ALLE ZEIT WACH 1842

Martin Hotine in 1956

Martin Hotine

Differential Geodesy

Edited with Commentaries by Joseph Zund

Contributions by
J. Nolton B.H. Chovitz C.A. Whitten

Springer-Verlag
Berlin Heidelberg New York
London Paris Tokyo
Hong Kong Barcelona
Budapest

Prof. Dr. Joseph Zund
New Mexico State University
Las Cruses, NM 88003-0001
USA

ISBN-13: 978-3-642-76498-1 e-ISBN-13: 978-3-642-76496-7
DOI: 10.1007/978-3-642-76496-7

Softcover reprint of the hardcover 1st edition 1991

Production Editor: Herta Böning, Heidelberg
Reproduction of the figures: Gustav Dreher GmbH, Stuttgart
Typesetting: K+V Fotosatz, 6124 Beerfelden
32/3145-543210 – Printed on acid-free paper

Contents

Foreword

After more than 20 years, the ideas expressed in Martin Hotine's treatise *Mathematical Geodesy* remain fresh and vibrant. Although geodesy has advanced remarkably over this span of time, this has been due mainly to increased precision of measurement and enlarged computing power. The bulk of the theory in *Mathematical Geodesy* still remains at the forefront of geodetic research. Examples of areas of current geodetic interest for which *Mathematical Geodesy* can provide a fruitful seedbed are precise coordinate systems linking various reference frames, and direct vector processing on modern computers.

We thus deem the appearance of this collection of Hotine's seminal papers which served as the basis for *Mathematical Geodesy* to be extremely providential. It is our hope that *Differential Geodesy* will motivate those interested in geodetic research to go forward, either by seeking out *Mathematical Geodesy*, or by continuing directly along the paths mapped out so elegantly by Hotine. We also consider it fortunate that the publication of this volume follows closely the issuance of *Intrinsic Geodesy* (Springer-Verlag 1985), which performed a similar service by making available the original research papers of Antonio Marussi, Hotine's friend and colleague. These two volumes are companions not only in time and subject matter, but also in spirit.

Martin Hotine inspired many persons during his lifetime, as we personally witnessed. But the true test of an outstanding scientist is his influence on those coming afterward. Joseph Zund, with a distinguished background in mathematical physics and differential geodesy, encountered *Mathematical Geodesy* by chance in 1981, and was immediately attracted by its creativity and richness of concepts. This happy interaction has yielded a series of publications advancing many of Hotine's ideas, and settling some long-standing problems posed by Hotine. We express our appreciation to Professor Zund for taking the time and effort to assemble, edit, and provide valuable comments to, the articles contained herein. We trust it was more a labour of love than a burden.

A debt of gratitude is also owed to John Nolton, who assembled the comprehensive bibliography from often quite remote sources. Finally, the willingness of Springer-Verlag to sponsor this undertaking deserves explicit recognition.

Charles A. Whitten Bernard H. Chovitz

Editorial Introduction

Joseph Zund

The purpose of this monograph is to make available to the geodetic community a selection of the largely unpublished papers of Martin Hotine (1898–1968). It is intended to be a companion volume to the collection of Antonio Marussi's papers in his *Intrinsic Geodesy* (Springer 1985). Together, these two volumes contain the essential and original ideas underlying the Marussi-Hotine approach to mathematical geodesy.

The collection contains nine papers:

1. Trends in mathematical geodesy. Bolletino di Geodesia e Scienze Affini, anno XXIV (1965) 607–622
2. Adjustment of triangulation in space. (I.A.G. Symposium on European Triangulation, Munich, 1956)
3. Metrical properties of the Earth's gravitational field. (I.A.G. General Assembly, Toronto, 1957)
4. Geodetic coordinate systems. (I.A.G. General Assembly, Toronto, 1957)
5. A primer on non-classical geodesy. (First Symposium on Three-Dimensional Geodesy, Venice, 1959)
6. The third dimension in geodesy. (I.A.G. General Assembly, Helsinki, 1960)
7. Harmonic functions. (Second Symposium on Three-Dimensional Geodesy, Cortina d'Ampezzo, 1962)
8. Downward continuation of the gravitational potential. (I.A.G. General Assembly, Lucerne-Zürich, 1967)
9. Curvature corrections in electronic distance measurements. (I.A.G. General Assembly, Lucerne-Zürich, 1967).

Apart from papers 1 and 7, none of these has previously been published in the open literature, or has been generally available to non-participants of these meetings. Actually, paper 7 appeared only as an appendix in P.L. Baetsle's comprehensive report (Baetsle 1963) of a meeting. In toto they represent the primary source material for Hotine's bold and dramatic approach to mathematical geodesy. They clearly show the evolution of his ideas, and furnish valuable insight into the contents of his posthumous treatise *Mathematical Geodesy* (U.S. Department of Commerce, Washington, D.C. 1969). Henceforth, since we will frequently refer to this book, it will be convenient to denote it by MG, and likewise IG designates Marussi's *Intrinsic Geodesy.* For each of these nine papers we have added an editorial commentary which contains some introductory remarks, indicates the relation of the paper to the corresponding discussions in MG, and gives more recent results dealing with Hotine's work.

The above nine items represent virtually all of Hotine's work on differential geodesy intended for presentation/publication, except for three items, which are readily available in the Bulletin Géodésique (Hotine 1966a, 1966b; Hotine and Morrison 1969). The contents of Hotine (1966a, b) are briefly discussed in our editorial commentary on paper 1, and Hotine and Morrison (1969) is an addendum to Hotine's work on satellite geodesy in Chapter 28 of MG.

Paper 1 is a delightful expository article and is essentially a companion to Marussi's *From Classical Geodesy to Geodesy in Three Dimensions*, which appeared as the first paper in IG (see IG pp. 2–12). Unquestionably, the most definitive and significant of these reports is item 5. It is simply a masterpiece, and Marussi commented (1969):

> "It is written in a style which makes it difficult to know which to admire most: its conciseness, or its eloquence."

In effect, it is the pièce de résistance of Hotine's formulation of mathematical geodesy. It refines and synthesizes, but does not totally supersede the contents of items 2–4, and 6 is a natural continuation of it. Papers 7–9 stand apart, but are introductions to the material in Chapter 21 of MG.

The viewpoint and content of these papers differ somewhat from that contained in MG. In military parlance, which we think Hotine might appreciate, we could characterize the former as discovery and reconnaissance, and the latter as conquest and occupation. MG aimed to present a comprehensive and unified deduction of the theoretical aspects of geodesy from basic principles in the spirit of mathematical physics. While this is admirable and suitable for the cognoscente, for someone wishing to sample the flavour and scope of Hotine's approach, the size and conciseness of MG is rather overpowering. We believe that the present monograph may be useful as a preliminary introduction to MG. It is by no means a replacement for MG, but rather a complement and guide to it.

As indicated in Chovitz's preview paper (reprinted in this book on pp. 157–170), Hotine was uncertain about what to choose as a suitable title for his treatise. We faced a similar dilemma in selecting a title for the present monograph. The best, and obvious, title had already been used, and to employ it again would have led to confusion with MG. Our decision, which we hope that Hotine would have found acceptable, was to use the title *Differential Geodesy*. This choice was a compromise, but one which is not radical, and which Hotine had previously considered for MG.

In preparing the material for publication, we have made minor changes of an editorial character, but nothing of substance has been altered. We have added a few missing equation numbers (without disturbing the original numbering of his equations) and Hotine's references have been made more precise. The format of the papers has been adjusted to conform to that employed in MG.

I would like to express my profound gratitude to Bernard Chovitz and Charles Whitten for their cooperation and permission to include their valuable articles in the monograph. I am also grateful to John Nolton for his assistance in compiling the Bibliography of Hotine's papers. Special thanks are also due to Erik Grafarend for his advice and enthusiastic encouragement of this project. Finally,

I am indebted to the Hotine family for their help and for supplying the photograph which appears as the frontispiece.

During part of the time of the preparation of this material, I was under contract to the Geophysics Laboratory of the Air Force Systems Command and I would like to express my appreciation to that agency for allowing me to devote part of my research time to this project.

Unfortunately, both Hotine and Marussi were taken from us too soon, but they have left us, as a challenge and legacy, the foundations of a beautiful and imaginative approach to geodesy. It is my hope that the present monograph and the companion volume of Marussi will serve to make their ideas better known in the geodetic community, and in this spirit the monograph is dedicated to the memory of Martin Hotine and his friend Antonio Marussi.

References

Baetsle PL (1963) Le deuxième symposium de géodésie à trois dimensions (Cortina d'Ampezzo, 1962). Bull Géod 67:27–62

Hotine M (1966a) Geodetic applications of conformal transformations in three dimensions. Bull Géod 80:123–140

Hotine M (1966b) Triply orthogonal coordinate systems. Bull Géod 81:195–222; and Note by writer of the paper on Triply orthogonal coordinate systems. Bull Géod 81:223–224

Hotine M, Morrison F (1969) First integrals of the equations of satellite motion. Bull Géod 91:41–45

Marussi A (1969) In memory of Martin Hotine. Surv Rev 152:58–59; also included in Whitten's memorial tribute in this monograph

1 Trends in Mathematical Geodesy[1]

Rutherford, a physicist, is reputed to have said that there are only two sciences: physics and stamp-collecting. He meant, I suppose, that all other sciences do nothing but collect and classify, which is not quite true, because some of them speculate where they cannot accumulate, whereas physics is subject to the sterner rational discipline of mathematics. Indeed, apart from a certain amount of gadgetry, physics nowadays seems to be nothing but mathematics.

What sort of mathematics? Inglis, sometime Professor of Mechanical Sciences at Cambridge, England – it would be called Engineering in any other University – used to say that mathematics is a can-opener; but whereas the engineer should use it to open cans of beef, the mathematician uses it to open cans of can-openers. The gulf between the two is much wider nowadays. The mathematicians have retired into a shell of formal logic and existence theorems and are no longer concerned with opening any sort of can; whereas the engineer is interested only in binary arithmetic and tends to feed the entire can to a computer.

We geodesists get the worst of both worlds. As "snappers-up of unconsidered trifles" (Shakespeare 1611) in the shape of deflections and anomalies, we are reviled by anti-philatelic physicists; and since we mostly have a basic engineering education we are abused by all other scientists, who classify engineers as half-educated morons in boiler suits. Yet in my time academic engineering consisted entirely of physics, that is mathematics, classical mechanics, electricity and magnetism, hydraulics and thermodynamics. I hope the only change has been to bring all that up to date and perhaps to add some wave mechanics. We need also to start up an engine occasionally, but only for the purpose of measuring its inefficiency. And in the later years of an engineering course, when the pure physicists go "boundless inward to the atom", we geodesists should go outwards into bounded or unbounded space – it is still possible to choose which – until we meet again in a stationary or receding nebula.

Meanwhile, we have only ourselves to blame for departing from our highest traditions. Men like Newton, Lagrange, Laplace, Gauss (Todhunter 1873, passim) – to name only a few – were the geodesists of their day in the sense that they were primarily interested in measuring the whole Earth, its attributes and its surroundings. In the process they made considerable contributions to mathematics

[1] Manuscript dated 15 September 1964 (Washington D.C.), appendix dated 15 November 1964 (Washington D.C.). Published in Bolletino di Geodesia e Scienze Affini, anno XXIV, 1965, pp. 607–622.

and physics. Gauss, for example, worked out a complete differential geometry of surfaces and founded the theory of errors in conjunction with the reduction and projection of a survey of Hanover. Since then we have mostly basked in Gaussian glory. It is true that we have vastly improved the accuracy of measurement and from time to time have helped to confirm or destroy some theories of the constitution of the Earth. But until recently we have done little on the mathematical side beyond elaborating the old, old formulae. We have not even kept abreast of the development by Ricci, Riemann and others of Gaussian differential geometry to three or more dimensions, which Gauss would probably have originated himself had he moved from Hanover to the Himalayas. We have imported some of the jargon of modern statistics into Gaussian least squares, but have done little or nothing to examine modern probability theory and adapt it to our use.

If we are to do our job properly, and incidentally to be less lightly considered by the scientific community, we must get back to a stricter mathematical discipline as the rational basis of our work; and in the process it may well be that we can once more be of greater use to other sciences than by merely supplying them with data. We have spent too long on re-designing reference spheroids and have almost come to regard this activity as an end in itself. We shall not have done much more to advance the breadth and the depth of human knowledge by the time we have put a few reliable contours on the geoid.

To avoid the charge that all this is just talk; that there is nothing more to be done in the theory of measurement of the Earth; that this is an age of specialization and if there is any work to be done on mathematical geodesy it will have to be done by pure mathematicians – to avoid all this and to illustrate my point, I will give you a couple of examples. These are not epoch-making discoveries. They may even be wrong in the same way that anything can be wrong until it has been independently checked several times. But they are new basic ideas in fields which have been ploughed, sown, reaped and gleaned for centuries. And they come up at the first scratch with new tools, new, that is, in the sense that geodesists have not used them before. They will be justified if they do no more than suggest to you that, in geodesy as in much else, the Canadian poet is still right who said: "The highest peaks haven't been climbed yet; the best work hasn't been done".

The Gravitational Field

All geodetic measurements are necessarily made in a gravitational field which provides us with our ideas of "horizontal" and "vertical" and it is natural that we should turn our attention to that first. For some time I have had an uncomfortable feeling that there is something missing, either in the classical theory of gravitation itself, or in the mathematical handling of it. For instance, it is well known that the form of an equipotential surface settles the value of Newtonian gravity on the surface, apart from constants, which suggests that gravity is in some way intrinsic to the surface, except for the nature – flat or curved – of the surrounding space. It should be possible to express gravity in terms of the first- and second-order magnitudes of the surface, yet the actual relation has been formulated in a very few simple cases only, by using particular coordinate systems. Gravity on a rotating spheroid is, for example, found by using spheroidal coordinates as $g = AN + B/N$, where A and B are constants over the surface and N is the principal radius of curvature perpendicular to the meridian. We can similarly obtain a relation for a triaxial ellipsoid, but nothing more general. This could be due to limited knowledge of how to manipulate more general harmonic functions, but when we try a different approach, independent of any particular coordinate system, we find that the law of gravitation itself offers no help. We can use the law in its harmonic form to determine the vertical variation of gravity by means of Brun's formula, but not the horizontal variations.

There is a similar element of indetermination in Newtonian dynamics. The artificial satellite is providing us with a wonderful new tool. But unless we are prepared to restrict its use to a light in the sky, an elevated triangulation beacon visible over long distances, we are forced to consider motion in orbits closer to the Earth than we have ever dealt with before. The actual motion must, of course, be regulated by the nature of the gravitational field, yet the Newtonian equations of motion in a free orbit do not contain the law of gravitation at all. We derive particular solutions of the equations by assuming a harmonic potential, but the equations themselves do not require this and would presumably apply equally well to other laws of gravitational attraction.

On the other hand, the corresponding law in four-dimensional theories of gravitation is woven into the metric of space-time from which the equations of motion are derived; it is impossible to obtain any solutions independent of the law of gravitation. The main object of these four-dimensional systems is to allow for very large velocities approaching the speed of light at cosmic distances. They all reduce to the Newtonian system for small velocities or for weak central fields, and indeed are made to do so. There would accordingly be no point in facing the formidable complications of using four-dimensional methods on, say, the orbits of near-Earth satellites. But nowadays there may well be a requirement for a *three*-dimensional dynamical system more rigidly connected to the actual law of gravitation.

Newton's law of gravitation was originally formulated to describe planetary motion in the solar system, and there can be no doubt that in these circumstances, of great distances between nearly spherically symmetrical masses, it has proved remarkably accurate. A pious opinion before Newton – not a scientific hypothesis or even an article of faith – was that the Earth is continuously steered in towards the life-giving Sun by a cohort of angels. Newton himself would have been the last to deny such a poetic description of Divine Providence, and indeed for all we know now it could be literally true. All Newton wanted to do was to describe the *results* of these angelic labours in a form which could be programmed for computation, and in the process he introduced notions of gravitational force and potential which have no more tangible existence than the angels. His inverse-square law of supposed force is nevertheless far more than inspired guess. No other feasible law would result in closed orbits in a central field, and it has stood up to numerical tests in the solar system in a remarkable way for centuries. But we have no right to argue from the particular to the general and assert that, because the law is necessary and has been amply verified in a central symmetrical field, it is therefore true of the field close to a large unsymmetrical mass. We can certainly say that whatever the law is near such a mass, it must degenerate to an inverse square at great distances, but the converse is by no means necessarily true. The usual way of overcoming the difficulty is to assert that the unsymmetrical mass consists of particles, each of which sets up a central field, just as it would if it were alone. The more complicated unsymmetrical fields is accordingly obtained by superimposing an infinite number of central fields. Mathematically this may be true, but it still has to be verified experimentally that matter does, in fact, behave like a conglomeration of independent particles in its gravitational effects. This has never been done, at any rate to the degree of accuracy we now seek.

To provide experimental verification we need an alternative gravitational theory, which we have never had, based on slightly different but equally plausible assumptions. This should be used on the same observations as the Newtonian system and the results compared statistically. The difference is not likely to be large enough for laboratory comparisons and the effect, if any, would mostly disappear at astronomical distances, but the artificial satellite, which provides us with the means of exploring the entire near-Earth gravitational field to a high degree of accuracy, should soon provide us with enough observations for a practical test. We geodesists may in that way be able to make a more significant contribution to human knowledge, more in keeping with our traditional role, than by arguing about whose value of an eighth harmonic is the best and what it means. We should first provide some assurance that an eighth Newtonian harmonic has any meaning at all.

Generalized Gravitational Equations

The ideas of "horizontal" and "vertical", which we can materialize with bubbles and plumb lines, must find some place in any gravitational theory, at any rate in the near-Earth field. We accordingly assume that there exists in Nature a family of surfaces, which we define as "level"; and a family of lines everywhere orthogonal to these surfaces, which we define as "vertical". We assume further that the "level" surfaces can be expressed mathematically by means of a continuous differentiable scalar function of position V, which is constant over any particular surface; the value of the constant being, of course, different for different surfaces.

Two neighbouring level surfaces, $V = \text{const.}$ and $V + dV = \text{const.}$, are separated by a distance (dn) measured along a vertical line. We define the limit

$$g = -dV/dn \tag{1}$$

as the "distance function".

In index notation for vectors, we write υ_r for the unit vector in the direction of the vertical and V_r for the gradient of V. All the above definitions are then included in the vector equation

$$V_r = -g\upsilon_r \; . \tag{2}$$

All this is pure geometry. If V were the Newtonian potential, then g would be the gravitational acceleration or force per unit mass, but for the present we do not say this.

We shall follow most gravitational theories (including both Newton's and Einstein's) by making the assumption that free orbits or trajectories are subject to a certain minimum condition, which appears in other natural laws. Such a minimum condition is generally reckoned to be an expression of the basic economy of Nature, which we occasionally see in ourselves as a resistance to change. The Duke of Cambridge, still Commander-in-Chief of the British Army at the age of 80, said he had finally come to the conclusion that "all change is for the worse", and in this he was merely voicing a slightly military over-simplification of a fundamental law of Nature: that some change is inevitable and does occur, but should be as little as possible.

In classical three-dimensional mechanics the law takes the form of the Principle of Least Action (McConnell 1931). We define the Action as

$$A = \int v\,ds = \int v^2 dt \ , \tag{3}$$

that is, the integral of the velocity (v) between two points on the path of a freely moving particle of unit mass, or the time integral of twice the "kinetic energy". We then assert that the Action is less along the actual path than it would be along any neighbouring path between the two points considered.

Here I must digress and introduce the notion of curved space. We are all familiar with flat and curved spaces of two dimensions. The flat or Euclidean space is distinguished by the fact that we can find coordinates (x, y) such that the line element (ds) is given in the Pythagorean form

$$ds^2 = dx^2 + dy^2 \ .$$

The line element on a curved surface cannot be expressed as simply, whatever coordinates we choose. Now suppose we set up a one-to-one correspondence between points in a plane and the corresponding points on a curved surface. The same numbers (x, y) will now serve to define the position of points on the curved surface as well as on the plane, but the line element on the curved surface will be given by the more complicated expression

$$ds^2 = E\,dx^2 + 2F\,dx\,dy + G\,dy^2 \ ,$$

in which E, F, G are functions of position, that is of x and y. Those functions and their first and second derivatives are related to the intrinsic or specific curvature of the surface, which is a property of the surface itself without regard to the surrounding space, by means of the Gauss characteristic equation. Gauss himself called it his "egregious theorem".

We can go further with Gauss into Hanover and set up a one-to-one correspondence with points on a plane such that the line element on the surface is given in the simpler form

$$ds^2 = m^2(dx^2 + dy^2) \ .$$

In other words, the ratio of the two line elements is a function of position m, which, of course, varies from point to point, but nevertheless implies that small corresponding figures around a point are similar. For this reason we call the transformation conformal, and we use it a great deal, as Gauss did, in the theory of map projections. In particular, it is evident that angles between corresponding directions around a point are preserved. We need not start with Cartesian coordinates x, y in the plane but can use any coordinates on either surface provided the two line elements are connected by the relation

$$d\bar{s}^2 = m^2 ds^2 \ .$$

But for such a transformation to be possible, the scale factor (m) must be related to the intrinsic curvature of the surface by means of the Gauss characteristic equation. Alternatively, we can use any continuous differentiable function of position we like for m and that will settle the curvature of the surface.

We now add a dimension to these ideas. We can have a flat space in which it is possible to find Cartesian coordinates x, y, z such that

$$ds^2 = dx^2 + dy^2 + dz^2$$

and a curved space whose line element is not expressible in this simple form. It will now contain six funtions instead of the three E, F, G. These functions and their derivatives provide, not one single function specifying the curvature as in two dimensions, but six distinct functions, known as the components of the curvature tensor. We can set up a conformal transformation from a flat space to a curved space, in which case there are six relations between derivatives of the scale factor and the components of the curvature tensor[2]. If we start with space of a defined curvature, then this drastically reduces the possible choice of scale factor. We can, however, start with any scale factor we like and this will settle all components of the curvature tensor.

The three-dimensional space in which we live is, so far as we know, flat. We cannot sense its flatness any more than a two-dimensional being, crawling along what it considers to be a straight line, could have any idea that this world surface is curved. Not so long ago (unfortunately, I have no reference), a room containing some simple furniture was actually constructed, using the geometry of a space of large constant curvature, and various observers were invited to look at it from the outside through two holes in the wall. None could see anything odd about it. However, if they had been allowed inside with a tape they would very soon have realized that what looked like a cube from the outside wasn't one. What a mess the photogrammetrists would make of things if they attempted to do without a floating mark!

For this reason we have started with flat space although we could almost as easily deal with two curved spaces. The curved space can be considered simply as a mathematical device without physical significance. In fact, we could obtain all the following results analytically by the calculus of variations, although not

[2] A first-order tensor in three dimensions is a vector having three components which transform in a particular way when we change coordinates. A second-order tensor is essentially a sum of products of vectors, and in general, has nine components. The curvature tensor is symmetrical and so has only six distinct components.

as simply. The approach via curved space moreover has other important applications as we shall see later.

We propose to set up a conformal transformation in which the scale factor is the velocity v, but before we can do this we must arrange for v to be defined not only along a particular trajectory but also throughout a region of space. We do this by taking a family of trajectories all having some feature in common; for example, all those trajectories starting from the same point in space with the same velocity, but in different directions. In a continuous field, one member of the family will then pass through every point in a region of space and we define velocity at the point as the velocity in that particular trajectory.

If we now set up the conformal transformation

$$d\bar{s} = v\,ds \ ,$$

integrate between two fixed points on a trajectory and require the Principle of Least Action [Eq. (3)] to hold, then we shall have

$$\bar{s} = A \ , \tag{4}$$

a minimum for the path, so that $\bar{s}$ or A is the length of the *geodesic* joining the two points in the curved space.

It can be shown (Levi-Civita 1926 p. 228, with some extensions) that the geodesics of the curved space correspond to curves in the original flat space which have the following properties:

(a) The arc rate of change of the logarithm of the scale factor in the direction of the principal normal to the curve is equal to the principal curvature of the curve, and
(b) There is no change in the scale factor in the direction of the binormal to the curve[3].

Now the gradient of the scale factor (v_r), like any other vector, is expressible in terms of three mutually perpendicular vectors which we shall take as l_r, m_r, n_r: respectively, the unit tangent to an trajectory of the family, its principal normal and binormal. Say

$$v_r = A l_r + B m_r + C n_r \ .$$

Multiplying this across in turn by l^r, m^r, n^r we find that $A = \partial v/\partial s$, if s is the arc length of the trajectory; B, from the condition (a) above is κv, where κ is the principal curvature of the trajectory; and C, from the condition (b) above is zero. So,

$$v_r = (\partial v/\partial s) l_r + (\kappa v) m_r \ , \tag{5}$$

[3] There is a similar proposition in two dimensions which is well known in the theory of conformal map projections. A geodesic on the curved surface corresponds to a curve on the plane surface whose curvature is similarly equal to the arc rate of change of the logarithm of the scale factor in the transverse direction. In two dimensions, a curve has no binormal.

which can also be written as

$$v_r = \frac{\delta(v l_r)}{\delta s} , \tag{6}$$

in which $\delta/\delta s$ implies intrinsic differentiation along the trajectory of the family which passes through the point under consideration. We may also multiply by $v = ds/dt$, t being the time, and write

$$\frac{\delta(v l_r)}{\delta t} = v v_r = \left(\frac{1}{2} v^2\right)_r . \tag{7}$$

Now the first member of this equation is a vector and it must also be the gradient of a scalar, say W, because the right-hand member is. So we can write further

$$\frac{\delta(v l_r)}{\delta t} = v v_r = \left(\frac{1}{2} v^2\right)_r = W_r . \tag{8}$$

This is the general differential vector equation of motion of all orbits or free trajectories satisfying the Principle of Least Action and depending on nothing but that principle. The vector $v l_r$ is evidently the velocity vector, in magnitude and direction, and is not to be confused with v_r the gradient of the scalar velocity. Equation (8) states that the time rate of change of the velocity vector, that is the acceleration vector, is equal to the gradient of $\frac{1}{2} v^2$.

There is one more property of a family of geodesics in curved space which we can use. It is well known (Eisenhart 1960 p. 57) that such a family cuts orthogonally a family of surfaces, known as geodesic parallels, and that these surfaces are parallel in the sense that any two of them cut off the same length measured along any geodesic of the family between them[4]. We can, accordingly, say that $\bar{s}$ or A measured from one of the surfaces is constant over any other of the surfaces. The geodesic parallels can accordingly be expressed as A = const., while the geodesics themselves are in the direction of the gradient $\bar{s}_r$ or A_r. Transforming the unit vector along the geodesics back to the flat space we can show without much difficulty that

$$v l_r = A_r . \tag{9}$$

This again is pure geometry resting solely on the Principle of Least Action. It is an integral of the equations of motion (8). It shows that the A = constant surfaces (corresponding to the geodesic parallels) contain the principal normal and binormal to l_r so that we have the following picture in the osculating plane of (l_r) (Fig. 1):

[4] This is a generalization of a theorem due to Gauss for parallel surfaces in flat space.

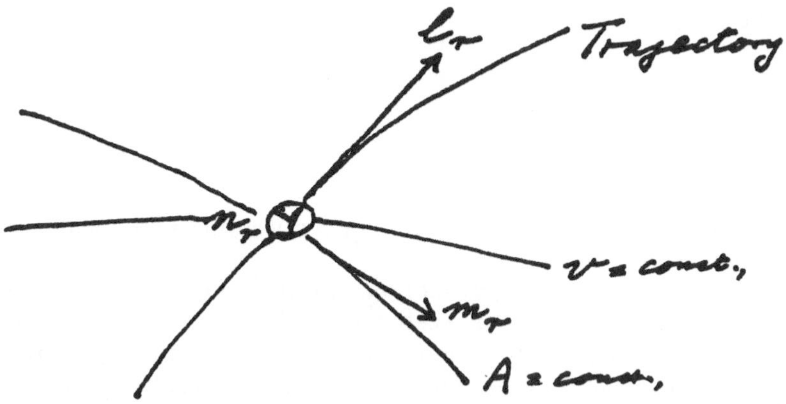

Fig. 1

From the fact that A, defined as above, is a scalar function of position, we can say that the curl of $v l_r$ is zero or that the curl of l_r is κn_r. This fact may sometimes be useful, although it conveys no more information than Eqs. (8) and (9).

Note that Eqs. (9) and (2) are entirely similar. We already have the differential geometry of the V-surface fully worked out without necessarily applying any restriction on V, such as requiring it to be a harmonic function (Hotine 1957a). All these properties can be applied to the A-surface by a simple change of notation.

We have said nothing so far about the gravitational field, except that bodies move in it in accordance with the Principle of Least Action. Nor have we provided any hook-up with the level surfaces and verticals. We can easily derive the Newtonian field from these Eq. (8). We then obtain a particular solution of these equations on the further assumption that V is a harmonic function.

There are, however, any number of other solutions. Suppose, for example, we make A equal to some function of V for a particular family of trajectories, which in that case would be the vertical lines. Then Eq. (9) tells us at cone that

$$v = g f(V) \,, \tag{10}$$

in which f(V) is some arbitrary function of V. This settles the velocity field for any family. By integrating the two last members of Eq. (8) (and this is a general result, corresponding to the Newtonian conservation of energy), we have

$$\tfrac{1}{2} v^2 - W = D \,, \tag{11}$$

in which D is an absolute constant for the family. We can further make the field Newtonian in the special case of spherical symmetry by putting

$$W = V = \mu/r \quad \text{and} \quad g = \mu/r^2 = V^2/\mu$$

so that the arbitrary function of V becomes

$$f(V) = \frac{\mu}{V^2}(2V+2D)^{1/2} \tag{12}$$

and

$$v = \frac{\mu g}{V^2}(2V+2D)^{1/2}\ . \tag{13}$$

In this equation g is the "distance function" for the V-surfaces and neither g nor V is Newtonian.

In solution for V of the equations of motion with this value of the velocity is certainly not as simple as the Newtonian case and would require the evaluation of more parameters from more observations, but nothing seems to daunt the binary arithmeticians these days. The resulting dynamical system is Newtonian in a central field. In a unsymmetrical field, it asserts that a body would fall along the physical vertical, whatever its velocity; whereas the Newtonians assert that the apple did not fall vertically but that the harmonic law of potential, derived from the central field, continues to hold. Which is nearer the facts and figures, we do not yet know.

This particular alternative to Newton is, however, merely an example. It is not my purpose to exhaust the entire field or even to pursue one solution to its numerical conclusion; I merely want to show how much remains to be done, not only in formulating alternative solutions but also in solving them. Equation (9), for instance, has not, so far as I know, entered into the calculations of geodesists and astronomers before. There is nevertheless a physical analogy in that, if the trajectories are considered as directions in which energy is propagated, then the surfaces A = const. are the wave-fronts. Perhaps we too should ride on the crest of a wave, with the help of all the work which has been done on wave propagation, instead of forever adding bits to a Kepler ellipse?

We have not yet dealt with the effect of diurnal rotation of the Earth, which, of course, affects the form of the V-surfaces as obtained from terrestrial observations. In any gravitational theory alternative to Newton's we must avoid such conceptions as centrifugal force and should treat the problem by means of rotating coordinate axes. The best way of doing so is probably to introduce a time dimension, although much more simply than Einstein does. The great advantage of the tensor methods outlined above is that they hold in any static coordinate system and are easily extended to more dimensions.

Static Measurements

We have so far considered measurements to a body in motion, relative to an observer on the Earth, with a view to using the orbital characteristics of the body; but important as that may be in the future for exploring the gravitational field, and even for position fixing, it is not yet an important method of geodetic measurement. Another geodetic use of artificial satellites, and perhaps at present the most important use, is to fix positions on the Earth by simultaneous observation of the satellite considered solely as a triangulation beacon, without requiring any accurate knowledge of its orbit. Satellite triangulation can bridge oceans and large desert areas on a single coordinate system, to greater accuracy and over longer distances than any previous system, and seems to be the best method in prospect for achieving a single world-wide datum and reference system. Present indications are that its accuracy is about the same as modern ground-to-ground measurements over land distances of about 1000 miles. Moreover, ground-to-ground methods will always be required – pace the advanced photogrammetrists – to provide a sufficient density of control for minor surveying over almost any land mass. Consequently, the sole use of satellite triangulation over most land masses is likely to be as over-all control on ground-to-ground networks and for this purpose it must be observed to the highest possible degree of accuracy.

It is hardly possible to use classical two-dimensional methods of reduction on the long sides and great altitudes of satellite triangulation, which must accordingly be reduced in three dimensions. There was, however, a move in this direction for ground-to-ground observations before the first sputnik went into orbit (Hotine 1956, 1957a, b) and three-dimensional methods are now becoming standard.

Geodetic measurements of length, angles, astronomical latitude, longitude and azimuth are affected by the direction of gravity – the vertical – but not by the magnitude of gravity and were the first to go into three dimensions (Hotine 1958 et ante). Spirit-levels, which are but little affected by the magnitude of gravity have been added since (Hotine 1960).

Most ground observations must necessarily be made along the paths of light or radio waves affected by atmospheric refraction and this, more than anything else, restricts the accuracy of our results. Essentially, this is a physical problem: actually to measure the refraction, or the variations which cause it, throughout the line. There is a good change that this will be done during the next few years, and meanwhile we should re-examine our basic mathematical treatment of the subject. So far as a ray of light in a medium of refractive index (μ) is concerned, Fermat's principle tells us that

$$\int \mu \, ds$$

must be a minimum along the path compared with any adjoining path between fixed ends, and this principle has as much solid philosophical and experimental support as the Principle of Least Action which it is so closely resembles. Indeed, we have only to write μ for v and take the A-surfaces as the wave-fronts in the above derivations from the Principle of Least Action to have the whole story. We shall then discover that Snell's Law of Refraction holds only across a surface of discontinuity in μ and is usually misapplied in books on geodesy and astronomy.

There is a growing tendency to include all geodetic operations which are affected only by the direction of gravity in a branch of the subject known as *geometrical geodesy.* Operations which are concerned with, or affected by, the magnitude of gravity are included in *physical geodesy.* In my view, such an artificial subdivision is a great mistake and will retard both sides of the subject. The magnitude of gravity is just as much a geometrical entity as its direction, although both are subject to a physical law, and to split the two is as retrograde as refusing to use even elementary vector analysis in our basic mathematical discipline. "Physical" geodesy should follow its geometrical twin into the reality of three dimensions. Instead, there is the same old desire to get onto a surface as soon as possible, and a nice smooth surface at that. This in turn requires certain smooth assumptions as to underlying densities, sometimes based on the rigid application of theories which are no longer accepted as more than half-true by other Earth sciences, or else the operation of a mammoth bulldozer. A three-dimensional system has, in fact, been proposed by Bjerhammar (1963), who indeed calls it a five-dimensional system because he includes the potential and the time as well as three space coordinates, but much more work remains to be done before it can be said that this is the only possible system, or the best one.

Above all, the "geometrical" and the "physical" geodesists must work on the same law of gravity, which is set for the rise and fall of us all, and that goes for the "satellite" geodesist, too. If it should eventually appear that the number V, which defines the level surfaces, is not after all exactly a harmonic function modified by diurnal rotation of the frame of reference, then the bottom will drop out of "physical" geodesy. It could be a profitable exercise to consider now what to do then, and who knows what else might come out of the exercise?

Coordinate Systems

I have time for only one other example of the interaction between mathematics and geodesy and for this we will choose the question of three-dimensional coordinate systems.

Fifteen years ago Marussi (1949) proposed the use of a coordinate system consisting of the astronomical latitude and longitude and the Newtonian geopotential and worked out the basic differential geometry of the system. Unfortunately, this is fairly complicated and there is something to be said for trying to find a simpler system which may also be based on directly measurable quantities. If, for instance, we could find a triply orthogonal system which fits the gravitational field, the consequences could be considerable, both in theory and in practical computing.

Now we have already seen in Eq. (10) that a conformal transformation to a curved three-dimensional space with a scale factor g results in the lines corresponding to the verticals becoming geodesics. From our definition of g and V in Eq. (2), this means, moreover, that an element of length along the geodesics in the curved space is dV. The line element in the curved space can accordingly be written in what is known as the geodesic form of the metric (Eisenhart 1960, p. 47) as

$$ds^2 = \bar{E}\,dx^2 + 2\bar{F}\,dx\,dy + \bar{G}\,dy^2 + dV^2$$

and transforming this back to the flat space in the same coordinates

$$ds^2 = E\,dx^2 + 2F\,dx\,dy + G\,dy^2 + (1/g^2)\,dV^2 \,, \tag{14}$$

in which E, F and G are components of the metric tensor of the level surfaces in the surface coordinates x, y. We do not yet know what these coordinates are, beyond the fact that they must be constant along the verticals; they are not the astronomical latitude and longitude. We have nevertheless shown that it is possible to find two coordinates, while retaining V as the third, which makes two of the six components of the metric tensor zero and reduces a third component to the directly measureable $1/g^2$. Neither this simplification nor any of the following argument depends on V being a harmonic function.

The question now arises whether we cannot also choose orthogonal surface coordinates, thereby making another component (F) of the metric tensor zero. In that case the coordinate surfaces (one family being the level surfaces V = const.) would cut one another everywhere at right angles. For this to be possible, the function which generates the level surfaces would have to satisfy a complicated third-order partial differential equation known as the Darboux equation.

It is well known that in a triply orthogonal system, each coordinate line would have to be a line of curvature of the two surfaces containing it; and since it would be normal to a coordinate surface, it would have to constitute what is known as a surface-normal field. Alternatively, we can say that one of the coordinates alone varies in the direction of each line of curvature and is constant in the two perpendicular directions, so that the lines of curvature must have the same directions as the gradient of some scalar, that is, of the appropriate coordinate.

If we express either of these conditions in the Marussi metric or in the simplified metric Eq. (14) we obtain the following simple formula:

$$\frac{\partial A}{\partial n} = \frac{\partial (\log g)}{\partial p} \tan \phi \ , \tag{15}$$

in which the first term is the vertical variation of the azimuth of the lines of curvature of the level surfaces: the second term is the variation of the logarithm of g in the direction of the astronomical parallel of latitude and ϕ is the astronomical latitude. It can be shown that this simple equation completely satisfies the complicated Darboux equation. It is a second-order differential equation in V and must therefore be an unsuspected integral of the third-order Darboux equation.

We have now to consider whether either Eq. (15) or the Darboux equation is a condition limiting the form of V, which would be the orthodox mathematical view, or whether it is true anyway in a space expressible in Marussi coordinates. For a start, Eq. (15) is automatically satisfied in an axially symmetrical field, which shows that any family of surfaces of revolution V = const. having a common axis of revolution is a possible member of a triply orthogonal system. This we already know.

We can, however, go further. *Each side* of Eq. (15) is zero in an axially symmetrical field, which means that the condition, if it is a condition, is over-filled in such a field and should therefore be satisfied in at any rate some unsymmetrical fields. The only way of resolving the matter beyond question is to find the other two coordinates, and this has not yet been done.

If Eq. (15) is generally true, it is a hitherto unsuspected relation between quantities in the field. It can be shown that it is equivalent to the following intrinsic level-surface tensor equation, true in any coordinates;

$$\left(\frac{1}{g}\right)_{\alpha\beta} n^{\alpha} t^{\beta} = 0 \ , \tag{16}$$

in which n^{α}, t^{β} are unit vectors in the direction of the lines of curvature. This may be, or may lead to, the missing relation between gravity and the form of the level surface, referred to earlier. It does not require the level surfaces to be Newtonian equipotentials.

If we are able to show by these means that the Darboux equation is automatically satisfied by any family of surfaces in Marussi space, then the mathematical consequences might well be very far-reaching. By Marussi space in this context I mean flat space containing a fixed axis to which the coordinates of latitude and longitude on the surfaces are referred; it does not have to be a gravitational field of any sort.

Equation (15) can be experimentally verified by torsion balance measurements over a range in altitude, say down a mineshaft. If it is satisfied over a sufficient range in the most gravitationally disturbed regions, we could accordingly use it in practice, and many of its consequences, whether it is rigorously true or not. This indicates a possible kickback to the data-production side of geodesy.

We must not underestimate the importance of data production, which is and must remain the bread and butter of geodesy. All I suggest is that we too use the data ourselves, primarily in furtherance of our own business of better data production, but also in order to see where it leads. There would, for instance, be far fewer advances in astronomy and in astrophysics if the observatories confined their activities to the production of ephemerides for the use of sailors and surveyors, important as that is. Never let us forget that Newton, Lagrange, Laplace, Gauss, among others who are now claimed as mathematicians, were geodesists who needed more and better mathematics to further their geodesy.

I hope therefore that I have said nothing to upset, much less to discourage stamp-collectors, or physicists, or those of us who need to be a little of both. The path to fame is beset by cut-throat bandits and it is no good being upset by them anyway.

I am grateful to Dr. John S. Rinehart of the Coast Survey, who gave me the basic idea for this paper quoting Rutherford, and much other encouragement to learn more physics; also to Mr. Bernard Chovitz of the Coast Survey who read the first draft and suggested several clarifications. He is not responsible for any errors and heresies which remain.

Appendix

Since this paper was written it has been found that the basic Eqs. (6) and (9) for free trajectories can be derived more simply.

The Principle of Least Action requires a particular family of trajectories in flat space to correspond to a family of geodesics in Riemannian space, a family of surfaces (geodesic parallels), orthogonal to the geodesics, therefore exists, on any one of which the action A is constant. It follows from the conformal properties of the transformation that corresponding constant A-surfaces exist in the flat space orthogonal to the trajectories. The gradient of A is accordingly in the direction of the trajectories, whose unit vector is l_r. We can therefore write

$$A_r = C l_r \ ,$$

in which C is a scalar function to be determined. Multiplying across by l_r, we have

$$A_r l^r = \partial A/\partial s = C \ ,$$

and therefore from Eq. (3), the definition of "action", $C = v$ and

$$v l_r = A_r \ , \tag{17}$$

which derives Eq. (9) without having to transform vectors between the two spaces.

Equation (6) can now be obtained by direct covariant differentiation of Eq. (17):

$$\begin{aligned}\delta(v l_r)/\delta s &= (v l_r)_s l^s \\ &= A_{rs} l^s \\ &= A_{sr} l^s\end{aligned}$$

(since A is a scalar function of position)

$$\begin{aligned}&= (v l_s)_r l^s \\ &= v_r l_s l^s + v l_{sr} l^s \\ &= v_r\end{aligned}$$

since l_s is a unit vector.

Equation (17), which as previously stated is an integral of Eq. (6), is accordingly a sufficient expression of the Principle of Least Action in this problem of free trajectories; it is unneccessary to use Eqs. (6) or (8) unless we wish to set up some relation between the acceleration vector and a scalar potential, as in the Newtonian theory.

The Cartesian form of Eq. (17) is

$$\frac{dx}{dt} = \frac{\partial A}{\partial x} \text{ etc.,}$$

and it might well be possible to solve these three equations numerically for A, expressed as a polynomial. The velocity is then given as the modulus of A_r. The function V which expresses the level surfaces, and which has to be compared with the Newtonian potential, would finally be obtained by inverse solution of Eq. (13).

Editorial Commentary

This paper represents Hotine's only expository work on mathematical geodesy. It is also unusual in that, unlike his other papers, it does not seem to have been prepared specifically in connection with a presentation at a geodetic conference or symposium. However, it was the basis of a talk that he gave an I.A.G. Symposium on Gravity Anomalies (Columbus 1964). Morever, it was written about a year after the I.A.G. General Assembly (Berkeley 1963), and it is also possible that it was an extended version of his presentation at that meeting. No text of his Berkeley report has been found, but this paper roughly seems to conform to the brief description of it given in Comptes rendus, Bull Geod (1963 p. 378). It offers an interesting companion to the expository paper *From Classical Geodesy to Geodesy in Three Dimensions* (1959) of Marussi, which is re-printed as the first paper in IG. Together these two papers admirably summarize the Marussi-Hotine approach to geodesy, and illustrate the subtle differences in their viewpoints.

Unlike Marussi's paper, which is largely historical, almost half of Hotine's paper is mathematical. In it, he suggested the use of conformal transformations, and the possibility of modelling the Earth's gravitational field on a triply orthogonal system of coordinates. Both of these were studied in more detail in his reports (Hotine 1966a, b) to the Third Symposium on Mathematical Geodesy (Turin 1965) which are not reprinted in this monograph.

Marussi regarded Hotine's ideas on the use of conformal transformations in geodesy as one of his most significant contributions. Indeed, in Parts V and VI of IG one can find five of Marussi's earlier papers on the same subject, and a valuable summary of his work is given in Bocchio (1978). The material in Hotine (1966a) was later presented in greater detail in MG: Chapter 10 was devoted to conformal transformations of space; Chapter 24 considered Fermat's principle in connection with atmospheric refraction. Nevertheless, the application of conformal techniques in three-dimensional geodesy is still in its nascent stage, and it is too early to know whether the great expectations of Hotine and Marussi will ultimately be fulfilled. A recent investigation dealing with the basic notions of conformal geometry is given in Zund (1987).

The final section of Hotine's paper was devoted to triply orthogonal coordinate systems. This was the subject of his publication (1966b) and Chapter 16 of MG. In particular, the former paper (1966b) contained his arguments for the existence of such coordinates, i.e the Hotine Conjecture, which was later shown to be invalid by Zund and Moore (1987).

Despite the informal and tentative nature of the material in this paper, it is rich in ideas – many of which have yet to be thought through to the end – and it vividly reveals the vibrant and exhuberant nature of Hotine's personality. As one of this anonymous obituary writers (see Memoir 1969) noted:

"Working with him was like dealing with the electric wiring of a house when the current has *not* been turned off at the main".

It is truly regrettable that we have so few examples of Hotine's engaging style and wit available to us. This paper is delightful, and deserves to be read and savoured. For the reader who has an interest in surveying, we recommend Hotine's five-part essay (Hotine 1952–53). An in-depth discussion of the mathematical background of the Marussi-Hotine approach to geodesy is given in Zund (1990).

References to Paper 1

Bjerhammar A (1963) A general world system without a reference surface. I. A. G. Berkeley

Eisenhart L (1960) Riemannian geometry. Princeton

Hotine M (1956) Adjustment of triangulation in space. I. A. G. Munich

Hotine M (1957a) Metrical properties of the earth's gravitational field. I. A. G. Toronto

Hotine M (1957b) Geodetic coordinate systems. I. A. G. Toronto

Hotine M (1958) A primer of non-classical geodesy. Venice Symposium

Hotine M (1960) The third dimension in geodesy. I. A. G. Helsinki

Levi-Civita T (1926) The absolute differential calculus. Blackie, London

Marussi A (1949) Fondements de géométrie différentielle absolue du champ Pontentiel Terrestre. Bull Géod 14

McConnell AJ (1931) Applications of the absolute differential calculus. Blackie, London

Shakespeare W (1611) A winter's tale.

Todhunter I (1873) A history of the mathematical theories of attraction and the figure of the earth.

References to Editorial Commentary

Bocchio F (1978) Some of Marussi's contributions in the field of two-dimensional and three-dimensional conformal representations. Boll Geod Sci Aff anno XXXVII:441–450

Comptes rendus resumés des séances des sections (1963) Section 1 Détermination géometrique de positions. Bull Géod 70:375–379

Hotine M (1952–53) Tales of a surveyor I–V. Geog Mag 25:198–200, 248–250, 307–309, 331–333, 480–482

Hotine M (1966a) Geodetic applications of conformal transformations in three dimensions. Bull Géod 80:123–140

Hotine M (1966b) Triply orthogonal coordinate systems. Bull Géod 81:195–224

Memoir: Brigadier Martin Hotine, CMG, CBE (1969). Roy Eng J 83:74–77

Zund JD (1987) The tensorial form of the Cauchy-Riemann equations. Tensor NS 44:281–290

Zund JD (1990) An essay on the mathematical foundations of the Marussi-Hotine approach to geodesy. Boll Geod Sci Aff, anno XLIX:133–179

Zund JD, Moore W (1987) Hotine's conjecture in differential geodesy. Bull Géod 61:209–222

2 Adjustment of Triangulation in Space[1]

Introduction

The science of geodesy bears unmistakeable marks of its two-dimensional origin. It is true that geodesists no longer consider that the Earth is flat, in the sense of being expressible in two Euclidian dimensions – we have, in the course of 2000 or more years, progressed to the stage of two non-Euclidian dimensions – but whenever a third dimension obtrudes, as in Nature it must, it is to be got rid of immediately by means of "corrections", or simply ignored, so that all calculations may be done on a surface.

As an example, consider the current methods of calculating geodetic triangulation. "Horizontal" angles are measured at a point in 3-space in the tangent plane to the gravitational equipotential surface passing through the point. These are assumed to be the same as angles between curves of normal section on the surface of a spheroid, which by hallowed convention has come to be considered as representing the Earth. Here at the outset is a large assumption – that the plumb line coincides with the normal to the spheroid – which has never been satisfactorily justified, although we hope in time to have sufficient knowledge of the gravitational field to apply "corrections" to the observed angles to bring the procedure more into line with reality. The curves of normal section, of which there are two corresponding to each straight line in 3-space, or six to a triangle, are then replaced by spheroidal geodesics through correction of the angles, leading to a triangular excess which is more readily calculable. Next we remove the excess in such a way – usually by the simple one-third approximation – as to convert the geodesic triangle to a plane triangle having the same side lengths. The plane triangles are then solved and adjusted by plane trigonometry.

On the face of it, this seems a very odd way of solving and adjusting triangles which, apart from the effects of atmospheric refraction, were plane and straight-sided in the first place, although not coplanar. It should be noted here that atmospheric refraction is a physical fact, to be allowed for as best as we can, before treating the residue as an error of observation; we do not escape it by a particular method of calculation. In the method which follows, residual errors of refraction will to some extent vitiate the result; they do so just as much in the classical method described above, even though they do not enter explicitly into the classical method.

[1] Report dated 24 April 1956 (Tolworth) and presented to an I.A.G. Symposium on European Triangulation (Munich 1956).

Laxity of scientific discipline in this matter is still more evident in the case of the so-called Laplace adjustment for azimuth, which has been the subject of so much unresolved controversy. In its usual form, this adjustment is based on the statement, which hardly rises above the level of a rule-of-thumb, that the difference between the astronomical and spheroidal azimuths at a point (usually without any awareness of the fact that the latter needs special definition at a point in space) is equal to the corresponding difference in longitudes multiplied by the sine of the latitude. As we shall see, this is a simplified – perhaps over-simplified – approximation to an equation of coordinate transformation, which in general applies only between different systems of coordinates of the same point in space. It is a large assumption, which should certainly be justified before the method is used further, that it holds at all between terminal azimuths etc. computed on a spheroid by the usual processes, and direct astronomical measures at a point in space, indefinitely related at most to the terminal point on the spheroid.

Again, measured base-lines are supposed to be reduced to the length which is obtained between corresponding points on the reference spheroid. This presupposes some system of one-to-one correspondence between points in physical space and calculated points on the spheroid which is not clearly defined, and the resulting slack is adjusted into the triangulation.

The time has come for critical examination of the classical methods and conceptions against the background of a method which frankly recognizes that geodetic measures are carried out in flat space of three dimensions, and which clearly defines the coordinate systems used to express the space, even if one coordinate is small. The following method, which is given in outline only and will require modification as it is put into practice, indicates how this may be done without any knowledge or assumption relating to the gravitational field. It requires a much greater volume of stereotyped computation which nowadays need not involve much delay or more than a fraction of the total cost, and in any case is a small price to pay for a rigorous control on the classical methods. If the latter stand up to the test, preferably a searching test in mountainous country, there will be no need to abandon them – any more than Newton's law of gravity, now known to be scientifically incorrect, has been abandoned for ordinary purposes of computation.

Although all the formulae in this particular paper can be obtained from spherical figures, but not as simply, it has been decided to use vector methods, whose introduction even into surface geodesy is long overdue. Index notation, exactly as in McConnell (1931) or most other modern textbooks on the tensor calculus, is used because it has certain advantages (and no disadvantages) in non-orthogonal coordinate systems and is in any case required for other applications, notably to the gravitational field.

Coordinate Systems

We take a right-handed orthogonal triad of fixed unit vectors λ^r, μ^r, ν^r defining the directions of Cartesian axes, x, y, z. The physical axis of rotation of the Earth is parallel to ν^r; and λ^r is perpendicular to ν^r in a plane which defines the origin of longitudes in *all* coordinate systems involving latitude and longitude.

Latitude and longitude are taken as two of the three curvilinear *space* coordinates and are defined as follows. At any point in space, there will be a unit vector ξ^r normal to the third coordinate surface passing through the point, viz. the surface on which the third coordinate has a constant value equal to its value at the point considered. Latitude ϕ (positive North) and Longitude ω (positive East) are then defined by the following scalar products expressing the direction cosines of ξ^r:

$$\xi^r \upsilon_r = \sin\phi$$

$$\xi^r \lambda_r = \cos\phi \cos\omega$$

$$\xi^r \mu_r = \cos\phi \sin\omega$$

from which we obtain the following vector equation:

$$\xi^r = \lambda^r \cos\phi \cos\omega + \mu^r \cos\phi \sin\omega + \upsilon^r \sin\phi \ . \tag{1}$$

We have now to consider the third coordinate. If the first two are to be *astronomical* latitude and longitude, then ξ^r will be the direction of the physical plumb line; the third-coordinate surfaces will be the gravitationally level or equipotential surfaces, and the third-coordinate *must* be the geo-potential or some function of the geo-potential, including the kinetic potential of rotation around the axis. The properties of this coordinate system have been established by Marussi (1949), who also deals (Marussi 1950) with the special case of an axially symmetrical gravitational field to provide standard or *geodetic* coordinates in the space surrounding an equipotential surface of revolution, such as a spheroid.

There is, however, another class of coordinate systems which possess certain advantages in dealing with the space surrounding a given surface, or base surface. If we measure an equal distance h along all the straight normals to this given surface, then the terminal points will lie on a surface which is said to be geodesically parallel to the given surface (see e.g. Eisenhart 1960 p. 57). The two surfaces have common normals. We can take h as the third coordinate, and the geodesically parallel surfaces as third-coordinate surfaces, including the given surface as $h = 0$. Latitude and longitude will then refer to the common normals of the surfaces $h =$ const. and the point considered will have the same latitude and longitude as the foot of the perpendicular on the base surface.

As a special case, take as base surface a spheroid whose principal curvatures are ϱ (meridian) and υ. If the origin of Cartesian coordinates is at the centre of symmetry of this spheroid, whose minor axis (b) is in the direction υ^r, then it can be shown without difficulty that the transformation equations to Cartesian coordinates are

$$\begin{aligned} x &= (\upsilon + h) \cos\phi \cos\omega \\ y &= (\upsilon + h) \cos\phi \sin\omega \\ z &= (k'^2 \upsilon + h) \sin\phi \\ (k'^2 &= b^2/a^2 = 1 - e^2) \end{aligned} \tag{2}$$

giving rise to the following triply orthogonal metric:

$$ds^2 = (\varrho + h)^2 d\phi^2 + (\upsilon + h)^2 \cos^2\phi \, d\omega^2 + dh^2 \ . \tag{3}$$

The choice of coordinate system should make no difference whatever to the final adjustment of a triangulation and if a curvilinear system should be required during the process, or for any purpose not requiring a close overall approximation to the geoid, then there is much to be said for choosing a base sphere ($\varrho = \upsilon$; $k' = 1$).

It is important to realize that each coordinate system is a distinct entity, which must satisfy certain conditions in flat space; we can transform from one system to another, and certain quantities will be invariant under such a transformation, but we may not mix them up, e.g. by attempting to make up a coordinate system from astronomical latitude and longitude and orthometric height.

Azimuth and Zenith Distance

The meridian plane at a point in space is defined as parallel to ξ^r and v^r and a unit vector u^r in the meridian (towards North) and in the third coordinate surface will be given by

$$\begin{aligned} u^r &= -\xi^r \tan\phi + v^r \sec\phi \\ &= -\lambda^r \sin\phi \cos\omega - \mu^r \sin\phi \sin\omega + v^r \cos\phi \ . \end{aligned} \tag{4}$$

A unit vector (towards East) perpendicular to u^r in the third coordinate surface is given by

$$v^r = -\lambda^r \sin\omega + \mu^r \cos\omega \ . \tag{5}$$

A unit vector in azimuth α (measured from u^r towards v^r) and zenith distance β (measured from ξ^r) is accordingly given by

$$\begin{aligned} \sigma^r &= (u^r \cos\alpha + v^r \sin\alpha) \sin\beta + \xi^r \cos\beta \\ &= a\lambda^r + b\mu^r + c\upsilon^r \end{aligned} \tag{6}$$

by substituting Eqs. (1), (4) and (5) where

$$\begin{aligned} a &= -\sin\phi \cos\omega \cos\alpha \sin\beta - \sin\upsilon \sin\alpha \sin\beta + \cos\phi \cos\omega \cos\beta \\ b &= -\sin\phi \sin\omega \cos\alpha \sin\beta + \cos\omega \sin\alpha \sin\beta + \cos\phi \sin\omega \cos\beta \\ c &= \cos\phi \cos\alpha \sin\beta + \sin\phi \cos\beta \end{aligned} \tag{7}$$

or, by forming scalar products with ξ^r, u^r, v^r

$$\begin{aligned} \cos\beta &= a\cos\phi\cos\omega + b\cos\phi\sin\omega + c\sin\phi \\ \cos\alpha\sin\beta &= -a\sin\phi\cos\omega - b\sin\phi\sin\omega + c\cos\phi \\ \sin\alpha\sin\beta &= -a\sin\omega + b\cos\omega \ . \end{aligned} \tag{8}$$

Note that a, b, c are not independent since $a^2 + b^2 + c^2 = 1$. It will, however, be advisable to carry the computation of all three to provide a check.

Now a, b, c are the Cartesian components – with respect to the fixed Cartesian axes λ^r, μ^r, υ^r – of the unit vector σ^r. They are accordingly invariant both for displacements along the straight line σ^r and for any change in the curvilinear coordinates ϕ, ω at a point, provided α, β are measured in relation to the new third-coordinate surface. We may, for instance, substitute direct astronomical measures of ϕ, ω, α, β in Eq. (7); the result would be exactly the same it geodetic coordinates in the system given by Eq. (2) were substituted, provided that the origin of latitudes and longitudes is the same (i.e. υ^r, λ^r are fixed) and so long as the geodetic values of α, β are taken in relation to the surface $h =$ const. through the point[2]. Again, we should obtain the same values for a, b, c at any other point on the straight line whose direction is σ^r; so that a, b, c have identical values at both ends of the side of a spatial triangle. Furthermore we can, at an initial point of a triangulation, substitute astronomical measures and use the resulting values of a, b, c to calculate differences of geodetic coordinates; for examnple, the changes in x, y, z along a straight line of length s are s a, s b, s c, which even though calculated from astronomical values can be used in such a geodetic system as Eq. (2).

To relate this result to the classical conception, differentiate Eq. (7) for small coordinate changes $\delta\phi$, $\delta\omega$ (whether changes in the coordinate system or due to displacements along the line) while retaining a, b, c fixed:

$$\delta\alpha = \sin\phi\,\delta\omega + \cos\beta(\sin\alpha\,\delta\phi - \cos\alpha\cos\phi\,\delta\omega)$$

$$\delta\beta = -\cos\alpha\,\delta\phi - \cos\phi\sin\alpha\,\delta\omega \ .$$

The first equation is the corrected form of the so-called Laplace azimuth equation for a non-horizontal azimuth line, both astronomical and geodetic azimuth at the same point in space being rigorously defined as above. The second equation has no counterpart in the classical methods of two-dimensional adjustment. We shall use both, but in the complete form (6) and (7).

Figural Adjustment

The best possible correction for atmospheric refraction is first made to observed zenith distances and thereafter it is assumed that, subject to observational error, we are dealing with straight lines between the observing stations.

Stations are numbered 1, 2, 3 etc., and we use the following notation:

α_{12} = Azimuth of 2 from 1.

β_{12} = Zenith distance of 2 from 1 corrected for refraction.

α_{123} = Measured "horizontal" angle at 1 between directions to 2 and 3.

[2] This azimuth is not the same as that of the projected direction on the base spheroid. If the latter azimuth is $\bar{\alpha}$, it can be shown that

$$\tan(\alpha-\bar{\alpha}) = \frac{h(\varrho-\upsilon)\sin\alpha\cos\alpha}{\varrho\upsilon + h(\varrho\cos^2\alpha + \upsilon\sin^2\alpha)} = \frac{h(\varrho-\upsilon)\sin\bar{\alpha}\cos\bar{\alpha}}{\varrho\upsilon + h(\varrho\sin^2\bar{\alpha} + \upsilon\cos^2\bar{\alpha})} \ .$$

We shall not, however, have occasion to apply this correction, since we shall not be working on the base spheroid.

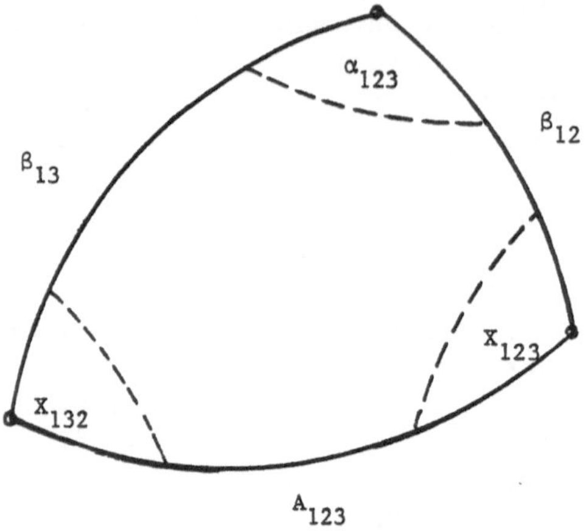

Fig. 1

A_{123} = "Facial" angle at 1 between directions to 2 and 3; that is the angle at 1 of the space triangle 123. This angle is shown in relation to the measured angles in the spherical Fig. 1, which also shows the base angles X_{123}, X_{132} of the spherical triangle.

A_{123}, X_{123}, X_{132} are first calculated between every pair of directions by a suitable formula for solving Fig. 1 from α_{123}, $\beta_{12}+\beta_{13}$. Note that a separate solution is required for every pair of directions. We cannot, for instance, add A_{123} to A_{134} to obtain A_{124} because these angles will not as a rule be in the same plane.

By differentiation of any such formula we relate errors or corrections to A_{123}, B_{12}, β_{13}, α_{123} as follows:

$$dA_{123} = \cos X_{123}\, d\beta_{12} + \cos X_{132}\, d\beta_{13} + \sin\beta_{12} \sin X_{123}\, d\alpha_{123} \ . \tag{9}$$

Angle and side equations are formed in the usual way, except that we incorporate closure corrections to the A_{123} angles instead of "horizontal" angles.

A typical angle equation would be:

$$dA_{123} + dA_{231} + dA_{312} = 180° - (A_{123} + A_{231} + A_{312}) \ ,$$

in which Eq. (9) should be substituted if it is desired to determine corrections to the actual observations; practical working alone can indicate whether this is necessary, or whether it would be sufficient to determine corrections to the facial angles. There are, of course, no corrections for spherical excess etc.

Side equations must be formed from one of the stations as pole – not the intersection of diagonals, which will usually be skew – and the sine of the A-angles of a particular spatial triangle must be substituted for the opposite sides of that same triangle before differentiation.

The X-angles will be nearly 90° and the coefficients of the $d\beta$'s in the condition equations will accordingly be small. This can be allowed for as usual by accurate computation of the X-angles and by altering the unit of the $d\beta$'s, which could also be weighted down to allow for the probably greater effect of residual refraction on these angles, if experience so indicates.

Adjustment for Scale

Measured base lengths are not reduced to sea level or to a spheroid but are computed as a traverse in a (vertical) plane to obtain the air-line distance between the terminals. Geodimeter measures are corrected solely for curvature of path and velocity; and radar measures are similarly reduced to provide the straight air-line distance between stations.

The base conditions are formed in the usual way by working through actual triangles between bases and expressing the resulting corrections to A-angles.

Adjustment of Unit Side Vectors

σ^r_{12} = Unit vector from 1 towards 2 = $-\sigma^r_{21}$ (Cartesian components a_{12}, b_{12}, c_{12})

τ^r_{123} = Vector product of σ^r_{12} and σ^r_{13} [Cartesian components $(b_{12}c_{13}-c_{12}b_{13})$, $(c_{12}a_{13}-a_{12}c_{13})$, $(a_{12}b_{13}-b_{12}a_{13})$].

A third side vector σ^r_{14} can be expressed as follows in terms of σ^r_{12}, σ^r_{13}, τ^r_{123} with only the facial angles in the coefficients. These will already have been computed and probably adjusted, and their use, in preference to formulae containing zenith distances, may serve to minimize the effect of residual refraction. It is assumed that σ^r_{12}, σ^r_{13}, σ^r_{14} are arranged in order of increasing azimuth, directions being reversed (and signs changed) as necessary to make this so.

$$
\begin{aligned}
\sin^2 A_{123}\sigma^r_{14} &= (\cos A_{124}-\cos A_{123}\cos A_{134})\sigma^r_{12} \\
&\quad +(\cos A_{134}-\cos A_{123}\cos A_{124})\sigma^r_{13}-2n\tau^r_{123} \\
n^2 &= \sin s \sin(s-A_{123})\sin(s-A_{124})\sin(s-A_{134}) \\
2s &= A_{123}+A_{124}+A_{134}\ .
\end{aligned} \tag{10}
$$

If the three side vectors are coplanar, in which case they could be parallel to the sides of a plane triangle, then $A_{124}=A_{123}+A_{134}$ and the formula reduces to:

$$
\sin A_{123}\sigma^r_{14} = -\sin A_{134}\sigma^r_{12}+\sin A_{124}\sigma^r_{13}\ . \tag{11}
$$

If σ^r_{14} is parallel to the third side σ^r_{23} of a plane triangle, then we may write $A_{134}=A_{312}$ and $A_{124}=180°-A_{231}$ and the equation becomes

$$
\sin A_{123}\sigma^r_{23} = -\sin A_{312}\sigma^r_{12}+\sin A_{231}\sigma^r_{13}\ . \tag{12}
$$

If we replace the sines of the angles by the lengths of the opposite sides, this equation correctly expresses the fact that the change in Cartesian coordinates round the triangle is zero.

By means of Eq. (12), which enables us to deduce the closing vector of a triangle, and Eq. (10), which determines a side vector in the next triangle not necessarily coplanar with the first, we can express all side vectors in terms of the first two. The Cartesian components of the first two vectors are determined by substituting astronomical measures of ϕ, ω, α and β in Eq. (7), using the value of β corrected for refraction, and we are accordingly able to carry Cartesian components throughout the triangulation. A direct determination will be provided by Eq. (7) whenever we again have astronomical measures; and the difference between the direct and extended components is to be disturbed throughout the triangulation on any reasonable basis. It will probably be sufficient to carry the components along the shortest traverse between astronomical stations and to distribute the closing error evenly along this traverse, followed by correction of the figural adjustment, much as is usually done in the two-dimensional Laplace azimuth adjustment, which is of course included in theoretically correct form in the present vector adjustment. Alternatively, it may well be advantageous, by differentiation of Eqs. (10) and (12), to form closure conditions for solution simultaneously with the figural and base conditions. The best method will no doubt emerge in the course of practical operation. The adjustment should finally provide consistent values of the Cartesian components for use in the calculation of geodetic coordinates; it is not necessary to evaluate corrections to the observed angles unless this is found to be the best method of effecting the adjustment.

Geodetic Coordinates

An origin can be taken from previous work, provided this has similarly been properly adjusted to astronomical measures. Otherwise, we must start with precise astronomical measures and use them in Eq. (7) to determine the initial pair of side vectors. We can use any geodetic starting elements *provided that when these are substituted in Eq. (7) they give the same components for the initial pair of vectors as the astronomical measures.* A glance at Eq. (8) shows that any alteration in initial latitude and longitude implies a corresponding alteration to both azimuths and zenith distances, whereas in current practice it is seldom enough that even the effect on azimuth is considered.

The choice of third coordinate will then settle the geodetic system. An axially symmetric potential, approximating to the actual physical potential, would seem to be a logical choice, of the same kind as the astronomical system, but the fact is that the calculation of finite differences of coordinates in such a system is not simple, and the main object of adopting a standard geodetic system is to simplify the calculation of position. At the same time we need a system which does not depart too far from the physical plumb line. From this point of view, the system of Eq. (2) is as good as any, if a suitable choice of base spheroid is made, and it is just as easy to calculate standard gravity in this system.

The initial value of the third coordinate at the origin can be taken from previous work. If geodetic latitude and longitude are also taken from previous work, subject to the above proviso, and the shape and size of base spheroid in Eq. (2) are the same, then the entire coordinate system will be the same. Failing any previous work, we can use a spirit-levelled height, or for that matter an arbitrary datum, but it should be appreciated that this settles the coordinate system and any later alteration will require recomputation, or at any rate correction, of all three geodetic coordinates.

The Cartesian coordinates of the origin $\bar{x}$, $\bar{y}$, $\bar{z}$ can now be computed from Eq. (2). Differences of Cartesian coordinates along a side of length s and adjusted vector (a, b, c) are then (s a, s b, s c), so that we can calculate the Cartesian coordinates of all the triangulation stations, and substitute in Eq. (2) to obtain geodetic longitudes and latitudes. Longitude presents no difficulty, and latitude would probably best be obtained by iteration, starting with the close approximation

$$\tan \phi = \frac{3}{k'^2 (x^2 + y^2)^{1/2}} \ .$$

Geodetic azimuth is then obtained from Eq. (8).

Alternatively, we can derive formulae in terms of differences of Cartesian coordinates, but these are not likely to present any advantage if sufficiently large-capacity machines are available to take the full coordinates, which are of the same order as υ.

Conversely, if we are given the geodetic coordinates of two points, we can calculate their Cartesian coordinates and hence their distance apart and direction (a, b, c) in space. Azimuths follow from Eq. (8).

Deflection of the Vertical

If the third vector σ^r_{14} in Eq. (10) is the unit normal ξ^r (in the direction of the plumb line), then $A_{134} = \beta_{13}$ and $A_{124} = \beta_{12}$ and (see Fig. 1) the formula can be reduced to:

$$\sin \alpha_{123} \csc X_{123} \csc X_{132} \xi^r = \cos X_{132} \sigma^r_{12} + \cos X_{123} \sigma^r_{13} - \tau^r_{123} \ . \tag{13}$$

If ϕ, ω are now the astronomical latitude and longitude, the Cartesian components of ξ^r are ($\cos \phi \cos \omega$, $\cos \phi \sin \omega$, $\sin \phi$). Substitution of these in Eq. (13) together with the corresponding (a, b, c) components of σ^r_{13}, τ^r_{123}, accordingly gives three equations (two of which are independent) to determine ϕ, ω. Comparison with geodetic coordinates calculated as above gives a measure of deflection at each point.

Editorial Commentary

This was Hotine's first formal paper which employed tensorial methods. Its content was straightforward and elementary in that it made only modest mathematical demands on the reader. Its content was re-considered in his *A Primer on Non-Classical Geodesy* (1959), and *The Third Dimension in Geodesy* (1960), both of which are included in this monograph. A more complete discussion on the line of observation, internal and external adjustment of networks in space can be found in Chapters 25–27 of MG.

Due to his extensive experience in photogrammetry and his pioneering work in aerial surveying, the topic of triangulation was a natural area for him to test his newly acquired mathematical skills, i.e. tensor analysis. Hotine's plea for a three-dimensional formulation was eloquently and forcibly made in the introduction of the paper. For those of us who grew up in the age of Sputnik or afterward, this plea is a non-issue. However, this was certainly not the case with people of Hotine's generation (see Levallois 1963). Indeed, the present Hotine-Marussi Symposia on Mathematical Geodesy were initiated as a Symposium on Three-Dimensional Geodesy (Venice 1959) with the express purpose of developing and popularizing such techniques. By 1965, this issue had been settled and the third symposium became the Symposium on Mathematical Geodesy (Turin 1965).

Note that on p. 24 Hotine's use of symbols λ^r, μ^r, υ^r for the Cartesian basis vectors was unfortunate, since he subsequently denoted these by A^r, B^r, C^r in *all* his other work. He then employed λ^r, μ^r, υ^r as a more general set of vectors, see paper 3, p. 33.

The contemporary literature in this area is extensive and it has grown into elaborate theories of linear and non-linear adjustment. Many geodesists have contributed to it, and many make serious use of tensor-theoretic methods (see References).

References to Paper 2

Eisenhart L (1960) Riemannian geometry. Princeton

Marussi A (1949) Fondements de géométrie différentielle absolue du champ potential terrestre. Bull Géod 14:411–439

Marussi A (1950) Principi die geodesia intrinseca applicati al campo di Somigliana. Boll Geod Sci Aff, anno IX:1–8; = Principles of intrinsic geodesy applied to the field of Somigliana, re-printed in IG, 101–108

McConnell AJ (1931) Applications of the absolute differential calculus. Blackie, London

References to Editorial Commentary

Blaha G (1984) Tensor structure applied to least squares revisited. Bull Géod 58, 1:1–30

Blaha G, Bessette R (1989) Non-linear least squares method by an isomorphic geometric setup. Bull Géod 63, 2:115–138

Brazier HH, Windsor LM (1957) A test triangulation, a report presented to Study Group No. 1 at the I. A. G. Toronto Assembly. This is a companion to item (4) in this monograph and Hotine's reference. A test triangulation (1957) in paper 3

Levallois JJ (1963) La réhabilitation de la géodésie classique et la géodésie tridimensionelle. Bull Géod 68:193–199

Reilly WI (1980) Three-dimensional adjustment of geodetic networks with incorporation of gravity field data. Report 160, Dep Sci Ind Res, Wellington, New Zealand

Teunissen P (1985) The geometry of geodetic inverse linear mapping and non-linear adjustment. Neth Geod Comm Publ Geod, New Ser 8; 1:177

Vanicek P (1979) Tensor structure and the least squares. Bull Géod 53:221–225

3   Metrical Properties of the Earth's Gravitational Field[1]

Introduction

Some, but not all, of the results in this paper have been obtained in an equivalent form by Professor Marussi (Marussi 1947, 1949) using a special coordinate system: the astronomical latitude and longitude, and the geopotential. The present treatment of the subject is independent of any coordinate system, and thereby achieves some simplification and completeness, although these three quantities, and others which can be directly measured, enter the analysis as scalar functions of position, defined either throughout space or along certain lines. All the properties of the field are shown to depend on five parameters, which are applied to such fundamental geodetic problems as variation of position and along refracted rays. Methods of measuring the parameters by torsion balance and by astro-geodetic observations are also considered. Finally, the results are used to derive briefly the properties of the Marussi metric as a special case of the general analysis and as an introduction to the similar derivation of geodetic coordinate systems in a further paper (Geodetic Coordinate Systems 1957).

The tensor notation used throughout is exactly as in McConnell (1947), and much the same, with obvious modification, as in Eisenhart (1949) and Levi-Civita (1926).

Base Vectors

1. We define as follows three mutually orthogonal unit vector fields A^r, B^r, C^r; right-handed in that order. C^r is parallel to the physical axis of rotation of the Earth; positive direction North. A^r is parallel to the plane determined by C^r and the zenith direction at the origin of longitudes, and is perpendicular to C^r; positive direction outwards from the Earth. B^r completes the right-handed orthogonal triad.

2. A second set of unit orthogonal vector fields λ^r, μ^r, υ^r, right-handed in that order, is defined as follows. υ^r is the direction of the zenith, or outward-drawn normal to the equipotential surface, or tangent to the line of force passing through the point in space under consideration. μ^r, to be known as the meridian

[1] Report dated 28 May 1957 (Tolworth) and presented to Study Group No. 1 at the I.A.G. Toronto Assembly 1957.

direction, lies in the equipotential surface and in the plane of υ^r and C^r; positive in a northerly direction. λ^r, also in the equipotential surface, and to be known as the parallel direction, completes the triad; positive easterly.

3. Astronomical longitude (ω) is the angle between the planes A^r, C^r and υ^r, μ^r, C^r; positive in the position rotation about C^r, that is from A^r to B^r, or East. Astronomical co-latitude ($\pi/2-\phi$) is the angle between C^r and υ^r; or latitude ϕ between υ^r and the plane A^r, B^r, positive North.

4. It should be noted that the direction of the meridian (parallel) as thus defined is not necessarily the same as the surface direction in which the longitude (latitude) is constant. We consider below the conditions for the two directions to coincide.

A curve on the equipotential surface at each point of which the tangent is in the meridian direction will be known as the meridian trace; and similarly for the parallel trace. Here again these curves are not necessarily the loci of points having the same longitude, or latitude.

5. Azimuth (α) is measured East from North, that is from μ^r towards λ^r. Accordingly a unit surface vector in azimuth α is given by ($\lambda^r \sin\alpha + \mu^r \cos\alpha$). A unit space vector in azimuth α and zenith distance β will be given by ($\lambda^r \sin\alpha \sin\beta + \mu^r \cos\alpha \sin\beta + \upsilon^r \cos\beta$).

6. The following covariant vector equations are easily verified for Cartesian axes parallel to A^r, B^r, C^r and are therefore true for the components in any coordinate system. The corresponding contravariant equations are obtained by simply raising the indices throughout.

$$\begin{aligned} \lambda_r &= -A_r \sin\omega + B_r \cos\omega \\ \mu_r &= -A_r \sin\phi\cos\omega - B_r \sin\phi\sin\omega + C_r\cos\phi \\ \upsilon_r &= A_r\cos\phi\cos\omega + B_r\cos\phi\sin\omega + C_r\sin\phi \ . \end{aligned} \tag{6.1}$$

If we consider the Cartesian coordinate x to be a scalar function of position, then the vector equation $x_r = A_r$ is true in Cartesian and therefore in any coordinates – and similarly for y and z – so that by solving the Eqs. (6.1) we have in any coordinate system:

$$\begin{aligned} x_r &= A_r = -\lambda_r\sin\omega - \mu_r\sin\phi\cos\omega + \upsilon_r\cos\phi\cos\omega \\ y_r &= B_r = \lambda_r\cos\omega - \mu_r\sin\phi\sin\omega + \upsilon_r\cos\phi\sin\omega \\ z_r &= C_r = \mu_r\cos\phi + \upsilon_r\sin\phi \ . \end{aligned} \tag{6.2}$$

7. If V is the geopotential (including the effect of uniform rotation, or of moving axes), then the resultant force (including centrifugal force) is the gradient of V and must act in the direction υ^r normal to the equipotential surface; its magnitude by definition is the gravity g. This gives rise to the vector equation

$$V_r = -g\upsilon_r \ , \tag{7.1}$$

in which the negative sign has been taken to make g positive when V decreases along the outward-drawn normal, or zenith.

Although mechanical definitions are given for V, g we are in fact dealing solely with the geometrical properties of a family of surfaces V = const. embedded in flat 3-space; V being any function of position and g the magnitude of its gradient, which is geometrically equivalent to what Weatherburn (1930, p. 32 et passim) called the "distance function." Later, V will be restricted to satisfy the modified Laplace or Poisson equation, but this also is essentially a geometrical consideration, analogous to Einstein's law of gravity obtained by contracting the curvature tensor of 4-space. The Laplacian of V, is indeed simply a contraction of the tensor V_{rs}.

First Derivatives of Base Vectors

8. Since A_r, B_r, C_r constitute parallel vector fields, defined throughout flat space, their intrinsic derivatives taken along *any* line are zero. Consequently, their covariant derivatives are zero.

Taking the covariant derivatives of Eqs. (6.1) we have, for example,

$$\lambda_{rs} = -(A_r \cos\omega + B_r \sin\omega)\omega_s = (\mu_r \sin\phi - \upsilon_r \cos\phi)\omega_s \tag{8.1}$$

on substituting for A_r, B_r from Eq. (6.2). In this tensor equation, true, of course, in any coordinate system, ω_s is the gradient of ω in space, that is $\partial\omega/\partial x^s$ in a space coordinate system x^s.

In the same way,

$$\mu_{rs} = -\sin\phi\,\lambda_r\omega_s - \upsilon_r\phi_s \tag{8,2}$$

$$\upsilon_{rs} = \cos\phi\,\lambda_r\omega_s + \mu_r\phi_s\ . \tag{8.3}$$

9. The covariant derivative of Eq. (7.1) gives

$$V_{rs} = -g_s\upsilon_r - g\upsilon_{rs} = -g\{\upsilon_{rs} + (\log g)_s\upsilon_r\} \tag{9.1}$$

leading to two important consequences. If we multiply by the metric tensor a^{rs} and contract, the left-hand side becomes the Laplacian $\Delta^2 V$ of V.

If the elements of arc in the directions λ^r, μ^r, υ^r are respectively dp, dm, dn then we have for example $a^{rs}\lambda_r\omega_s = \lambda^s\omega_s = (\partial\omega/\partial p)$, and by substituting Eq. (8.3), the final contracted equation becomes

$$-\frac{1}{g}\Delta^2 V = \cos\phi\left(\frac{\partial\omega}{\partial p}\right) + \frac{\partial\phi}{\partial m} + \frac{\partial \log g}{\partial n}\ . \tag{9.2}$$

10. The other consequence of (9.1) is that since V_r is the gradient of a scalar, V_{rs} is symmetrical, i.e. $V_{rs} = V_{sr}$. Using Eq. (8.3), we have the following tensor equation;

$$(\log g)_r\upsilon_s - (\log g)_s\upsilon_r = \cos\phi\,\lambda_r\omega_s + \mu_r\phi_s - \cos\phi\,\lambda_s\omega_r - \mu_s\phi_r\ , \tag{10.1}$$

which can conveniently be split into three vector equations by multiplying by λ^r, μ^r, υ^r in turn and contracting. Since the elements of arc in the directions λ^r, μ^r, υ^r are respectively dp, dm, dn, we have after some slight rearrangement the following equations for the gradient of ω, ϕ and g.

$$\omega_s \cos\phi = (\cos\phi\, \partial\omega/\partial p)\lambda_s + (\partial\phi/\partial p)\mu_s + (\partial \log g/\partial p)\upsilon_s \tag{10.2}$$

$$\phi_s = (\cos\phi\, \partial\omega/\partial m)\lambda_s + (\partial\phi/\partial m)\mu_s + (\partial \log g/\partial m)\upsilon_s \tag{10.3}$$

$$(\log g)_s = (\cos\phi\, \partial\omega/\partial n)\lambda_s + (\partial\phi/\partial n)\mu_s + (\partial \log g/\partial n)\upsilon_s\ . \tag{10.4}$$

Multiplying each equation in turn by λ^s, μ^s, υ^s we have, apart from identities, the following scalar equations:

$$\cos\phi(\partial\omega/\partial m) = (\partial\phi/\partial p) \tag{10.5}$$

$$\cos\phi(\partial\omega/\partial n) = (\partial \log g/\partial p) = \gamma_1 \quad \text{say} \tag{10.6}$$

$$(\partial\phi/\partial n) = (\partial \log g/\partial m) = \gamma_2 \quad \text{say}\ . \tag{10.7}$$

The last equation is often used for a spheroidal equipotential, for which case (10.5) and (10.6) are satisfied identically, but is now seen to be true in a wider context.

The Five Parameters of the Field

11. Consider a geodesic of the equipotential surface in the meridian direction μ^r. The geodesic torsion (τ) and normal curvature (χ_2) of the surface in this direction are by definition the torsion and principal curvature of this geodesic, whose principal normal is minus[2] υ^r and whose binormal is minus λ^r. The covariant form of the second Frenet formula for this geodesic is accordingly

$$-\upsilon_{rs}\mu^s = -\tau\lambda_r - \chi_2\mu_r\ ,$$

whence

$$\tau = \upsilon_{rs}\lambda^r\mu^s = \cos\phi(\partial\omega/\partial m) \quad \text{from (8.3)} \tag{11.1}$$

and

$$\chi_2 = \upsilon_{rs}\mu^r\mu^s = \partial\phi/\partial m \quad \text{from (8.3)}\ . \tag{11.2}$$

12. In the same way, the geodesic torsion (τ') and normal curvature (χ_1) in the parallel direction λ^r are given by the following equation, (the binormal to the geodesic in this case being plus μ^s):

$$-\upsilon_{rs}\lambda^s = +\tau'\mu_r - \chi_1\lambda_r\ ,$$

---

[2] By taking the principal normal of the geodesic in the opposite sense to the outward-drawn surface normal, we make the curvatures positive in the usual case of an equipotential convex to the outward-drawn normal.

whence

$$\chi_1 = \upsilon_{rs}\lambda^r\lambda^s = \cos\phi(\partial\phi/\partial p) \quad \text{from (8.3)} \tag{12.1}$$

and

$$\begin{aligned}\tau' &= -\upsilon_{rs}\mu^r\lambda^s \\ &= -(\partial\phi/\partial p) \quad \text{from (8.3)} \\ &= -\cos\phi(\partial\omega/\partial m) \quad \text{from (10.5)} \\ &= -\tau\,;\end{aligned} \tag{12.2}$$

that is the negative of the geodesic torsion in the meridian direction; a well-known result connecting the geodesic torsions in any orthogonal directions, which enables us to use the same symbol for both.

13. These three parameters χ_1, χ_2, τ, together with the parallel and meridian components γ_1, γ_2 of the gradient of (log g) [Eqs. (10.6 and 10.7)], are fundamental in this treatment of the subject. Collecting results so far;

$$\begin{aligned}&\cos\phi(\partial\omega/\partial m) = \partial\phi/\partial p = \tau \quad \text{(geodetic torsion of meridian)} \\ &\cos\phi(\partial\omega/\partial n) = (\partial\log g)/\partial p = \gamma_1 \\ &\partial\phi/\partial n = (\partial\log g)/\partial m = \gamma_2 \\ &\cos\phi(\partial\omega/\partial p) = \chi_1 \quad \text{(normal curvature of parallel)} \\ &\partial\phi/\partial m = \chi_2\;.\end{aligned} \tag{13.1}$$

14. The divergence of λ^r is now given by

$$\begin{aligned}\lambda^r_r &= (\mu^r\omega_r)\sin\phi - (\upsilon^r\omega_r)\cos\phi \\ &= \sin\phi(\partial\omega/\partial m) - \cos\phi(\partial\omega/\partial n) \\ &= (\tau\tan\phi - \gamma_1)\;.\end{aligned} \tag{14.1}$$

Similarly,

$$\mu^r_r = -(\kappa_1\tan\phi + \gamma_2) \tag{14.2}$$

and

$$\upsilon^r_r = \chi_1 + \chi_2 = 2H\;, \tag{14.3}$$

if H is the mean curvature; a well-known result.

15. The *surface* divergence λ^α_α obtained by dropping the υ_r component of λ_{rs} is $\tau\tan\phi$; and similarly $\mu^\alpha_\alpha = -\chi_1\tan\phi$.

16. We are also able to rewrite Eqs. (10.2) and (10.3) as

$$\cos\phi(\omega_r) = \chi_1\lambda_r + \tau\mu_r + \gamma_1\upsilon_r \tag{16.1}$$

and

$$\phi_r = \tau\lambda_r + \chi_2\mu_r + \gamma_2\nu_r \ . \tag{16.2}$$

We can also rewrite Eqs. (9.2) and (10.4) as

$$(\partial \log g)/\partial n = -\chi_1 - \chi_2 - (\Delta^2 V)/g = -2H - (\Delta^2 V)/g \tag{16.3}$$

$$(\log g)_r = \gamma_1\lambda_r + \gamma_2\mu_r - \{\chi_1 + \chi_2 + (\Delta^2 V)/g\}\nu_r \ . \tag{16.4}$$

17. If $\tilde{\omega}$ is the angular rotation of the Earth (in radians per sidereal second), ϱ the density of matter at the point considered (zero for a point in free air), and k the gravitational constant, the modified Poisson equation enables us to write in conjunction with Eq. (16.3)

$$\Delta^2 V = 2\tilde{\omega}^2 - 4\pi k\varrho \ . \tag{17.1}$$

18. We can easily express the symmetrical tensor

$$V_{rs} = -g_s\nu_r - g\nu_{rs} = -g_s\nu_r - (g\cos\phi)\lambda_r\omega_s - g\mu_r\phi_s$$

in terms of the five parameters by forming such scalar products as $V_{rs}\lambda^r\lambda^s = -(g\cos\phi)(\partial\omega/\partial p)$, the final result being

$$\begin{aligned} -V_{rs}/g = {} & \chi_1\lambda_r\lambda_s + \tau\lambda_r\mu_s + \gamma_1\lambda_r\nu_s \\ & + \tau_1\mu_r\lambda_s + \chi_2\mu_r\mu_s + \gamma_2\mu_r\nu_s \\ & + \gamma_1\nu_r\lambda_s + \gamma_2\nu_r\mu_s - \{\chi_1 + \chi_2 + (\Delta^2 V)/g\}\nu_r\nu_s \ . \end{aligned} \tag{18.1}$$

The five parameters are accordingly components of the Marussi tensor W_{rs} (Marussi 1949 p. 433).

Properties of the Lines of Force

19. We now derive the usually required elements of the field in terms of the five parameters, starting with the properties of the lines of force. Their vector curvature is

$$\begin{aligned} \nu_{rs}\nu^s &= \cos\phi(\partial\omega/\partial n)\lambda_r + (\partial\phi/\partial n)\mu_r \quad \text{from (8.3)} \\ &= \gamma_1\lambda_r + \gamma_2\mu_r \quad \text{from (13.1)} \ , \end{aligned} \tag{19.1}$$

whence we conclude that the principal curvature is $(\gamma_1^2 + \gamma_2^2)^{1/2}$ and the principal normal is a surface vector in azimuth $\tan^{-1}(\gamma_1/\gamma_2)$ which from Eq. (16.4) is also the surface gradient of (log g). The binormal of the line of force is accordingly a surface vector along which (log g) is constant.

20. If $\alpha = \tan^{-1}(\gamma_1/\gamma_2)$, the principal normal is the unit vector $(\lambda_r \sin\alpha + \mu_r\cos\alpha)$ and the binormal is $(-\lambda_r\cos\alpha + \mu_r\sin\alpha)$. Writing τ_0 for the torsion of the line of force, we have from the third Frenet formula:

$$(-\lambda_r\cos\alpha + \mu_r\sin\alpha)_s\nu^s = -\tau_0(\lambda_r\sin\alpha + \mu_r\cos\alpha) \ ,$$

which, with the help of Eqs. (8.1) and (8.2) works out at

$$\tau_0 = \sin\phi(\partial\omega/\partial n) - (\partial\alpha/\partial n)$$

$$= \gamma_1 \tan\phi - \partial\alpha/\partial n \ . \tag{20.1}$$

We investigate $\partial\alpha/\partial n$ later in § 38.

21. If the principal curvature is χ_0 and α is still the azimuth of the principal normal, we have also

$$\gamma_1 = \chi_0 \sin\alpha \ ; \quad \gamma_2 = \chi_0 \cos\alpha \ . \tag{21.1}$$

If the element of arc in the direction of the principal normal is du and in the direction of the binormal is dv, then

$$(\partial \log g)/\partial u = \gamma_1 \sin\alpha + \gamma_2 \cos\alpha = \chi_0 \tag{21.2}$$

$$(\partial \log g)/\partial n = -\gamma_1 \cos\alpha + \gamma_2 \sin\alpha = 0 \ , \tag{21.3}$$

agreeing with the remark at the end of § 19. These simple equations are often useful.

Properties of the Equipotential Surfaces

22. The normal curvature in any azimuth α, on the same lines as Eqs. (11.2) and (12.1, is

$$v_{rs}(\lambda^r \sin\alpha + \mu^r \cos\alpha)(\lambda^s \sin\alpha + \mu^s \cos\alpha)$$

$$= \chi_1 \sin^2\alpha + 2\tau \sin\alpha \cos\alpha + \chi_2 \cos^2\alpha \ ,$$

a generalized form of Euler's theorem.

23. The geodesic torsion in any azimuth α, on the same lines as Eq. (11.1) and remembering that the orthogonal surface vector is now in azimuth $(\pi/2+\alpha)$, is

$$v_{rs}(\lambda^r \cos\alpha - \mu^r \sin\alpha)(\lambda^s \sin\alpha + \mu^s \cos\alpha)$$

$$= (\chi_1 - \chi_2) \sin\alpha \cos\alpha + \tau(\cos^2\alpha - \sin^2\alpha) \ .$$

24. Since the geodesic torsion is zero in the principal directions, the azimuths of the latter are given by

$$\tan 2A = 2\tau/(\chi_2 - \chi_1) \ .$$

25. If κ_1, κ_2 are the principal curvatures of the surface, occurring respectively in azimuths $(\pi/2+A)$, A, we have from § 22

$$\kappa_1 = \chi_1 \cos^2 A - 2\tau \sin A \cos A + \chi_2 \sin^2 A$$

$$\kappa_2 = \chi_1 \sin^2 A + 2\tau \sin A \cos A + \chi_2 \cos^2 A$$

$$\kappa_1 + \kappa_2 = 2H = \chi_1 + \chi_2 \ , \quad \text{a well-known result} \ .$$

$$\kappa_1 - \kappa_2 = (\chi_1 - \chi_2)\cos 2A - 2\tau \sin 2A$$
$$= (\chi_1 - \chi_2)\sec 2A = -2\tau \csc 2A \ , \quad \text{using § 24}$$
$$\kappa_1 \kappa_2 \quad = \text{Gauss curvature K}$$
$$= \tfrac{1}{4}(\chi_1 + \chi_2)^2 - \tfrac{1}{4}(\chi_1 - \chi_2)^2 \sec^2 2A = \chi_1 \chi_2 - \tau^2 \ .$$

The formulae may be inverted as

$$\chi_1 = \kappa_1 \cos^2 A + \kappa_2 \sin^2 A$$
$$\chi_2 = \kappa_2 \sin^2 A + \kappa_2 \cos^2 A$$
$$\tau \ = (\kappa_2 - \kappa_1)\cos A \sin A \ .$$

26. If $\varrho^s = \lambda^s \sin\alpha + \mu^s \cos\alpha$ is a unit surface vector in whose direction the longitude ω is constant, then multiplying Eq. (16.1) by ϱ^r and contracting, we have

$$0 = \chi_1 \sin\alpha + \tau \cos\alpha \ ,$$

so that ω is constant in azimuth $\tan^{-1}(-\tau/\chi_1)$; and similarly ϕ is constant in azimuth $\cos^{-1}(-\tau/\chi_2)$.

27. Similarly, the isozenithal direction in space, along which both ω and ϕ are constant, occurs in azimuth α, zenith distance β, if

$$0 = \chi_1 \sin\alpha \sin\beta + \tau \cos\alpha \sin\beta + \gamma_1 \cos\beta \ ,$$

and

$$0 = \chi_2 \cos\alpha \sin\beta + \tau \sin\alpha \sin\beta + \gamma_2 \cos\beta \ ,$$

whence

$$\tan\alpha = (\gamma_2 \tau - \gamma_1 \chi_2)/(\gamma_1 \tau - \gamma_2 \chi_1) \ ,$$

and

$$\tan\beta = -(\gamma_1 \sin\alpha + \gamma_2 \cos\alpha)/(\text{Normal curvature in azimuth } \alpha)$$
$$= -(\gamma_1 \cos\alpha - \gamma_2 \sin\alpha)/(\text{Geodesic torsion in azimuth } \alpha) \ .$$

Alternatively, the isozenithal line must lie in a direction perpendicular to both ω_s and ϕ_s and is accordingly given by the vector equation

$$(\cos\phi)\varepsilon^{rst}\omega_s\phi_t = \varepsilon^{rst}(\chi_1\lambda_s + \tau\mu_s + \gamma_1 \nu_s)(\tau\lambda_t + \chi_2\mu_t + \gamma_2\nu_t)$$
$$= (\gamma_2\tau - \gamma_1\chi_2)\lambda^r + (\gamma_1\tau - \gamma_2\chi_1)\mu^r + K\nu^r \ , \tag{27.1}$$

whence

$$\sin\alpha \tan\beta \ = (\gamma_2\tau - \gamma_1\chi_2)/K$$
$$\cos\alpha \tan\beta = (\gamma_1\tau - \gamma_2\chi_1)/K \ .$$

28. The geodesic curvature (σ_1) of the parallel trace λ^r is by definition the component in the direction μ^r of its vector $\lambda_{rs}\lambda^s$, so that

$$\sigma_1 = \lambda_{rs}\lambda^s\mu^r = \sin\phi(\partial\omega/\partial p) \quad \text{from (8.1)}$$
$$= \chi_1 \tan\phi \quad \text{from (13.1)} .$$

Similarly the geodesic curvature (σ_2) of the meridian trace is given by

$$\sigma_2 = -\mu_{rs}\lambda^r\mu^s = \sin\phi(\partial\omega/\partial m) = \tau\tan\phi .$$

29. If $\varrho^r = (\lambda^r \cos\alpha - \mu^r \sin\alpha)$ is a unit surface vector in azimuth $(\pi/2+\alpha)$, the geodesic curvature in any azimuth α (arc length s) is given by

$$-(\lambda_r \sin\alpha + \mu_r \cos\alpha)_s \varrho^r (\lambda^s \sin\alpha + \mu^s \cos\alpha) ,$$

which an expansion and substitution of Eqs. (8.1) and (8.2) becomes

$$\sin\phi \sin\alpha(\partial\omega/\partial p) + \sin\phi\cos\alpha(\partial\omega/\partial m) - (\partial\alpha/\partial s)$$
$$= \sigma_1 \sin\alpha + \sigma_2 \cos\alpha - (\partial\alpha/\partial s)$$
$$= (\chi_1 \sin\alpha + \tau\cos\alpha)\tan\phi - (\partial\alpha/\partial s)$$

and since

$$(\partial\omega/\partial s) = \omega_r(\lambda^r \sin\alpha + \mu^r \cos\alpha)$$
$$= \sin\alpha(\partial\omega/\partial p) + \cos\alpha(\partial\omega/\partial m) ,$$

the geodesic curvature may also be written

$$\sin\phi(\partial\omega/\partial s) - (\partial\alpha/\partial s) .$$

30. The equation of the surface geodesic in azimuth α is accordingly

$$(\partial\alpha/\partial s) = (\chi_1 \sin\alpha + \tau\cos\alpha)\tan\phi = \sin\phi(\partial\omega/\partial s) .$$

The latter form of the equation is often used for spheroidal geodesics, but is here seen to be true for any surface V = const.

31. If τ is zero over a particular surface, then the meridian and parallel traces are lines of curvature (§ 24), and are also the surface directions in which ω and ϕ respectively are constant (§ 26); χ_1 and χ_2 are the principal curvatures (§ 25); the meridian traces are surface geodesics ($\sigma_2 = 0$) and are also plane curves, since their torsion as space curves is zero. Conversely, if any one of these statements is generally true, the others are true and $\tau = 0$. But these properties apply to an equipotential surface which is a surface revolution (such as a spheroid) symmetrical about the axis C^r. Accordingly, the parameter τ is a measure of a departure from this state.

Torsion Balance Measures

32. For the sake of completeness, the theory of the inclined torsion balance is reworked in the present notation, although this does not lead to any fresh results.

The line joining the two masses is of length 2l, azimuth α, zenith distance β, unit vector l^r. A unit surface vector in azimuth α is b^r and in $(3\pi/2+\alpha)$ is n^r.

The force acting on one mass m is $m V_r$; its resolved part perpendicular to the plane of v_r and l_r is $m V_r n^r$ and its turning moment about the suspension is $(ml \sin\beta) V_r n^r$. The force acting on the other mass is $m \bar{V}_r$ and the resultant moment of the two is

$$(ml \sin\beta)(V_r - \bar{V}_r) n^r \ .$$

But $(V_r - \bar{V}_r)$ is the intrinsic change in V_r over the short length 2l in the unit direction l^r so that

$$(V_r - \bar{V}_r) = (2l) V_{rs} l^s = -2gl\{\upsilon_{rs} + (\log g)_s \upsilon_r\} l^s \quad \text{from (9.1)} \ ,$$

and remembering that $n^r \upsilon_r = 0$, since these two vectors are perpendicular, the resultant moment is

$$-2mgl^2 \sin\beta\, \upsilon_{rs} n^r l^s = -2mgl^2 \sin\beta\, \upsilon_{rs} n^r (v^s \cos\beta + b^s \sin\beta) \ . \qquad (3.21)$$

Now from Eq. (11.1), minus $\upsilon_{rs} n^r b^s$ is the geodesic torsion in the direction b^s, i.e. in azimuth α (§ 23); $\upsilon_{rs} v^s$ is given by Eq. (19.1); and

$$n^r = -\lambda^r \cos\alpha + \mu^r \sin\alpha \ ,$$

so that the turning moment is finally

$$+2mgl^2 \sin^2\beta [(\chi_1 - \chi_2) \sin\alpha \cos\alpha + \tau(\cos^2\alpha - \sin^2\alpha)$$
$$+ \gamma_1 \cos\alpha \cos\beta - \gamma_2 \sin\alpha \cos\beta] \ . \qquad (32.2)$$

33. Measurements in at least four azimuths will accordingly determine the parameters $(\chi_1 - \chi_2), \tau, \gamma_1, \gamma_2$ and with them many of the other local characteristics of the field, but this instrument will not separate χ_1 and χ_2. These two parameters could conceivably be obtained separately by measuring the turning moment about the *horizontal* axis n^r as suggested by Marussi (1947, p. 25). In that case, if m^r is a unit vector perpendicular to the balance arm in the vertical plane containing the balance arm [i.e. in azimuth α, zenith distance $(\pi/2+\beta)$], then the turning moment is

$$ml(\bar{V}_r - V_r) m^r = 2mgl^2 \{\upsilon_{rs} + (\log g)_s \upsilon_r\} l^s m^r$$

which, with

$$m^r = l^r \sin\alpha \cos\beta + \mu^r \cos\alpha \cos\beta - \upsilon^r \sin\beta \ ,$$

works out at

$$2mgl^2 \sin\beta \cos\beta \{(\Delta^2 V)/g + \chi_1 (1 + \sin^2\alpha) + 2\tau \sin\alpha \cos\alpha$$
$$+ \chi_2 (1 + \cos^2\alpha)\} + 2mgl^2 \cos 2\beta (\gamma_1 \sin\alpha + \gamma_2 \cos\alpha) \qquad (33.1)$$

with, of course, $\Delta^2 V = 2\,\tilde{\omega}^2$ for a point in free air [Eq. (17.1)].

Derivatives of the Parameters

34. The conditions of integrability, corresponding to the Mainardi-Codazzi equations of coordinate surfaces, are easily obtained in this notation. If F is any scalar function of position,

$$\frac{\partial}{\partial p}\left(\frac{\partial F}{\partial n}\right)-\frac{\partial}{\partial n}\left(\frac{\partial F}{\partial p}\right)=(F_r\upsilon^r)_s\lambda^s-(F_r\lambda^r)_s\upsilon^s$$

$$=F_{rs}\upsilon^r\lambda^s+F_r\upsilon^r_{,s}\lambda^s-F_{rs}\lambda^r\upsilon^s-F_r\lambda^r_{,s}\upsilon^s$$

$$=F_r(\upsilon^r_{,s}\lambda^s-\lambda^r_{,s}\upsilon^s)\quad\text{(since } F_{rs} \text{ is symmetrical)}$$

$$=\chi_1\frac{\partial F}{\partial p}+(\tau-\gamma_1\tan\phi)\frac{\partial F}{\partial m}+\gamma_1\frac{\partial F}{\partial n}\ ,$$

on substituting from Eqs. (8.3) and (8.1).

In the same way:

$$\frac{\partial}{\partial m}\left(\frac{\partial F}{\partial n}\right)-\frac{\partial}{\partial n}\left(\frac{\partial F}{\partial m}\right)=(\tau+\gamma_1\tan\phi)\frac{\partial F}{\partial p}+\chi_2\frac{\partial F}{\partial m}+\gamma_2\frac{\partial F}{\partial n}$$

$$\frac{\partial}{\partial m}\left(\frac{\partial F}{\partial p}\right)-\frac{\partial}{\partial p}\left(\frac{\partial F}{\partial m}\right)=\chi_1\tan\phi\frac{\partial F}{\partial p}+\tau\tan\phi\frac{\partial F}{\partial m}\ .$$

Writing $2H=(\chi_1+\chi_2)$ and putting $F=\phi$ we have

$$(\partial\gamma_2/\partial p)-(\partial\tau/\partial m)=2H\tau-\gamma_1\chi_2\tan\phi+\gamma_1\gamma_2\ , \tag{34.1}$$

$$(\partial\gamma_2/\partial m)-(\partial\chi_2/\partial n)=\tau^2+\chi_2^2+\gamma_2^2+\gamma_1\tau\tan\phi\ , \tag{34.2}$$

$$(\partial\tau/\partial m)-(\partial\chi_2/\partial p)=2H\tau\tan\phi \tag{34.3}$$

and the following three equations for $F=\omega$

$$(\partial\gamma_1/\partial p)-(\partial\chi_1/\partial n)=\chi_1^2+\gamma_1^2+\tau^2-2\gamma_1\tau\tan\phi+\chi_1\gamma_2\tan\phi \tag{34.4}$$

$$(\partial\gamma_1/\partial m)-(\partial\tau/\partial n)=2H\tau+(\chi_1+\chi_2)\gamma_1\tan\phi+\gamma_1\gamma_2\tau\tan\phi \tag{34.5}$$

$$(\partial\chi_1/\partial m)-(\partial\tau/\partial p)=(\chi_1^2-\chi_1\chi_2+2\tau^2)\tan\phi\ . \tag{34.6}$$

These six equations are equivalent to the Codazzi equations in (ω,ϕ,V) coordinates, and we may similarly form corresponding equations for any other coordinates or for any other scalar. The following three equations, obtained from $F=(\log g)$ for displacements in free air for which Δ^2V is constant, are, for instance, frequently useful.

$$(\partial\gamma_1/\partial n)+[\partial(2H)/\partial p]=2\gamma_1\{(\Delta^2V)/g\}+\gamma_1\chi_2-\gamma_2\tau+\gamma_1\gamma_2\tan\phi \tag{34.7}$$

$$(\partial\gamma_2/\partial n)+[\partial(2H)/\partial m]=2\gamma_2\{(\Delta^2V)/g\}+\gamma_2\chi_1-\gamma_1\tau-\gamma_1^2\tan\phi \tag{34.8}$$

$$(\partial\gamma_1/\partial m)-(\partial\gamma_2/\partial p)=(\chi_1\gamma_1+\gamma_2\tau)\tan\phi\ , \tag{34.9}$$

the last equation being equivalent to Eq. (34.5) minus Eq. (34.1).

35. By means of these equations we can readily find the Laplacians ($\Delta^2\phi$ etc.) of the main scalars. For example, covariant differentiation of Eq. (16.2) gives us

$$\phi_{rs} = (\tau)_s\lambda_r + (\chi_2)_s\mu_r + (\gamma_2)_s\upsilon_r + \tau\lambda_{rs} + \chi_2\mu_{rs} + \gamma_2\upsilon_{rs} ,$$

which becomes after multiplication by the metric tensor a^{rs} and substitution of Eq. (8.1) etc.

$$\begin{aligned}\Delta^2\phi &= \frac{\partial\tau}{\partial p} + \frac{\partial\chi_2}{\partial m} + \frac{\partial\gamma_2}{\partial n} + \chi_1\gamma_2 - \gamma_1\tau + \tau^2\tan\phi - \chi_1\chi_2\tan\phi \\ &= -(\chi_1^2 + \tau^2 + \gamma_1^2)\tan\phi + 2\gamma_2\{(\Delta^2 V)/g\} + 2\chi_1\gamma_2 - 2\gamma_1\tau\end{aligned}$$

on substituting Eqs. (34.6) and (34.8). And if we write

$$\nabla\omega = a^{rs}\omega_r\omega_s \quad \text{and} \quad \nabla(\log g, \phi) = a^{rs}(\log g)_r\phi_s ,$$

this can be expressed alternatively as

$$\Delta^2\phi = -\sin\phi\cos\phi\nabla\omega - 2\nabla(\log g, \phi) .$$

The equipotential *surface* Laplacian of ϕ is

$$\frac{\partial\tau}{\partial p} + \frac{\partial\chi_2}{\partial m} + \tau\lambda^\alpha_{,\alpha} + \chi_2\mu^\alpha_{,\alpha} = \frac{\partial(2H)}{\partial m} - (\chi_1^2 + \tau^2)\tan\phi$$

on substituting Eq. (34.6) and § 15.

36. In the same way, we have

$$\begin{aligned}(\cos\phi)\Delta^2\omega &= 2\gamma_1\{(\Delta^2 V)/g\} + 2\gamma_1\chi_2 - 2\gamma_2\tau + 2\tan\phi(\chi_1\tau + \chi_2\tau + \gamma_1\gamma_2) \\ &= 2\sin\phi\nabla(\phi, \omega) - 2\cos\phi\nabla(\log g, \omega)\end{aligned}$$

and the *surface* Laplacian of ω is

$$\sec\phi\frac{\partial(2H)}{\partial p} + 4H\tau\sec\phi\tan\phi .$$

37. We have also by the same means

$$\Delta^2(\log g) = -2K - 4H\{(\Delta^2 V)/g\} - \{(\Delta^2 V)/g\}^2 ,$$

which together with

$$\Delta^2(\log g) = a^{rs}(g_r/g)_s = (\Delta^2 g)/g - \nabla(\log g)$$

and

$$\nabla(\log g) = \gamma_1^2 + \gamma_2^2 + (\chi_1 + \chi_2)^2 + 2(\chi_1 + \chi_2)\{(\Delta^2 V)/g\} + \{(\Delta^2 V)/g\}^2$$

gives

$$(\Delta^2 g)/g = \gamma_1^2 + \gamma_2^2 + 4H^2 - 2K ,$$

which is equivalent to the sum of the squares of the principal curvatures of the equipotential surface and of the line of force – a rather remarkable result.

The *surface* Laplacian of (log g) is

$$\frac{\partial\gamma_1}{\partial p}+\frac{\partial\gamma_2}{\partial m}+\gamma_1\tau\tan\phi-\gamma_2\chi_1\tan\phi \;,$$

which with Eqs. (34.4), (34.2) and § 25 can be transformed to

$$\frac{\partial(2H)}{\partial n}+(\gamma_1^2+\gamma_2^2)+(4H^2-2K) \;.$$

38. By means of the above equations, we can also find the derivatives of other scalars along the line of force. For instance, if α is the azimuth of the principal normal to the line of force, we have by differentiating Eq. (21.1) and using Eqs. (34.7), (34.8)

$$\begin{aligned}(\gamma_1^2+\gamma_2^2)(\partial\alpha/\partial n) &= \gamma_2(\partial\gamma_1/\partial n)-\gamma_1(\partial\gamma_2/\partial n)\\ &= \gamma_1\gamma_2(\chi_2-\chi_1)+\tau(\gamma_1^2-\gamma_2^2)+\gamma_1\tan\phi(\gamma_1^2+\gamma_2^2)\\ &\quad -\gamma_2\{\partial(2H)/\partial p\}+\gamma_1\{\partial(2H)/\partial m\} \;,\end{aligned}$$

or, if $\chi_0=(\gamma_1^2+\gamma_2^2)^{1/2}$ is the curvature of the line of force,

$$\begin{aligned}(\partial\alpha/\partial n) &= (\chi_2-\chi_1)\sin\alpha\cos\alpha+\tau(\sin^2\alpha-\cos^2\alpha)+\gamma_1\tan\phi\\ &= (2H)_r(-\lambda^r\cos\alpha+\mu^r\sin\alpha)/\chi_0 \;.\end{aligned}$$

But $(-\lambda^r\cos\alpha+\mu^r\sin\alpha)$ is the binormal to the line of force (§ 20) and if dv is the element of arc in this direction the last term is $\{\partial(2H)/\partial v\}\chi_0$. From Eq. (20.1) the torsion τ_0 of the line of force is

$$\tau_0=(\chi_1-\chi_2)\sin\alpha\cos\alpha+\tau(\cos^2\alpha-\sin^2\alpha)-\{\partial(2H)/\partial v\}/\chi_0 \;,$$

in which the first two terms are the geodesic torsion of the surface in azimuth α.

39. We can similarly derive an expression for $(\partial\chi_0/\partial n)$ from Eqs. (34.7), (34.8) together with $\gamma_1=\chi_0\sin\alpha$; $\gamma_2=\chi_0\cos\alpha$, as in Eq. (21.1)

$$\begin{aligned}\chi_0(\partial\chi_0/\partial n) &= \gamma_1(\partial\gamma_1/\partial n)+\gamma_2(\partial\gamma_2/\partial n)\\ &= 4\chi_0^2\tilde{\omega}^2/g+\chi_0^2(\chi_1\cos^2\alpha-2\tau\sin\alpha\cos\alpha+\chi_2\sin^2\alpha)\\ &\quad -\chi_0(2H)_r(\lambda^r\sin\alpha+\mu^r\cos\alpha) \;;\end{aligned}$$

and if du is the element of arc in the direction of the principal normal, this can be written

$$\begin{aligned}(\partial\log\chi_0/\partial n) &= 4\tilde{\omega}^2/g+(\chi_1\cos^2\alpha-2\tau\sin\alpha\cos\alpha+\chi_2\sin^2\alpha)\\ &\quad -\{\partial(2H)/\partial u\}/\chi_0 \;.\end{aligned}$$

The second term is the normal curvature of the equipotential surface in azimuth $(\pi/2+\alpha)$ or $(3\pi/2+\alpha)$, i.e. in the direction of the binormal to the line of force.

Variation Along Lines

40. We now proceed to find the total change ΔF in a scalar function of position F over a finite length of a line whose unit tangent vector is l^r, principal normal m^r, binormal n^r, curvature $\bar{\chi}$, torsion $\bar{\tau}$ from the ordinary expansion

$$\Delta F = \left(\frac{\partial F}{\partial s}\right)_0 s + \frac{1}{2}\left(\frac{\partial^2 F}{\partial s^2}\right)_0 s^2 + \frac{1}{6}\left(\frac{\partial^3 F}{\partial s^3}\right)_0 s^3 + \ldots \text{etc.} \tag{40.1}$$

for which we require to know the values of $(\partial F/\partial s)$ etc. at the initial point. By taking successive covariant derivatives, we have $(\partial F/\partial s) = F_r l^r$, F_r being as usual the space gradient of F,

$$(\partial^2 F/\partial s^2) = (F_r l^r)_s l^s = F_{rs} l^r l^s + \bar{\chi} F_r m^r \ ,$$

$$(\partial^3 F/\partial s^3) = F_{rst} l^r l^s l^t + 3\bar{\chi} F_{rs} m^r l^s + (\partial\bar{\xi}/\partial s) F_r m^r + \bar{\chi} F_r (\bar{\tau} n^r - \bar{\chi} l^r) \ ,$$

etc. using the Frenet formulae.

41. It may be noted in passing that changes in Cartesian coordinates (e.g. x) along a line are particularly easy to compute from these formulae, since all components of the tensors x_{rs}, x_{rst} etc. are zero in Cartesian coordinates and are therefore zero in any coordinates. The other, non-zero, invariants can, of course, be computed in any coordinates. For example, we take a Cartesian system with origin at one end of the line, x (respectively y) in the direction of λ^r (respectively μ^r) at the same end, z normal to the initial equipotential surface, and consider the change in these coordinates along the line of force. If α is the azimuth of the principal normal to the line of force, we have the following Cartesian components at the origin.

$$l^r = (0, 0, 1)$$

$$m^r = (\sin\alpha, \cos\alpha, 0)$$

$$n^r = (-\cos\alpha, \sin\alpha, 0)$$

and evaluating the above invariants at the origin in the Cartesian system,

$$(\partial x/\partial s)_0 = 0 \ ; \quad (\partial^2 x/\partial s^2)_0 = \chi_0 \sin\alpha = \gamma_1 \quad \text{as in (21.1)} \ ;$$

$$(\partial^3 x/\partial s^3)_0 = (\partial\chi_0/\partial n)\sin\alpha - \chi_0 \tau_0 \cos\alpha = (\partial\gamma_1/\partial n) - \gamma_1\gamma_2 \tan\phi \ ,$$

by differentiating $\gamma_1 = \chi_0 \sin\alpha$ along n and using Eq. (20.1). The values of γ_1, γ_2 are obtained from Eq. (34.7) in terms of the parameters and the form of the initial equipotential surface. We have finally

$$\Delta x = \tfrac{1}{2}\gamma_1 s^2 + \tfrac{1}{6}s^3(\partial\gamma_1/\partial n - \gamma_1\gamma_2\tan\phi) + \ldots \quad \text{and similarly} \ ,$$

$$\Delta y = \tfrac{1}{2}\gamma_2 s^2 + \tfrac{1}{6}s^3(\partial\gamma_2/\partial n + \gamma_1^2\tan\phi) + \ldots$$

$$\Delta z = s - \tfrac{1}{6}s^3(\gamma_1^2 + \gamma_2^2) + \ldots \ .$$

Variation Along Refracted Rays

42. Of particular importance in geodesy is the path of light refracted by the atmosphere, since all triangulation observations are made along such lines. We assume that the air density is a function of potential, since otherwise the model atmosphere would introduce lateral refraction. In setting up such a mathematical model, we do not, of course, assume a complete absence of lateral refraction, but merely that over a large number of lines it will occur as often to one side as the other.

43. Subject to this one assumption, it can be shown[3] that the principal normal m^r to the refracted ray (l^r, azimuth α, zenith distance β) lies in the plane of l^r and υ^r, the normal to the equipotential surface; so that m^r is in azimuth α, zenith distance $(\pi/2+\beta)$. The binormal n^r is a surface vector in azimuth $(3\pi/2+\alpha)$.

$$\begin{aligned} l^r &= \lambda^r \sin\alpha \sin\beta + \mu^r \cos\alpha \sin\beta + \upsilon^r \cos\beta \;, \\ m^r &= \lambda^r \sin\alpha \cos\beta + \mu^r \cos\alpha \cos\beta - \upsilon^r \sin\beta \;, \\ n^r &= -\lambda^r \cos\alpha + \mu^r \sin\alpha \;. \end{aligned} \tag{43.1}$$

44. If μ is the refracting index, the principal curvature of the ray is given by

$$\bar{\chi} = (\log\mu)_r m^r = -\sin\beta\,\partial(\log\mu)/\partial n \;, \tag{44.1}$$

which is calculable for a standard atmosphere (Bomford 1952, p. 156).

45. To find the torsion $\bar{\tau}$, we multiply the third Frenet formula $n_{rs} l^s = -\bar{\tau} m_r$ by υ^r.

$$\begin{aligned} \bar{\tau} &= (\csc\beta) n_{rs} \upsilon^r l^s = -(\csc\beta) \upsilon_{rs} n^r l^s \\ &= (\chi_1 - \chi_2) \sin\alpha \cos\alpha + \tau(\cos^2\alpha - \sin^2\alpha) \\ &\quad + \gamma_1 \cos\alpha \cos\beta - \gamma_2 \sin\alpha \cos\beta \quad \text{[cf. (32.1), (32.2)]} \;. \end{aligned} \tag{45.1}$$

Note that if the ray were straight, we could not put the latter expression equal to zero since the Frenet formula from which it is derived would not apply. Formulae derived for a curved ray in which the latter expression is substituted for $\bar{\tau}$ will nevertheless apply equally to straight rays.

46. Multiplying Eqs. (16.1), (16.2), (16.4) and (7.1) across by l^r we have the following first-order formulae for changes in ω etc. along the line

$$\cos\phi(\partial\omega/\partial s) = \chi_1 \sin\alpha \sin\beta + \tau \cos\alpha \sin\beta + \gamma_1 \cos\beta \tag{46.1}$$

[3] A rapid way of obtaining these results is to make a conformal transformation to a curved space $d\bar{s}^2 = \mu^2 ds^2$ (Levi-Civita 1926, p. 228) and to use Fermat's principle (Levi-Civita 1926, p. 334) by expressing the transformed rays as geodesics of the curved space.

$$\partial\phi/\partial s = \tau \sin\alpha \sin\beta + \chi_2 \cos\alpha \sin\beta + \gamma_2 \cos\beta \tag{46.2}$$

$$\partial(\log g)/\partial s = \gamma_1 \sin\alpha \sin\beta + \gamma_2 \cos\alpha \sin\beta - \{\chi_1 + \chi_2 + (\Delta^2 V)/g\} \cos\beta \tag{46.3}$$

$$\partial V/\partial s = -g \cos\beta\ . \tag{46.4}$$

Notice that, identically,

$$\bar{\tau} \sin\beta = \cos\alpha \cos\phi (\partial\omega/\partial s) - \sin\alpha (\partial\phi/\partial s)\ . \tag{46.5}$$

47. It will also be convenient to have equations for $(\partial\alpha/\partial s)$, $(\partial\beta/\partial s)$. Since n^r is defined along the line, we can differentiate it intrinsically as

$$n_{rs} l^s = (-\lambda_{rs} \cos\alpha + \mu_{rs} \sin\alpha) l^s + (\lambda_r \sin\alpha + \mu_r \cos\alpha)(\partial\alpha/\partial s)\ .$$

Now $b_r = (\lambda_r \sin\alpha + \mu_r \cos\alpha)$ is a surface vector in the plane of l_r and υ_r, so that it is also equal to $(l_r \csc\beta - \upsilon_r \cos\beta)$, and multiplying the last equation across by b^r, we have after substituting for λ_{rs} etc.

$$\begin{aligned} \partial\alpha/\partial s &= \sin\phi (\partial\omega/\partial s) + n_{rs}(l^r \csc\beta - \upsilon^r \cos\beta) l^s \\ &= \sin\phi (\partial\omega/\partial s) - \bar{\tau} \cos\beta \end{aligned} \tag{47.1}$$

[using Eq. (45.1) and noting that $n_{rs} l^r l^s = -\bar{\tau} m_r l^r = 0$] finally we have

$$\partial\alpha/\partial s = (\sin\phi - \cos\phi \cos\alpha \cos\beta)(\partial\omega/\partial s) + \sin\alpha \cos\beta (\partial\phi/\partial s) \tag{47.2}$$

from Eq. (46.5).

48. The simplest way to find $(\partial\beta/\partial s)$ is to differentiate $l^r \upsilon_r = \cos\beta$ intrinsically along l^s. We then have

$$\partial\beta/\partial s = -\upsilon_{rs} l^r l^s \csc\beta - (\bar{\chi} m^r) \upsilon_r \csc\beta\ ,$$

and writing

$$l^r = \upsilon^r \cos\beta + b^r \sin\beta$$

(b^r as above a unit surface vector in azimuth α), this is

$$\begin{aligned} \partial\beta/\partial s &= \bar{\chi} - \upsilon_{rs} \upsilon^s b^r \cos\beta - \upsilon_{rs} b^r b^s \sin\beta \\ &= \bar{\chi} - (\gamma_1 \cos\alpha + \gamma_2 \sin\alpha) \cos\beta \\ &\quad - (\chi_1 \sin^2\alpha + 2\tau \sin\alpha \cos\alpha + \chi_2 \cos^2\alpha) \sin\beta \end{aligned}$$

[using Eq. (19.1) and § 22] finally we have

$$\partial\alpha/\partial s = \bar{\chi} - \sin\alpha \cos\phi (\partial\omega/\partial s) - \cos\alpha (\partial\phi/\partial s) \tag{48.1}$$

from Eqs. (46.1) and (46.2).

49. The second differentials may be obtained from the invariant equations (§ 40) or, now that we have expressions for $(\partial\alpha/\partial s)$, $(\partial\beta/\partial s)$, by direct differentiation of Eq. (46.1) etc.

$$\cos\phi(\partial^2\omega/\partial s^2) = (\partial\chi_1/\partial s)\sin\alpha\sin\beta+(\partial\tau/\partial s)\cos\alpha\sin\beta+(\partial\gamma_1/\partial s)\cos\beta$$
$$+(\partial\omega/\partial s)\{(\chi_1+\chi_2)\sin\phi\cos\alpha\sin\beta-\chi_1\cos\phi\cos\beta$$
$$+\gamma_1\cos\phi\sin\alpha\sin\beta+\gamma_2\sin\phi\cos\beta\}$$
$$+(\partial\phi/\partial s)(-\tau\cos\beta+\gamma_1\cos\alpha\sin\beta)$$
$$+\bar{\chi}(\chi_1\sin\alpha\cos\beta+\tau\cos\alpha\cos\beta-\gamma_1\sin\beta)$$

$$(\partial^2\phi/\partial s^2) = (\partial\tau/\partial s)\sin\alpha\sin\beta+(\partial\chi_2/\partial s)\cos\alpha\sin\beta+(\partial\gamma_2/\partial s)\cos\beta$$
$$+(\partial\omega/\partial s)\{\tau(\sin\phi\cos\alpha\sin\beta-\cos\phi\cos\beta)-\chi_2\sin\phi\sin\alpha\sin\beta$$
$$+\gamma_2\cos\phi\sin\alpha\sin\beta\}+(\partial\phi/\partial s)\{-\chi_2\cos\beta+\gamma_2\cos\alpha\sin\beta\}$$
$$+\bar{\chi}(\tau\sin\alpha\cos\beta+\chi_2\cos\alpha\cos\beta-\gamma_2\sin\beta)$$

$$\partial^2(\log g)/\partial s^2 = -\cos\beta\,\partial(\chi_1+\chi_2)/\partial s+(\partial\gamma_1/\partial s)\sin\alpha\sin\beta$$
$$+(\partial\gamma_2/\partial s)\cos\alpha\sin\beta+2\tilde{\omega}^2\cos\beta(\partial\log g/\partial s)/g$$
$$+(\partial\omega/\partial s)\{\gamma_1(\sin\phi\cos\alpha\sin\beta-\cos\phi\cos\beta)$$
$$-\gamma_2\sin\phi\sin\alpha\sin\beta-(\chi_1+\chi_2+2\tilde{\omega}^2/g)\cos\phi\sin\alpha\sin\beta\}$$
$$+(\partial\phi/\partial s)\{-\gamma_2\cos\beta-(\chi_1+\chi_2+2\tilde{\omega}^2/g)\cos\alpha\sin\beta\}$$
$$+\bar{\chi}\{\gamma_1\sin\alpha\cos\beta+\gamma_2\cos\alpha\cos\beta+(\chi_1+\chi_2+2\tilde{\omega}^2/g)\sin\beta\}$$

$$\partial^2 \upsilon/\partial s^2 = -g\cos\beta(\partial\log g/\partial s)+g\sin\beta(\partial\beta/\partial s)\ .$$

By using Eqs. (47.2) and (48.1), we can in the above expressions have terms in $(\partial\alpha/\partial s)$, $(\partial\beta/\partial s)$ instead of $(\partial\omega/\partial s)$, $(\partial\phi/\partial s)$. Many terms will in practice be negligible, but are given in full for the sake of completeness. If we know the five parameters at both ends of the line, say by torsion balance measures, we can calculate $(\partial\chi_1/\partial s)$ etc. on the assumption that the parameters vary uniformly over the line, and so the second differentials. Substitution of the values of the first and second differentials for the initial point in Eq. (40.1) will then give us to a second order the changes in ω, ϕ, (log g) and V over the line. The third-order terms will, however, contain $(\partial^2\chi_1/\partial s^2)$ etc., and there is no way of evaluating these over a single line. Nor is it likely that they could be determined with worthwhile accuracy from differences in a network necessarily measured at points on different equipotential surfaces. For these reasons it is unlikely that astronomical coordinates can ever be used to calculate geodetic triangulation to sufficient accuracy, even though the measured azimuths and zenith distances are in this gravitational field in which the measurements are necessarily made, fundamental to the whole subject, and in the course of another paper (Geodetic Coordinate Systems 1957) will be applied to geodetic calculations.

Astro-Geodetic Measures

50. It remains to consider conversely whether measures of astronomical latitude etc. over the lines of a triangulation can be used to determine the parameters as

a contribution to knowledge of the form of the field. The procedure would be to determine approximate values of the parameters from the first-order equations [(46.1) etc.] and use these to derive the second-order terms; then re-solving the first-order equations after applying the second-order terms as a correction. For the first-order solution we use observed differences for $(\partial\omega/\partial s)$ etc. and mean values of the coefficients computed from the values at both ends, azimuth, for instance, at the far end being the azimuth of the line produced. The results should give mean values of the parameters over the line, that is the values at some intermediate point on the line.

In the first place, it should be appreciated from Eqs. (47.2) and (48.1) that the azimuth and zenith distance equations are not independent of the latitude and longitude equations; two only of these equations should be used, probably latitude and azimuth, which are more easily measured precisely. These equations will not determine γ_1, γ_2 with any accuracy, although the terms containing γ_1, γ_2 are not necessarily negligible in any but flat country. We might, however, reckon on gravimeter observations at the stations to obtain (46.3) as a third equation. Accordingly, we need at least two lines in different azimuths to determine the five parameters. But the equations for each line will refer to the values at some intermediate point on that line, so that the problem in relation to one isolated point is soluble only on the implicit assumption that the parameters are sensibly constant over an area extending to something like half the length of the lines. If several lines radiating from a point are used in the solution, the residuals will indicate how far this assumption is justified.

51. Current methods of determining "geoidal sections" by similar means achieve a certain simplicity by assuming in addition that the stations are all on the same equipotential surface. They could hardly be expected to produce worthwhile results if the more rigorous methods discussed here do not. True, they do not use large differences of direct astronomical observations, but small "deflections", and that may lend an air of precision. At best, this procedure would amount to forming another first-order equation akin to e.g. Eq. (46.2) in a different set of coordinates and subtracting from (46.2), the precision of which is not thereby increased at all. The assumption that deflections measured and calculated at a point well above the geoid would remain the same at a point vertically below the geoid is probably very far from justifiable.

52. To test these matters, trial computations have been carried out at two points of a test triangulation which is free from observational error, one point, D, being a very disturbed area, which would be unlikely to occur in practice, and one, I, a much less severe case. In both cases the lines are long. Details of the two points and of the lines radiating from them are given in the annexed Table 1. A least-squares solution of the first-order latitude, longitude and gravity equations for all lines radiating from a point has been made and the residuals of each equation are given in Table 2, together with the magnitude of the second-order terms (in brackets). Residuals for each azimuth equation (not included in the solution) are also given. Even if allowance is made for such second-order terms as can be computed, it is apparent that the method does not give results commensurate with

Table 1.

Point	Approximate					Deflection		Gravity anomaly (milligals)
	Latitude (N)		Longitude (E)		Height (m)	Meridian	Prime vertical	
D	37°	00′	20°	40′	1300	−13.982″	−22.141″	−5876
B	36	35	20	25	0	−31.675	−49.379	−6115
C	36	48	19	41	1200	−24.753	+21.541	−6003
E	37	10	20	00	3900	−23.756	+1.051	−5913
G	37	40	20	11	4000	−13.367	−1.458	−5791
F	37	20	20	50	1100	−11.856	−8.861	−5801
I	38	00	20	40	3800	−4.617	−4.747	−5736
G	37	40	20	11	4000	−13.367	−1.458	−5791
F	37	20	20	50	1100	−11.856	−8.861	−5801
H	37	40	21	10	1000	−10.662	−3.430	−5768
J	38	00	21	50	1400	−12.368	−15.598	−5782
K	38	21	21	10	3500	+0.910	−8.611	−5727

the attainable accuracy of observation. The implicit assumption that the parameters are sensibly constant over the length of the lines is evidently too drastic.

53. Apart from uneconomic shortening of the lines, the only other possibility would seem to be simultaneous determination of the parameters over a network of triangulation on the assumption that they vary uniformly along each line. In that case the unknowns in the first-order observational equations [(46.1) etc.] might be written as the mean of the parameters at the two ends. Each additional point of the triangulation thus adds five unknowns, but since it must be connected to at least two points already included, it would provide at least six more equations, so that ultimately the system would be determinable or even redundant. Further consideration needs to be given, however, to the possibility of dependent equations on closing lines.

The Marussi Metric

54. If we take (ω, ϕ, V) as coordinates $(1, 2, 3)$ the contravariant components of the base vectors, using Eq. (13.1), will be:

$$\begin{aligned}
\lambda^r &= \{(\partial\omega/\partial p)\,,\quad (\partial\phi/\partial p)\,,\quad (\partial V/\partial p)\} = (\chi_1 \sec\phi, \tau, 0) \\
\mu^r &= \{(\partial\omega/\partial m)\,,\quad (\partial\phi/\partial m)\,,\quad (\partial V/\partial m)\} = (\tau \sec\phi, \chi_2, 0) \\
\nu^r &= \{(\partial\omega/\partial n)\,,\quad (\partial\phi/\partial n)\,,\quad (\partial V/\partial n)\} = (\gamma_1 \sec\phi, \gamma_2, -g)\ .
\end{aligned} \tag{54.1}$$

55. By solving the nine equations

$$\lambda^r \lambda_s + \mu^r \mu_s + \nu^r \nu_s = \delta^r_s$$

Table 2.

Line	Longitude		Latitude		Gravity		Azimuth	
	Difference	Residual (Comp. − Obs.)	Difference	Residual (Comp. − Obs.)	Difference (Milligals)	Residual (Comp. − Obs.)	Difference	Residual (Comp. − Obs.)
DB	−913.776″	+23.01″ (+4.77″)	−1467.693″)	+6.45″ (−0.93″)	+127	+88 (−61)	−546.914″	+13.42″ (+5.39″)
DC	−3575.378″	−33.86″ (+9.30″)	−710.771″	−19.44″ (−14.71″)	−113	−64 (−203)	−2146.662″	−20.31″ (+10.81″)
DE[a]	−2400.959″	−1.12″ (−5.01″)	+630.226″	−5.79″ (−6.78″)	−823	+3 (−95)	−1447.768″	+0.42″ (−5.92″)
DG	−1764.120″	+18.54″ (−14.99″)	+2400.616″	−6.26″ (−3.88″)	−689	+44 (−179)	−1070.511″	+10.97″ (−17.39″)
DF	+596.576″	−6.20″ (+2.77″)	+1242.126″	+5.34″ (−0.28″)	+167	+44 (−40)	+360.480″	−3.78″ (+3.08″)
IG	−1775.818″	−0.68″ (+8.22″)	−1238.750″	−3.19″ (−3.63″)	−147	−29 (−81)	−1089.234″	−0.41″ (+9.30″)
IF	+584.879″	−2.29″ (−5.13″)	−2397.239″	+2.87″ (−0.76″)	+709	+12 (−137)	+357.730″	−1.36″ (−5.87″)
IH[a]	+1781.691″	+7.91″ (−7.60″)	−1196.045″	−0.90″ (−3.93″)	+802	+16 (−79)	+1093.398″	+4.68″ (−8.77″)
IJ	+4156.230″	−0.30″ (+0.41″)	−27.751″	−8.15″ (−20.26″)	+693	−21 (−250)	+2559.257″	+0.06″ (+0.04″)
IK	+1805.046″	+0.01″ (+8.27″)	+1205.526	+0.07″ (−3.76″)	+131	−16 (−80)	+1115.578	+0.01″ (+9.31″)

[a] Not included in solution.

we can readily obtain the covariant components, beginning with the three equations $-g\upsilon_s = \delta^3_s$, which give υ_s at once. Writing $K = (\chi_1\chi_2 - \tau^2)$, the Gauss curvature of the equipotential surface, we have

$$\begin{aligned} K\lambda_r &= \{\chi_2 \cos\phi, -\tau, (\chi_2\gamma_1 - \tau\gamma_2)/g\} \\ K\mu_r &= \{-\tau\cos\phi, \chi_1, (\chi_1\gamma_2 - \tau\gamma_1)/g\} \\ \upsilon_r &= \{0, 0, -1/g\} \ . \end{aligned} \tag{55.1}$$

56. From Eq. (16.4), we have

$$\partial(\log g)/\partial\omega = \gamma_1\lambda_1 + \gamma_2\mu_1 = (\chi_2\gamma_1 - \tau\gamma_2)\cos\phi/K \tag{56.1}$$

and

$$\partial(\log g)/\partial\phi = \gamma_1\lambda_2 + \gamma_2\mu_2 = (\chi_1\gamma_2 - \tau\gamma_1)/K \ , \tag{56.2}$$

so that we can write alternatively

$$\lambda_3 = -\sec\phi\{\partial(1/g)/\partial\omega\} \tag{56.3}$$

and

$$\mu_3 = -\partial(1/g)/\partial\phi$$

and then

$$\begin{aligned} \partial g/\partial V &= g\gamma_1\lambda_3 + g\gamma_2\mu_3 - g\{\chi_1 + \chi_2 + (\Delta^2 V)/g\}\upsilon_3 \quad (16.4) \\ &= (\chi_1\gamma_2^2 + \chi_2\gamma_1^2 - 2\tau\gamma_1\gamma_2)/K + \{\chi_1 + \chi_2 + (\Delta^2 V)/g\} \ . \end{aligned} \tag{56.4}$$

57. Components of the metric tensor a_{rs} or a^{rs} follow at once by substitution in the formulae:

$$a_{rs} = \lambda_r\lambda_s + \mu_r\mu_s + \upsilon_r\upsilon_s \ ,$$

or

$$a^{rs} = \lambda^r\lambda^s + \mu^r\mu^s + \upsilon^r\upsilon^s \ ,$$

true for any triad of orthogonal unit vectors.

58. If F is any scalar

$$\begin{aligned} (\partial F/\partial p) &= F_r\lambda^r = \chi_1\sec\phi(\partial F/\partial\omega) + \tau(\partial F/\partial\phi) \\ (\partial F/\partial m) &= F_r\mu^r = \tau\sec\phi(\partial F/\partial\omega) + \chi_2(\partial F/\partial\phi) \\ (\partial F/\partial n) &= F_r\upsilon^r = \gamma_1\sec\phi(\partial F/\partial\omega) + \gamma_2(\partial F/\partial\phi) - g(\partial F/\partial V) \ . \end{aligned} \tag{58.1}$$

The Codazzi relations, which in the case of these coordinates reduce to five independent equations, can be obtained as derivatives of the five parameters χ_1 etc. with respect to ω, ϕ, V by means of the above equations and § 34, after any desired manipulation.

By solving the last set of equations or by evaluating the equation $F_r = (\partial F/\partial p)\lambda_r + (\partial F/\partial m)\mu_r + (\partial F/\partial n)\upsilon_r$ in (ω, ϕ, V) coordinates, we have also

$$K \sec\phi(\partial F/\partial\omega) = \chi_2(\partial F/\partial p) - \tau(\partial F/\partial m)$$
$$K(\partial F/\partial\phi) = -\tau(\partial F/\partial p) + \chi_1(\partial F/\partial m)$$
$$Kg(\partial F/\partial V) = (\chi_2\gamma_1 - \gamma_2\tau)(\partial F/\partial p) + (\chi_1\gamma_2 - \gamma_1\tau)(\partial F/\partial m) - K(\partial F/\partial n) \ . \tag{58.2}$$

59. The remaining metrical properties of the space follow in the usual way from the metric tensor, but the foregoing general analysis will enable us to take several short-cuts. For instance, to find the Christoffel symbols, we have

$$V_{rs} = \frac{\partial^2 V}{\partial x^r \partial x^s} - \begin{Bmatrix} t \\ r\,s \end{Bmatrix} V_t = -\begin{Bmatrix} 3 \\ r\,s \end{Bmatrix}$$

so that, using Eqs. (9.1) and (8.3), the six symbols with superscript 3 are given by

$$\frac{1}{g}\begin{Bmatrix} 3 \\ r\,s \end{Bmatrix} = \upsilon_{rs} + (\log g)_s \upsilon_r = \cos\phi\,\lambda_r\omega_s + \mu_r\phi_s + \upsilon_r(\log g)_s$$
$$= \cos\phi\,\lambda_r\delta^1_s + \mu_r\delta^2_s - \delta^3_r(\log g)_s/g \ .$$

For example,

$$\begin{Bmatrix} 3 \\ 1\,2 \end{Bmatrix} = -\frac{g\tau\cos\phi}{K} \ ; \quad \begin{Bmatrix} 3 \\ 3\,3 \end{Bmatrix} = -\frac{\partial \log g}{\partial V} \quad \text{etc.}$$

60. In the same way, we have, by differentiating Eq. (10.2),

$$-\cos\phi \begin{Bmatrix} 1 \\ r\,s \end{Bmatrix} = \cos\phi\,\omega_{rs}$$
$$= \phi_s\omega_r\sin\phi + (\chi_1)_s\lambda_r + (\tau)_s\mu_r + (\gamma_1)_s\upsilon_r + \chi_1\lambda_{rs} + \tau\mu_{rs} + \gamma_1\upsilon_{rs} \ ;$$

and by substituting Eq. (8.1) etc. this is

$$-\cos\phi \begin{Bmatrix} 1 \\ r\,s \end{Bmatrix} = \sin\phi\,\delta^1_r\delta^2_s + \lambda_r(\chi_1)_s + \mu_r(\tau)_s + \upsilon_r(\gamma_1)_s$$
$$+ \chi_1(\mu_r\sin\phi - \upsilon_r\cos\phi)\delta^1_s - \tau(\sin\phi\,\lambda_r\delta^1_s + \upsilon_r\delta^2_s)$$
$$+ \gamma_1(\cos\phi\,\lambda_r\delta^1_s + \mu_r\delta^2_s)$$

so that, for example

$$\begin{Bmatrix} 1 \\ 1\,2 \end{Bmatrix} = \frac{\tau}{K}\frac{\partial\tau}{\partial\phi} = \frac{\chi_2}{K}\frac{\partial\chi_1}{\partial\phi} + \frac{\gamma_1\tau}{K} - \tan\phi$$

or

$$\begin{Bmatrix} 1 \\ 2\,1 \end{Bmatrix} = \frac{\tau\sec\phi}{K}\frac{\partial\chi_1}{\partial\omega} - \frac{\chi_1\sec\phi}{K}\frac{\partial\tau}{\partial\omega} - \frac{\chi_1^2\tan\phi}{K} - \frac{\tau^2\tan\phi}{K} + \frac{\gamma_1\tau}{K} \ ,$$

the two being shown to be identical by means of § 58 and § 34. Also,

$$\left\{ {2 \atop rs} \right\} = -\phi_{rs} = -(\tau)_s \lambda_r - (\chi_2)_s \mu_r - (\gamma_2)_s \upsilon_r$$
$$-\tau(\mu_r \sin\phi - \upsilon_r \cos\phi)\delta_s^1 + \chi_2(\sin\phi\,\lambda_r \delta_s^1 + \upsilon_r \delta_s^2)$$
$$-\gamma_2(\cos\phi\,\lambda_r \delta_s^1 + \mu_r \delta_s^2) \ .$$

61. If we take ω, ϕ as coordinates (1,2) on the equipotential *surface*, the components of the base vectors, considered now as surface vectors in two dimensions, are

$$\lambda^\alpha = \{(\partial\omega/\partial p), (\partial\phi/\partial p)\} = (\chi_1 \sec\phi, \tau)$$
$$\mu^\alpha = \{(\partial\omega/\partial m), (\partial\phi/\partial m)\} = (\tau \sec\phi, \chi_2)$$

and by solving the tensor equation

$$\lambda^\alpha \lambda_\beta + \mu^\alpha \mu_\beta = \delta_\beta^\alpha$$

we have the covariant components

$$K\lambda_\alpha = (\chi_2 \cos\phi, -\tau)$$
$$K\mu_\alpha = (-\tau\cos\phi, \chi_1) \ ,$$

all of which are the same as the (1,2) space cmponents.

62. The surface metric tensor follows from

$$\bar{a}^{\alpha\beta} = \lambda^\alpha \lambda^\beta + \mu^\alpha \mu^\beta \ ; \quad \bar{a}_{\alpha\beta} = \lambda_\alpha \lambda_\beta + \mu_\alpha \mu_\beta \ .$$

63. In these coordinates the second-order magnitudes $b_{\alpha\beta}$ of the surface are given by

$$b_{\alpha\beta} = -a_{mn} \upsilon^m_{,\alpha} (\partial x^n / \partial x^\beta) = -\upsilon_{\beta\alpha}$$

(taken in relation to the space metric)

$$= -\cos\phi\,\lambda_\beta \lambda_\alpha^1 - \mu_\beta \delta_\alpha^2 \quad \text{from (8.3)}$$
$$= (-\chi_2 \cos^2\phi/K, \tau\cos\phi/K, -\chi_1/K)$$

and the following tensor equation, true in any coordinates, is easily verified

$$b_{\alpha\beta} = -\chi_1 \lambda_\alpha \lambda_\beta - \tau(\lambda_\alpha \mu_\beta + \mu_\alpha \lambda_\beta) - \chi_2 \mu_\alpha \mu_\beta \ .$$

64. The third-order magnitudes of the surface in these coordinates are

$$c_{\alpha\beta} = a_{mn} \upsilon^m_{,\alpha} \upsilon^n_{,\beta} = \cos^2\phi\,\delta_\alpha^1 \delta_\beta^1 + \delta_\alpha^2 \delta_\beta^2$$
$$= (\cos^2\phi, 0, 1)$$

with the tensor equation

$$c_{\alpha\beta} = (\chi_1^2 + \tau^2)\lambda_\alpha \lambda_\beta + 2H\tau(\lambda_\alpha \mu_\beta + \mu_\alpha \lambda_\beta) + (\chi_2^2 + \tau^2)\mu_\alpha \mu_\beta \ ,$$

showing that with the sign conventions used throughout this paper (to make χ_1, χ_2 positive),

$$c_{\alpha\beta} + 2H b_{\alpha\beta} + K a_{\alpha\beta} = 0 \ .$$

65. If barred quantities refer to the surface metric, and unbarred to the space metric, notice that

$$a_{\alpha\beta} = \bar{a}_{\alpha\beta} \quad \text{and} \quad a^{\alpha\beta} = \bar{a}^{\alpha\beta} + \upsilon^{\alpha}\upsilon^{\beta} (\alpha, \beta = 1, 2) \tag{65.1}$$

and since the (1,2) covariant components are the same, so are the Christoffel symbols of the first kind;

$$[\alpha\beta, \gamma] = \overline{[\alpha\beta, \gamma]} \ . \tag{65.2}$$

As regards the Christoffel symbols of the second kind, we have

$$\begin{aligned} \begin{Bmatrix} \gamma \\ \alpha\beta \end{Bmatrix} &= a^{\gamma m}[\alpha\beta, m] = a^{\gamma 3}[\alpha\beta, 3] + a^{\gamma\delta}[\alpha\beta, \delta] \\ &= \bar{a}^{\gamma\delta}\overline{[\alpha\beta, \delta]} + \upsilon^{\gamma}\upsilon^{3}[\alpha\beta, 3] + \upsilon^{\gamma}\upsilon^{\delta}[\alpha\beta, \delta] \ , \end{aligned} \tag{65.3}$$

$$\begin{aligned} \begin{Bmatrix} \gamma \\ \alpha\beta \end{Bmatrix} - \overline{\begin{Bmatrix} \gamma \\ \alpha\beta \end{Bmatrix}} &= \upsilon^{\gamma}\upsilon^{m}[\alpha\beta, m] = \upsilon^{\gamma}\upsilon_{k} a^{km}[\alpha\beta, m] = \upsilon^{\gamma}\upsilon_{k} \begin{Bmatrix} k \\ \alpha\beta \end{Bmatrix} \\ &= -\frac{1}{g}\upsilon^{\gamma} \begin{Bmatrix} 3 \\ \alpha\beta \end{Bmatrix} = -\upsilon^{\gamma}\upsilon_{\alpha\beta} \ , \quad \text{by § 59} \ , \end{aligned} \tag{65.4}$$

a tensor equation which is true in any coordinates, provided only that the coordinates of the surface and of the space are the same. Subject to the same proviso, the (1,2) components of any surface vector ϱ^{α}, which must be a linear combination of λ^{α} and μ^{α} are the same, viz. $\varrho^{\alpha} = \bar{\varrho}^{\alpha}$ and $\varrho_{\alpha} = \bar{\varrho}_{\alpha}$. As regards the covariant derivative

$$\begin{aligned} \varrho_{\alpha\beta} &= \frac{\partial \varrho_{\alpha}}{\partial x^{\beta}} - \begin{Bmatrix} m \\ \alpha\beta \end{Bmatrix} \varrho_{m} = \frac{\partial \varrho_{\alpha}}{\partial x^{\beta}} - \overline{\begin{Bmatrix} \gamma \\ \alpha\beta \end{Bmatrix}} \bar{\varrho}_{\gamma} + \upsilon^{\gamma}\upsilon_{\alpha\beta}\varrho_{\gamma} - \begin{Bmatrix} 3 \\ \alpha\beta \end{Bmatrix} \varrho_{3} \\ &= \bar{\varrho}_{\alpha\beta} + \upsilon^{\gamma}\upsilon_{\alpha\beta}\varrho_{\gamma} + \upsilon^{3}\upsilon_{\alpha\beta}\varrho_{3} = \bar{\varrho}_{\alpha\beta} + (\upsilon^{m}\varrho_{m})\upsilon_{\alpha\beta} \\ &= \bar{\varrho}_{\alpha\beta} \ , \quad \text{since } \varrho_{m} \text{ is a surface vector and } \upsilon^{m}\varrho_{m} = 0 \ . \end{aligned}$$

As a special case, we have from Eqs. (8.1) and (8.2)

$$\begin{aligned} \bar{\lambda}_{\alpha\beta} &= \sin\phi\,\mu_{\alpha}\delta^{1}_{\beta} \\ \bar{\mu}_{\alpha\beta} &= -\sin\phi\,\lambda_{\alpha}\delta^{1}_{\beta} \ , \end{aligned} \tag{65.6}$$

which are frequently useful in considering the differential geometry of the equipotential surface.

Variation of Position

66. We next investigate changes in the length (ds), azimuth ($d\alpha$), and zenith distance ($d\beta$) of a light ray in space due to changes in the coordinates dx^r and $d\bar{x}^r$ at both ends of the line. This leads to certain formulae which are required for the rigorous adjustment of triangulation, in whatever coordinates this may be carried out.

Quantities referring to the far end of the line are barred, e.g. $\bar{\lambda}_r$ is the base vector λ_r at the far end of the line. Azimuth and zenith distance at the far end refer to the production of the line, not to the back azimuth etc.

67. For the present purpose it will be sufficient to suppose that the refracted ray has been replaced by its chord, or in other words to suppose that the chord suffers the same changes as the ray. We denote the length of the chord by S and its unit vector by

$$l^r = \lambda^r \sin\alpha \sin\beta + \mu^r \cos\alpha \sin\beta + \nu^r \cos\beta \ . \tag{67.1}$$

As before, we denote a unit surface vector in azimuth α by

$$b^r = \lambda^r \sin\alpha + \mu^r \cos\alpha \ ; \tag{67.2}$$

a unit surface vector in azimuth $3\pi/2+\alpha$ perpendicular to l^r by

$$n^r = -\lambda^r \cos\alpha + \mu^r \sin\alpha \ ; \tag{67.3}$$

also a unit vector in azimuth α perpendicular to l^r by m^r

$$m^r = \lambda^r \sin\alpha \cos\beta + \mu^r \cos\alpha \cos\beta - \nu^r \sin\beta \ . \tag{67.4}$$

68. If the position vector [Cartesian components (x, y, z)] is ϱ^r, we have in any coordinates

$$s\,l^r = \bar{\varrho}^r - \varrho^r$$

and differentiating this for a change in coordinate dx^r (the barred end being for the moment held fixed), we have

$$l^r ds + s\,l^r_s dx^s = -\varrho^r_{,s} dx^s \ . \tag{68.1}$$

But $\varrho^r_{,s}$ in Cartesian coordinates is $\partial x^r/\partial s^s = \delta^r_s$ so that in any coordinates we have the tensor equation

$$\varrho^r_{,s} = \delta^r_s = \bar{\varrho}^r_s$$

and multiplying (68.1) by l_r, we find the variation in length of the line to be

$$ds = -\delta^r_s l_r dx^s = -b_s dx^s \ .$$

Substituting in (68.1) we obtain

$$s\,l^r_{,s} dx^s = (l^r l_s - \delta^r_s) dx^s \ . \tag{68.2}$$

Similarly for changes $d\bar{x}^r$ in coordinates at the barred end,

$$ds = \bar{l}_s d\bar{x}^s$$

and

$$s\bar{l}^r_{,s} d\bar{x}^s = (\delta^r_s - \bar{l}^r \bar{l}_s) d\bar{x}^s \tag{68.3}$$

so that the total variation in length of the line is

$$ds = \bar{l}_s d\bar{x}^s - l_s dx^s \; . \tag{68.4}$$

Note that $\bar{l}_s$, although the same vector as l_s, has not the same coordinates at the barred end, except in Cartesian coordinates.

69. Now consider changes in $\sin\alpha \sin\beta = l^r \lambda_r$ and in $\cos\alpha \sin\beta = l^r \mu_r$ due to $d\bar{x}^s$ and dx^s at the two ends of the line, and write e.g. (λ_r) for a unit vector parallel to λ_r at the barred end of the line:

$$\cos\alpha \sin\beta d\alpha + \sin\alpha \cos\beta d\beta = (l^r_s \lambda_r + l^r \lambda_{rs}) dx^s + \bar{l}^r_{,s} (\lambda_r) d\bar{x}^s \tag{69.1}$$

$$-\sin\alpha \sin\beta d\alpha + \cos\alpha \cos\beta d\beta = (l^r_{,s} \mu_r + l^r \mu_{rs}) dx^s + \bar{l}^r_{,s} (\mu_r) d\bar{x}^s \; , \tag{69.2}$$

in which (λ_r), (μ_r) have remained fixed in direction during the variation $d\bar{x}^s$.

Eliminating $d\beta$ and using Eqs. (67.3), (68.2), (68.3) and the usual expressions for λ_{rs} etc.

$$\begin{aligned} s \sin\beta d\alpha &= -s l^r_{,s} n_r dx^s - s\bar{l}^r_{,s} (n_r) d\bar{x}^s + s' l^r (\lambda_{rs} \cos\alpha - \mu_{rs} \sin\alpha) dx^s \\ &= +n_s dx^s - (n_s) d\bar{x}^s + s(\sin\phi \sin\beta - \cos\alpha \cos\beta \cos\phi)\omega_s dx^s \\ &\quad + s(\sin\alpha \cos\beta)\phi_s dx^s \; . \end{aligned} \tag{69.3}$$

70. This equation is true in any coordinates, provided that we take the components n_s etc., in the same coordinates. In the (ω, ϕ, V) system, the components of $n_r = -\lambda_r \cos\alpha + \mu_r \sin\alpha$ using § 56 are

$$\begin{aligned} K n_1 &= -(\chi_2 \cos\alpha + \tau \sin\alpha) \cos\phi \\ K n_2 &= (\tau \cos\alpha + \chi_1 \sin\alpha) \\ g K n_3 &= -\gamma_1 (\chi_2 \cos\alpha + \tau \sin\alpha) + \gamma_2 (\tau \cos\alpha + \chi_1 \sin\alpha) \; . \end{aligned}$$

To evaluate (n_r) note that its Cartesian components are the same as those of n_r so that

$$(n_r) = (n_s A^s) \bar{A}_r + (n_s B^s) \bar{B}_r + (n_s C^s) \bar{C}_r$$

which on using Eq. (6.2) becomes

$$(n_r) = P\bar{\lambda}_r + Q\bar{\mu}_r + R\bar{\nu}_r$$

with

$$\begin{aligned} P &= -\cos\alpha \cos(\bar{\omega} - \omega) + \sin\phi \sin\alpha \sin(\bar{\omega} - \omega) \\ Q &= \sin\bar{\phi} \sin\phi \sin\alpha \cos(\bar{\omega} - \omega) + \sin\bar{\phi} \cos\alpha \sin(\bar{\omega} - \omega) + \cos\phi \cos\bar{\phi} \sin\alpha \\ R &= -\sin\phi \cos\bar{\phi} \sin\alpha \cos(\bar{\omega} - \omega) - \cos\bar{\phi} \cos\alpha \sin(\bar{\omega} - \omega) + \cos\phi \sin\bar{\phi} \sin\alpha \end{aligned}$$

and $P^2 + Q^2 + R^2 = 1$ as a check, since (n_r) is a unit vector.

The change in azimuth for changes in (ω, ϕ, V) coordinates is then given by

$$\begin{aligned} s \sin\beta\, d\alpha = {} & d\omega(n_1 + s\sin\phi\sin\beta - s\cos\alpha\cos\beta\cos\phi) \\ & + d\phi(n_2 + s\sin\alpha\cos\beta) + n_3 dN \\ & + d\bar{\omega}(-P\bar{\chi}_2\cos\bar{\phi}/\bar{K} + Q\bar{\tau}\cos\bar{\phi}/\bar{K}) \\ & + d\bar{\phi}(+P\bar{\tau}/\bar{K} - Q\bar{\chi}_1/\bar{K}) \\ & - d\bar{N}\{P(\bar{\chi}_2\bar{\gamma}_1 - \bar{\tau}\bar{\gamma}_2)/\bar{g}\bar{K} + Q(\bar{\chi}_1\bar{\gamma}_2 - \bar{\tau}\bar{\gamma}_1)/\bar{g}\bar{K} - R/\bar{g}\} \;. \end{aligned}$$

We can, of course, use changes in dynamic height $dh = -dN/g$ (and $d\bar{h} = -d\bar{N}/\bar{g}$) instead of dN (and $d\bar{N}$).

71. To obtain the change in zenith distance we eliminate $d\alpha$ between Eqs. (69.1) and (69.2) and use (67.2):

$$s\cos\beta\, d\beta = s\, l^r_{,s} b_r dx^s + s\bar{l}^r_s(b_r) d\bar{x}^s + s\, l^r(\lambda_{rs}\sin\alpha + \mu_{rs}\cos\alpha) dx^s \;,$$

which, by means of (68.2) and (68.3), reduces to

$$s\, d\beta = -m_s dx^s + (m_s) d\bar{x}^s - s\cos\phi\sin\alpha\,\omega_s dn^s - s\cos\alpha\,\phi_s dx^s \;, \tag{71.1}$$

where (m_s) is a unit vector at the barred end parallel to m_s [Eq. (67.4)]. Again this equation is true in any coordinates.

72. In (ω, ϕ, V) the components of m_s are given by:

$$\begin{aligned} K m_1 &= \chi_2\cos\phi\sin\alpha\cos\beta - \tau\cos\phi\cos\alpha\cos\beta \\ K m_2 &= -\tau\sin\alpha\cos\beta + \chi_1\cos\alpha\cos\beta \\ gK m_3 &= (\chi_2\gamma_1 - \tau\gamma_2)\sin\alpha\cos\beta + (\chi_1\gamma_2 - \tau\gamma_1)\cos\alpha\cos\beta + K\sin\beta \end{aligned}$$

and

$$(m_s) = S\bar{\lambda}_s + T\bar{\mu}_s + U\bar{\nu}_s \;,$$

where

$$\begin{aligned} S &= \sin\alpha\cos\beta\cos(\bar{\omega}-\omega) + \{\sin\phi\cos\alpha\cos\beta + \cos\phi\sin\beta\}\sin(\bar{\omega}-\omega) \\ &= \sin\bar{\alpha}\sin\bar{\beta}\cos\beta + \cos\phi\csc\beta\sin(\bar{\omega}-\omega) \\ T &= \{\sin\phi\cos\alpha\cos\beta + \cos\phi\sin\beta\}\sin\bar{\phi}\cos(\bar{\omega}-\omega) - \sin\phi\cos\bar{\phi}\sin\beta \\ &\quad - \sin\alpha\cos\beta\sin\bar{\phi}\sin(\bar{\omega}-\omega) + \cos\alpha\cos\beta\cos\phi\cos\bar{\phi} \\ &= \cos\bar{\alpha}\sin\bar{\beta}\cos\beta + \cos\phi\sin\bar{\phi}\csc\beta\cos(\bar{\omega}-\omega) - \sin\phi\cos\bar{\phi}\csc\beta \\ U &= -\{\sin\phi\cos\alpha\cos\beta + \cos\phi\sin\beta\}\cos\bar{\phi}\cos(\bar{\omega}-\omega) - \sin\phi\sin\bar{\phi}\sin\beta \\ &\quad + \sin\alpha\cos\beta\cos\bar{\phi}\sin(\bar{\omega}-\omega) + \cos\alpha\cos\beta\cos\phi\sin\bar{\phi} \\ &= \cos\bar{\beta}\cos\beta - \cos\phi\cos\bar{\phi}\csc\beta\cos(\bar{\omega}-\omega) - \sin\phi\sin\bar{\phi}\csc\beta \end{aligned}$$

the alternative expressions for S, T, U having been obtained from the fact that the Cartesian components of l^r are the same at both ends of the line, which leads to the following three useful formulae (two independent) for reverse azimuth and zenith distance.

$$\begin{aligned}
\sin\bar{\alpha}\sin\bar{\beta} &= \sin\alpha\sin\beta\cos(\bar{\omega}-\omega)\\
&\quad+\{\cos\alpha\sin\beta\sin\phi-\cos\beta\cos\phi\}\sin(\bar{\omega}-\omega)\\
\cos\bar{\alpha}\sin\bar{\beta} &= \{\cos\alpha\sin\beta\sin\phi-\cos\phi\cos\beta\}\sin\bar{\phi}\cos(\bar{\omega}-\omega)\\
&\quad-\sin\alpha\sin\beta\sin\bar{\phi}(\bar{\omega}-\omega)+(\cos\alpha\sin\beta\cos\phi+\sin\phi\cos\beta)\cos\bar{\phi}\\
\cos\bar{\beta} &= -\{\cos\alpha\sin\beta\sin\phi-\cos\phi\cos\beta\}\cos\bar{\phi}\cos(\bar{\omega}-\omega)\\
&\quad+\sin\alpha\sin\beta\cos\bar{\phi}\sin(\bar{\omega}-\omega)\\
&\quad+(\cos\alpha\sin\beta\cos\phi+\sin\phi\cos\beta)\sin\phi\ .
\end{aligned}$$

The following identities are easily obtained from various scalar and vector products and are frequently useful.

$$PS+QT+RU=0\ ,$$

$$PT-QS=\cos\bar{\beta}\ ,$$

$$QU-RT=\sin\bar{\alpha}\sin\bar{\beta}\ ,$$

$$RS-PU=\cos\bar{\alpha}\sin\bar{\beta}\ ,$$

$$PQ+ST=-\sin\bar{\alpha}\cos\bar{\alpha}\sin^2\bar{\beta}\ ,$$

$$QR+TU=-\cos\bar{\alpha}\cos\bar{\beta}\sin\bar{\beta}\ ,$$

$$RP+US=-\sin\bar{\alpha}\cos\bar{\beta}\sin\bar{\beta}\ ,$$

$$P^2+Q^2+R^2=1$$

$$S^2+T^2+U^2=1\ ,$$

$$P^2+S^2=\sin^2\bar{\alpha}\cos^2\bar{\beta}+\cos^2\bar{\alpha}\ ,$$

$$Q^2+T^2=\cos^2\bar{\alpha}\cos^2\bar{\beta}+\sin^2\bar{\alpha}\ ,$$

$$R^2+U^2=\sin^2\bar{\beta}\ ,$$

$$P\sin\bar{\alpha}\sin\bar{\beta}+Q\cos\bar{\alpha}\sin\bar{\beta}+R\cos\bar{\beta}=0\ ,$$

$$S\sin\bar{\alpha}\sin\bar{\beta}+T\cos\bar{\alpha}\sin\bar{\beta}+U\cos\bar{\beta}=0\ ,$$

$$S=R\cos\bar{\alpha}\sin\bar{\beta}-Q\cos\bar{\beta}\ ,$$

$$T=P\cos\bar{\beta}-R\sin\bar{\alpha}\sin\bar{\beta}\ ,$$

$$U=Q\sin\bar{\alpha}\sin\bar{\beta}-P\cos\bar{\alpha}\sin\bar{\beta}\ ,$$

$$P=T\cos\bar{\beta}-U\cos\bar{\alpha}\sin\bar{\beta}\ ,$$

$$Q=U\sin\bar{\alpha}\sin\bar{\beta}-S\cos\bar{\beta}\ ,$$

$$R=S\cos\bar{\alpha}\sin\bar{\beta}-T\sin\bar{\alpha}\sin\bar{\beta}\ ,$$

$$\bar{P}\sin\bar{\beta}=R\sin\alpha\cos\beta-U\cos\alpha\ ,$$

$$\bar{Q} \sin \bar{\beta} = R \cos \alpha \cos \beta + U \sin \alpha \ ,$$

$$\bar{R} \sin \bar{\beta} = -R \sin \beta \ ,$$

$$\bar{S} \sin \bar{\beta} = -R \cos \alpha - U \sin \alpha \cos \beta \ ,$$

$$\bar{T} \sin \bar{\beta} = R \sin \alpha - U \cos \alpha \cos \beta \ ,$$

$$\bar{U} \sin \bar{\beta} = U \sin \beta \ .$$

The change in zenith distance due to changes in (ω, ϕ, V) coordinates is finally given by

$$\begin{aligned} S\,d\beta = {} & d\omega(-m_1 - S \cos\phi \sin\alpha) + d\phi(-m_2 - S\cos\alpha) - m_3 dV \\ & + d\bar{\omega}(S\bar{\chi}_2 \cos\bar{\phi}/\bar{K} - T\bar{\tau}\cos\bar{\phi}/\bar{K}) + d\phi(-S\bar{\tau}/\bar{K} + T\bar{\chi}_1/\bar{K}) \\ & + d\bar{V}\{S(\bar{\chi}_2\bar{\gamma}_1 - \bar{\tau}\bar{\gamma}_2)/\bar{g}\bar{K} + T(\bar{\chi}_1\bar{\gamma}_2 - \bar{\tau}\bar{\gamma}_1)/\bar{g}\bar{K} - U/\bar{g}\} \ . \end{aligned}$$

In all the above equations we can interchange the bars, that is bar the unbarred quantities and unbar the unbarred, provided only that we change the sign of S.

Index of Main Symbols

A_r, B_r, C_r	Cartesian vectors (§ 1)
λ_r	Unit vector in parallel direction (§ 2)
μ_r	Unit vector in meridian direction (§ 2)
υ_r	Unit vector in zenithal direction (§ 2)
ω	Astronomical longitude direction (§ 3)
ϕ	Astronomical latitude direction (§ 3)
α	Astronomical azimuth direction (§ 5)
β	Astronomical zenith distance (§ 5)
V	Potential (§ 7)
g	Gravity (§ 7)
dp, dm, dn	Elements of arc in direction λ_r, μ_r, υ_r (§ 9)
γ_1	Arc rate of change of (log g) along parallel (§ 10)
γ_2	Arc rate of change of (log g) along meridian (§ 10)
χ_1	Normal curvature of equipotential along parallel (§ 11)
χ_2	Normal curvature of equipotential along meridian (§ 11)
τ	Geodesic torsion in meridian direction (§ 11)
H	Mean curvature of equipotential surface (§ 14)
K	Gauss curvature of equipotential surface (§ 25)
$\tilde{\omega}$	Angular velocity of the Earth (§ 17)
χ_0	Curvature of line of force (§ 19, § 21)
τ_0	Torsion of line of force (§ 20)

Editorial Commentary

This was Hotine's first major paper on differential geodesy. In it he made full use of not only tensorial methods, but also classical differential geometry. Its goal, which he admirably attained, was to recast the original work of Marussi [see Hotine's references Marussi (1947, 1949), which were not included in IG] into tensor form. This was necessary, since many of Marussi's seminal ideas in these papers, as well as in Marussi (1951, 1988) were formulated in special coordinate systems and obtained by using the homographic calculus of Burali-Forti and Marcolongo. The valuable article of Reilly (Appendix IG) gives a brief discussion of this formalism and a guide of how homographic expressions may be translated into vector and tensor notation. Hotine never made any use of the homographic calculus, whereas it played a major role in Marussi's thinking.

Hotine's treatment of this material was a veritable tour de force and a striking example of his mathematical ingenuity and ability. It was quite remarkable, since it was the work of someone who some 9 years earlier had set out – at the age of 50 – to master the intricacies of tensor calculus and differential geometry on his own. As Whitten's memorial lecture (see the final paper in this monograph) relates, this task was done while he held responsible full-time administrative positions in the service of his country. Hotine's mastery of tensorial and geometrical methods, while not breaking new ground mathematically, transcended the standard techniques appearing in his references (McConnell 1947; Eisenhart 1949; Levi-Civita 1926; Weatherburn 1930). As discussed in some detail in Zund (1990), although his mathematical reading was selective, it was essentially adequate for his purposes. Indeed, he anticipated – but did not explicitly formalize – many of the ideas which form the basis of the contemporary theory.

Hotine was not content merely to transcribe Marussi's work into tensor notation, he also embarked on his own formulation of the theory and in doing so he obtained a number of new results. In particular, the contents of this report cut across much of the material in Chapters 7, 12, 20, 24, 25 and 26 of Parts II and III of MG. It contained a general covariant theory of the Marussi tensor, and demonstrated how the metrical properties of the Earth's gravitational field can be described in terms of five parameters (in Chap. 12 of MG these were called curvature parameters).

The notation employed in the report differs from that in MG. In particular the basic quantities:

$$\mathrm{N},\ \mathrm{n},\ \mathrm{k}_1,\ \mathrm{k}_2 \text{ and } \mathrm{t}_1$$

in MG correspond respectively to

$$-\mathrm{V},\ -\mathrm{g},\ \chi_1,\ \chi_2 \text{ and } \tau$$

in the report. Likewise, the arc length elements

$$\mathrm{d}\lambda,\ \mathrm{d}\mu \quad\text{and}\quad \mathrm{ds}$$

in MG, appear in the report as

$$\mathrm{dp},\ \mathrm{dm} \text{ and } \mathrm{dn}$$

respectively. The sign difference in N and V, and n and g results in numerous sign differences between equations in MG and the report. However, these pose no difficulties to anyone seeking to establish a correspondence between the discussions.

A few other minor discrepancies occur between the notations, but these are obvious and require no comment.

Generally speaking, the presentation in the report is clearcut and, being less comprehensive and exhaustive than that in MG, it is easier to assimilate on a first reading. Almost all of the equations and results of the report occur *somewhere* in MG; however, two noteworthy exceptions have been detected.

The first of these is the expression in Eq. (18.1) for the Marussi tensor in terms of products of the covariant components of the vector fields $\underline{\lambda}$, $\underline{\mu}$, $\underline{\nu}$ (the Hotine triad, or 3-leg, of MG) and the five parameters (called the curvature parameters in MG). This is an important representation of the Marussi tensor which is not repeated in MG. There it appears only in contracted form, viz. [MG Eq. (12.162)], and although this is equivalent to Eq. (18.1), it is useful to have this representation in hand. In particular, by the definition of V_{rs} given in Eqs. (9.1), (18.1) also gives an expression for ν_{rs} which only occurs in contracted form like Eq. (19.1) in MG [i.e. MG Eq. (12.021)]. Reilly (1981, 1982, 1985) has given some elegant extensions of the notion of the Marussi tensor V_{rs} to include the case of a time-varying gravity field, and applied it to crustal deformations.

The second of these concerns the material in § 34 dealing with the Mainardi-Codazzi, or more simply the Codazzi, equations. This is a substantive topic, and Hotine considered it on no less than *three* occasions in MG: in Chapters 8 §§ 8–13; in Chapter 12 §§ 84–94, and again from an alternate viewpoint in §§ 95–97. None of these discussions precisely duplicates the contents of § 34 of the report. In particular, neither the general integrability conditions for the arbitrary function F, nor the specialized Eqs. (34.1)–(34.3) when $F = \phi$, or (34.4)–(34.6) when $F = \omega$, or (34.7)–(34.9) when $F = \log g$, appear in MG. The entire matter is one of some mathematical delicacy, since it concerns the existence and imbedding of the equipotential surfaces in Euclidean 3-space. A critical examination of this material was recently undertaken by the editor (Zund 1990).

A Test Triangulation (1957), cited in the report, was compiled by H. H. Brazier and L. M. Windsor and presented at the Toronto Assembly (Brazier and Windsor 1957).

A valuable report of the Toronto Assembly is given in Marussi (1958). Hotine's work was reported in Section I: *Triangulations*, under the presidency of C. A. Whitten, and with A. Marussi as secretary. Hotine himself was president of the special study group No. 1: *Problèmes théoretiques intéressant le calcul et la compensation des grandes triangulations, en prenant en considération la forme due géoïde*, which met on September 7, 1957. Its activities were discussed on pages 62–64 of Marussi (1958).

References to Paper 3

Bomford G (1952) Geodesy. Clarendon Press, Oxford

Brazier HH, Windsor LM *A test triangulation.* (1957) Paper to be submitted to the Toronto Assembly of I. A. G.

Eisenhart L (1949) Riemannian geometry. Princeton edn

Hotine M *Geodetic coordinate systems.* (1957) Paper to be submitted to the Toronto Assembly of I. A. G.

Levi-Civita T (1926) The absolute differential calculus (trans). Blackie, London
Marussi A (1949) Fondements de géométrie différentielle absolue du champ potential terrestre. Bull Géod Dec
Marussi A (1957) Sulla struttura locale del geoide, e sui mezzi geometrici e meccanici atti a determinarla. Univ Trieste
McConnell AJ (1947) Applications of the absolute differential calculus. Blackie, London edn
Weatherburn CE (1930) Differential geometry II. Cambridge

References to Editorial Commentary

Brazier HH, Windsor LM (1957) A test triangulation, a report submitted to Study Group No. 1, I.A.G. Toronto Assembly, unpubl
Marussi A (1951) Fondementi di Geodesia intrinseca. Pubbl Comm Geod Ital (Terza Serie) Mem 7:1–47 = Foundations of intrinsic geodesy, reprinted in IG, pp 13–58
Marussi A (1988) Intrinsic geodesy (a revised and edited version of his 1952 lectures prepared by J.D. Zund) Rep 390 Dep Geod Sci Survey. Ohio State Univ, Columbus 137 pp
Marussi A (1958) Resumé des procès-verbaux des séances des sections: Section I Triangulations. Bull Géod 47:59–70
Reilly WI (1981) Complete determination of local crustal deformation from geodetic observations. Tectonophys 71:111–123
Reilly WI (1982) Three-dimensional kinematics of Earth deformation from geodetic observations. Proc Int Symp Geodetic Networks and Computations, Vol 5 (I.A.G. Munich, 1981) Deutsche Geodätische Kommission, Reihe B. *258 V*:207–221
Reilly WI (1985) Differential geometry of a time-varying gravity field. Boll Geod Sci Aff anno XLIV:283–293
Reilly WI: Notations of vector analysis – the vectorial homographies of Burali-Forti and Marcolongo. Appendix in IG pp 190–195
Zund JD (1990a) An essay on the mathematical foundations of the Marussi-Hotine approach to geodesy. Boll Geod Sci Aff anno XLIX:133–179
Zund JD (1990b) The assertation of Hotine on the integrability conditions for his general (ω, ϕ, N) coordinate system. Man Geod 15:362–372

4 Geodetic Coordinate Systems[1]

Introduction

1. For the present purpose, a coordinate system is defined as a set of three continuous, single-valued, differentiable functions of a Cartesian system (x, y, z) within a certain region of flat 3-space. Other means of defining position (e.g. Square X 56, or "follow the normal to the spheroid as far as the geoid and thence along the line of force") are better described as reference systems and are not considered here; they are not amenable to the ordinary processes of analysis.

Two coordinates in all the systems considered are the longitude and latitude, in relation to fixed Cartesian axes, of the normal to the third coordinate surface. Moreover, the directions of the Cartesian axes are arranged to be the same, in all four systems considered, by a suitable choice of origin. Azimuth (and zenith distance) are in all cases considered to be measured about (and from) the normal to the third coordinate surface. The origin of azimuth is the plane defined by the normal and the Cartesian axis C^r parallel to the axis of rotation of the Earth.

2. The following quantities, all scalar functions of position, are taken as third coordinate in each case:

i) The geopotential, in which case the normal is the astronomical zenith direction, and the latitude and longitude are directly measurable astronomically. The Cartesian axes (which will be common to all other systems) are parallel to unit vectors A^r, B^r, C^r, right-handed in that order.

This case has already been considered in a previous paper (Hotine 1957), to which frequent reference will be made and the notation of which, summarized in a list of symbols at the end of this paper, will be used for each coordinate system in turn. It was concluded (Hotine 1957, § 49) that this system is unsuitable for accurate geodetic calculations, but it does nevertheless provide the basis of all other systems, which must in any case be related to it, since most of the field measurements are necessarily made in this system.

ii) A standard potential, giving rise to a field symmetrical about the axis of rotation. The equipotentials of this field are surfaces of revolution and one of them may be a spheroid. It will be shown that this system is unsuitable for

[1] Report dated 29 June 1957 (Tolworth), and presented to Study Group No. 1 at the I. A. G. Toronto Assembly 1957. Hotine noted that it was to be read in conjunction with *Metrical Properties of the Earth's Gravitational Field*, which was also presented to the Study Group at the same time.

geodetic calculation of position, distances etc., and indeed has little to commend it for any purpose.

iii) The distance between third-coordinate surfaces, measured along their common normals, which are necessarily straight. The third-coordinate surfaces are geodesic parallels to a base surface which can be of any form. This system may be of use in problems involving reduction to a base surface.

iv) The same as (iii) with a spheroid whose minor axis is parallel to the axis of rotation, as base surface. This system leads to simple closed formulae which are very suitable for geodetic calculation, not differing violently from previous results obtained by classical methods. The rigorous three-dimensional adjustment of triangulation is considered in some detail in this system, together with the assimilation, without successive approximation and in a single process, of astronomical measurements, spirit-levels and measured bases. It is suggested that this provides a complete and rigorous answer to the fundamental problem posed to I. A. G. Study Group No. 1, although it does not require any explicit knowledge of the form of the geoid. The amount of computation is considerably greater than by classical methods but still costs only a small fraction of the cost of the field work, even without electronic aids.

Relations Between Coordinate Systems

3. We first consider a few general propositions relating any two of the above four coordinate systems, one of which is denoted by barred notation.

The fact that the Cartesian vectors A^r, B^r, C^r (Hotine 1957, § 1) are the same in all systems [we consider below (§ 7) how this can be ensured], enables us to relate the parallel, meridian and zenith vectors (Hotine 1957, § 2) at a point as follows [Hotine 1957, Eqs. (6.1, 6.2)]:

$$\begin{aligned}\bar{\lambda}_r &= -A_r \sin\bar{\omega} + B_r \cos\bar{\omega}\\ &= \lambda_r \cos(\bar{\omega}-\omega) + \mu_r \sin\phi \sin(\bar{\omega}-\omega) - \nu_r \cos\phi \sin(\bar{\omega}-\omega)\end{aligned} \tag{3.1}$$

and similarly,

$$\begin{aligned}\bar{\mu}_r = &-\lambda_r \sin\bar{\phi} \sin(\bar{\omega}-\omega) + \mu_r\,[\cos\phi \cos\bar{\phi} + \sin\phi \sin\bar{\phi} \cos(\bar{\omega}-\omega)]\\ &+ \nu_r\,[\sin\phi \cos\bar{\phi} - \cos\phi \sin\bar{\phi} \cos(\bar{\omega}-\omega)]\end{aligned} \tag{3.2}$$

$$\begin{aligned}\bar{\nu}_r = &\lambda_r \cos\bar{\phi} \sin(\bar{\omega}-\omega) + \mu_r\,[\cos\phi \sin\bar{\phi} - \sin\phi \cos\bar{\phi} \cos(\bar{\omega}-\omega)]\\ &+ \nu_r\,[\sin\phi \sin\bar{\phi} + \cos\phi \cos\bar{\phi} \cos(\bar{\omega}-\omega)]\ .\end{aligned} \tag{3.3}$$

4. Since the Cartesian components of a vector are the same as those of a parallel vector, these equations are also true *in Cartesian coordinates* if μ_r etc. is the meridian etc. vector at some other point in the same space whose position is given by the barred coordinates. Alternatively, if we define $\bar{\mu}_r$ etc. as parallel to the meridian etc. vector at the barred point and transported to the unbarred point,

then these vector equations are true at the unbarred point in Cartesian coordinates and therefore in any coordinates.

5. We next consider a unit vector l^r in azimuth α zenith distance β ($\bar{\alpha}, \bar{\beta}$ in the barred system) so that either of the following expressions represents the same vector at the same point in space:

$$\begin{aligned} l^r &= \lambda^r \sin\alpha \sin\beta + \mu^r \cos\alpha \sin\beta + \upsilon^r \cos\beta \\ &= \bar{\lambda}^r \sin\bar{\alpha} \sin\bar{\beta} + \bar{\mu}^r \cos\bar{\alpha} \sin\bar{\beta} + \bar{\upsilon}^r \cos\bar{\beta} \ . \end{aligned}$$

Multiplying Eq. (3.1) etc. across by whichever expression is appropriate gives us the following equations (two independent) for azimuth and zenith distance resulting from a change of coordinates.

$$\begin{aligned} \sin\bar{\alpha} \sin\bar{\beta} = {} & \sin\alpha \sin\beta \cos(\bar{\omega}-\omega) + \cos\alpha \sin\beta \sin\phi \sin(\bar{\omega}-\omega) \\ & -\cos\beta \cos\phi \sin(\bar{\omega}-\omega) \ . \end{aligned} \tag{5.1}$$

$$\begin{aligned} \cos\bar{\alpha} \sin\bar{\beta} = {} & -\sin\alpha \sin\beta \sin\phi \sin(\bar{\omega}-\omega) \\ & +\cos\alpha \sin\beta\,[\cos\phi \cos\bar{\phi} + \sin\phi \sin\bar{\phi} \cos[\bar{\omega}-\omega)] \\ & +\cos\beta\,[\sin\phi \cos\bar{\phi} - \cos\phi \sin\bar{\phi} \cos(\bar{\omega}-\omega)] \end{aligned} \tag{5.2}$$

$$\begin{aligned} \cos\bar{\beta} = {} & \sin\alpha \sin\beta \cos\bar{\phi} \sin(\bar{\omega}-\omega) \\ & +\cos\alpha \sin\beta\,[\cos\phi \sin\bar{\phi} - \sin\phi \cos\bar{\phi} \cos(\bar{\omega}-\omega)] \\ & +\cos\beta\,[\sin\phi \sin\bar{\phi} + \cos\phi \cos\bar{\phi} \cos(\bar{\omega}-\omega)] \ . \end{aligned} \tag{5.3}$$

It will be apparent from § 4 that these equations hold if the barred quantities refer to any other point on the straight line l^r in the same space. They are accordingly true as between two points on this line, whether there is also a change of coordinates or not.

6. If the changes are small and e.g. $\delta\phi = (\bar{\phi}-\phi)$, then the Eqs. (5.1) etc. reduce to the following first-order formulae:

$$\delta\alpha = \sin\phi\,\delta\omega + \cos\beta\,(\sin\alpha\,\delta\phi - \cos\alpha \cos\phi\,\delta\omega) \tag{6.1}$$

$$\delta\beta = -\cos\alpha\,\delta\phi - \cos\phi \sin\alpha\,\delta\omega \ . \tag{6.2}$$

These will usually be sufficient in practice for a change in coordinates, of the same order as "deflections of the vertical", but the full equations will be required to obtain reverse azimuth etc. at the other end of a long line. The equations contain the so-called Laplace azimuth equation, generalized for a line in 3-space and given a rigorous interpretation.

7. Conversely, if the barred and unbarred quantities, referring to a particular line at a particular point, satisfy two of the Eqs. (5.1)–(5.3), then the Cartesian vectors will be parallel in the two systems and the equations will hold at any other point. The simplest way of ensuring this is to make the barred and unbarred quan-

tities the same at a particular point, or origin. [It is not sufficient merely to make the barred and unbarred latitudes, longitudes and azimuths the same, since this might leave Eq. (6.2) unsatisfied.]

Relations Between Parameters

8. If we form barred and unbarred equations from [Hotine 1957, Eq. (16.2)], subtract, write $\delta\phi = \bar{\phi} - \phi$ and use Eqs. (3.1)–(3.3), we obtain the following vector equation:

$$
\begin{aligned}
(\delta\phi)_r = {} & \lambda_r\{\bar{\tau}\cos\delta\omega - \bar{\chi}_2\sin\bar{\phi}\sin\delta\omega + \bar{\gamma}_2\cos\bar{\phi}\sin\delta\omega - \tau\} \\
& + \mu_r\{\bar{\tau}\sin\phi\sin\delta\omega + \bar{\alpha}_2(\cos\phi\cos\bar{\phi} + \sin\phi\sin\bar{\phi}\cos\delta\omega) \\
& + \bar{\gamma}_2(\cos\phi\sin\bar{\phi} - \sin\phi\cos\bar{\phi}\cos\delta\omega) - \chi_2\} \\
& + \upsilon_r\{-\bar{\tau}\cos\phi\sin\delta\omega + \bar{\chi}_2(\sin\phi\cos\bar{\phi} - \cos\phi\sin\phi\cos\delta\omega) \\
& + \bar{\gamma}_2(\sin\phi\sin\bar{\phi} + \cos\phi\cos\bar{\phi}\cos\delta\omega) - \gamma_2\}
\end{aligned}
\tag{8.1}
$$

and forming scalar products with each of the base vectors,

$$\tau = \bar{\tau}\cos\delta\omega - \bar{\chi}_2\sin\bar{\phi}\sin\delta\omega + \bar{\gamma}_2\cos\bar{\phi}\sin\delta\omega - \partial(\delta\phi)/\partial \mathrm{p} \ . \tag{8.2}$$

$$
\begin{aligned}
\chi_2 = {} & \bar{\tau}\sin\phi\sin\delta\omega + \bar{\chi}_2(\cos\phi\cos\bar{\phi} + \sin\phi\sin\bar{\phi}\cos\delta\omega) \\
& + \bar{\gamma}_2(\cos\phi\sin\bar{\phi} - \sin\phi\cos\bar{\phi}\cos\delta\omega) - \partial(\delta\phi)/\partial \mathrm{m} \ .
\end{aligned}
\tag{8.3}
$$

$$
\begin{aligned}
\gamma_2 = {} & -\bar{\tau}\cos\phi\sin\delta\omega + \bar{\chi}_2(\sin\phi\cos\bar{\phi} - \cos\phi\sin\bar{\phi}\cos\delta\omega) \\
& + \bar{\gamma}_2(\sin\phi\sin\bar{\phi} + \cos\phi\cos\bar{\phi}\cos\delta\omega) - \partial(\delta\phi)/\partial \mathrm{n} \ .
\end{aligned}
\tag{8.4}
$$

These equations can be considerably simplified by a suitable choice of geodetic coordinates. For instance, if the barred system is that described in § 21 et seq., then both $\bar{\tau}$ and $\bar{\gamma}_2$ are zero and $\bar{\chi}_2 = 1/(\varrho + h)$. We then have some very simple equations for directly determining, say, the astronomical (unbarred) parameters from deflections, without requiring the deflections to be small. We should need to make the assumption that the deflections vary uniformly along finite lines, in order to evaluate e.g. $\partial(\delta\phi)/\partial \mathrm{m}$, the change in latitude deflection per unit of meridian distance, or $\partial(\delta\phi)/\partial \mathrm{n}$, the change in deflection per unit of height. The mean of longitude deflections and of geodetic parameters at the two ends of the line should be used in the coefficients, and we should expect the result to give a value for the parameter at some intermediate point along the line. The method is more practicable than the absolute method in (Hotine 1957, § 50) but suffers from much the same assumptions; at the moment of writing it has not been tried out. It should be noted, moreover, that the equations hold only for coordinate systems related as in § 7; any departure from this relation could have very serious effects.

9. In exactly the same way, from a pair of longitude equations [Hotine 1957, Eq. (16.1)], we have

$$\chi_1 \sec\phi = \bar{\chi}_1 \sec\bar{\phi} \cos\delta\omega - \bar{\tau}\tan\bar{\phi}\sin\delta\omega + \bar{\gamma}_1 \sin\delta\omega - \partial(\delta\omega)/\partial p \ , \qquad (9.1)$$

$$\tau \sec\phi = \bar{\chi}_1 \sec\bar{\phi} \sin\phi \sin\delta\omega + \bar{\tau}\sec\phi\,(\cos\phi\cos\bar{\phi} + \sin\phi\sin\bar{\phi}\cos\delta\omega)$$
$$+ \bar{\gamma}_1 \sec\bar{\phi}\,(\cos\phi\sin\bar{\phi} - \sin\phi\cos\bar{\phi}\cos\delta\omega) - \partial(\delta\omega)/\partial m \ ,$$

$$\gamma_1 \sec\phi = -\bar{\chi}_1 \sec\bar{\phi}\cos\phi\sin\delta\omega + \bar{\tau}\sec\bar{\phi}\,(\sin\phi\cos\bar{\phi} - \cos\phi\sin\bar{\phi}\cos\delta\omega)$$
$$+ \bar{\gamma}_1 \sec\bar{\phi}\,(\sin\phi\sin\bar{\phi} + \cos\phi\cos\bar{\phi}\cos\delta\omega) - \partial(\delta\omega)/\partial n \ , \qquad (9.3)$$

for the change in longitude deflections. For the geodetic coordinate system (§ 21 et seq.), $\bar{\tau}$ and $\bar{\gamma}_1$ are zero and $\bar{\chi}_1 = 1/(\nu + h)$. In the case of an oblique line, the two values of τ [from Eqs. (8.2) and (9.2)] should indicate how far the basic assumptions are justified. A line mainly in the meridian (or parallel) direction will not of course give a good determination of χ_1 (or χ_2) and γ_1, γ_2 will not be well-determined in view of the smallness of dn on most terrestrial lines. Better values of γ_1, γ_2 would be obtained by measuring gravity at the two ends of the line and evaluating the parameters directly from the equations [Hotine 1957, Eq. (16.4)]

$$\gamma_1 = \partial(\log g)/\partial p$$
$$\gamma_2 = \partial(\log g)/\partial m \ .$$

10. We can obtain vector relations involving the third coordinates from Eq. (3.3). For instance, if the barred system is astronomical, we can substitute $-\bar{V}_r/\bar{g}_r$ for $\bar{\nu}_r$ [Hotine 1957, Eq. (7.1)]. The change in potential along a line of length ds (geodetic azimuth α, geodetic zenith distance β) whose unit vector is $(\lambda^r \sin\alpha\sin\beta + \mu^r\cos\alpha\sin\beta + \nu^r\cos\beta)$ is then given by

$$-(1/\bar{g})(\partial\bar{V}/\partial s) = \cos\bar{\phi}\sin\delta\omega\sin\alpha\sin\beta$$
$$+\cos\alpha\sin\beta\,[\cos\phi\sin\bar{\phi} - \sin\phi\cos\bar{\phi}\cos\delta\omega]$$
$$+\cos\beta\,[\sin\phi\sin\bar{\phi} + \cos\phi\cos\bar{\phi}\cos\delta\omega] \ . \qquad (10.1)$$

For small deflections this is

$$-(1/\bar{g})(\partial\bar{V}/\partial s) = \cos\phi\sin\alpha\sin\beta\,\delta\omega + \cos\alpha\sin\beta\,\delta\phi + \cos\beta \ . \qquad (10.2)$$

Symmetrical Field Coordinates

11. If the gravitational field is symmetrical about the axis of rotation, all meridian sections will be plane and similar. Any line of force will lie in a meridian plane and its principal normal will be in the meridian direction μ_r. Consequently, (Hotine 1957, § 19) $\gamma_1 = 0$; and since (log g) will also be independent of longitude, $\tau = 0$ (Hotine 1957, § 56). All the properties in (Hotine 1957, § 31) follow; there will be no change in curvatures in the parallels direction; and the Codazzi etc., equations (Hotine 1957, § 34) reduce to four;

$$\frac{\partial\gamma_2}{\partial m} - \frac{\partial\chi_2}{\partial n} = \chi_2^2 + \gamma_2^2 \ , \qquad (11.1)$$

$$-\frac{\partial \chi_1}{\partial n} = \chi_1^2 + \chi_1 \gamma_2 \tan\phi \;, \tag{11.2}$$

$$\frac{\partial \chi_1}{\partial m} = \chi_1 (\chi_1 - \chi_2) \tan\phi \;, \tag{11.3}$$

$$\frac{\partial(\chi_1+\chi_2)}{\partial m} + \frac{\partial \gamma_2}{\partial n} = \frac{4\tilde{\omega}^2}{g}\gamma_2 + \chi_2\gamma_2 \quad \text{in free air} \;. \tag{11.4}$$

12. These equations enable us to determine the three non-zero parameters χ_1, χ_2 and γ_2. Suppose we have a spheroid as base equipotential surface on which $\chi_1 = 1/\upsilon$; $\chi_2 = 1/\varrho$ (ϱ, υ being as usual the principal curvatures of the spheroid). The theoretical value of gravity on such a surface is known (§ 35 et seq.) in terms of latitude. Accordingly, we know

$$\gamma_2 = \partial(\log g_0)/(\varrho\partial\phi)$$

on the spheroid. Then at a distance n from the spheroid measured along the line of force, we have to a first order:

$$\chi_1 = \frac{1}{\upsilon} + n\left(\frac{\partial\chi_1}{\partial n}\right)_0 = \frac{1}{\upsilon} - n\left\{\frac{1}{\upsilon^2} + \frac{\tan\phi}{\varrho\upsilon}\frac{\partial \log g_0}{\partial\phi}\right\} \quad \text{from (11.2)} \tag{12.1}$$

$$\chi_2 = \frac{1}{\varrho} - n\left\{\frac{1}{\varrho^2} + \frac{1}{\varrho^2}\left(\frac{\partial \log g_0}{\partial\phi}\right)^2 - \frac{\partial}{\varrho\partial\phi}\left(\frac{\partial \log g_0}{\varrho\partial\phi}\right)\right\} \quad \text{from (11.1)} \tag{12.2}$$

$$\gamma_2 = \frac{\partial \log g_0}{\varrho\partial\phi} + n\left\{\frac{4\tilde{\omega}^2}{\varrho g_0}\frac{\partial \log g_0}{\partial\phi} + \frac{1}{\varrho\upsilon}\frac{\partial \log g_0}{\partial\phi} - \frac{\partial}{\varrho\partial\phi}\left(\frac{1}{\varrho}+\frac{1}{\upsilon}\right)\right\} \quad \text{from (11.4)} \tag{12.3}$$

Gravity at the same point is given by Hotine [1957, Eq. (16.3)] as

$$(\log g) = (\log g_0) - n\left\{\frac{1}{\varrho} + \frac{1}{\upsilon} + \frac{2\tilde{\omega}^2}{g_0}\right\} \;, \tag{12.4}$$

in which the last term corresponds to the "free air" reduction.

The latitude at a distance n along the line of force is greater than the latitude on the spheroid by $n\,(\partial\phi/\partial n)_0 = n(\gamma_2)_0 = (n/\varrho)(\partial \log g_0/\partial\phi)$ [Hotine 1957, Eq. (13.1)] and allowance for this should properly be made before looking up the spheroidal functions of latitude.

In all these formulae, to a comparable accuracy, n may be taken as the "orthometric" geodetic height.

The third Codazzi equation [Eq. (11.3)] becomes on the spheroid

$$\partial\upsilon/\partial\phi = (\upsilon - \varrho)\tan\phi \;, \tag{12.5}$$

a well-known result.

13. Any other required metrical properties of the field now follow by straight substitution in the corresponding formulae of Hotine (1957). For instance, the

components of the base vectors in this system of coordinates are (Hotine 1957, § 54);

$$\lambda^r = (\chi_1 \sec\phi,\ 0,\ 0)\ ,$$
$$\mu^r = (0,\ \chi_2,\ 0)\ , \qquad (13.1)$$
$$\upsilon^r = (0,\ \gamma_2,\ -g)\ ,$$

$$\lambda_r = (\cos\phi/\chi_1,\ 0,\ 0)\ ,$$
$$\mu_r = (0,\ 1/\chi_2,\ \gamma_2/(g\chi_2))\ , \qquad (13.2)$$
$$\upsilon_r = (0,\ 0,\ -1/g)\ ,$$

whence we can easily calculate the components of the metric tensor etc. in a form equivalent to that already given by Marussi (1950) for this system of coordinates.

14. As regards the expansion of coordinates along a line (Hotine 1957, § 42 et seq.), we need first to determine the variation of the three parameters along the line. We have, for example,

$$\frac{\partial\chi_1}{\partial s} = \frac{\partial\chi_1}{\partial m}\cos\alpha\sin\beta + \frac{\partial\chi_1}{\partial n}\cos\beta \text{ (since } \chi_1 \text{ is independent of } \omega\text{)}$$
$$= \frac{\partial\chi_1}{\partial\phi}\chi_2\cos\alpha\sin\beta + \frac{\partial\chi_1}{\partial n}\cos\beta\ , \qquad (14.1)$$

which can be evaluated from Eqs. (11.2) or (12.1), and similarly for $(\partial\chi_2/\partial s)$, $(\partial\gamma_2/\partial s)$. We can therefore calculate the second-order terms (Hotine 1957, § 49) and could theoretically obtain terms of still higher order. Accurate geodetic calculation is accordingly possible in this system, but by no means simple.

15. It is often supposed that this system leads to greater facility in handling gravity anomalies, but it is very doubtful if this is so. Exactly as in § 8, we can form a difference equation from Eq. (16.4) (Hotine 1957). If $\delta g = (\bar{g} - g)$ we have

$$\{\log(\bar{g}/g)\}_r = \{\log(1+\delta g/g)\}_r = (\delta g/g)_r \quad \text{nearly}$$
$$= \bar{\gamma}_1\bar{\lambda}_r + \bar{\gamma}_2\bar{\mu}_r - \{\bar{\chi}_1 + \bar{\chi}_2 + 2\bar{\omega}^2/\bar{g}\}\bar{\upsilon}_r$$
$$- \gamma_1\lambda_r - \gamma_2\mu_r + \{\chi_1 + \chi_2 + 2\bar{\omega}^2/g\}\upsilon_r$$

so that, for example

$$\gamma_1 = \bar{\gamma}_1\cos\delta\omega - \bar{\gamma}_2\sin\bar{\phi}\sin\delta\omega - (\bar{\chi}_1 + \bar{\chi}_2 + 2\bar{\omega}^2/\bar{g})\cos\bar{\phi}\sin\delta\omega - \frac{\partial}{\partial\varrho}\left(\frac{\delta g}{g}\right)\ .$$

If we substitute for the barred quantities the geodetic values in this coordinate system ($\bar{\gamma}_1 = 0$) from Eq. (12.1) etc. together with the change in gravity anomaly over a line, we should accordingly have equations to determine the astronomical parameters γ_1, γ_2 and $(\chi_1 + \chi_2)$. The difficulty of accurately computing geodetic positions in this system would, however, vitiate the accuracy of terms containing

$\delta\omega$ etc. and it is very doubtful if the method offers any advantage over § 8 and § 9, which are equally applicable in much simpler coordinate systems.

It is no easier, in fact it is more difficult, to calculate standard gravity accurately (§ 35 et seq.) at points whose positions are given in these coordinates.

Geodesic Parallel Coordinates

16. The next special case of the general gravitational metric is obtained by putting $g = -1$. In that case $\gamma_1 = \gamma_2 = 0$ and the normals are straight [Hotine 1957, Eq. (19.1)]. The contravariant unit normal (Hotine 1957, § 54) is (0, 0, 1), confirming that there is no change in ω or ϕ along the normal, and showing that the third coordinate can be interpreted as the distance h measured along the outward normal from some one of the third coordinate surfaces which we shall call the *base surface*. Since the third coordinate surfaces are equidistant along the straight normals, they are known as geodesic parallels (Eisenhart 1949, p. 57); they are no longer gravitational equipotentials, although the base surface can be chosen as such, e.g. we can take it to be the geoid.

17. The Codazzi equations (Hotine 1957, § 34) reduce to five [Eqs. (34.7)–(34.9), Hotine (1957) being no longer applicable]:

$$(\partial\chi_1/\partial h) = -\chi_1^2 - \tau^2 , \qquad (17.1)$$

$$(\partial\chi_2/\partial h) = -\chi_2^2 - \tau^2 , \qquad (17.2)$$

$$(\partial\tau/\partial h) = -2H\tau , \qquad (17.3)$$

$$(\partial\tau/\partial m) - (\partial\chi_2/\partial p) = 2H\tau \tan\phi , \qquad (17.4)$$

$$(\partial\chi_1/\partial m) - (\partial\tau/\partial p) = (\chi_1^2 + 2\tau^2 - \chi_1\chi_2) \tan\phi . \qquad (17.5)$$

18. We next solve the first three Codazzi equations. By substracting the first two, we have

$$\frac{\partial \log(\chi_1 - \chi_2)}{\partial h} = -(\chi_1 + \chi_2) = -2H = \frac{\partial \log \tau}{\partial h} \quad \text{from (17.3)} .$$

Consequently $\tau/(\chi_1 - \chi_2)$ is constant along the normal, which implies that the azimuth A of the principal directions of the h-coordinate surfaces remains constant along the normal (Hotine 1957, § 24). Substituting the principal curvatures κ_1, κ_2 (Hotine 1957, § 25) in the first two Codazzi equations,

$$(\partial\kappa_1/\partial h)\cos^2 A + (\partial\kappa_2/\partial h)\sin^2 A = -\kappa_1^2 \cos^2 A - \kappa_2^2 \sin^2 A ,$$

$$(\partial\kappa_1/\partial h)\sin^2 A + (\partial\kappa_2/\partial h)\cos^2 A = -\kappa_1^2 \sin^2 A - \kappa_2^2 \cos^2 A ,$$

so that $(\partial\kappa_1/\partial h) = -\kappa_1^2$, and $(\partial\kappa_2/\partial h) = -\kappa_2^2$, or $\partial(1/\kappa_1)/\partial h = 1$, and $\partial(1/\kappa_2)/\partial h = 1$.

If the quantities on the base surface at the foot of the normal are barred, the integrated equations are

$$(1/\kappa_1) = (1/\bar{\kappa}_1) + h \quad \text{and} \quad (1/\kappa_2) = (1/\bar{\kappa}_2) + h \, .$$

Substituting back in the formulae at the end of (Hotine 1957, § 25), and writing K for the Gauss curvature of the h-coordinate surface, we have finally

$$(\chi_1/\mathrm{K}) = (\bar{\chi}_1/\bar{\mathrm{K}}) + h \, , \tag{18.1}$$

$$(\chi_2/\mathrm{K}) = (\bar{\chi}_2/\bar{\mathrm{K}}) + h \, , \tag{18.2}$$

$$(\tau/\mathrm{K}) = (\bar{\tau}/\bar{\mathrm{K}}) \, . \tag{18.3}$$

Also

$$(1/\mathrm{K}) = (1/\bar{\mathrm{K}} + (2\bar{\mathrm{H}}/\bar{\mathrm{K}})h + h^2 \tag{18.4}$$

and from (18.1) and 18.2)

$$(2\mathrm{H}/\mathrm{K}) = (2\bar{\mathrm{H}}/\bar{\mathrm{K}}) + 2h \, . \tag{18.5}$$

These formulae enable us to determine the parameters, and in consequence any metric property, at any point in space from values on the base surface for the same latitude and longitude, or vice versa. Notice that in any formula we can interchange the bars provided that we change the sign of h, since this procedure merely amounts to changing the base surface. For example, Eq. (18.4) becomes

$$(1/\bar{\mathrm{K}}) = (1/\mathrm{K}) - (2\mathrm{H}/\mathrm{K})h + h^2 \, ,$$

which can easily be verified from the other equations.

19. The base vectors (Hotine 1957, § 54) in these coordinates are

$$\begin{aligned} &\lambda^r = (\chi_1 \sec\phi, \ \tau, \ 0) \, , \quad &&\mathrm{K}\lambda_r = (\chi_2 \cos\phi, \ -\tau, \ 0) \, , \\ &\mu^r = (\tau \sec\phi, \ \chi_2, \ 0) \, , \quad &&\mathrm{K}\mu_r = (-\tau\cos\phi, \ \chi_1, \ 0) \, , \\ &\nu^r = (0, \ 0, \ 1) \, , \quad &&\nu_r \ = (0, \ 0, \ 1) \, , \end{aligned} \tag{19.1}$$

and if $\bar{\lambda}_r$ etc. are components of the base vectors at the foot of the normal, these can be combined with (18.1) etc. to give

$$\begin{aligned} \lambda_r &= \bar{\lambda}_r + h \cos\phi \, \delta^1_r \\ \mu_r &= \bar{\mu}_r + h \, \delta^2_r \\ \nu_r &= \bar{\nu}_r \, . \end{aligned} \tag{19.2}$$

The first two are not vector equations, true in any other coordinates, since λ_r, $\bar{\lambda}_r$ etc. are not defined at the same point in space. The space and surface metric tensors follow, together with all other results by straight substitution in the general formuale of Hotine (1957); except that, although we can write $\mathrm{V} = h$ for the third coordinate, $\Delta^2 h = 2\mathrm{H}$, which unlike $\Delta^2\mathrm{V}$ is not a constant; equations containing $\Delta^2\mathrm{V}$, or obtained by differentiating $\Delta^2\mathrm{V}$, should accordingly be re-worked. The following equations (from Hotine 1957, § 62, § 63, § 64) are of particular interest, giving the three fundamental forms of an h-coordinate surface in terms of those of the base surface at the foot of the normal

$$\begin{aligned} a_{\alpha\beta} &= \bar{a}_{\alpha\beta} - 2h\,\bar{b}_{\alpha\beta} + h^2 \bar{c}_{\alpha\beta} \quad (\alpha, \beta = 1,2) \\ b_{\alpha\beta} &= \bar{b}_{\alpha\beta} - h\bar{c}_{\alpha\beta} \\ c_{\alpha\beta} &= \bar{c}_{\alpha\beta} = (\cos^2 \phi, 0, 1) \ . \end{aligned} \tag{19.3}$$

The Laplacian of a scalar F in these coordinates can be found without difficulty as

$$\Delta^2 F = (\Delta^2 F) + 2H \frac{\partial F}{\partial h} + \frac{\partial^2 F}{\partial h^2} \ , \tag{19.4}$$

in which $(\Delta^2 F)$ is the *surface* Laplacian taken with respect to the metric of the h-coordinate surface passing through the point.

20. If the axes of Cartesian coordinates are as usual parallel to A^r, B^r, C^r (Hotine 1957, § 1), we have the Cartesian coordinates of a point distant h along the straight normal from $(\bar{x}, \bar{y}, \bar{z})$ on the base surface as

$$\begin{aligned} x &= \bar{x} + h \cos\phi \cos\omega \\ y &= \bar{y} + h \cos\phi \sin\omega \\ z &= \bar{z} + h \sin\phi \ , \end{aligned} \tag{20.1}$$

in which $\bar{x}, \bar{y}, \bar{z}$ are functions of (ω, ϕ) only. From Hotine 1957, Eq. (6.2) we have

$$\begin{aligned} \bar{x}_\alpha &= -\bar{\lambda}_\alpha \sin\omega - \bar{\mu}_\alpha \sin\phi \cos\phi \cos\omega \quad (\alpha = 1,2) \\ \bar{y}_\alpha &= \bar{\lambda}_\alpha \cos\omega - \bar{\mu}_\alpha \sin\phi \sin\omega \\ \bar{z}_\alpha &= \bar{\mu}_\alpha \cos\phi \ , \end{aligned}$$

where

$$\begin{aligned} \bar{\lambda}_\alpha &= -\bar{x}_\alpha \sin\omega + \bar{y}_\alpha \cos\omega \\ \bar{\mu}_\alpha &= -\bar{x}_\alpha \csc\phi \cos\omega - \bar{y}_\alpha \csc\phi \sin\omega = \bar{z}_\alpha \sec\phi \ , \end{aligned}$$

and substituting the components of § 19 of the vectors, these give

$$(\bar{\chi}_2 \cos\phi)/\bar{K} = -\sin\omega\,(\partial\bar{x}/\partial\omega) + \cos\omega\,(\partial\bar{y}/\partial\omega) \tag{20.2}$$

$$\begin{aligned} \bar{\tau}/\bar{K} &= \sin\omega\,(\partial\bar{x}/\partial\phi) - \cos\omega\,(\partial\bar{y}/\partial\phi) \quad \text{or} \ , \\ &= \sec\phi \csc\phi \cos\omega\,(\partial\bar{x}/\partial\omega) + \sec\phi \csc\phi \sin\omega\,(\partial\bar{y}/\partial\omega) \quad \text{or} \\ &= -\sec^2\phi\,(\partial\bar{z}/\partial\omega) \ , \end{aligned} \tag{20.3}$$

$$\begin{aligned} \bar{\chi}_1/\bar{K} &= -\csc\phi \cos\omega\,(\partial\bar{x}/\partial\phi) - \csc\phi \sin\omega\,(\partial\bar{y}/\partial\phi) \ , \quad \text{or} \\ &= \sec\phi\,(\partial\bar{z}/\partial\phi) \ . \end{aligned} \tag{20.4}$$

Starting from Eq. (6.2) in Hotine (1957), the same equations obviously hold without the bars for any other third-coordinate surface and together with

Eq. (20.1) this enables us to verify (18.1) etc. These equations, together with $\bar{K} = \bar{\kappa}_1 \bar{\kappa}_2 - \bar{\tau}^2$, will enable us to determine the parameters on the base surface [and therefore anywhere else from (18.1) etc.] in cases where the base surface is given as $\bar{x} = f(\omega, \phi)$ etc.

Geodesic Parallels to a Spheroid

21. If the base surface is a surface of revolution about an axis parallel to C^r (or z-axis), then the Cartesian coordinates of points on the base surface can take the form

$$\bar{x} = f(\phi) \cos \omega$$
$$\bar{y} = f(\phi) \sin \omega$$
$$\bar{z} = g(\phi) ,$$

which, on substitution in Eqs. (20.3), (20.2) and (20.4), gives

$$\bar{\tau} = 0 \; ; \quad \bar{K} = \bar{\kappa}_1 \bar{\kappa}_2 \; ; \quad f(\phi) = \cos \phi / \bar{\kappa}_1 \; ; \quad f'(\phi) = -\sin \phi / \bar{\kappa}_2 \; ;$$
$$g'(\phi) = \cos \phi / \bar{\kappa}_2 \; . \tag{21.1}$$

The fact that we have derived separate equations for $f(\phi)$ and $f'(\phi)$ does not introduce any limitation on $f(\phi)$, since these are reconciled by the Codazzi equation (17.5), which in this case takes the form (Hotine 1957, § 58)

$$\chi_2 (\partial \chi_1 / \partial \phi) = (\chi_1^2 - \chi_1 \chi_2) \tan \phi \; . \tag{21.2}$$

The other Codazzi equation [17.4] is satisfied identically.

22. If the base surface is a spheroid of eccentricity e(with $\bar{e}^2 = 1 - e^2$), then $1/\bar{\chi}_1 = \upsilon$, $1/\bar{\chi}_2 = \varrho$, and the last equation is

$$\partial \upsilon / \partial \phi = (\upsilon - \varrho) \tan \phi \tag{22.1}$$

and since (21.1) $f(\phi) = \upsilon \cos \phi$; $f'(\phi) = -\varrho \sin \phi$; $g'(\phi) = \varrho \cos \phi$; $g(\phi) = \bar{e}^2 \upsilon \sin \phi$ the coordinate equations (20.1) are

$$x = (\upsilon + h) \cos \phi \cos \omega$$
$$y = (\upsilon + h) \cos \phi \sin \omega \tag{22.2}$$
$$z = (\bar{e}^2 \upsilon + h) \sin \phi \; .$$

The non-zero parameters at any point in space are (18.1) etc.:

$$\chi_1 = 1/(\upsilon + h) \; ; \quad \chi_2 = 1/(\varrho + h) \; . \tag{22.3}$$

Non-zero components of the base vectors are

$$\lambda^1 = 1/\{(\upsilon + h) \cos \phi\}$$
$$\mu^2 = 1/(\varrho + h) \tag{22.4}$$
$$\upsilon^3 = 1 ,$$

$$\lambda_1 = (\upsilon + h)\cos\phi$$
$$\mu_2 = (\varrho + h) \tag{22.5}$$
$$\upsilon_3 = 1$$

and the metric is

$$ds^2 = (\upsilon + h)^2 \cos^2\phi\, d\omega^2 + (\varrho + h)^2 d\phi^2 + dh^2 \ . \tag{22.6}$$

The second-order magnitudes of the h = const. surfaces (Hotine 1957, § 64) are

$$b_{\alpha\beta} = [(-\varrho\upsilon + h)\cos^2\phi, 0, -(\varrho - h)] \tag{22.7}$$

and the third-order magnitudes are

$$c_{\alpha\beta} = (\cos^2\phi, 0, 1) \ . \tag{22.8}$$

23. The equations for variation of coordinates along a line which is either straight or refracted in a medium whose density is now a function of h are obtained straight from (Hotine 1957, § 42) et seq. as

$$\cos\phi(\partial\omega/\partial s) = \sin\alpha\sin\beta/(\upsilon + h) \ ,$$
$$(\partial\phi/\partial s) = \cos\alpha\sin\beta/(\varrho + h) \ ,$$
$$(\partial h/\partial s) = \cos\beta \ , \tag{23.1}$$
$$(\partial\alpha/\partial s) = \tan\phi\sin\alpha\sin\beta/(\upsilon + h)$$
$$\qquad - \sin\alpha\cos\alpha\sin\beta\cos\beta(\varrho - \upsilon)/\{(\upsilon + h)(\varrho + h)\} \ ,$$
$$(\partial\beta/\partial s) = \bar{\chi} - \sin\beta\sin^2\alpha/(\upsilon + h) - \sin\beta\cos^2\alpha/(\varrho + h) \ ,$$

(in which $\bar{\chi}$ is the curvature of the refracted ray).

Since, in this case, we know the parameters $\chi_1 = 1/(\upsilon + h)$, $\chi_2 = 1/(\varrho + h)$, $\tau = 0$ as functions of the coordinates, we can compute $(\partial\chi_1/\partial s)$ etc. and therefore determine the second-order terms (Hotine 1957, § 49), but this operation is still by no means simple. Whenever the line can be considered straight, as it can be for azimuth calculations, and for others by removal of a standard refraction, it is better to use closed formulae obtained as follows.

24. It is evident from Eq. (22.2), that the position vector ϱ^r can be written as

$$\varrho^r = xA^r + yB^r + zC^r$$
$$= \{(a^2/\upsilon) + h\}\upsilon^r - (e^2\upsilon\sin\phi\cos\phi)\mu^r \quad \text{[Eq. (6.2), Hotine 1957]} \tag{24.1}$$

a vector equation which is true in any coordinates. Now suppose that all quantities at the far end of a line of length s (unit vector $l^r = \lambda^r\sin\alpha\sin\beta + \mu^r\cos\alpha\sin\beta + \upsilon^r\cos\beta$) are barred, and that $\bar{\varrho}^r$ represents either the Cartesian components of the position vector at the barred end, or else a vector parallel to the latter at the unbarred end, then

$$\bar{\varrho}^r - \varrho^r = s l^r \tag{24.2}$$

with

$$\bar{\varrho}^r = \{(a^2/\bar{\upsilon}) + \bar{h}\}\bar{\upsilon}^r - (e^2\bar{\upsilon}\sin\bar{\phi}\cos\bar{\phi})\bar{\mu}^r ,$$

in which we can immediately substitute Eqs. (3.3) and (3.2) for $\bar{\upsilon}^r$ and $\bar{\mu}^r$.

Multiplying Eq. (24.2) across by λ^r

$$\begin{aligned} s\sin\alpha\sin\beta &= \{(a^2/\bar{\upsilon}) + \bar{h}\{\cos\bar{\phi}\sin(\bar{\omega}-\omega) \\ &\quad + (e^2\bar{\upsilon}\sin^2\bar{\phi}\cos\bar{\phi}\sin(\bar{\omega}-\omega) = (\bar{\upsilon}-\bar{h})\cos\bar{\phi}\sin(\bar{\omega}-\omega) \end{aligned} \tag{24.3}$$

and similarly through multiplication by μ^r and υ^r

$$\begin{aligned} s\cos\alpha\sin\beta &= -(\bar{\upsilon}+\bar{h})\sin\phi\cos\bar{\phi}\cos(\bar{\omega}-\omega) \\ &\quad + (\bar{e}^2\bar{\upsilon}+\bar{h})\cos\phi\sin\bar{\phi} + e^2\upsilon\sin\phi\cos\phi \end{aligned} \tag{24.4}$$

or

$$\begin{aligned} &= (\bar{\upsilon}+\bar{h})\{\cos\phi\sin\bar{\phi} - \sin\phi\cos\bar{\phi}\cos(\bar{\omega}-\omega)\} \\ &\quad - e^2\cos\phi(\bar{\upsilon}\sin\bar{\phi} - \upsilon\sin\phi) . \end{aligned}$$

$$\begin{aligned} s\cos\beta &= (\bar{\upsilon}+\bar{h})\cos\phi\cos\bar{\phi}\cos(\bar{\omega}-\omega) \\ &\quad + (\bar{e}^2\bar{\upsilon}+\bar{h})\sin\phi\sin\bar{\phi} - \{(a^2/\upsilon)+h\} \end{aligned} \tag{24.5}$$

or

$$\begin{aligned} &= (\bar{\upsilon}+\bar{h})\{\sin\phi\sin\bar{\phi} + \cos\phi\cos\bar{\phi}\cos(\bar{\omega}-\omega)\} \\ &\quad - (\upsilon+h) - e^2\sin\phi(\bar{\upsilon}\sin\bar{\phi} - \upsilon\sin\phi) . \end{aligned}$$

These three equations directly determine the length, azimuth and zenith distance of a line whose end geodetic coordinates are given. In these and all similar formulae we can interchange the bars provided only that we change the sign of s, assuming as usual that barred azimuth etc. refers to the line produced and not to the back azimuth. For example,

$$s\sin\bar{\alpha}\sin\bar{\beta} = (\upsilon+h)\cos\phi\sin(\bar{\omega}-\omega) \tag{24.6}$$

and combining this with (24.3) we find that

$$(\upsilon+h)\cos\phi\sin\alpha\sin\beta \tag{24.7}$$

is a constant along the line, an analogous result in space to Clairaut's theorem for a geodesic on the spheroid.

25. For the inverse problem, given s, α, β and the coordinates of the unbarred end, we can reverse the above formulae as

$$\begin{aligned} (\bar{\upsilon}+\bar{h})\cos\bar{\phi}\sin(\bar{\omega}-\omega) &= s\sin\alpha\sin\beta \\ (\bar{\upsilon}+\bar{h})\cos\bar{\phi}\cos(\bar{\omega}-\omega) &= -s\cos\alpha\sin\beta\sin\phi + s\cos\beta\cos\phi + (\upsilon+h)\cos\phi \\ (\bar{e}^2\bar{\upsilon}+\bar{h})\sin\bar{\phi} &= s\cos\alpha\sin\beta\cos\phi + s\cos\beta\sin\phi + (\bar{e}^2\upsilon+h)\sin\phi , \end{aligned}$$

which could have been obtained from the difference of Cartesian components over the line, using a temporary origin for longitude at the unbarred end. The difference in longitude follows at once by division of the first two equations. We then have to solve by iteration

$$\begin{cases}(\bar{\nu}+\bar{h})\cos\bar{\phi}=A\\(\bar{e}^2\bar{\nu}+\bar{h})\sin\bar{\phi}=B\end{cases}$$

for the latitude and height, starting with an approximate latitude given by $\bar{e}^2\tan\bar{\phi}=B/A$.

Adjustment of Triangulation in Space

26. As in Hotine (1957, § 66 et seq.), we consider the changes in azimuth α and zenith distance β of a line l^r due to changes in the positions of the ends. We use geodesic parallel coordinates (ω, ϕ, h) with a spheroidal base.

The components of the vector n_s in these coordinates (Hotine 1957, § 70) and § 22 are:

$$n_s=\{-(\nu+h)\cos\phi\cos\alpha, (\varrho+h)\sin\alpha, 0\}$$

so that the change in azimuth is given by [Eq. (70.1), Hotine 1957]:

$$\begin{aligned}\sin\beta\, d\alpha &= d\omega\{-\cos\phi\cos\alpha(\nu+h)/s+\sin\phi\sin\beta-\cos\phi\cos\alpha\cos\beta\}\\&\quad+d\phi\{\sin\alpha(\varrho+h)/s+\sin\alpha\cos\beta\}\\&\quad+d\bar{\omega}\{-P\cos\bar{\phi}(\bar{\nu}+\bar{h})/s\}\\&\quad+d\bar{\phi}\{-Q(\bar{\varrho}+\bar{h})/s\}\\&\quad+d\bar{h}\{-R/s\}\end{aligned}$$

with

$$\begin{aligned}P &= -\cos\alpha\cos(\bar{\omega}-\omega)+\sin\phi\sin\alpha\sin(\bar{\omega}-\omega)\ ,\\Q &= +\sin\phi\sin\bar{\phi}\sin\alpha\cos(\bar{\omega}-\omega)+\sin\bar{\phi}\cos\alpha\sin(\bar{\omega}-\omega)\\&\quad+\cos\phi\cos\bar{\phi}\sin\alpha\\R &= -\sin\phi\cos\bar{\phi}\sin\alpha\cos(\bar{\omega}-\omega)-\cos\bar{\phi}\cos\alpha\sin(\bar{\omega}-\omega)\\&\quad+\cos\phi\sin\bar{\phi}\sin\alpha\end{aligned}$$

and $P^2+Q^2+R^2=1$ as in check.

In these axially symmetrical coordinates (but not in general), the coefficients of $d\omega$ and $d\bar{\omega}$ in the above equation should obviously be equal and opposite in sign, so that there may be a single term in $(d\omega-d\bar{\omega})$, since it must be immaterial if the same quantity is added to both ω and $\bar{\omega}$. That this is so can easily be verified with the help of Eqs. (24.3), (24.6) and (5.1). This fact provides another check, or lessens the computation.

An alternative expression for R in these coordinates is

$$e^2(\bar{\nu}\sin\bar{\phi}-\nu\sin\phi)\cos\phi\sin\alpha/(\bar{\nu}+\bar{\hbar})$$

obtainable from Eqs. (24.4) and (24.3).

If we change coordinates to the astronomical system by adding the (astronomic minus geodetic) deflections $\delta\omega$, $\delta\phi$ (at present unknown), there will be a further change in azimuth [Eq. (6.1)] of

$$\delta\alpha = (\sin\phi-\cos\phi\cos\alpha\cos\beta)\,\delta\omega+\sin\alpha\cos\beta\,\delta\phi \ ,$$

and if we then add a station correction $\Delta\alpha$ (at present unknown) to all rays at a station, the result should be the observed astronomical azimuth. The final observation equation, starting with a geodetic azimuth computed from provisional values of the geodetic coordinates, is accordingly:

$$\begin{aligned}
&\text{Observed Azimuth} - \text{Computed Azimuth}\\
&= \Delta\alpha + d\omega\{-(\nu+\hbar)\cos\phi\cos\alpha\csc\beta/s\} \qquad (26.1)\\
&\quad + d\phi\{(\varrho+\hbar)\sin\alpha\cos\beta/s\}\\
&\quad + d\bar{\omega}\{-P\cos\bar{\phi}\csc\beta(\bar{\nu}+\bar{\hbar})/s\}\\
&\quad + d\bar{\phi}\{-Q\csc\beta(\bar{\varrho}+\bar{\hbar})/s\}\\
&\quad + d\bar{\hbar}\{-R\csc\beta/s\}\\
&\quad + (d\omega+\delta\omega)(\sin\phi-\cos\phi\cos\alpha\cos\phi)\\
&\quad + (d\phi+\delta\phi)(\sin\alpha\cos\beta) \ .
\end{aligned}$$

If the deflections $\delta\omega$, $\delta\phi$ are unknown, it is permissible to take $(d\omega+\delta\omega)$, $(d\phi+\delta\phi)$ as composite unknowns and later to evaluate the deflections by subtracting the position corrections $d\omega$, $d\phi$. If, however, astronomical latitude and/or longitude has been measured, then $(d\phi+\delta\phi)$ and/or $(d\omega+\delta\omega)$ are the astronomical values minus the *preliminary* geodetic values and should be substituted in the equations before solution, thereby reducing the number of unknowns by two. Similarly, if astronomical azimuth has been measured at a station and it has accordingly not been necessary to assume an azimuth in order to orient the observed directions, then the $\Delta\alpha$ term should be dropped.

27. If the triangulation has been computed in Cartesian coordinates, some simplification is possible by determining the position corrections (dn, etc.) in Cartesian coordinates. In that case the vectors

$$n_r = -\lambda_r\cos\alpha+\mu_r\sin\alpha$$

and (n_s) [Eq. (69.3), Hotine 1957] both have the same Cartesian components (Hotine 1957, § 6):

$$\begin{aligned}
n_1 &= \cos\alpha\sin\omega-\sin\phi\sin\alpha\cos\omega\\
n_2 &= -\cos\alpha\cos\omega-\sin\phi\sin\alpha\sin\omega\\
n_3 &= \cos\phi\sin\alpha
\end{aligned}$$

and [Eq. (69.3), Hotine 1957] the (Observed minus Computed) azimuth is

$$
\begin{aligned}
&\Delta\alpha + (d\bar{n} - dn)(-\cos\alpha\sin\omega + \sin\phi\sin\alpha\cos\omega)\csc\beta/s \\
&\quad + (d\bar{y} - dy)(\cos\alpha\cos\omega + \sin\phi\sin\alpha\sin\omega)\csc\beta/s \\
&\quad - (d\bar{z} - dz)(\cos\phi\sin\alpha\csc\beta)/s \qquad (27.1) \\
&\quad + (d\omega + \delta\omega)(\sin\phi - \cos\phi\cos\alpha\cos\beta) \\
&\quad + (d\phi + \delta\phi)(\sin\alpha\cos\beta)\ .
\end{aligned}
$$

28. In § 21 coordinates, the auxiliary vector m_r (Hotine 1957, § 71), required for the zenith distance equation, has the following components:

$$
\begin{aligned}
m_1 &= (\nu + \hbar)\cos\phi\sin\alpha\cos\beta \\
m_2 &= (\varrho + \hbar)\cos\alpha\cos\beta \\
m_3 &= -\sin\beta
\end{aligned}
$$

and the change in zenith distance due to change of coordinates to the astronomical system [Eq. (6.2)] is

$$\delta\beta = -\cos\phi\sin\alpha\,\delta\omega - \cos\alpha\,\delta\phi\ .$$

The (Observed minus Computed) zenith distance (Hotine 1957, § 71) is accordingly

$$
\begin{aligned}
&-d\omega\{(\nu + \hbar)\cos\phi\sin\alpha\cos\beta/s\} \\
&-d\phi\{(\varrho + \hbar)\cos\alpha\cos\beta/s\} \\
&+d\hbar\,(\sin\beta/s) \\
&+d\bar{\omega}\{S\cos\bar{\phi}(\bar{\nu} + \bar{\hbar})/s\} \qquad (28.1) \\
&+d\bar{\phi}\{T(\bar{\varrho} + \bar{\hbar})/s\} \\
&+d\bar{\hbar}\,\{U/s\} \\
&-(d\omega - \delta\omega)\cos\phi\sin\alpha \\
&-(d\phi + \delta\phi)\cos\alpha
\end{aligned}
$$

with S, T, U as in Hotine (1957, § 71).

Here again, in these symmetrical coordinates, the coefficients of $d\omega$ and $d\bar{\omega}$ must be equal and opposite in sign, so that

$$S\cos\bar{\phi}(\bar{\nu} + \bar{\hbar}) = (\nu + \hbar)\cos\phi\sin\alpha\cos\beta + S\cos\phi\sin\alpha\ ,$$

which is easily verified from Eqs. (24.3) and (24.6).

29. Again, if the triangulation has been computed in Cartesian, the Cartesian components of m_s or (m_s) are

$$
\begin{aligned}
m_1 &= -\sin\alpha\cos\beta\sin\omega - \sin\phi\cos\alpha\cos\beta\cos\omega - \cos\phi\sin\beta\cos\omega \\
m_2 &= \sin\alpha\cos\beta\cos\omega - \sin\phi\cos\alpha\cos\beta\sin\omega - \cos\phi\sin\beta\sin\omega \\
m_3 &= \cos\phi\cos\alpha\cos\beta - \sin\phi\sin\beta
\end{aligned}
$$

and the (Observed minus Computed) zenith distance takes the alternative form [Eq. (71.1) Hotine 1957]:

$$\begin{aligned}
&+(d\bar{n}-dn)(m_1/s)\\
&+(d\bar{y}-dy)(m_2/s)\\
&+(d\bar{z}-dz)(m_3/s)\\
&-(d\omega+\delta\omega)\cos\phi\sin\alpha\\
&-(d\phi+\delta\phi)\cos\alpha\ .
\end{aligned}\qquad (29.1)$$

30. The observed zenith distance should be freed from refraction as far as possible before forming all the above equations, whether the zenith distance occurs in the absolute term or in the coefficients of any of the observation equations. Even so, the observation equations for zenith distances should be weighted down, since zenith distances will certainly not be measured as accurately as the horizontal angles, but there seems to be no justification for omitting them altogether. If zenith distances are omitted, the adjustment would be weak (although still stable) and would tend not to alter the provisional values.

As shown in § 26, the adjustment can be stiffened by measuring astronomical azimuths and positions, and these are very easily assimilated wherever they have been measured. Current methods of adjustment in two dimensions, besides assuming that the third dimension can be treated quite independently, usually reduce the number of unknowns in effect by simply ignoring $\delta\phi$, $\delta\omega$, but this does nothing to increase the reliability of the result.

31. For adjustment to a known base length [Eq. (68.4), Hotine 1957], we require the components of the vector l_s in the § 21 system, which are in § 22:

$$\begin{aligned}
l_1 &= (\nu+h)\cos\phi\sin\alpha\sin\beta\\
l_2 &= (\rho+h)\cos\alpha\sin\beta\\
l_3 &= \cos\beta\\
\bar{l}_1 &= (\bar{\nu}+\bar{h})\cos\bar{\phi}\sin\bar{\alpha}\sin\bar{\beta}\ \text{etc.}
\end{aligned}$$

The (Observed minus Computed) length *of the air-line* is then [Eq. (68.4), Hotine 1957]

$$-l_1\,d\omega-l_2\,d\phi-l_3\,dh+\bar{l}_1\,d\bar{\omega}+\bar{l}_2\,d\bar{\phi}+\bar{l}_3\,d\bar{h}\ .\qquad (31.1)$$

Notice once more that the symmetricality of the system requires $l_1 = \bar{l}_1$, which is verified from Eq. (24.7).

A direct measure by invar, geodimeter or tellurometer (which in any current adjustment would be held fixed) would require this equation to be heavily weighted. (It needs, of course, to be divided by the length of the line to reduce it to the same terms as the azimuth and zenith distance equations).

Alternatively, it can be held fixed by treating Eq. (31.1) as a condition equation and using it to eliminate one of the unknowns from the other observation equations before solution, just as was proposed in § 26 for astronomical measures.

32. To obtain a length equation to solve with Eqs. (27.1) and (29.1), we need the Cartesian components of l_r, which are

$$l_1 = -\sin\alpha \sin\beta \sin\omega - \sin\phi \cos\alpha \sin\beta \cos\omega + \cos\phi \cos\beta \cos\omega$$

$$l_2 = \sin\alpha \sin\beta \cos\omega - \sin\phi \cos\alpha \sin\beta \sin\omega + \cos\phi \cos\beta \sin\omega$$

$$l_3 = \cos\phi \cos\alpha \sin\beta + \sin\phi \cos\beta$$

and the (Observed minus Computed) length becomes

$$+(d\bar{n} - dn)l_1 + (d\bar{y} - dy)l_2 + (d\bar{z} - dz)l_3 \ . \tag{32.1}$$

33. We have now obtained rigorous first-order observation equations to take care of all the quantities usually measured in triangulation, and we have done so without requiring any knowledge of the form of the geoid. In particular the method of § 31, § 32 completely eliminates any "reduction to sea level", for which the depth of the geoid in relation to the spheroid might be required. The only way in which a knowledge of the geoid could assist the adjustment would be to provide means of deducing $\delta\phi$ and $\delta\omega$ instead of making direct astronomical measures, and it will certainly be more accurate and economical to make these measures as required, rather than undertake the much greater programme of measurement necessary to determine the geoid. It should be noted, moreover, that what would be required is an accurate knowledge of the geoidal depth and slope at particular points and not a deliberately smoothed first-order result as currently proposed.

34. The only additional measurement, not as a rule undertaken at present, which would stiffen the adjustment seems to be spirit-levelling between stations. Combined with occasional measures of gravity, this in effect would give a direct measure of difference of potential V over the line. Spirit-levelled "heights" by themselves are not independent of the path; they are not point-functions like potential, or geodetic heights rigorously defined as in § 16, and we cannot properly speak of the "height" of a point obtained by spirit-levelling without also specifying the route by which it was measured. Much futile argument would be saved if this were more generally recognized.

For a rigorous treatment, we should accordingly have a relation between physical potential V and geodetic height h, analogous to the differences $\delta\phi$, $\delta\omega$ between the other coordinates in the two systems, and it would need to be more exact over a finite line than § 10. This might in time be provided by a knowledge at each station of N (the separation between geoid and spheroid or between geop and spherop) obtainable from the Stokes' theory, as recently modified by Levallois and de Graaff Hunter, but it is doubtful whether this is necessary to a first order, which is all the Stokes' treatment would give.

The change in potential due to a change in position dx^r is $V_r\, dx^r = -g(\upsilon_r)dx^r$, where (υ_r) is the astronomical zenith. But (υ_r) differs from υ_r, the normal to the $h = \text{const.}$ surface in the geodetic coordinates by a vector whose modulus is of the order $\delta\phi$ or $\delta\omega$, so that to a first order we can say the change of potential is $-g\upsilon_r dx^r = -g\,dh$ in § 21 coordinates. If we start with the spirit-levelled heights (perhaps after a circuit adjustment), as approximate geodetic coordinates (and use these for calculating geodetic azimuths and zenith distances), the observation equation is accordingly

$$0 = g\,dh - \bar{g}\,d\bar{h} \ , \qquad (34.1)$$

which should be divided by mean gravity and by the side-length to reduce it to the same dimensions as the azimuth etc. equations, and then heavily weighted, as in the case of a directly measured base length. Alternatively, Eq. (34.1) can be used as a condition equation to eliminate one of the unknowns from the other observation equations before solution, as proposed for astronomical measures and bases. The final values will, as a result, be close to the spirit-levelling and cannot, of course, be considered as true coordinates in the geodetic system for any such purpose as determining the geoid. They may also affect derived values of $\delta\phi$, $\delta\omega$ for such purposes, but so far as the geometrical adjustment of the triangulation is concerned, these matters are of little consequence. At the time of writing, this method of absorbing the results of spirit-levelling has not yet been tried out in practice.

Standard Gravity

35. For certain purposes we use gravity "anomalies", defined as the difference between measured gravity and the value of gravity in some standard field at the same point. We now define the standard field rigorously as one in which the base spheroid of the geodetic coordinate system is an equipotential surface enclosing a mass M equal to that of the physical Earth; the field to rotate about the minor axis of the spheroid with the same angular velocity $\tilde{\omega}$ as the Earth does about its physical axis. The two axes will be parallel in the case of any coordinate system proposed in this paper, and will be made so in setting up the origin conditions of the coordinate system (§ 7); they will not necessarily coincide.

36. To determine the standard potential, we use (for this purpose only) ordinary spheroidal coordinates related to the usual Cartesian coordinates by the equations

$$x = c(1+u^2)^{1/2} \sin U \cos \omega$$

$$y = c(1+u^2)^{1/2} \sin U \sin \omega$$

$$z = cu \cos U \ .$$

The u = const. surfaces are the spheroids

$$\frac{x^2+y^2}{c^2(1+u^2)} + \frac{z^2}{c^2u^2} = 1$$

confocal with the base spheroid

$$\frac{x^2+y^2}{a^2} + \frac{z^2}{b^2} = 1 \ .$$

The constant F is half the inter-focal distance, or $c = ae = (a^2 - b^2)^{1/2}$; and the value of u for the base spheroid is

$$b/c = b/(ae) = \bar{e}/e \ .$$

The coordinate U is the *reduced* co-latitude on the u-spheroid passing through the point.

37. The normal solution, independent of longitude, of the Laplace equation in these coordinates (e.g. MacRobert 1947, p. 215) is easily found in terms of Legendre functions as

$$A_n Q_n(iu) P_n(\cos U) \ ,$$

so that the potential of the standard field is

$$\begin{aligned} V &= \sum_{n=0}^{\infty} A_n Q_n(iu) P_n(\cos U) + \tfrac{1}{2}\tilde{\omega}^2(x^2+y^2) \\ &= \sum_{n=0}^{\infty} A_n Q_n(iu) P_n(\cos U) + \tfrac{1}{3}\tilde{\omega}^2 c^2(1+u^2)\{1 - P_2(\cos U)\} \ . \end{aligned}$$

On the base spheroid $u = b/c = \bar{e}/e$ and

$$V_0 = \sum_{n=0}^{\infty} A_n Q_n(i\bar{e}/e) P_n(\cos U) + \tfrac{1}{3}\tilde{\omega}^2 a^2 - \tfrac{1}{3}\tilde{\omega}^2 a^2 P_2(\cos U)$$

for all values of U, so that the coefficients of the P_n's may be equated to zero, giving

$$V_0 = A_0 Q_0(i\bar{e}/e) + \tfrac{1}{3}\tilde{\omega}^2 a^2 \qquad (37.1)$$

and

$$A_2 Q_2(i\bar{e}/e) = \tfrac{1}{3}\tilde{\omega}^2 a^2 \qquad (37.2)$$

and all other A's zero, whence finally

$$V = A_0 Q_0(iu) + A_2 Q_2(iu) P_2(\cos U) + \tfrac{1}{2}\tilde{\omega}^2(x^2+y^2) \ . \qquad (37.3)$$

38. So far the treatment is equivalent to Somigliana's and others, but the coordinates (U, u), although valuable for this particular purpose, are of little use for other geodetic purposes and do not convert very easily to any of the systems proposed in this paper. Accordingly, we obtain the potential from Eq. (37.3) in ordinary spherical harmonics which are more readily computed from Cartesians and hence from (ω, ϕ, h) coordinates [Eq. (22.2)].

39. On the z-axis of symmetry, Eq. (37.3) gives the potential as

$$A_0 Q_0(iz/c) + A_2 Q_2(iz/c) \ ,$$

which can easily be expanded as

$$\sum_{n=1}^{\infty} \frac{(2n+1)iA_0+(2n-2)iA_2}{(2n-1)(2n+1)} (-1)^n \frac{(ae)^{2n-1}}{z^{2n-1}} . \tag{39.1}$$

To find the external attraction potential off the axis of symmetry, all we need do is to replace $(1/z^{2n-1})$ by $(1/r^{2n-1})P_{2n-2}(\cos\theta)$ (Ramsey 1952, p. 132), in which r is the radius vector and θ is the "geocentric" co-latitude. We must also restore the rotation term and obtain finally

$$V = \sum_{n=1}^{\infty} \frac{(2n+1)iA_0+(2n-2)iA_2}{(2n-1)(2n+1)} (-1)^n \frac{(ae)^{2n-1}}{r^{2n-1}} P_{2n-2}(\cos\theta)+\tfrac{1}{2}\tilde{\omega}^2(x^2+y^2) \tag{39.2}$$

with

$$r^2 = x^2+y^2+z^2 \; ; \quad \cos\theta = 3/r .$$

40. For large values of z, Eq. (39.1) reduces to $-iA_0/(ae/z)$ and at the same large distance we can consider the mass M condensed at the origin to give rise to a potential kM/z so that

$$iA_0 = -kM/(ae) . \tag{40.1}$$

The other constant, iA_2, is obtained from Eq. (37.2). If $e = \sin\alpha$, this expands as

$$iA_2 = \tfrac{2}{3}\tilde{\omega}^2 a^2/\{(1+3\cos^2\alpha)\alpha-3\cos\alpha\} . \tag{40.2}$$

41. If λ is the geographical *co*-latitude of the line of force, we have, by differentiating Eq. (39.2) with respect to z, after some simplification

$$g\cos\lambda = \sum_{n=1}^{\infty} \frac{(2n+1)iA_0+(2n-2)iA_2}{(2n+1)} (-1)^n \frac{(ae)^{2n-1}}{r^{2n}} P_{2n-1}(\cos\theta) \tag{41.1}$$

and by differentiation with respect to x or y

$$g\sin\lambda = \sum_{n=1}^{\infty} \frac{(2n+1)iA_0+(2n-2)iA_2}{(2n-1)(2n+1)} (-1)^n \frac{(ae)^{2n-1}}{r^{2n}} P'_{2n-1}(\cos\theta)\sin\theta - \tilde{\omega}^2 r\sin\theta . \tag{41.2}$$

These equations may be combined in the alternative forms:

$$g\cos(\lambda-\theta) = \sum_{n=1}^{\infty} \frac{(2n+1)iA_0+(2n-2)iA_2}{(2n+1)} (-1)^n \frac{(ae)^{2n-1}}{r^{2n}} P_{2n-2}(\cos\theta) - \tilde{\omega}^2 r\sin^2\theta \tag{41.3}$$

$$g\sin(\lambda-\theta) = \sum_{n=1}^{\infty} \frac{(2n+1)iA_0+(2n-2)iA_2}{(2n-1)(2n+1)} (-1)^n \frac{(ae)^{2n-1}}{r^{2n}} P'_{2n-2}(\cos\theta) - \tilde{\omega}^2 r\sin\theta\cos\theta . \tag{41.4}$$

42. Gravity on the equator of the base spheroid ($\lambda = \theta = \pi/2$; $r = a$) is from Eq. (41.3)

$$g_e = \sum_{n=1}^{\infty} \frac{(2n+1)iA_0 + (2n-2)iA_2}{(2n+1)} (-1)^n \frac{e^{2n-1}}{a} P_{2n-2}(0) \ ,$$

which relates g_e to M, or determines iA_0 in terms of g_e instead of M if required. An alternative formula, which can be shown to be equivalent, is obtained by differentiating Eq. (37.3) for $U = \pi/2$ with respect to

$$(x^2 + y^2)^{1/2} = c(1 + u^2)^{1/2} \ .$$

With $e = \sin\alpha$, the result after some manipulation is

$$a g_e = -iA_0 \tan\alpha + iA_2(\alpha - \tan\alpha) - \tfrac{2}{3}\bar{\omega}^2 a^2 \ .$$

43. All the above formulae are rigorous and rapidly convergent and may often be required as a control on more approximate computation and interpolation. Gravity on the international spheroid tabulated in the Geodetic Tables (1956) agrees with results from these formulae, but the application of ordinary first-order height reductions can introduce errors up to 7 mg.

44. We can, however, expand gravity, or any other scalar, in terms of geodetic height h, by the method of Hotine (1957, § 40). Notice that since the geodetic normal $\bar{\upsilon}^r$ in § 22 coordinates is straight

$$\partial^2 F/\partial h^2 = F_{rs}\bar{\upsilon}^r\bar{\upsilon}^s \quad \text{and} \quad \partial^3 F/\partial h^3 = F_{rst}\bar{\upsilon}^r\bar{\upsilon}^s\bar{\upsilon}^t \ \text{etc.}$$

For the line of force υ^r we have (from Hotine 1957, § 19) the following for substitution in the formulae of Hotine (1957, § 40)

$$\chi m^r = \gamma_1 \lambda^r + \gamma_2 \mu^r \ .$$

If we take an equipotential of the gravitational field as base surface ($h = 0$) of § 22 coordinates, than *at a point on this surface* we have from Hotine (1957, § 40)

$$\partial F/\partial h = \partial F/\partial n \tag{44.1}$$

$$\begin{aligned} \partial^2 F/\partial h^2 &= \partial^2 F/\partial n^2 - F_r(\gamma_1\lambda^r + \gamma_2\mu^r) \\ &= \partial^2 F/\partial n^2 - \gamma_1(\partial F/\partial p) - \gamma_2(\partial F/\partial m) \end{aligned} \tag{44.2}$$

and the third differential follows from the substitution of results already given in Hotine (1957, § 38 and § 39). For example,

$$\frac{\partial \log g}{\partial n} = -2H - 2\bar{\omega}^2/g \ \text{[Eq. (16.3), Hotine 1957]}$$

$$\frac{\partial^2 \log g}{\partial n^2} = -\frac{\partial(2H)}{\partial n} + \frac{2\bar{\omega}^2}{g}(-2H - 2\bar{\omega}^2/g)$$

$$\frac{\partial^2 \log g}{\partial n^2} = -\frac{\partial(2H)}{\partial n} - 2H\left(\frac{2\bar{\omega}^2}{g}\right) - \left(\frac{2\bar{\omega}^2}{g}\right)^2 - \gamma_1^2 - \gamma_2^2$$

[from Eq. (44.2)]. In the last equation, which is true only for a point on the base surface, all quantities on the right have base surface values. Next, we have from Eqs. (34.2) and (34.4) from Hotine (1957)

$$\frac{\partial(2H)}{\partial n}=\frac{\partial\gamma_1}{\partial\varrho}+\frac{\partial\gamma_2}{\partial m}-\chi_1^2-\chi_2^2-2\tau^2-\gamma_1^2-\gamma_2^2+(\gamma_1\tau-\chi_1\gamma_2)\tan\phi$$

so that

$$\frac{\partial^2\log g}{\partial n^2}=(\chi_1^2+\chi_2^2+2\tau^2)-(\gamma_1\tau-\chi_1\chi_2)\tan\phi$$
$$-2H\left(\frac{2\tilde{\omega}^2}{g}\right)-\left(\frac{2\tilde{\omega}^2}{g}\right)^2-\partial\gamma_1/\partial p-\partial\gamma_2/\partial m\ , \qquad (44.3)$$

which is calculable for any base surface if the form of the base surface, and gravity on it, are known. For a spheroidal base surface

$$\chi_1=1/\upsilon\ ,\quad \chi_2=1/\varrho\ ,\quad \tau=0\ ,\quad \gamma_2=\partial\log g/(\varrho\partial\phi)$$

etc. and the expression reduces to

$$\frac{\partial^2\log g}{\partial h^2}=\left(\frac{1}{\upsilon^2}+\frac{1}{\varrho^2}\right)+\frac{\tan\phi}{\varrho\upsilon}\frac{\partial\log g}{\partial\phi}-\left(\frac{1}{\upsilon}+\frac{1}{\varrho}\right)\left(\frac{2\tilde{\omega}^2}{g}\right)-\left(\frac{2\tilde{\omega}^2}{g}\right)^2$$
$$-\frac{\partial}{\varrho\partial\phi}\left(\frac{\partial\log g}{\varrho\partial\phi}\right).$$

Since

$$\frac{\partial^2\log g}{\partial h^2}=-\left(\frac{\partial\log g}{\partial h}\right)^2+\frac{1}{g}\frac{\partial^2 g}{\partial h^2}\ ,$$

we have finally the following expression as far as terms of the second-order for gravity at a geodetic height h:

$$g-h\left\{\left(\frac{1}{\upsilon}+\frac{1}{\varrho}\right)g+2\tilde{\omega}^2\right\}+\frac{h^2}{2}\left[2g\left(\frac{1}{\upsilon^2}+\frac{1}{\varrho\upsilon}+\frac{1}{\varrho^2}\right)\right.$$
$$\left.+\left(\frac{1}{\upsilon}+\frac{1}{\varrho}\right)(2\tilde{\omega}^2)+\frac{\tan\phi}{\varrho\upsilon}\frac{\partial g}{\partial\phi}-\frac{g}{\varrho}\frac{\partial}{\partial\phi}\left(\frac{1}{\varrho g}\frac{\partial g}{\partial\varrho}\right)\right]\ , \qquad (44.4)$$

in which g and all spheroidal functions are to have their values on the base spheroid at the foot of the normal ($h=0$) viz. at a point on the base spheroid having the same geodetic latitude.

We can similarly expand potential and latitude of the line of force ($\bar{\phi}$) as far as necessary, starting with

$$\frac{\partial V}{\partial n}=-g$$

and

$$\frac{\partial \bar{\phi}}{\partial n} = \gamma_2 \, .$$

These expansions will not have to be used often in practice, until the characteristics of the field are required at much greater altitudes than exist on the surface of the Earth, but they do provide yet another illustration of the power of the present theory, which can derive quite general expressions like Eq. (44.3).

Index of Main Symbols

A^r, B^r, C^r	Cartesian vectors, right-handed in that order (see Hotine 1957, § 1)
ω, ϕ	Longitude, latitude of υ^r in relation to A^r, B^r, C^r (two of the three coordinates in all systems)
υ^r	Outward-drawn unit normal to the third-coordinate surface
μ^r	Unit "meridian" vector, lying towards the North in the third-coordinate surface, coplanar with υ^r and C^r
λ^r	Unit "parallel" vector, perpendicular to υ^r and μ^r. λ^r, μ^r, υ^r is a right-handed set in that order
α	Azimuth, measured eastwards from μ^r
β	Zenith distance, measured from υ^r
V	Potential; the third coordinate in the astronomical system or in the standard field geodetic system
h	Geodetic height; the third coordinate in geodesic parallel systems
g	Gravity; physical in the astronomical system; standard or notional in other systems
dp, dm, dn	Elements of arc in directions λ^r, μ^r, υ^r
γ_1	Arc rate of change of (log g) in direction λ^r
γ_2	Arc rate of change of (log g) in direction μ^r
χ_1	Normal curvature of third-coordinate surface in direction λ^r
χ_2	Normal curvature of third-coordinate surface in direction μ^r
τ	Geodesic torsion of third-coordinate surface in direction μ^r
H	Mean curvature of third-coordinate surface $= (\chi_1 + \chi_2)/2$
K	Gauss curvature of third-coordinate surface $= \chi_1 \chi_2 - \tau^2$
ϖ	Angular velocity; physical in the astronomical system; notional, same value but anout the axis of symmetry, in the case of the standard field (see § 11 and § 35)
ϱ, υ	Principal curvatures of base spheroid in systems § 12 and § 22
e	Eccentricity of spheroid. $\bar{e}^2 = (1 - e^2)$.

(*Note:* ω, ϕ, α, β etc. of the same point or line etc., will not of course be equal in different coordinate systems. Whenever two coordinate systems are considered together, one is denoted by barred notation, e.g. $\bar{\omega}$, $\bar{\phi}$, $\bar{\alpha}$, $\bar{\beta}$. In some cases, clear from the context, barred notation is also used for a different point in the same coordinate system.)

Editorial Commentary

This report is a direct continuation of the previous report to the Toronto Assembly. It addressed the complicated question of the construction of geodetic coordinate systems, and represented Hotine's first response to the challenge posed by Marussi's ideas on differential geodesy (see Marussi 1988; Zund 1989). As such is represents a preliminary account to the material in Part II of MG, in particular that in Chapters 12–15; and Part IIII of MG, viz. Chapters 22–27.

It furnishes a valuable, albeit admittedly brief account of this material, much of which is reconsidered in more depth and detail in the later two reports *A Primer of Non-Classical Geodesy* (Venice, 1959). and *The Third Dimension in Geodesy* (Helsinki, 1960). Nevertheless, one is pleased to have it, even if it is only of historical interest as indicative of the evolution of Hotine's ideas.

The wealth of material touched on in this report, i.e. not only the construction of geodetic coordinate systems (§§ 1–25), but the adjustment of triangulation in space (§§ 25–34), and standard gravity (§§ 35–44), is rather overpowering. It must have been – at least momentarily – overwhelming even to one of Hotine's ability, since it suggested the magnitude of the challenge awaiting him.

A particularly noteworthy derivation of the equations for the components of the potential and gravity is contained in §§ 37–40. This includes the higher order flattening expressions subsequently given by Cook (1959) and Lambert (1961). The corresponding discussion in MG is contained in Chapter 23.

References to Paper 4

Eisenhart L (1949) Riemannian geometry. Princeton edn

Geodetic Tables, International Ellipsoid (1956) Danish Geodetic Institute, Copenhagen

Hotine M (1957) Metrical properties of the earth's gravitational field. I. A. G. Toronto

MacRobert TM (1947) Spherical harmonics. Methuen, London edn

Marussi A (1950) Principi di geodesia intrinseca applicati al campo di Somigliana. Boll Geod Sci Aff

Ramsey AS (1952) Newtonian attraction. Cambridge

References to Editorial Commentary

Cook AH (1959) External gravity field of a rotating spheroid to the order e^3. Geophys J Roy Astron Soc 2:199–214

Lambert WD (1961) The gravity field of an ellipsoid of revolution as a level surface. Publ Inst Geod Photogram Cartogr No. 14. Ohio State Univ, Columbus

Marussi A (1988) Intrinsic geodesy (a revised and edited version of his 1952 lectures prepared by J. D. Zund) Rep 390, Dept Geod Sci Survey. Ohio State Univ, Columbus, 137 pp

Zund JD (1989) A mathematical appreciation of Antonio Marussi's contributions to geodesy. Geophysics Laboratory, GL-TR-89-0309 24 pp

5 A Primer of Non-Classical Geodesy[1]

Introduction

1. This paper simplifies and extends some of the contents of two earlier papers[2] presented to the International Geodetic Association at Toronto in 1957. It is intended as a new approach to the basic mathematical theory of geodetic measurement, leading to rigorous methods of reduction and adjustment, unrestricted by the length of observed lines and suited to modern electronic computation. It is submitted as a complete answer to the problem posed to Study Group No. 1 of the International Association of Geodesy (Toronto, 1957), viz. the adjustment of large triangulation networks "taking into account the form of the geoid". The basic material theory contained in this paper shows that it is unnecessary for this purpose to know the form of the geoid at all.

2. The basis of the classical method is to project the observing stations orthogonally onto a spheroid. Observed directions, necessarily measured in relation to the plumb line, are (in some cases) corrected for "geoidal tilt" to derive directions which would be obtained if measurable in relation to the spheroidal normal. The actual line of observation – an optical or radio path – is considered to lie in a plane containing the spheroidal normal, and is projected to the spheroid as a curve of normal section. By applying a correction to the initial direction, this curve is then shifted to the normal section curve passing through the projection of the other station. A different curve in general results from observation at the other end of the line, and the two curves are replaced (through "correction" of the end directions) by the (unique) spheroidal geodesic joining the projected terminals. Precise calculation of the elements of this geodesic is often described as the fundamental problem of geodesy, despite the fact that it has no direct correspondence throughout with the actual line of observation. Geodesic triangles are solved by "correcting" their angles to those of a plane triangle having the same side-lengths. Two coordinates of the observing stations are finally computed either on the spheroid or on a plane-conformal transformation of the spheroid. For purposes of adjustment, astronomical measures, particularly of longitude and azimuth, made at points in physical space are compared with geodetic values derived at different points on the spheroid, without laying down a complete and unequivocal correspondence between the two systems.

---

[1] Presented at the first symposium on Three-Dimensional Geodesy (Venice 1959).

[2] See papers 3 and 4 in this monograph.

3. Determination of geodetic heights in the classical method receives scant theoretical attention. As a rule, not even corrections for "geoidal tilt" are applied to observed zenith distances, with the result that "trigonometric heights" are given in no defined system independent of the particular line of observation. It is usually assumed that heights should be provided by spirit-levelling. These are no doubt the best heights to show on maps of countries fortunate enough to have a network of precise levels, but they are related more to the astronomical system of coordinates and are not directly related to any possible analytical geodetic system.

4. Measured bases should be reduced to the length of the spheroidal geodesic between their projected terminals if they are to be in harmony with other classically reduced measures. This presupposes a knowledge of heights above the spheroid, which the classical method does not provide. If spirit-levels above mean sea level are used instead, the base length would be wrongly reduced to the general level of the geoid underlying it, not to the spheroid. This leads to an attempt to establish the relative depths of spheroid and geoid by integrating the effect of deflections along a section. The method rests on several not always clearly specified approximations, and assumptions relating to underground densities, whose validity may fairly be questioned. Determination of the form of the geoid in relation to a reference spheroid from gravity measures is another method which does not yet command universal acceptance in detail; it requires worldwide measures, which are not likely to be available for many years, to give a smoothed first-order result.

5. It may be said at once that the classical method does usually give as numerically satisfactory results as the rigorous methods described in this paper, at any rate over the normal side lengths of geodetic triangulation between terrestrial stations. So, at the time, did pre-Copernican methods of astronomical computation give satisfactory numerical results. The classical methods of azimuth and base adjustment have not yet been compared but they too may well be numerically adequate. There can be no doubt, however, that the theoretical basis of the classical method leads to quite unnecessary complication and confusion and should be replaced by more modern conceptions, which are vital to the further development of the subject, particularly the utilization of much longer lines and the incorporation of more frequent astronomical measures and measures of length.

6. The methods developed in this paper, and fully illustrated by application to most geodetic processes, are three-dimensional throughout. A base spheroid appears in the adopted geodetic 3-coordinate system, mainly to ensure results in fair sympathy with antecedent work[3], but it is not used for two-dimensional com-

[3] For consideration of other possible systems, see paper 4 this Volume, where it is argued that the system adopted in the present paper is the most advantageous. The astronomical system is fully considered in paper 3. On these matters, the earlier papers are still current until they can be replaced.

putation on classical lines. No apology is offered for the inclusion of vector methods in index notation, which are much easier to acquire than to avoid[4].

7. The paper can be read and understood without prior knowledge of vector methods from the following very brief description. The notation l^r does not mean a number raised to the r-th power; it is shorthand for three numbers l^1, l^2, l^3 specifying a length in a certain direction, known as a *vector*, whose three *contravariant components* are l^1, l^2, l^3. If changes dx^r in the three coordinates x^1, x^2, x^3 occur in the direction of the vector, then these changes dx^1, dx^2, dx^3 are proportional to l^1, l^2, l^3 respectively. If the changes occur over a length ds (which may be infinitesimal) of the vector and if $l^r = dx^r/ds$, then l^r is known as a *unit* vector. A vector L^r of magnitude L in the direction l^r would have as its contravariant components Ll^1, Ll^2, Ll^3.

8. A vector l^r also has *covariant components*, denoted by l_r, which are the same as the contravariant components in Cartesian coordinates but are not as easy to define in more general curvilinear coordinates (such as latitude and longitude); wherever necessary their derivation will be illustrated in the text.

9. A product of two vectors $L^r M_s$ is a second-order *tensor* having nine components (e.g. L^1M_1, L^1M_2, L^1M_3, L^2M_1, etc.). If the superscript and subscript are the same, the *summation convention* applies and the three remaining components of the tensor are summed to form a pure number or *scalar*, e.g. $L^rM_r = L^1M_1 + L^2M_2 + L^3M_3$. This is known as the *scalar product* of the two vectors and is equal to the product of their magnitudes and the cosine of the angle between them. If the two vectors are the same, or are parallel and of equal length, then their scalar product is equal to the square of their common magnitude. If they are perpendicular, their scalar product is zero. The scalar product is the same, or *invariant*, whatever the coordinate system associated with the vector components.

10. If a linear relation between vectors at a point in space is true in one coordinate system, then it is also true in any other coordinate system. A linear relation exists between any three *coplanar* vectors. Any vector can be expressed linearly in terms of three other non-coplanar vectors. If l^r, m^r, n^r are three mutually orthogonal (perpendicular) vectors of non-zero length and $Al^r + Bm^r + Cn^r = 0$, then by multiplying across by l_r, we find that $A = 0$, and similarly $B = 0$, $C = 0$.

11. Apologists for the classical methods have suggested that they were deliberately adopted in order to escape the effects of atmospheric refraction, and that this makes them superior to three-dimensional methods. Neither claim is valid. Any uncertainty in the form of the line of observation is bound to affect the same

---

[4] Two simple books on the subject are (1) McConnell's *Application of the Absolute Differential Calculus* and (2) Weatherburn's *Introduction to Riemannian Geometry and the Tensor Calculus*. The first 50 pages plus Chapter XI of the former are ample for the present application; or the first 40 pages of the latter.

results to the same extent, whatever method of computation is adopted. For instance, the classical method assumes implicitly that the line of observation lies in both normal section planes, which in general is impossible unless it is straight, so that the effect of refraction on the calculation of latitude and longitude is not overcome; it is simply ignored. True, the effect is small, but it is equally small in the three-dimensional method. The effect on heights is more serious but again does not favour one method over the other: equivalent assumptions have to be made in both methods as regards the form of the line, and the only difference between the two methods is that the three-dimensional method makes no other assumptions. Approximations equivalent to, or fewer than, those made or implied in the classical method can be made in the three-dimensional method, which then becomes just as easy to compute. In short, the classical method has no real comparative advantages. Historically, it has evolved from the flat Earth in two Euclidean dimensions to the round Earth in two non-Euclidean dimensions; it has made much use of Gaussian differential geometry of surfaces, but no use of the later extension of Gauss' methods by Ricci and others to three (or more) dimensions. Since Gauss, geodesy and geometry have gone their separate ways, to the disadvantage of geodesy.

Definitions

A diagram (Fig. 1) and list of symbols is provided at the end of the paper.

12. Three mutually orthogonal unit vector fields A^r, B^r, C^r – right-handed in that order – are defined as follows.

C^r at any point in space is parallel to the physical axis of rotation of the Earth; positive direction North.

A^r is parallel to the plane determined by C^r and the plumb line or astronomic zenith at the origin of the survey, and is perpendicular to C^r; positive direction outwards from the centre of the Earth.

B^r completes the triad; positive direction eastwards.

13. These three vector fields A^r, B^r, C^r are common to both the astronomic and geodetic systems of coordinates. In the geodetic system in § 31 C^r is parallel to the minor axis of the base spheroid. The plane A^r, C^r remains parallel to the astronomic meridian at the origin; it is not parallel to the geodetic meridian at the origin unless the astronomic and geodetic longitudes of the origin are made equal. Means of orienting the geodetic system to ensure parallelism of the three fundamental vectors are considered below (§ 24 et seq.).

The three vector fields A^r, B^r, C^r are also the base vectors of a Cartesian system of coordinates whose origin in the geodetic system is the geometrical centre of the base spheroid.

14. A second set of (variable) orthogonal unit vector fields λ^r, μ^r, ν^r – right handed in that order – is defined at any point in space as follows.

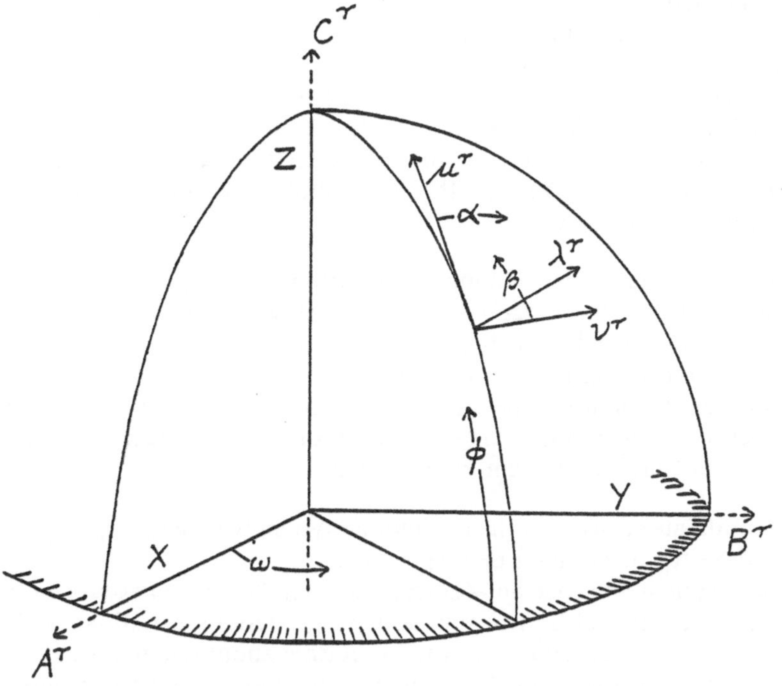

Fig. 1.

υ^r is the direction of the astronomic zenith, or outward-drawn normal to the equipotential surface (or tangent to the line of force).

μ^r in the direction of the astronomic meridian, lies in the equipotential surface and in the plane υ^r, C^r; positive northwards.

λ^r in the direction of the astronomic parallel, lies in the equipotential surface and completes the triad; positive eastwards.

A similar set of vectors, not in general parallel to the corresponding astronomic vectors, is defined in the geodetic system. In this case, υ^r is the straight outward-drawn normal to the base spheroid passing through the point in space under consideration; μ^r is in the plane υ^r, C^r; and λ^r completes the triad. Whenever the two sets are used together, the notation for one specified set will be barred, viz. $\bar{\lambda}^r$, $\bar{\mu}^r$, $\bar{\upsilon}^r$.

15. In both the astronomic and geodetic systems, longitude and latitude are defined as follows, using the appropriate vectors υ^r, μ^r.

Longitude (ω) is the angle between the planes A^r, C^r and υ^r, μ^r, C^r; positive in the positive rotation about C^r, that is from A^r to B^r, or *East*. This differs from the normal astronomic convention (and from some geodetic systems), but is necessary to preserve the ordinary analytical conventions. The sign of longitudes listed as positive to the West should be changed before entering any of the formulae in this paper and should also be abstracted by subtracting the astronomic longitude of the origin. The plane A^r, C^r thus becomes the zero of both astronomic and geodetic longitudes. The *geodetic* longitude of the origin will not necessarily be zero, unless it had been made so during orientation of the geodetic system (§ 24 et seq.).

Latitude (ϕ) is the angle between ν^r and the plane A^r, B^r; positive North. Co-latitude ($\pi/2-\phi$) is the angle between ν^r and C^r.

In the astronomic system, the direction of the meridian (or parallel) as defined in § 14 is not necessarily the same as the horizontal direction in which the longitude (or latitude) remains constant.

16. The third coordinate in the astronomic system will be the *geopotential*. In the geodetic system it will be the *geodetic height* (h), defined as the length of the straight outward-drawn normal to the base spheroid, intercepted between the base spheroid and the point in space under consideration.

Both systems can be shown to be analytical coordinate systems (as distinct from mere reference systems) in the sense of continuous, single-valued, differentiable functions of a Cartesian system in flat 3-space.

17. The third-coordinate surfaces in the astronomic system will be the equipotential or level surfaces, as determined by a spirit-level.

In the geodetic system, the constant-h surfaces are parallel to the base spheroid but are not themselves spheroids. The standard (or geodetic) gravitational field will be defined, apart from constants, as a field rotating about the minor axis of the base spheroid with the same angular velocity as the Earth; the base spheroid to be an equipotential surface of this field. The other equipotential surfaces are neither spheroids nor constant-h surfaces.

18. In both astronomic and geodetic systems, *azimuth* (α) is measured East from North, that is from μ^r towards λ^r, using the vectors appropriate.

A unit space vector in azimuth α and zenith distance β will accordingly be given by

$$l^r = \lambda^r \sin\alpha \sin\beta + \mu^r \cos\alpha \sin\beta + \nu^r \cos\beta \ , \tag{18.1}$$

which can easily be verified by forming scalar products.

19. The *declination* (D) of a space vector l^r is the angle between l^r and the plane A^r, B^r; positive North. This follows the usual astronomic convention if we consider the vector l^r prolonged to meet the "celestial sphere" in a "star".

The *origin hour angle* (H) of a space vector l^r is the angle between the planes[5] A^r, C^r and l^r, C^r; positive in the positive rotation about C^r, that is from A^r towards B^r, or East. This reverses the sign of the usual astronomic convention. With the normal conventions for right ascension and sidereal time, H is the right ascension of the direction l^r *minus* the local sidereal time at the origin, both expressed in angular measure. The local sidereal time at the origin is the Greenwich sidereal time *plus* the astronomic longitude of the origin relative to Greenwich (measured as stated above positive eastwards).

[5] Although l^r, C^r do not intersect, lines parallel to them define the plane in question.

A unit space vector l^r in declination D and origin hour angle H is given by

$$l^r = A^r \cos H \cos D + B^r \sin H \cos D + C^r \sin D \ . \tag{19.1}$$

Relations Between Fundamental Vectors

20. We denote a general set of coordinates by x^r. If these are either the astronomic or the geodetic system, then they are to be taken in the order x^1 = longitude, x^2 = latitude, x^3 = geodetic height or the geopotential; if they are Cartesian coordinates, $x^1 = x$ measured in the direction A^r; $x^2 = y$ in the direction B^r; $x^3 = z$ in the direction C^r.

x, y, z can be considered as scalar functions of position, in which case such covariant vectors as x_r are, as usual, given by $\partial x/\partial x^r$. The vector equation $x_r = A_r$ is immediately verifiable in Cartesian coordinates and is accordingly true in any coordinate system. Similarly, $y_r = B_r$; $z_r = C_r$.

21. The following covariant vector equations, true in any coordinate system, are verified straight from the definitions by forming scalar produces (e.g. if we multiply the first equation across by B^r, we have $\lambda_r B^r = \cos \omega$ and can easily verify that this is so).

$$\begin{aligned} \lambda_r &= -A_r \sin \omega + B_r \cos \omega \\ \mu_r &= -A_r \sin \phi \cos \omega - B_r \sin \phi \sin \omega + C_r \cos \phi \\ \upsilon_r &= A_r \cos \phi \cos \omega + B_r \cos \phi \sin \omega + C_r \sin \phi \end{aligned} \tag{21.1}$$

and the reverse equations:

$$\begin{aligned} A_r = x_r &= -\lambda_r \sin \omega - \mu_r \sin \phi \cos \omega + \upsilon_r \cos \phi \cos \omega \\ B_r = y_r &= \lambda_r \cos \omega - \mu_r \sin \phi \sin \omega + \upsilon_r \cos \phi \sin \omega \\ C_r = z_r &= \mu_r \cos \phi + \upsilon_r \sin \phi \ . \end{aligned} \tag{21.2}$$

The corresponding contravariant equations, e.g.

$$\lambda^r = -A^r \sin \omega + B^r \cos \omega \ , \tag{21.3}$$

are obtained by simply raising the indices throughout.

22. In the above equations, λ_r, μ_r, υ_r, ϕ, ω naturally all refer to the same system, astronomic or geodetic. If the corresponding functions in the other system, *but at the same point in space*, are denoted by bars (e.g. $\bar{\lambda}_r$), then we should also have

$$\bar{\lambda}_r = -A_r \sin \bar{\omega} + B_r \cos \omega$$

etc. since the vectors A_r, B_r, C_r are common to both systems; and by eliminating A_r etc. with Eq. (21.2) we have the following relations applicable to a transformation between the two systems

$$\bar{\lambda}_r = -A_r \sin\bar{\omega} + B_r \cos\bar{\omega}$$
$$= \lambda_r \cos(\bar{\omega}-\omega) + \mu_r \sin\phi \sin(\bar{\omega}-\omega) - \nu_r \cos\phi \sin(\bar{\omega}-\omega) \tag{22.1}$$

$$\bar{\mu}_r = -\lambda_r \sin\bar{\phi} \sin(\bar{\omega}-\omega) + \mu_r [\cos\phi \cos\bar{\phi} + \sin\phi \sin\bar{\phi} \cos(\bar{\omega}-\omega)]$$
$$+ \nu_r [\sin\phi \cos\bar{\phi} - \cos\phi \sin\bar{\phi} \cos(\bar{\omega}-\omega)] \tag{22.2}$$

$$\bar{\nu}_r = \lambda_r \cos\bar{\phi} \sin(\bar{\omega}-\omega) + \mu_r [\cos\phi \sin\bar{\phi} - \sin\phi \cos\bar{\phi} \cos(\bar{\omega}-\omega)]$$
$$+ \nu_r [\sin\phi \sin\bar{\phi} + \cos\phi \cos\bar{\phi} \cos(\bar{\omega}-\omega)] \; . \tag{22.3}$$

These last equations [(22.1), (22.2), (22.3)] should also be true if the barred quantities refer to the same coordinate system but to a different point in space, since we have merely used the constancy of the vectors A_r, B_r, C_r in order to derive them. Nevertheless, such vector equations apply in general only between the components of vectors defined at the same point in space. Accordingly, we define e.g. $\bar{\lambda}_r$ in the usual way at the barred point and then consider it translated parallel to itself to the unbarred point. Since the Cartesian components of equal and parallel vectors are the same, the vector Eqs. (22.1), (22.2), (22.3) will then hold in Cartesian coordinates at the unbarred point; and therefore, on this interpretation of $\bar{\lambda}_r$ etc. they will also hold at the unbarred point in any other coordinate system, even though the components of parallel vectors at different points are not the same in other coordinate systems. In the same way, we could consider λ_r etc. as defined in the usual way at the unbarred point and then translated parallel to themselves to the barred point; in which case the vector equations would hold in any coordinate system at the barred point.

23. If the Cartesian components of a unit vector l^r – in azimuth α, zenith distance β, declination D, origin hour angle H – are (a, b, c) then

$$l^r = aA^r + bB^r + cC^r \; , \tag{23.1}$$

and combining this with Eqs. (19.1) and (18.1), after substituting in the latter the contravariant form of (21.1), we have

$$\cos H \cos D = a = -\sin\phi \cos\omega \cos\alpha \sin\beta - \sin\omega \sin\alpha \sin\beta$$
$$+ \cos\phi \cos\omega \cos\beta \; .$$
$$\sin H \cos D = b = -\sin\phi \sin\omega \cos\alpha \sin\beta + \cos\omega \sin\alpha \sin\beta \tag{23.2}$$
$$+ \cos\phi \sin\omega \cos\beta \; .$$
$$\sin D = c = \cos\phi \cos\alpha \sin\beta + \sin\phi \cos\beta \; .$$

Conversely, by substituting Eqs. (21.2) in (23.1) and comparing the result with § 18 we have

$$\sin\alpha \sin\beta = -a \sin\omega + b \cos\omega$$
$$= \cos D \sin(H-\omega)$$
$$\cos\alpha \sin\beta = -a \sin\phi \cos\omega - b \sin\phi \sin\omega + c \cos\phi \tag{23.3}$$
$$= \sin D \cos\phi - \sin\phi \cos D \cos(H-\omega)$$

$$\cos\beta = a\cos\phi\cos\omega + b\cos\phi\sin\omega + c\sin\phi$$
$$= \sin D\sin\phi + \cos\phi\cos D\cos(H-\omega)\ .$$

These equations [(23.2) and (23.3)] clearly hold if α, β, ω, ϕ all refer either to the astronomic or to the geodetic system; since a, b, c, H, D are common to both systems. If we denote the other system by bars, we have, for example

$$\sin\bar{\alpha}\sin\bar{\beta} = -a\sin\bar{\omega} + b\cos\bar{\omega}$$

from Eq. (23.3) and substituting for a, b, c from (23.2) we have the following transformation equations:

$$\begin{aligned}
\sin\bar{\alpha}\sin\bar{\beta} &= \sin\alpha\sin\beta\cos(\bar{\omega}-\omega) + \cos\alpha\sin\beta\sin\phi\sin(\bar{\omega}-\omega) \\
&\quad -\cos\beta\cos\phi\sin(\bar{\omega}-\omega)\ , \\
\cos\bar{\alpha}\sin\bar{\beta} &= -\sin\alpha\sin\beta\sin\bar{\phi}\sin(\bar{\omega}-\omega) \\
&\quad +\cos\alpha\sin\beta\,[\cos\phi\cos\bar{\phi} + \sin\phi\sin\bar{\phi}\cos(\bar{\omega}-\omega)] \\
&\quad +\cos\beta\,[\sin\phi\cos\bar{\phi} - \cos\phi\sin\bar{\phi}\cos(\bar{\omega}-\omega)] \\
\cos\bar{\beta} &= \sin\alpha\sin\beta\cos\bar{\phi}\sin(\bar{\omega}-\omega) \\
&\quad +\cos\alpha\sin\beta\,[\cos\phi\sin\bar{\phi} - \sin\phi\cos\bar{\phi}\cos(\bar{\omega}-\omega)] \\
&\quad +\cos\beta\,[\sin\phi\sin\bar{\phi} + \cos\phi\cos\bar{\phi}\cos(\bar{\omega}-\omega)]\ .
\end{aligned} \tag{23.4}$$

Moreover, these equations [(23.4)] will also hold for a vector parallel to l^r at another point in the same space, denoted by bars, since a, b, c are the same for such a parallel vector. In particular, they hold for the straight line joining the unbarred and barred points and enable us to compute azimuth and zenith distance at the far end of such a line. But is should be noted that $\bar{\alpha}$, $\bar{\beta}$ refer to the same sense of the line as α, β; $\bar{\beta}$ should be subtracted from 180° and 180° should be added to $\bar{\alpha}$, if they are to refer to the back direction. The equations are true as between barred and unbarred points on the same straight line if $\bar{\alpha}$, $\bar{\beta}$, α, β, $\bar{\omega}$, $\bar{\phi}$, ω, ϕ all refer either to the astronomic or to the geodetic system, regardless of irregularities or anomalies affecting the plumb line.

Only two of the equations in each set [(23.2), (23.3) and (23.4)] are independent, since $a^2+b^2+c^2=1$; the third equation provides a check.

Origin Conditions

24. We now consider means of ensuring that the Cartesian vectors A_r, B_r, C_r are parallel in the astronomic and geodetic systems, since all the analysis so far depends on this.

It is evident that the Eqs. (23.4), considered as transformation equations from the astronomic system (unbarred) to the geodetic system (barred), must necessarily be satisfied at the origin, as at any other point in space. If we adopt measured astronomic values for α, β, ϕ, ω then we must choose initial geodetic values $\bar{\alpha}$, $\bar{\beta}$, $\bar{\phi}$, $\bar{\omega}$ to satisfy Eq. (23.4). The Cartesian components of the vector l^r (an actual physical direction in space between survey stations) will then be equal in the two systems. [This can easily be verified by forming the geodetic (barred) equations corresponding to (23.2) and using (23.4) to prove that $a = \bar{a}$ etc.] But this alone is not sufficient to ensure parallelism of the Cartesian vectors in the two systems, since it would still be possible to rotate the geodetic system complete with its Cartesian vectors about l^r. To ensure parallelism we must also equate the Cartesian components of a second direction.

25. This question is of such importance as to justify alternative and less intuitive consideration. Suppose that the Cartesian vectors of the geodetic system are *not* parallel to those of the astronomic system, but are denoted by $\bar{A}^r$ etc. yet nevertheless the Cartesian components of the unit vector l^r are the same in both systems. Then

$$l^r = aA^r + bB^r + cC^r = a\bar{A}^r + b\bar{B}^r + c\bar{C}^r$$

or

$$a(A^r - \bar{A}^r) + b(B^r - \bar{B}^r) + c(C^r - \bar{C}^r) = 0 \ ,$$

which merely requires the three vectors $(A^r - \bar{A}^r)$ etc. to be coplanar (and incidentally perpendicular to l^r, since $l^r\,(A_r - \bar{A}_r) = l^r A_r - l^r \bar{A}_r = a - a = 0$ etc.). But if, in addition, the Cartesian components (a^1, b^1, c^1) of a second, independent unit vector are the same in both systems, then we must also have

$$a^1(A^r - \bar{A}^r) + b^1(B^r - \bar{B}^r) + c_1(C^r - \bar{C}^r) = 0 \ ,$$

so that *either* $a/a^1 = b/b^1 = c/c^1$, in which case the second vector would coincide with the first *or* the three vectors $(A^r - \bar{A}^r)$ etc. are null and $A^r = \bar{A}^r$, $B^r = \bar{B}^r$, $C^r = \bar{C}^r$.

26. The difference between the astronomic and geodetic systems will usually be small. If we write $\bar{\phi} = (\phi + \delta\phi)$ etc. then to a first order the Eqs. (23.4) reduce to:

$$\begin{aligned} \delta\alpha &= \sin\phi\,\delta\omega + \cos\beta\,(\sin\alpha\,\delta\phi - \cos\alpha\,\cos\phi\,\delta\omega) \\ \delta\beta &= -\cos\phi\,\sin\alpha\,\delta\omega - \cos\alpha\,\delta\phi \ . \end{aligned} \tag{26.1}$$

If, in addition, β is nearly 90°, then the first equation reduces to $\delta\alpha = \sin\phi\,\delta\omega$. This is the so-called Laplace azimuth equation, which alone is used in the classical method to orient the spheroid. It is probably sufficiently accurate in terrestrial triangulation to satisfy the first equation, but the classical method makes no attempt to satisfy the second equation, much less to do so for the necessary two distinct lines radiating from the origin.

The extra conditions in the present method do not arise from any special requirement in orienting the geodetic system. If astronomic and geodetic measures are to be used together in calculating and adjusting an extensive survey, then the relation between the two systems must be specified completely and the present specification is no more and no less exacting than is necessary for the purpose.

It may be objected that the second of the two equations (26.1) is vitiated by errors of atmospheric refraction in the measured (astronomic) zenith distances and ought not therefore to be used. Indeed, there is everything to be said for careful reciprocal observation of these angles over several lines to minimize the effect of refraction, but there is nothing to be said for omitting these observations altogether. We do not overcome the difficulty by simply ignoring the second equations of (26.1) without making any attempt to satisfy them.

27. The simplest way of satisfying the origin conditions is to accept, as the geodetic starting elements, astronomic measures of latitude, longitude and azimuth, plus two measured zenith distances corrected for refraction. This procedure will in effect have been adopted, without full consideration of the underlying theory, in choosing the origin conditions for some triangulations. In other cases, where the influence of vertical angles may have been ignored and where perhaps not even the Laplace azimuth equation has been satisfied at the origin, the matter should be investigated before the results are extended or further adjusted in conjunction with astronomic measures.

Means of correcting for refraction and of minimizing the effect of observational error are considered below in the section on triangulation adjustment.

28. The initial geodetic height may be assumed or be taken as a spirit-levelled height of the origin. It will settle the position of the base spheroid in relation to the actual point on the Earth's surface taken as origin. Whatever initial height is taken for the origin will not, however, affect the orientation of the spheroid and will therefore be without effect on any of the preceding formulae.

Differentiation of the Fundamental Vectors

29. Throughout sections § 29 and § 30, we assume that we are working in Cartesian coordinates, so that the components of the vectors A^r, B^r, C^r have the same values at all points in space. Accordingly, small changes in the components of the vector λ^r, during which the components of A^r and B^r remain constant, are from Eq. (21.3) given by

$$d(\lambda^r) = -(A^r \cos \omega + B^r \sin \omega) d\omega$$

and by substituting the contravariant forms of Eq. (21.2), this becomes

$$d(\lambda^r) = (\mu^r \sin \phi - \upsilon^r \cos \phi) d\omega \ . \tag{29.1}$$

Similarly,

$$d(\mu^r) = -\lambda^r \sin \phi \, d\omega - \upsilon^r d\phi \tag{29.2}$$

and

$$d(\upsilon^r) = \lambda^r \cos \phi \, d\omega + \mu^r d\phi \ . \tag{29.3}$$

If in the same way and using these last results we differentiate a general unit vector l^r in azimuth α and zenith distance β given by Eq. (18.1), that is

$$l^r = \lambda^r \sin \alpha \sin \beta + \mu^r \cos \alpha \sin \beta + \upsilon^r \cos \beta \ , \tag{29.4}$$

we have

$$\begin{aligned} d(l^r) = &(\lambda^r \cos\alpha \sin\beta - \mu^r \sin\alpha \sin\beta)ds \\ &+(\lambda^r \sin\alpha \cos\beta + \mu^r \cos\alpha \cos\beta - \upsilon^r \sin\beta)d\beta \\ &+\sin\alpha \sin\beta(\mu^r \sin\phi - \upsilon^r \cos\phi)d\omega \\ &-\cos\alpha \sin\beta(\lambda^r \sin\phi\, d\omega + \upsilon^r d\phi) \\ &+\cos\beta(\lambda^r \cos\phi\, d\omega + \mu^r d\phi) \;. \end{aligned} \tag{29.5}$$

Denote a unit vector in azimuth α, zenith distance $(\pi/2+\beta)$ by m^r so that

$$m^r = \lambda^r \sin\alpha \cos\beta + \mu^r \cos\alpha \cos\beta - \upsilon^r \sin\beta \;; \tag{29.6}$$

and a unit vector in azimuth $(3\pi/2+\alpha)$, zenith distance $\pi/2$ by

$$n^r = -\lambda^r \cos\alpha + m^r \sin\alpha \;. \tag{29.7}$$

The vectors l^r, m^r, n^r constitute a right-handed orthogonal set in that order. Substituting in (29.5) and simplifying, we have finally

$$\begin{aligned} d(l^r) = &m^r\{d\beta + \cos\phi \sin\alpha\, d\omega + \cos\alpha\, d\phi\} \\ &+n^r\{-\sin\beta\, d\alpha + (\sin\phi \sin\beta - \cos\phi \cos\alpha \cos\beta)d\omega + \sin\alpha \cos\beta\, d\phi\} \;. \end{aligned} \tag{29.8}$$

This equation shows that $d(l^r)$ is a vector in the plane of m^r, n^r and therefore at right-angles to l^r. The equation holds whether the changes in ω, ϕ, α, β are due to a change in position or a change from the astronomic to the geodetic system. Note that if there is no change in the vector l^r despite changes in ω, ϕ, α, β then the coefficients of m^r, n^r must both be zero and we have Eq. (26.1).

30. Next suppose that l^r is expressed in terms of declination D and origin hour angle H as Eq. (19.1), that is

$$l^r = A^r \cos H \cos D + B^r \sin H \cos D + C^r \sin D \tag{30.1}$$

and write

$$M^r = A^r \cos H \sin D + B^r \sin H \sin D - C^r \cos D \tag{30.2}$$

for a unit vector in origin hour angle H, declination $(D-\pi/2)$; also

$$N^r = -A^r \sin H + B^r \cos H \tag{30.3}$$

for a unit vector in origin hour angle $(\pi/2+H)$, declination zero. Here again l^r, M^r, N^r form a right-handed orthogonal set in that order and by differentiation of (30.1) we have at once, in view of the constancy of A^r, B^r, C^r,

$$d(l^r) = -M^r dD + N^r \cos D\, dH \;. \tag{30.4}$$

The Geodetic Coordinate System

31. The principal radii of curvature of the base spheroid are as usual denoted by ϱ, the meridian curvature; and υ, the length of the normal intercepted by the

minor axis. From the ordinary geometry of the elliptic meridian section of eccentricity e, we know that the normal strikes the minor axis at a distance $(\upsilon e^2 \sin\phi)$ below the centre of the spheroid, which we take as the origin of Cartesian coordinates (x, y, z). If the normal is extended a distance h, the geodetic height, above the base spheroid then we have at once:

$$
\begin{aligned}
x &= (\upsilon + h)\cos\phi\cos\omega \\
y &= (\upsilon + h)\cos\phi\sin\omega \\
z &= (\upsilon + h)\sin\phi - \upsilon e^2 \sin\phi \\
&= (\upsilon \bar{e}^2 + h)\sin\phi \;,
\end{aligned}
\qquad (31.1)
$$

writing $\bar{e}^2$ for $(1-e^2)$.

32. The *position vector* ϱ^r whose Cartesian components are (x, y, z) is evidently given by the following vector equation:

$$
\begin{aligned}
\varrho^r &= xA^r + yB^r + zC^r \\
&= (\upsilon + h)\upsilon^r - (\upsilon e^2 \sin\phi)C^r \\
&= (h + a^2/\upsilon)\upsilon^r - (\upsilon e^2 \sin\phi\cos\phi)\mu^r \;.
\end{aligned}
\qquad (32.1)
$$

33. Next, we find the components of the fundamental vectors λ^r, μ^r, υ^r in this system taken in the order $(\omega, \phi, h) = (1, 2, 3)$. The longitude only will vary in the direction λ^r along which an element of length is $(\upsilon + h)\cos\phi\, d\omega$, so that the only non-zero component of λ^r is $d\omega/ds = 1/\{(\upsilon + h)\cos\phi\}$. Proceeding in the same way, we find that the only non-zero contravariant components of the triad are

$$
\begin{aligned}
\lambda^1 &= 1/\{(\upsilon + h)\cos\phi\} \\
\mu^2 &= 1/(\varrho + h) \\
\upsilon^3 &= 1 \;.
\end{aligned}
\qquad (33.1)
$$

By evaluating such expressions as $\lambda^r\lambda_r = 1, \mu^r\lambda_r = 0, \upsilon^r\lambda_r = 0$, we find that the only non-zero covariant components are

$$
\begin{aligned}
\lambda_1 &= (\upsilon + h)\cos\phi \\
\mu_2 &= (\varrho + h) \\
\upsilon_3 &= 1 \;.
\end{aligned}
\qquad (33.2)
$$

The metric of the space follows as

$$
ds^2 = (\upsilon + h)\cos^2\phi\, d\omega^2 + (\varrho + h)^2 d\phi^2 + dh^2 \;. \qquad (33.3)
$$

34. By substitution in Eq. (21.2), we find that the components of the Cartesian vectors are:

$$\mathrm{A}^{r} = \left(-\frac{\sin\omega}{(\nu+h)\cos\phi};\ -\frac{\sin\phi\cos\omega}{(\varrho+h)};\ \cos\phi\cos\omega \right)$$

$$\mathrm{B}^{r} = \left(\frac{\cos\omega}{(\nu+h)\cos\phi};\ -\frac{\sin\phi\sin\omega}{(\varrho+h)};\ \cos\phi\sin\omega \right) \tag{34.1}$$

$$\mathrm{C}^{r} = \left(0; \frac{\cos\phi}{(\varrho+h)};\ \sin\phi \right)$$

$$\mathrm{A}_{r} = (-(\nu+h)\cos\phi\sin\omega;\ -(\varrho+h)\sin\phi\cos\omega;\ \cos\phi\cos\omega)$$

$$\mathrm{B}_{r} = ((\nu+h)\cos\phi\cos\omega;\ -(\varrho+h)\sin\phi\sin\omega;\ \cos\phi\sin\omega) \tag{34.2}$$

$$\mathrm{C}_{r} = (0;\ (\varrho+h)\cos\phi;\ \sin\phi)\ .$$

35. From Eq. (18.1), we find that the components of a unit vector l^r in azimuth α, zenith distance β are:

$$\mathrm{l}^{r} = \left(\frac{\sin\alpha\sin\beta}{(\nu+h)\cos\phi};\ \frac{\cos\alpha\sin\beta}{(\varrho+h)};\ \cos\beta \right)$$

$$\mathrm{l}_{r} = [(\nu+h)\cos\phi\sin\alpha\sin\beta;\ (\varrho+h)\cos\alpha\sin\beta;\ \cos\beta]\ . \tag{35.1}$$

36. We shall need the covariant components of a parallel vector, which are the same in Cartesian coordinates but not in geodetic coordinates, at a different point in space. If L_r is a unit vector in azimuth A, zenith distance B at an unbarred point, then Eq. (35.1) combined with (23.4) shows that the components of the parallel vector at a barred point are as follows:

$$\begin{aligned}\bar{\mathrm{L}}_1 = (\bar{\nu}+\bar{h})\cos\bar{\phi}\,\{&\sin\mathrm{A}\sin\mathrm{B}\cos(\bar{\omega}-\omega)\\ &+\cos\mathrm{A}\sin\mathrm{B}\sin\phi\sin(\bar{\omega}-\omega)\\ &-\cos\mathrm{B}\cos\phi\sin(\bar{\omega}-\omega)\}\end{aligned} \tag{36.1}$$

$$\begin{aligned}\bar{\mathrm{L}}_2 = (\bar{\varrho}+\bar{h})\{&-\sin\mathrm{A}\sin\mathrm{B}\sin\bar{\phi}\sin(\bar{\omega}-\omega)\\ &+\cos\mathrm{A}\sin\mathrm{B}[\cos\phi\cos\bar{\phi}+\sin\phi\sin\bar{\phi}\cos(\bar{\omega}-\omega)]\\ &+\cos\mathrm{B}[\sin\phi\cos\bar{\phi}-\cos\phi\sin\bar{\phi}\cos(\bar{\omega}-\omega)]\}\end{aligned} \tag{36.2}$$

$$\begin{aligned}\bar{\mathrm{L}}_3 = &\sin\mathrm{A}\sin\mathrm{B}\cos\bar{\phi}\sin(\bar{\omega}-\omega)\\ &+\cos\mathrm{A}\sin\mathrm{B}[\cos\phi\sin\bar{\phi}-\sin\phi\cos\bar{\phi}\cos(\bar{\omega}-\omega)]\\ &+\cos\mathrm{B}[\sin\phi\sin\bar{\phi}+\cos\phi\cos\bar{\phi}\cos(\bar{\omega}-\omega)]\ .\end{aligned} \tag{36.3}$$

The Straight Line in Space

37. If (a, b, c) are the Cartesian components of the unit vector l^r in the direction of the line from (x, y, z) to $(\bar{x}, \bar{y}, \bar{z})$ and s is the length of the line, then

$$(\bar{x}-x) = s\,a$$
$$(\bar{y}-y) = s\,b \tag{37.1}$$
$$(\bar{z}-z) = s\,c\ .$$

Substitute in (23.3) and use (31.1). Then we have after slight simplification,

$$\begin{aligned} s \sin\alpha \sin\beta &= -(\bar{x}-x)\sin\omega + (\bar{y}-y)\cos\omega \\ &= (\bar{\nu}+h)\cos\bar{\phi}\sin(\bar{\omega}-\omega) \end{aligned}$$

$$\begin{aligned} s \cos\alpha \sin\beta &= -(\bar{x}-x)\sin\phi\cos\omega - (\bar{y}-y)\sin\phi\sin\omega + (\bar{z}-z)\cos\phi \\ &= (\bar{\nu}+h)\{\sin\bar{\phi}\cos\phi - \cos\bar{\phi}\sin\phi\cos(\bar{\omega}-\omega)\} \\ &\quad - e^2\cos\phi(\bar{\nu}\sin\bar{\phi} - \nu\sin\phi) \end{aligned} \tag{37.3}$$

$$\begin{aligned} s\cos\beta &= (\bar{x}-x)\cos\phi\cos\omega + (\bar{y}-y)\cos\phi\sin\omega + (\bar{z}-z)\sin\phi \\ &= (\bar{\nu}+\bar{h})\{\sin\bar{\phi}\sin\phi + \cos\bar{\phi}\cos\phi\cos(\bar{\omega}-\omega)\} \\ &\quad -(\nu+h) - e^2\sin\phi(\bar{\nu}\sin\bar{\phi} - \nu\sin\phi)\ . \end{aligned} \tag{37.4}$$

These three equations directly determine the length, azimuth and zenith distance of a line whose end coordinates are given. The azimuth ($\bar{\alpha}$) and zenith distance ($\bar{\beta}$) at the other end of the line (produced through the barred point, not the back azimuth and back zenith distance) are obtained by interchanging the bars and changing the sign of s, since the length will now be measured in the direction opposite to l^r: e.g. Eq. (37.2) becomes

$$s\sin\bar{\alpha}\sin\bar{\beta} = (\nu+h)\cos\phi\sin(\bar{\omega}-\omega)$$

from which, by combining with Eq. (37.2) and (35.1), we infer that

$$(\nu+h)\cos\phi\sin\alpha\sin\beta = l_1 \tag{37.5}$$

is constant along the line.

38. The inverse problem, of finding the coordinates of the barred point given those of the unbarred together with the length and initial azimuth and zenith distance, is solved at once from Eq. (23.2). By taking a temporary origin for geodetic longitude at the unbarred point, we have

$$(\bar{\nu}+\bar{h})\cos\bar{\phi}\cos(\bar{\omega}-\omega) = (\nu+h)\cos\phi - s\sin\phi\cos\alpha\sin\beta + s\cos\phi\cos\beta \tag{38.1}$$

$$(\bar{\nu}+\bar{h})\cos\bar{\phi}\sin(\bar{\omega}-\omega) = s\sin\alpha\sin\beta \tag{38.2}$$

$$(\bar{e}^2\bar{\nu}+h)\sin\bar{\phi} = (\bar{e}^2\nu+h)\sin\phi + s\cos\phi\cos\alpha\sin\beta + s\sin\phi\cos\beta\ . \tag{38.3}$$

The difference of longitude follows at once by division of the first two equations. We then have to solve by iteration

$$(\nu+h)\cos\bar{\phi} = A$$

and

$$(\bar{e}^2\bar{\nu}+\bar{h})\sin\bar{\phi} = B$$

for the latitude and height, starting with an approximate latitude given by $\bar{e}^2 \tan \bar{\phi}\, B/A$.

The Plane Triangle in Space

39. For the present we assume that deflections are zero at the vertices of the triangle, that is, the geodetic normal coincides with the plumb line at each vertex.

If l^r, $\bar{l}^r$ are the unit vectors in the directions of two adjacent sides of a plane triangle (astronomical azimuths and zenith distances α, β; $\bar{\alpha}, \bar{\beta}$), then the included angle L is given by Eq. (35.1) as

$$\begin{aligned}\cos L &= l^r \bar{l}_r \\ &= \sin\alpha \sin\beta \sin\bar{\alpha} \sin\bar{\beta} + \cos\alpha \sin\beta \cos\bar{\alpha} \sin\bar{\beta} + \cos\beta \cos\bar{\beta} \\ &= \cos\beta \cos\bar{\beta} + \sin\beta \sin\bar{\beta} \cos(\bar{\alpha} - \alpha)\ ,\end{aligned} \tag{39.1}$$

in which $(\bar{\alpha} - \alpha)$ is simply the "horizontal" angle as measured between the two directions. The other angles of the triangle are found similarly. One side of the triangle will always be known by direct measurement or from antecedent work and the other sides can next be computed from the ordinary sine formula. In the case of a triangle containing the origin of a triangulation, the geodetic position of the origin together with a geodetic azimuth and two zenith distances will have been adopted to satisfy the conditions in § 24 et seq. and we can now compute the coordinates and azimuths at the other vertices of the triangle from the formulae in the last section. In all other cases we shall know the geodetic coordinates etc. of two vertices from antecedent work and can compute the coordinates etc. of the third point from either of these known points.

40. If angular measurements have been made at all three vertices, then before computing the sides, we should adjust the sum of the angles L to 180° by subtracting one-third of the excess from each. This, of course, has nothing to do with the "spherical excess" of the classical theory and the one-third rule arises simply from the fact that it is the most probable distribution of error among angles supposed to be measured with equal precision at the three vertices. In ordinary terrestrial triangulation, the β's are so nearly 90° that the resulting correction to the angles of the space triangle can be thrown straight into the difference of geodetic azimuths before calculating geodetic coordinates etc. round the triangle.

41. Observations are actually made, not along straight lines or the sides of plane triangles in space, but tangential to slightly curved optical paths refracted by the atmosphere. Before applying the formulae in this and the preceding section, we should accordingly add to observed zenith distances the angle or refraction as computed from whatever data may be available. Any error or defect of data in this operation will not affect the final calculation of geodetic coordinates any more than it does in the classical theory. The question is considered further in the section on triangulations adjustment below, where improved means of solving triangles are proposed in § 53 and the effect of refraction considered in more

detail in § 49. This more advanced method also enables full account to be taken of deflections at the vertices.

Variation of Geodetic Coordinates

42. We now proceed to find the changes in length, azimuth and zenith distance resulting from a change in the end coordinates of the straight line whose unit vector is l^r.

If for the moment we suppose we are working in Cartesian coordinates x^r, ($\bar{x}^r$ at the far end of the line), then

$$s\,l^r = \bar{x}^r - x^r \ ,$$

and differentiating this we have

$$s\,d(l^r) + l^r ds = d\bar{x}^r - dx^r \ . \tag{42.1}$$

Now we know from Eq. (29.8) that $d(l^r)$ is a vector perpendicular to l^r, so that by forming the scalar product of Eq. (42.1) with l_r we have

$$ds = (d\bar{x}^r - dx^r)l_r = \bar{l}_r d\bar{x}^r - l_r dx^r \ , \tag{42.2}$$

an invariant equation true in any coordinates provided we interpret $\bar{l}_r$ as a vector parallel to l_r at the barred point and take out its components in the same coordinate system. Using (35.1) and (37.5), the full equation in geodetic coordinates becomes

$$\begin{aligned} ds = {} & (\nu + h)\cos\phi \sin\alpha \sin\beta (d\bar{\omega} - \omega) \\ & + (\bar{\varrho} + \bar{h})\cos\bar{\alpha}\sin\bar{\beta}\,d\bar{\phi} - (\varrho + h)\cos\alpha\sin\beta\,d\phi \\ & + \cos\bar{\beta}\,d\bar{h} - \cos\beta\,dh \ . \end{aligned}$$

In this equation, $\bar{\alpha}, \bar{\beta}$ refer, as usual, to the prolongation of the line through the barred point.

43. Eliminating ds between Eqs. (42.2) and (42.1), we have

$$\begin{aligned} s\,d(l^r) &= (d\bar{x}^r - dx^r) - (d\bar{x}^s - dx^s)l_s l^r \\ &= (d\bar{x}^s - dx^s)(\delta^r_s - l_s l^r) \\ &= (d\bar{x}^s - dx^s)(m_s m^r + n_s n^r) \ , \end{aligned} \tag{43.1}$$

where δ^r_s is the Kronecker delta, which is unity if $r = s$ and is otherwise zero. If l_r, m_r, n_r are *any* mutually orthogonal triad of unit vectors which can be taken as temporary Cartesian axes, then the tensor equation

$$\delta^r_s = l_s l^r + m_s m^r + n_s n^r$$

is easily verified in Cartesian coordinates and is therefore true in any coordinates. In this case we take m^r, n^r as the same vectors as (29.6) and (29.7), substitute (43.1) in (29.8), equate coefficients of m^r, n^r and find that

$$s\,d\beta = \bar{m}_s d\bar{x}^s - m_s dx^s - s\cos\phi\sin\alpha\, d\omega - s\cos\alpha\, d\phi \tag{43.2}$$

$$s\sin\beta\, d\alpha = -\bar{n}_s d\bar{x}^s + n_s dx^s + s(\sin\phi\sin\beta - \cos\beta\cos\phi\cos\alpha)d\omega + s\cos\beta\sin\alpha\, d\phi\ . \tag{43.3}$$

Here again, $\bar{m}_s, \bar{n}_s$ must be interpreted as parallel to m_s, n_s respectively but drawn through the barred point, and their components in the geodetic system taken out accordingly. Substituting the appropriate azimuths and zenith distances in § 36 (for $m_s, A = \alpha, B = \pi/2$; for $n_s, A = 3\pi/2 + \alpha, B = \pi/2$), we have

$$\begin{aligned}
\bar{m}_1 &= (\bar{\nu} + \bar{h})\cos\phi\{\sin\alpha\cos\beta\cos(\bar{\omega}-\omega) + \cos\alpha\cos\beta\sin\phi\sin(\bar{\omega}-\omega) \\
&\quad + \sin\beta\cos\phi\sin(\bar{\omega}-\omega)\} \quad \text{or} \\
&= (\bar{\nu} + \bar{h})\cos\bar{\phi}\{\sin\bar{\alpha}\sin\bar{\beta}\cos\beta + \cos\phi\csc\beta\sin(\bar{\omega}-\omega)\} \quad \text{from (23.4)} \\
&= (\nu + h)\cos\phi\sin\alpha\cos\beta + s\cos\phi\sin\alpha \quad \text{from (37.5) and (37.2)} \\
&= m_1 + s\cos\phi\sin\alpha \quad \text{from § 35}
\end{aligned}$$

$$\begin{aligned}
\bar{m}_2 &= (\bar{\rho} + \bar{h})\{-\sin\alpha\cos\beta\sin\bar{\phi}\sin(\bar{\omega}-\omega) \\
&\quad + \cos\alpha\cos\beta[\cos\phi\cos\bar{\phi} + \sin\phi\sin\bar{\phi}\cos(\bar{\omega}-\omega)] \\
&\quad - \sin\beta[\sin\phi\cos\bar{\phi} - \cos\phi\sin\bar{\phi}\cos(\bar{\omega}-\omega)]\} \quad \text{or} \\
&= (\bar{\rho} + \bar{h})\{\cos\bar{\alpha}\sin\bar{\beta}\cot\beta - \sin\phi\cos\bar{\phi}\csc\beta \\
&\quad + \cos\phi\sin\bar{\phi}\csc\beta\cos(\bar{\omega}-\omega)\} \quad \text{from (23.4)}
\end{aligned}$$

$$\begin{aligned}
\bar{m}_3 &= \sin\alpha\cos\beta\cos\bar{\phi}\sin(\bar{\omega}-\omega) \\
&\quad + \cos\alpha\cos\beta[\cos\phi\sin\bar{\phi} - \sin\phi\cos\bar{\phi}\cos(\bar{\omega}-\omega)] \\
&\quad - \sin\beta[\sin\phi\sin\bar{\phi} + \cos\phi\cos\bar{\phi}\cos(\bar{\omega}-\omega)] \quad \text{or} \\
&= \cos\bar{\beta}\cot\beta - \cos\phi\cos\bar{\phi}\csc\beta\cos(\bar{\omega}-\omega) - \sin\phi\sin\bar{\phi}\csc\beta
\end{aligned}$$

$$\begin{aligned}
m_1 &= (\nu + h)\cos\phi\sin\alpha\cos\beta \\
m_2 &= (\rho + h)\cos\alpha\cos\beta \\
m_3 &= -\sin\beta
\end{aligned}$$

$$\begin{aligned}
\bar{n}_1 &= (\bar{\nu} + \bar{h})\cos\bar{\phi}\{-\cos\alpha\cos(\bar{\omega}-\omega) + \sin\alpha\sin\phi\sin(\bar{\omega}-\omega)\} \quad \text{or} \\
&= -(\nu + h)\cos\phi\cos\alpha + s\sin\phi\sin\beta - s\cos\beta\cos\phi\cos\alpha \\
&\quad \text{using (38.1) and (38.2)} \\
&= n_1 + s\sin\phi\sin\beta - s\cos\beta\cos\phi\cos\alpha\ .
\end{aligned}$$

$$\begin{aligned}
\bar{n}_2 &= (\bar{\rho} + \bar{h})\{\cos\alpha\sin\bar{\phi}\sin(\bar{\omega}-\omega) \\
&\quad + \sin\alpha[\cos\phi\cos\bar{\phi} + \sin\phi\sin\bar{\phi}\cos(\bar{\omega}-\omega)]
\end{aligned}$$

$$\begin{aligned}
\bar{n}_3 &= -\cos\alpha\cos\bar{\phi}\sin(\bar{\omega}-\omega) \\
&\quad + \sin\alpha[\cos\phi\sin\bar{\phi} - \sin\phi\cos\bar{\phi}\cos(\bar{\omega}-\omega)]
\end{aligned}$$

$$n_1 = -(\upsilon + h)\cos\phi\cos\alpha$$
$$n_2 = (\varrho + h)\sin\alpha$$
$$n_3 = 0 \ .$$

As a check, note that

$$\frac{(\bar{m}_1)^2}{(\bar{\upsilon}+\bar{h})^2\cos^2\bar{\phi}} + \frac{(\bar{m}_2)^2}{(\bar{\varrho}+\bar{h})^2} + (\bar{m}_3)^2 = 1$$

and similarly for the components of the other vectors.

44. We could form similar equations giving the result of changes in the Cartesian coordinates (x, y, z) provided that the Cartesian components of the corresponding vectors are used. These components can be calculated from Eq. (23.2) using the appropriate azimuth and zenith distance of the vector; and any parallel vector at another point will have the same Cartesian components. Covariant and contravariant components will also be the same in Cartesian coordinates. Thus, for example, Eq. (42.2) expanded in Cartesian coordinates would be

$$ds = a(d\bar{x} - dx) + b(d\bar{y} - dy) + c(d\bar{z} - dz) \ .$$

The coordinates dx, $d\bar{x}$ etc. can still be varied quite independently at the two ends of the line, but will have the same coefficients; whereas in geodetic coordinates the coefficients of e.g. $d\phi$, $d\bar{\phi}$ are not the same (see § 48).

Adjustment of Triangulation and Traverse

45. We start with approximate geodetic positions (ω, ϕ, h) of the stations computed as in § 39 and § 53 and use these to compute accurately $s, \alpha, \beta, \bar{\alpha}, \bar{\beta}$ for each direction. Note once again that $\bar{\alpha}, \bar{\beta}$ refer to the same direction produced through the far station and not the back azimuth etc. We are then able to compute the coefficients of Eqs. (43.2) and (43.3), giving the changes in α and β resulting from corrections $dx^s, d\bar{x}^s$ (as yet unknown) which are to be applied to the initial approximate coordinates. In forming the equations for the reverse directions, we take $(180° + \bar{\alpha})$ and $(180° - \bar{\beta})$ as the initial unbarred α, β; and $(180° + \alpha)$, $(180° - \beta)$ as the far (barred) $\bar{\alpha}, \bar{\beta}$. The formulae then remain exactly the same, whereas confusion could result from writing down separate formulae applicable to the reverse direction[6].

46. Next we apply further changes to α, β given by Eq. (26.1) in order to transform to the astronomical system, so that we may have azimuths and zenith distances measured, as they are in practice, in relation to the plumb line. In these equations $\delta\omega, \delta\phi$ are the *deflection* at the station from which the direction is observed, with the sign (astronomic minus geodetic) longitude or latitude. Again note that longitude is positive eastwards. We may, or may not yet know, these deflections.

---

6 For instance, new initial vectors m_r, n_r at what was the far end are not the same as $\bar{m}_r, \bar{n}_r$.

47. We may not have measured an astronomic azimuth at the station, in which case it will be necessary to assume a direction for the astronomic meridian. On this account we must add a *station correction* $\Delta\alpha$ to the assumed astronomic azimuths; or subtract it from the calculated azimuths. To reduce the observed zenith distance to the straight line joining the two end stations, we must also add the *angle of refraction* $\Delta\beta$; or subtract if from the calculated β.

48. We then have finally the following two observation equations for each observed direction; the coefficients $\bar{m}_1$ etc. being as given in § 43: (Observed minus Computed) Zenith Distance is

$$\begin{aligned}
&-\Delta\beta+\bar{m}_1\,d\bar{\omega}/s+\bar{m}_2\,d\bar{\phi}/s+\bar{m}_3\,d\bar{h}/s\\
&-m_1\,d\omega/s-m_2\,d\phi/s-m_3\,dh/s\\
&-(d\omega+\delta\omega)\cos\phi\sin\alpha\\
&-(d\phi+\delta\phi)\cos\alpha
\end{aligned}\tag{48.1}$$

(Observed minus Computed) Azimuth is

$$\begin{aligned}
&-\Delta\alpha-\bar{n}_1\,d\bar{\omega}\csc\beta/s-\bar{n}_2\,d\bar{\phi}\csc\beta/s-\bar{n}_3\,d\bar{h}\csc\beta/s\\
&+n_1\,d\omega\csc\beta/s+n_2\,d\phi\csc\beta/s\\
&+(d\omega+\delta\omega)(\sin\phi-\cos\alpha\cot\beta\cos\phi)\\
&+(d\phi+\delta\phi)\sin\alpha\cot\beta\ .
\end{aligned}\tag{48.2}$$

It is evident that $(d\omega+\delta\omega)$ is the astronomic minus the *initial* approximate geodetic longitude and is known if the astronomic longitude has been measured. If not, then it is permissible to treat $(d\omega+\delta\omega)$ as a single unknown and later to subtract $d\omega$, the correction to geodetic longitude, in order to determine the deflection $\delta\omega$. The same applies to $(d\phi-\delta\phi)$. For this reason it will usually be better to work in geodetic coordinates throughout, despite the fact that working in Cartesian coordinates as in § 44 for final conversion to geodetic coordinates otherwise offers some simplification.

49. We cannot treat $\Delta\beta$ as completely unknown at both ends of all lines if the β-equations are to contribute to the solution. There are two main possibilities at present, until more work has been done on atmospheric refraction, leading to an assurance that it can be accurately evaluated from physical data. We can assume that the refraction of all lines radiating from a particular station is the same per unit length, treat this as a single unknown for the station and express the $\Delta\beta$'s in terms of it. This procedure would be indicated where, as usual in practice, all vertical angles at a station are measured together, even though some complete measures may be taken at a different time. If, for instance, several complete sets were measured by day and several by night, then the refraction determined by this procedure would result from some intermediate atmospheric condition; it would have no physical significance, but the accuracy of the adjustment would be unaffected by thus including complete sets taken at such different times.

The other alternative would be to assume that $\Delta\beta$ is the same at the two ends of a line, in which case it would be eliminated by subtracting the β-equations in pairs before solving. This procedure would be indicated where reciprocal observations have been made.

50. The residual refraction must then be treated as a random error and for this reason the equations should properly be given less weight. There is, however, so little interaction between the two sets of equations *in normal terrestrial triangulation* that weighting has little effect, and indeed the α- and β-equations might well be solved separately. The coefficients $d\omega, d\phi, d\bar{\omega}, d\bar{\phi}$ in the β-equations are all small and these terms could be omitted in a first solution. The main function of the β-equations, assisted by frequent astronomic observations, is to determine $dh, d\bar{h}, (d\omega+\delta\omega)$ and $(d\phi+\delta\phi)$. The coefficients of these unknowns in the α-equations are, however, small, except the term $(d\omega+\delta\omega)\sin\phi$, which might be considered as incorporated in $\Delta\alpha$, so that fairly large errors in them would have little effect on the determination of $d\omega, d\phi, d\bar{\omega}, d\bar{\phi}$ from the α-equations.

51. If, by making assumptions equivalent to those normally made in classical methods, we were to solve the α-equations separately – ignore the effect of deflections, in view of the impossibility of measuring them at every station – and drop the small $d\bar{h}$ term, then the α-equation boils down to

$$-\Delta\alpha+\bar{n}_1 \csc\beta(d\omega-d\bar{\omega})/s+n_2 \csc\beta\, d\phi/s-\bar{n}_2 \csc\beta\, d\bar{\phi}/s\ , \qquad (51.1)$$

which is just as easy to compute as the usual classical equation for variation of position, whether on the spheroid or via the spheroid on a plane projection, without making as many other assumptions.

52. Incorporation of astronomic measures is simple. If azimuth is measured, the $\Delta\alpha$ term is dropped. If longitude is measured, we evaluate the $(d\omega+\delta\omega)$ terms and throw them into the absolute terms; and the same for latitude. This does not, of course, mean that astronomic measures are treated as error-free; errors in them will appear in the residuals of the observation equations after adjustment. It is not necessary to measure longitude and azimuth at the same station as for the classical Laplace adjustment. Indeed, except for the purpose of fixing an origin, it is not necessary to measure longitude at all; the full equations will amply serve to bridge gaps in the deflections between frequent latitude and azimuth measures, which are, of course, much easier to observe precisely.

53. Approximate geodetic coordinates to serve as initial data for the adjustment could be obtained by solving the azimuth (48.2) and zenith distance (48.1) equations for selected triangles taken one at a time. The two zenith distance equations for each line should be subtracted as proposed in § 49 to eliminate the refraction term. An electronic computer programmed to iterate the computation would itself require only very rough initial coordinates and would also deliver accurate coefficients of the azimuth and zenith-distance equations for the selected lines to use in the final adjustment.

To determine corrections to the three coordinates, the station correction and two deflections at an apex from two previously fixed points, we should have four azimuth and two (composite) zenith distance equations, so that the calculation is just determinate without astronomic observations at the apex.

54. In the case of rays radiating from the origin $d\omega, d\phi, dh$ are all zero and astronomic longitude, latitude and azimuth should all be measured so that the $\delta\omega, \delta\phi, \Delta\alpha$ terms can be evaluated. (If astronomic values are accepted as the initial geodetic elements then $\delta\omega, \delta\phi, \Delta\alpha$ are zero). The effect of this is to ensure that the conditions of § 24 et seq. are held satisfied (on at least two lines), apart from observational error. The remainder of these observational equations are left in the adjustment so as to minimize the effect of observational error by the inclusion of more lines and to assist the determination of $d\bar{\omega}, d\bar{\phi}, d\bar{h}$.

55. Measures of length are likely to be much more frequent in future through use of the tellurometer. Wherever a side has been measured, an observational equation is formed from Eq. (42.3), in which ds is the measured minus the computed length. The equation should be divided by a constant of the same order as the average side-lengths in the triangulation in order to reduce it to the same dimensions as the α- and β-equations; it is then taken straight into the general adjustment. Experience so far indicates that tellurometer measures (or for that matter invar measures, at any rate after base extension) are of much the same order of accuracy as the horizontal angular measures, both being about 1/300000. Consequently, there is no object in relative weighting. Sectionally measured bases are not reduced to the spheroid or to "sea level" but are "reduced" upwards to the air line between the terminals.

Flare Triangulation

56. The formulae derived in the last section can be used as they stand wherever angular observations of altitude and azimuth, or difference of azimuth are made in relation to the astronomic zenith. Such observations are made to parachute flares dropped from aircraft in order to provide geodetic connections between the non-intervisable shores of wide water gaps, as originally proposed by W. E. Browne.

57. A usual arrangement is to observe flares from three stations of known geodetic coordinates on one side of the gap, and from three stations whose geodetic coordinates are required in the same system on the far side of the gap. Observations at all six stations are synchronized by radio signals. In some cases the circle readings are photographed by cameras whose shutters are operated by the radio signals, so that all the observers need do is to keep the flare intersected. Flares, which can be dropped at different times, are required in at least two, preferably three, widely separated positions over the water, making reasonably well-conditioned figures with the ground stations.

58. Two observation equations [(48.1) and (48.2)] are formed for each line from approximate positions of the flares and of the unknown ground stations. Between six ground stations and three flares, there are then 36 equations (or 44 if connections between the three unknown ground stations are observed in both directions) to determine corrections to six sets of three coordinates, if full astronomical observations are taken at all ground stations.

In present practice, vertical angles would not be measured, and corrections to latitudes and longitudes only would be determined. Between six ground stations and three flares, 18 equations (22 if directions are observed both ways between the three unknown ground stations) would then be available to determine 12 corrections. Theoretically, there would be enough equations to determine heights but the equations would be too ill-conditioned for the purpose. The result would be an approximation whose validity should be demonstrated numerically from the full equations. It should be possible to measure vertical angles and to obtain reasonable refraction corrections from reciprocal observations between the ground stations, thereby obtaining a theoretically correct solution and carrying geodetic heights across the gap. If the flare is roughly mid-way between known and unknown ground stations, residual effects of refraction would tend to cancel as between the heights of ground stations, although perhaps seriously affecting the flare heights.

59. To minimize pointing errors, it is usual to make several observations to the same flare as it falls. We have no knowledge of the path of the flare and must accordingly treat its position as unknown at the time of every such additional observation. An extra set of observations accordingly introduces three more unknown corrections to the position of the flare and two extra equations for each ground station observing it. If the extra observations are made rapidly, the same coefficients would probably suffice. If vertical angles are not observed and heights are not to be determined, then an extra set of observations introduces two more unknowns in the flare position and one extra equation per ground station observing it.

Shoran or Hiran Measures

60. Radar measures of slant-ranges are made between each of two ground stations (S_1, S_2) and an aircraft (A) flying a straight and level course between them. The minimum sum of the radar ranges is used to determine the distance between the two ground stations, and hence geodetic positions by trilateration. In addition to the usual approximations inherent in classical geodesy, a particular assumption is made that the minimum position occurs when the plane S_1, A, S_2 is vertical at A. The limitations of this assumption can be seen at once by considering the aircraft course as tangential to a prolate spheroid whose foci are S_1, S_2. The sum $(S_1P + PS_2)$ is the same for any point on this spheroid and less than for any point on the straight aircraft course external to the spheroid, so that the minimum position occurs at the point of contact of the course with the spheroid. In theory, the assumption is accordingly justified only if (a) the aircraft course is perpendicular

to S_1, S_2, which it usually is not, or (b) the aircraft crosses in the mid-way position. The problem can, however, be solved simply and rigorously in three dimensions without making any such assumptions.

61. Appropriate corrections, necessary for any method of reduction, should first be made to the radar readings to obtain the straight-line slant ranges.

Coordinates of S_1, A, S_2 respectively are denoted by $x^r, \bar{x}^r, \bar{\bar{x}}^r$. Unit vectors in the directions S_1A, AS_2 are p^r, q^r and the slant-ranges S_1A, AS_2 are u, v. The unit vectors for the aircraft course is $\bar{a}^r$. Parallel vectors at the three points are denoted by appropriate bars, e.g. a vector parallel to $\bar{a}^r$ at S_2 is $\bar{\bar{a}}^r$; and at S_1 is a^r.

Equations (42.2) for variation of the two slant-ranges are

$$du = \bar{p}_r d\bar{x}^r - p_r dx^r \tag{61.1}$$

$$dv = \bar{\bar{q}}_r d\bar{\bar{x}}^r - \bar{q}_r d\bar{x}^r \; . \tag{61.2}$$

62. To establish the minimum position, we assume first that S_1, S_2 are fixed and the aircraft only moves by $d\bar{x}^r$, while $dx^r = d\bar{\bar{x}}^r = 0$. At the minimum position, we have also $du + dv = 0$, and therefore from Eqs. (61.1) and (61.2)

$$\bar{p}_r d\bar{x}^r = \bar{q}_r d\bar{x}^r \; .$$

But $d\bar{x}^r$ is proportional to the (contravariant) course vector $\bar{a}^r$, so that

$$\bar{p}_r \bar{a}^r = \bar{q}_r \bar{a}^r \; . \tag{62.1}$$

This is the minimum condition and implies that the aircraft course makes equal angles with the directions S_1A, AS_2 when the minimum occurs, say $\cos P = \cos Q$.

63. Now suppose that the direction of the aircraft (a^r) is fixed but that the coordinates of S_1, A and S_2 are all varied by dx^r etc. from approximate initial positions to final positions, which final positions satisfy the minimum condition. The corresponding changes in cos P and cos Q are the given as follows, [see Eq. (42.1)].

$$\begin{aligned} u\times\{\text{Final } (\cos P) \text{ minus Initial } (\cos P)\} &= u\, d(\cos P) = u\, d(a_r p^r) = u\, a_r\, d(p^r) \\ &= a_r (d\bar{x}^r - dx^r) - du \cos P \end{aligned}$$

and similarly

$$\begin{aligned} v\times\{\text{Final } (\cos Q) \text{ minus Initial } (\cos Q)\} &= v\, d(\cos Q) \\ &= a_r (d\bar{\bar{x}}^r - d\bar{x}^r) - dv \cos Q \; . \end{aligned}$$

Substracting these two equations, and remembering that the final values of cos P and cos Q are to be equal, we have

$$\begin{aligned} \text{Initial } (\cos Q - \cos P) = {} & -\frac{1}{u} a_r dx^r + \left(\frac{1}{u} + \frac{1}{v}\right) \bar{a}_1 d\bar{x}^r - \frac{1}{v} \bar{\bar{a}}_r d\bar{\bar{x}}^r \\ & -\frac{du}{u} \cos P + \frac{dv}{v} \cos Q \; , \end{aligned} \tag{63.1}$$

in which P, Q are initial values computed from the initial assumed coordinates and du (or dv) is the final observed u (or v) minus the initial computed u (or v). This equation and (61.1) and (61.2) can be used either as condition equations, or as observation equations in conjunction with any other measurements which may have been made by triangulation or traverse etc., to fix S_1, S_2. As before in § 42, these equations are true in any coordinates provided that the components of $\bar{a}_r$ etc. and the parallel vectors $a_r, \bar{\bar{a}}_r$ are taken out in the same coordinates. P, Q, etc. must, of course, be accurately computed even though the aircraft course has been only roughly measured.

64. In the usual case, the azimuth of the aircraft course is $\bar{A}$ and the zenith distance for level flight is 90°. From § 35 and § 36 we then have for the coefficients in Eq. (63.1):

$$\bar{a}_1 = (\bar{\upsilon}+\bar{\hslash})\cos\bar{\phi}\sin\bar{A}$$

$$\bar{a}_2 = (\bar{\varrho}+\bar{\hslash})\cos\bar{A}$$

$$\bar{a}_3 = 0\ .$$

$$a_1 = (\upsilon+\hslash)\cos\phi\{\sin\bar{A}\cos(\omega-\bar{\omega})+\cos\bar{A}\sin\bar{\phi}\sin(\omega-\bar{\omega})\}$$

$$a_2 = (\varrho+\hslash)\{-\sin\bar{A}\sin\phi\sin(\omega-\bar{\omega})$$
$$+\cos\bar{A}[\cos\bar{\phi}\cos\phi+\sin\bar{\phi}\sin\phi\cos(\omega-\bar{\omega})]\}$$

$$a_3 = \sin\bar{A}\cos\phi\sin(\omega-\bar{\omega})$$
$$+\cos\bar{A}[\cos\bar{\phi}\sin\phi-\sin\bar{\phi}\cos\phi\cos(\omega-\bar{\omega})]$$

$$\bar{\bar{a}}_1 = (\bar{\bar{\upsilon}}+\bar{\bar{\hslash}})\cos\bar{\bar{\phi}}\{\sin\bar{A}\cos(\bar{\bar{\omega}}-\bar{\omega})+\cos\bar{A}\sin\bar{\phi}\sin(\bar{\bar{\omega}}-\bar{\omega})\}$$

$$\bar{\bar{a}}_2 = \bar{\bar{\varrho}}+\bar{\bar{\hslash}})\{-\sin\bar{A}\sin\bar{\bar{\phi}}\sin(\bar{\bar{\omega}}-\bar{\omega})$$
$$+\cos\bar{A}[\cos\bar{\phi}\cos\bar{\bar{\phi}}+\sin\bar{\phi}\sin\bar{\bar{\phi}}\cos(\bar{\bar{\omega}}-\bar{\omega})]\}$$

$$\bar{\bar{a}}_3 = \sin\bar{A}\cos\bar{\bar{\phi}}\sin(\bar{\bar{\omega}}-\bar{\omega})$$
$$+\cos\bar{A}[\cos\bar{\phi}\sin\bar{\bar{\phi}}-\sin\bar{\phi}\cos\bar{\bar{\phi}}\cos(\bar{\bar{\omega}}-\bar{\omega})]$$

and the term in e.g. dx_r in Eq. (63.1) expands as

$$-\frac{1}{u}a_1\,d\omega-\frac{1}{u}a_2\,d\phi-\frac{1}{u}a_3\,d\hslash.$$

The coefficients need to be accurately computed even though $\bar{A}$ has been only roughly measured.

65. If α, β are the azimuth and zenith distance from S_1 to A as computed from the assumed initial coordinates of S_1 and A; and $\bar{\alpha}, \bar{\beta}$ refer as usual to the same direction at A and in the same sense, then the first Eq. (61.1) expands exactly as (42.3) (see also § 35 and § 64),

$$\cos\pi = \bar{p}^r\bar{a}_r = \sin\bar{A}\sin\bar{\alpha}\sin\bar{\beta}+\cos\bar{A}\cos\bar{\alpha}\sin\bar{\beta} = \sin\bar{\beta}\cos(\bar{A}-\bar{\alpha})\ .$$

Similarly for the second Eq. (61.2) and $\cos Q = \bar{q}^r\bar{a}_r$.

66. If there are no other measures besides Shoran connecting the ground stations, then it will certainly be impossible to determine corrections to heights and the terms containing $d\mathscr{h}$, $d\bar{\mathscr{h}}$, $d\bar{\bar{\mathscr{h}}}$ must be dropped. For instance, in a simple trilateration, where a third ground point is to be fixed from two previously known points, there would be only the above three equations for each side, and these could do no more than determine six unknowns, viz. corrections to latitude and longitude of the third point and of the two aircraft positions. Even though the latter may not be required, they must, of course, be left in the equations. The result is not very sensitive to height changes, but nevertheless the difficulty of height determination is a weakness of Shoran – not of the method of computation – since even if heights are not required the omission of the $d\mathscr{h}$ terms must to some extent affect the determination of latitude and longitude.

67. If the initial assumed positions are within 15 seconds of the truth, which can usually be arranged by rough spherical computation and by placing the aircraft along the line in simple proportion to the measured ranges, then a single solution provides results correct to about two feet. In a test case of a single trilateral, deliberately rough values were assumed which turned out to be 8 minutes and 5 minutes adrift in latitude and longitude respectively of the unknown station, and 3 minutes adrift to the final aircraft position. The first solution averaged about 14 seconds adrift, the largest difference being 47 seconds in longitude of the unknown ground station. A repeat computation gave results within 0.025 second of correct values. Movements of the aircraft are not very sensitive; Eq. (63.1) is soon satisfied, and when that occurs, the corrections to the aircraft position have the same coefficients in the remaining equations and can be eliminated. An electronic computer programmed to iterate the whole computation would soon satisfy all six equations in a trilateral, however rough the initial values, and at the same time produce coefficients of the observation equations to be used in a final net adjustment.

68. Results obtained by the classical method which have so far been checked work out very nearly the same, which indicates that the aircraft does in practice usually cross nearly mid-way. Nevertheless, it would be desirable in future to measure the aircraft course roughly and apply the rigorous method, at least as a final check.

Geodetic Positions from Rocket Flashes

69. This method – suggested in 1946 by Väisälä and more recently by R. d'E. Atkinson independently and in greater practical detail – of fixing geodetic positions over long distances, possibly across the oceans, is similar in principle to flare triangulation. Instead of observing a flare instrumentally, photographs of an "instantaneous" flash produced by a powerful rocket are taken in equatorially mounted long-focus cameras against a background of stars. Measurement of the plates provides the apparent right ascension (reducible to origin hour angle H by § 19), and declination D, of the flash from each of the ground observing stations. Accordingly, we need observation equations to give changes in H and D, rather

than changes in azimuth and zenith distance, arising from variation in position of the flash and of a ground station.

70. The required observation equations are obtained at once from Eqs. (30.4) and (43.1) as:

$$s\,dD = -\bar{M}_s d\bar{x}^s + M_s dx^s \;;$$

and

$$s \cos D\, dH = \bar{N}_s d\bar{x}^s - N_s dx^s \;, \tag{70.1}$$

in which the unit vectors M_s, N_s (and the parallel vectors $\bar{M}_s, \bar{N}_s$ at the barred point, which we shall assume to be the flash) are given by Eqs. (30.2) and (30.3). These equations are true in any coordinate system.

As usual, the displacements $d\bar{x}^s, dx^s$ are assumed to be made from initial (assumed) positions of the flash and the ground station, to final (observed) positions. We also compute initial values of H, D and s from Eqs. (23.2), (37.1) and (31.1) and use these in the coefficients of the observation equations. Thus:

$$\begin{aligned} s \cos H \cos D &= (\bar{x}-x) = (\bar{\nu}+\bar{h}) \cos\bar{\phi} \cos\bar{\omega} - (\nu+h) \cos\phi \cos\omega \\ s \cos D \cos H &= (\bar{y}-y) = (\bar{\nu}+\bar{h}) \cos\bar{\phi} \sin\bar{\omega} - (\nu+h) \cos\phi \sin\omega \\ s \sin D &= (\bar{z}-z) = (\bar{\nu}\bar{e}^2+\bar{h}) \sin\bar{\phi} - (\nu\bar{e}^2+h) \sin\phi \;. \end{aligned} \tag{70.2}$$

Finally we assume an approximate value of local sidereal time *at the origin* of the survey when the flash occurs; we use this to give us the computed right ascension from § 19, and we shall seek to correct it by determining an additive correction dt to the origin sidereal time of the flash, so that dt will be the same for all observed lines to the same flash. It will be clear from § 19 that we then have: (Observed minus Computed) Right ascension of the flash = dH + dt, all expressed in arc.

71. In geodetic coordinates the components of the vectors $M_2, \bar{M}_s$ etc., are obtained straight from Eqs. (30.2), (30.3) and (34.2) as:

$$\begin{aligned} \bar{M}_1 &= (\bar{\nu}+\bar{h}) \cos\bar{\phi} \sin D \sin(H-\bar{\omega}) \\ \bar{M}_2 &= -(\bar{\rho}+\bar{h})\{\cos\bar{\phi} \cos D + \sin\bar{\phi} \sin D \cos(H-\bar{\omega})\} \\ \bar{M}_3 &= -\sin\bar{\phi} \cos D + \cos\bar{\phi} \sin D \cos(H-\bar{\omega}) \\ M_1 &= (\nu+h) \cos\phi \sin D \sin(H-\omega) \\ M_2 &= -(\rho+h)\{\cos\phi \cos D + \sin\phi \sin D \cos(H-\omega)\} \\ M_3 &= -\sin\phi \cos D + \cos\phi \sin D \cos(H-\omega) \\ \bar{N}_1 &= (\bar{\nu}+\bar{h}) \cos\bar{\phi} \cos(H-\bar{\omega}) \\ \bar{N}_2 &= (\bar{\rho}+\bar{h}) \sin\bar{\phi} \sin(H-\bar{\omega}) \\ \bar{N}_3 &= -\cos\bar{\phi} \sin(H-\bar{\omega}) \end{aligned}$$

$$N_1 = (\nu + h)\cos\phi\cos(H-\omega)$$

$$N_2 = (\varrho + h)\sin\phi\sin(H-\omega)$$

$$N_3 = -\cos\phi\sin(H-\omega)$$

leading to the expanded observation equations:
(Observed minus Computed) Declination is

$$= -\bar{M}_1 d\bar{\omega}/s - \bar{M}_2 d\bar{\phi}/s - \bar{M}_3 d\bar{h}/s$$
$$+ M_1 d\omega/s + M_2 d\phi/s + M_3 dh/s \tag{71.1}$$

(Observed minus Computed) Right Ascension is

$$= dt + \bar{M}_1 \sec D\, d\bar{\omega}/s + \bar{N}_2 \sec D\, d\bar{\phi}/s + \bar{N}_3 \sec D\, d\bar{h}/s$$
$$- N_1 \sec D\, d\omega/s - N_2 \sec D\, d\phi/s - N_3 \sec D\, dh/s \tag{71.2}$$

72. To fix the position of a single unknown station, at least two known stations and two flashes (which need not be simultaneous) are required in suitably different positions to provide reasonably well-conditioned intersections. There are then 12 equations to determine 11 unknown viz. two dt's for the two flashes and three each of $d\omega$'s, $d\phi$'s and dh's for the two flashes and for the unknown station. Any number of other unknown stations can be fixed off the same two flashes, since each additional unknown station adds four equations and only three unknowns. Additional known stations and flashes will, of course, strengthen the network: an additional flash would add six equations and four unknowns, while an extra known station observing two flashes would add four equation and no unknowns.

It should be clear from the foregoing analysis that deflections can have no effect since the plumb line does not appear in the working, except at the origin of the survey.

73. If the geodetic coordinates of only one unknown station across a wide gap are fixed by this method, this single station can still be used to extend the survey in the same geodetic system by measuring astronomical latitude, longitude and azimuth at it, and by adopting geodetic azimuths and zenith distances in at least two directions radiating from it to satisfy Eqs. (23.4) or (26.1). The effect of refraction and of observational error is minimized by adopting the procedure given in § 54 during the adjustment of the extended survey.

74. The method is virtually free from geometrical errors of refraction. The flash will usually occur outside the refracting atmosphere and will therefore be subject to the same refraction as a star, so that its right ascension and declination will be determinable to the same degree of accuracy on this account as the apparent places of stars are determined by photographic methods. Most of the effect of refraction is taken out in determining the plate constants from the background star images, whose apparent places are known.

Changing refraction would affect the position of the "instantaneous" flash image relative to the star images, the effect of refraction on which is meaned over a longer exposure. Clear weather over such long distances would probably mean steady atmospheric conditions, but to minimize the effect of unsteady refraction, Atkinson suggests arranging the flash to appear at zenith distances not exceeding 70°. He also suggests the use of repeating flashes.

75. Instead of a rocket, it has been suggested that a flashing artificial satellite might be used as a beacon, observations from several ground stations being synchronized by means of the flashes. This probably offers the best chance of using an artificial satellite for fixing geodetic positions. It would move too fast to be used instead of a flare for accurate altazimuth measurements and its orbit is unlikely to be well enough known for use in the lunar methods now to be described.

Position Fixing by Lunar Methods

76. We consider next the method of position-fixation by photography of the moon against a background of stars, recently proposed and worked out in detail by W. Markowitz for wide use during the International Geophysical Year. The camera itself is mounted equatorially to hold the exposure of the stellar background. It also carries a parallel-plate filter (analogous to the parallel-plate micrometer of a precise level) which can be rotated to hold the photographic image of the moon fixed in relation to the stars. The time of the observation is considered to be when the rotating filter introduces no relative displacement between moon and stars. Measurement of the plate, which can be corrected for irregularities in the limb by referring all lunar profiles to a single datum, provides the apparent right ascension and declination of the moon's centre from the known positions of several stars. The observational data is accordingly the same as in Atkinson's method, with the moon in place of the flash, except that the observation must be accurately timed. Moreover, the position of the moon in space, unlike that of the flash, is reckoned to be known from the Lunar Ephermeris, so that it need not be fixed by photography from known ground stations. Photography of the moon in at least two different positions from the same unknown station will fix the position of that station in all three coordinates.

77. As in the case of Atkinson's method, we reduce right ascensions to origin hour angles H by § 19. The observation equations are then exactly the same as Eq. (70.1), with the moon at the barred point. For a reason which will presently appear, we retain the corrections $d\bar{x}^s$ to the position of the moon, even though this is supposed to be known.

78. The Ephermeris gives the position of the moon's centre by its right ascension (which we shall reduce to origin hour angle by § 19 and then call γ to distinguish it from H, the origin hour angle of the line joining the ground station to the moon); by its declination (which we shall call δ to distinguish it similarly from D, the declination of the line joining the ground station to the moon); and by its parallax (directly related to the radius vector r) – all of the line joining the centre

of the moon to the centre of mass of the Earth. Accordingly, we propose to use $(\gamma, \delta, r) = (1, 2, 3)$ as coordinates and to derive the position of the ground station in the same coordinates; δ will then be the geocentric latitude (the complement of the angle between the axis and the radius vector drawn from the physical centre of mass to the ground station); and γ will be the geocentric longitude (the angle between the axial plane containing the radius vector and the plane of the *astronomic* meridian of the origin). Note that these geocentric coordinates have nothing to do with the geodetic system and its base spheroid; the radius vector is drawn from the centre of mass of the Earth, not from the centre of the spheroid.

79. As usual, we assume approximate coordinates $(\bar{\gamma}, \bar{\delta}, \bar{r})$ for the moon and for the ground station (γ, δ, r) and compute the length S, declination D and origin hour angle H of the line from the following formulae, which are obtained by projection on the three Cartesian axes:

$$\begin{aligned} S\cos H\cos D &= \bar{r}\cos\bar{\gamma}\cos\bar{\delta} - r\cos\gamma\cos\delta \\ S\sin H\cos D &= \bar{r}\sin\bar{\gamma}\cos\bar{\delta} - r\sin\gamma\cos\delta \\ S\sin D &= \bar{r}\sin\bar{\delta} - r\sin\delta\ . \end{aligned} \tag{79.1}$$

80. Components of the fundamental Cartesian vectors in the geocentric coordinate system can be written down at once from (34.2) – even though the Cartesian origin is not the same – by substituting $\gamma = \omega$, $\delta = \phi$, $(\nu + h) = (\varrho + h) = r$, viz.

$$\begin{aligned} A_r &= (-r\cos\delta\sin\gamma\ ;\quad -r\sin\delta\cos\gamma\ ;\quad \cos\delta\cos\gamma) \\ B_r &= (r\cos\delta\cos\gamma\ ;\quad -r\sin\delta\sin\gamma\ ;\quad \cos\delta\sin\gamma) \\ C_r &= (0\ ;\quad r\cos\delta\ ;\quad \sin\delta)\ . \end{aligned} \tag{80.1}$$

The components of the auxiliary vectors M_r, N_r follow by making the same substitutions in the formulae of § 71 or direct from Eqs. (30.2), (30.3) and (80.1), viz.

$$\begin{aligned} \bar{M}_1 &= \bar{r}\cos\bar{\delta}\sin D\sin(H-\bar{\gamma}) \\ \bar{M}_2 &= -\bar{r}\{\cos\bar{\delta}\cos D + \sin\bar{\delta}\sin D\cos(H-\bar{\gamma})\} \\ \bar{M}_3 &= -\sin\bar{\delta}\cos D + \cos\bar{\delta}\sin D\cos(H-\bar{\gamma}) \\ M_1 &= r\cos\delta\sin D\sin(H-\gamma) \\ M_2 &= -r\{\cos\delta\cos D + \sin\delta\sin D\cos(H-\gamma)\} \\ M_3 &= -\sin\delta\cos D + \cos\delta\sin D\cos(H-\gamma) \\ \bar{N}_1 &= \bar{r}\cos\bar{\delta}\cos(H-\bar{\gamma}) \\ \bar{N}_2 &= \bar{r}\sin\bar{\delta}\sin(H-\bar{\gamma}) \\ \bar{N}_3 &= -\cos\bar{\delta}\sin(H-\bar{\gamma}) \\ N_1 &= r\cos\delta\cos(H-\gamma) \end{aligned}$$

$$N_2 = r \sin\delta \sin(H-\gamma)$$
$$N_3 = -\cos\delta \sin(H-\gamma) \ ,$$

whence by substitution in (70.1), the expanded observational equations become: (Observed minus Computed) Declination of the line ground station to Moon's centre

$$= -M_1\, d\bar{\gamma}/s - \bar{M}_2\, d\bar{\delta}/s - \bar{M}_3\, d\bar{r}/s$$
$$+ M_1\, d\gamma/s + M_2\, d\delta/s + M_3\, dr/s \qquad (80.2)$$

(Observed minus Computed) Right Ascension (or Origin Hour Angle) of the same line

$$= \bar{N}_1 \sec D\, d\bar{\gamma}/s + \bar{N}_2 \sec D\, d\bar{\delta}/s + \bar{N}_3 \sec D\, d\bar{r}/s$$
$$- N_1 \sec D\, d\gamma/s - N_2 \sec D\, d\delta/s - N_3 \sec D\, dr/s \ . \qquad (80.3)$$

81. If the position of the moon really were known, we could put $d\bar{\gamma}$, $d\bar{\delta}$, $d\bar{r}$ equal to zero in these equations without further ado. Unfortunately, we cannot be sure that the Universal Time of the observation is the same as the Ephermeris Time used as argument in the Lunar Ephermeris; there is a difference between the two which is not tabulated and which varies slowly. The simplest way of overcoming the difficulty is to envisage a correction dt to the time of the observation, to find $(d\bar{\gamma}/dt)$ etc. from the tabular differences and to replace $d\bar{\gamma}$ etc. in the observational equation by $(d\bar{\gamma}/dt)\,dt$. This reduces the number of unknowns by two. The equations are still soluble from two separate positions of the moon, provided that dt is considered to be the same for both; but in practice many positions will be photographed from several stations, and it may also be possible to derive corrections to the elements of the orbit as well as by expressing $d\bar{\gamma}$ etc. in terms of these elements.

82. If π is the tabulated parallax, then $\bar{r} = \bar{S} \csc \pi$, in which the constant $\bar{S}$ is an assumed equatorial radius of the Earth. It may accordingly be necessary to write $\bar{r} = \csc \pi \ \{d\bar{r} = -\cos \pi \csc \pi \, d\pi\}$ and thereby in effect to determine $S/\bar{S}$ and $r/\bar{S}$ in a reduced scale model. In that case, $\bar{S}$ would be determined later from measured terrestrial distances between stations fixed by this method.

83. It has been suggested that a radar distance to the moon should also be measured, in which case there would be an additional observational equation obtainable from (42.2). If l_r is the unit vector in the direction of the moon from the ground station, then

$$l_r = (\cos H \cos D) A_r + (\sin H \cos D) B_r + (\sin D) C_r$$

and the components of this vector at the two ends of the line in $(\gamma, \bar{\delta}, r)$ coordinates are accordingly found from Eq. (80.1) to be:

$$\bar{l}_1 = \bar{r} \cos\bar{\delta} \cos D \sin(H-\bar{\gamma})$$
$$\bar{l}_2 = \bar{r}\{\cos\bar{\delta} \sin D - \sin\bar{\delta} \cos D \cos(H-\bar{\gamma})\}$$
$$\bar{l}_3 = \sin\bar{\delta} \sin D + \cos\bar{\delta} \cos D \cos(H-\bar{\gamma})$$

$$l_1 = r \cos\delta \cos D \sin(H-\gamma)$$
$$l_2 = r\{\cos\delta \sin D - \sin\delta \cos D \cos(H-\gamma)\}$$
$$l_3 = \sin\delta \sin D + \cos\delta \cos D \cos(H-\gamma)$$

and the required observational equation by substitution in Eq. (42.2) is: (Observed minus Computed) Distance

$$= \bar{l}_1 d\bar{\gamma} + \bar{l}_2 d\bar{\delta} + \bar{l}_3 d\bar{r}$$
$$-l_1 d\gamma - l_2 d\delta - l_3 dr \ . \tag{83.1}$$

With the help of (79.1) we find that $\bar{l}_1 = l_1$.

84. We have ensured that the Cartesian axes of the geocentric coordinates (γ, δ, r) are parallel to those of the geodetic system (ω, ϕ, h). If we know the position of a station in both systems, we can accordingly find the Cartesian coordinates (dx_0, dy_0, dz_0) of the centre of mass of the Earth relative to the centre of the spheroid from Eqs. (31.1) and (79.1);

$$dx_0 = (\nu + h)\cos\phi \cos\omega - r\cos\delta\cos\gamma$$
$$dy_0 = (\nu + h)\cos\phi \sin\omega - r\cos\delta\sin\gamma \tag{84.1}$$
$$dz_0 = (\nu\bar{e}^2 + h)\sin\phi - r\sin\delta \ .$$

The accuracy of the result will depend largely on the lunar theory on which the Ephermeris is based, and also it must be admitted on the method of computing the geodetic survey, but in time we may reasonably expect to obtain consistent values for dx_0, dy_0, dz_0 from a number of common stations. In that case, we can apply Eq. (84.1) in reverse to derive the geodetic coordinates of points whose geocentric coordinates have been obtained off the moon. We can then extend the geodetic survey on the same system (ultimately all round the Earth) as already proposed in § 73.

This is the only rigorously valid method of mixing lunar with other geodetic measurements.

85. Interest in methods involving stellar occultation has been revived in recent years by improvements in timing and photo-electric recording. All we get out of one timed occultation is an observation at a ground station for right ascension and declination of a particular point on the moon's limb where the occultation occurred. If these observations are reduced to right ascension and declination of the moon's centre, then we can form two observation equations [(80.2) and (80.3)] just as in the photographic method. Other equations can be added from different occultations observed at the same station, leading in exactly the same way to a solution for the geocentric coordinates of that station.

Unfortunately, the reduction to the moon's centre usually involves serious inaccuracies, owing to irregularities in the limb, which are largely avoided in the photographic method by locating the centre from several points on the limb. To overcome this difficulty, J. A. O'Keefe chooses two stations, one known and one unknown, where the same star is occulted by the same part of the limb. From the unknown coordinates of one station he is able in effect to provide data for the reduction to the moon's centre in this particular case, including any other uncertainty in lunar data which can be considered constant between the two observations, and to use this data in reducing the observations at the unknown station. Another such pair of occultations, including the same unknown station, provides two more observational equations and therefore the geocentric coordinates of the unknown station.

86. Solar eclipse observations could be reduced in much the same way, including the further reduction from the sun's centre to the point of contact with the moon. A single eclipse will not provide a complete fix, but the observational equations could nevertheless be of use in conjunction with other data.

Spirit-Levelling and the Geoid

87. Comparison of latitudes and longitudes in the astronomic and geodetic coordinate systems enters the argument at the outset, because all terrestrial angular measures are necessarily made in relation to the astronomic zenith, but we have not so far compared the third coordinates in the two systems.

The third coordinate in the astronomic system would naturally be the geopotential, or some function of the geopotential, since this is an independently variable function of position associated with the astronomic latitude and longitude, by reason of the fact that the astronomic zenith, whose direction defines latitude and longitude, is also the direction of the gradient of the geopotential. Moreover, the geopotential would be the best definition of natural "height" to use for practical purposes: water will not flow between two points of equal potential; starting from the same point, the same work would have to be done to climb two mountains whose summits are at the same potential; and the only physical notion of "horizontal" or "level" we can have is perpendicular to the astronomic zenith, in a direction in which the potential does not change.

88. If g is measured gravity and dl the difference in spirit-levelled heights between two near points, then the corresponding difference in potential is $g\,dl$, which can be integrated over a long line as $\int g\,dl$. The steps into which this integral is divided for numerical calculation can be quite large, and gravity measures correspondingly infrequent, unless the terrain is very hilly or geologically disturbed.

The potentials at junction-points in a level network should properly be obtained in this manner and used in the adjustment. Neat spirit-levels are not functions of position and are not independent of the path along which they are measured. Different values for the level of junction points can accordingly be obtained from different lines of levelling connecting them, even if the levelling is free of all error. The actual numerical difference depends on the extent to which gravity can be considered constant over the area of the network.

89. Unfortunately, the idea of potential is alien to most users of maps and survey data. We can, however, convey the same information in a form which looks more

like the conventional idea of height if we divide differences of potential by a constant having the dimensions of mean gravity. If we want to compare potentials over the whole Earth, then only one constant can be used for the whole Earth. Otherwise we can divide by mean gravity taken over a particular region and avoid comparison with "heights" obtained by the use of a different constant for another region. It is obvious that the more often the divisor is changed, the less the extent of the region over which comparisons can be made. To adopt a continuously variable divisor, as is sometimes suggested, rules out the possibility of making any valid "height" comparisons and stultifies the whole process. The important point to realize is that whatever system is used amounts to no more than a convention, which should be clearly recorded. Such expressions as "orthometric" and "dynamic" heights resulting from a particular convention and wrongly suggesting some measurable physical entity, are merely misleading and should be discontinued.

90. Suppose that we have determined the potential at a survey station by spirit-levelling in relation to the potential at mean sea level. If we sink a shaft, having depth but no other significant dimensions, under the station and make occasional gravity measurements down the shaft, we shall clearly be able to calculate the change in potential down the shaft, just as in surface levelling, and can thus find the depth at which the potential of mean sea level is recovered. By definition this will be the depth of the geoid below the station. The difference between this and the geodetic h-coordinate of the station gives us at once the separation of the spheroid and geoid under the station, usually denoted $N = (d - h)$, subject to a cosine error of the deflection.

Since we cannot sink shafts under the station, some approximation is necessary. We could measure gravity at the surface and apply half the *free-air* correction (to simulate the conditions in a shaft) for an assumed depth to derive a mean value of gravity) half way down; then divide the difference in potential between the station and sea level by this mean value of gravity to obtain a better value of t. Except in very hilly country, it would probably be sufficient to ignore gravity altogether and to take d as the neat spirit-levelled height of the station above sea level.

91. The classical method of taking *geoidal sections* claims to obtain a difference of N between two stations as a product of the mean deflection in the direction of the line joining the two stations and the distance apart of the stations. Since neither d nor h enters the method at all, it is evident that some additional assumption is being made to determine their difference. Apart from the over-simplified plane geometry used in deriving the method, the main implicit assumption is that a deflection measured on the surface has the same value underground at the level of the spheroid (or can be shown that the change in deflection down a line of force is proportional to the *horizontal* gradient of gravity. This gradient is certainly subject to rapid local variation on the surface and there is no reason to suppose it would be any more predictable underground. There would also be discontinuity in the deflection on crossing dipping strata of different densities. The method may nevertheless be numerically adequate over short lines in flat country, despite its

theoretical defects. It should be tested in hilly country against the simpler and more rigorous method given above.

Change of Geodetic Coordinates

92. We next investigate the effect of changing the geodetic coordinate system, while retaining fixed the positions of points in space. Parallelism of the fundamental Cartesian vectors A^r, B^r, C^r is also retained, so that the basis of comparison between astronomic and geodetic measures may remain unimpaired. Accordingly, we seek first-order changes in the coordinates (ω, ϕ, h) resulting from changes (dx_0, dy_0, dz_0) in the Cartesian coordinates of the centre of the base spheroid; and changes da and dt ($= -d\bar{e}$) in the major semi-axis and the flattening $f = (1-\bar{e})$ of the base spheroid.

93. We first change a and f without altering the Cartesian origin, which means that the position vector p^r remains constant. If we write $p = h + a^2/\upsilon$ and $q = -\upsilon e^2 \sin\phi \cos\phi$, it follows from Eq. (32.1) that

$$p^r = p\upsilon^r + q\mu^r .$$

Differentiating this vector equation and substituting Eqs. (29.3) and (29.2), we have

$$0 = (p\cos\phi - q\sin\phi)\,d\omega\,\lambda^r + (dq + p\,d\phi)\mu^r + (dp - q\,d\phi)\upsilon^r ,$$

in which we can equate to zero the coefficients of the mutually orthogonal vectors λ^r, μ^r, υ^r. The coefficient of λ^r shows that $d\omega$ is zero since $(p\cos\phi - q\sin\phi) = (\upsilon + h)\cos\phi$ is not zero; in other words, as might have been anticipated, no change in geodetic longitude results from changes in the shape and size of the base spheroid. We are left with the two conditions

$$dq + p\,d\phi = 0 \tag{93.1}$$

$$dp - q\,d\phi = 0 .$$

Considering p, q as functions of a, f, ϕ and h we find after some simplification that $\partial p/\partial\phi = q$ and $\partial q/\partial\phi = \varrho - a^2/\upsilon$, so that these two equations (93.1) reduce on expansion to

$$(\varrho + h)\,d\phi + (\partial q/\partial h)\,dh + (\partial q/\partial a)\,da + (\partial q/\partial f)\,df = 0$$

and

$$(\partial p/\partial h)\,dh + (\partial p/\partial a)\,da + (\partial p/\partial f)\,df = 0$$

which on differentiation of p+q, substitution and simplification become

$$dh = -(a/\upsilon)\,da + (\upsilon\bar{e}\sin^2\phi)\,df \tag{93.2}$$

$$(\varrho + h)\,d\phi = (\upsilon e^2 \sin\phi\cos\phi/a)\,da + (\upsilon\bar{e} + \varrho/\bar{e})\sin\phi\cos\phi\,df . \tag{93.3}$$

94. Next we introduce a change (dx_0, dy_0, dz_0) in the Cartesian origin, involving a corresponding translation of the spheroid. The effect will be the same if we keep the Cartesian origin and the spheroid fixed and alter the Cartesian coordinates of the point in space under consideration by $(-dx_0, -dy_0, -dz_0)$. The corresponding changes in geodetic coordinates are given by

$$\begin{aligned} d\omega &= -\frac{\partial\omega}{\partial x}dx_0 - \frac{\partial\omega}{\partial y}dy_0 - \frac{\partial\omega}{\partial z}dz_0 \\ &= -A^1 dx_0 - B^1 dy_0 - C^1 dz_0 \ , \end{aligned}$$

which on substituting (34.1) becomes

$$(\nu + h)\cos\phi\, d\omega = \sin\omega\, dx_0 - \cos\omega\, dy_0 \ . \tag{94.1}$$

In the same way, we have

$$\begin{aligned} d\phi &= -A^2 dx_0 - B^2 dy_0 - C^2 dz_0 \\ dh &= -A^3 dx_0 - B^3 dy_0 - C^3 dz_0 \ , \end{aligned}$$

which on substituting (34.1) and including the terms (93.2) and (93.3), arising from change of spheroid, give us finally:

$$\begin{aligned} (\varrho + h)\, d\phi = {} & \sin\phi\cos\omega\, dx_0 + \sin\phi\sin\omega\, dy_0 - \cos\phi\, dz_0 \\ & + (\nu e^2 \sin\phi\cos\phi/a)\, da + (\nu\bar{e} + \varrho/\bar{e})\sin\phi\cos\phi\, df \end{aligned} \tag{94.2}$$

$$\begin{aligned} dh = {} & -\cos\phi\cos\omega\, dx_0 - \cos\phi\sin\omega\, dy_0 - \sin\phi\, dz_0 \\ & -(a/\beta)\, da + (\nu\bar{e}\sin^2\phi)\, df \ . \end{aligned} \tag{94.3}$$

Corresponding changes in azimuth and zenith distance are obtained by substituting (94.2) and (94.3) in (26.1), which apply just as much to a change of geodetic coordinates as to a change to astronomic coordinates. The resulting equations are not, of course, independent and may not be used as well as (94.2) and (94.3) when required as observational equations. Either can, however, be used instead of (94.2) and (94.3).

95. The three equations (94.1), (94.2) and (94.3) enable us to bring adjacent surveys, say P and Q, into sympathy, through the geodetic coordinates of their common points. If dx_0, dy_0, dz_0, da, df are "corrections" to the Q system of coordinates, then $d\omega$ is the longitude of a point in the P system minus the longitude of the same point in the Q system; and similarly for $d\phi$, dh. Three observational equations are thus obtained for each common point to determine the five unknowns[7], dx_0 etc. When dx_0 etc. have been found, the equations can be used to determine corrections $d\omega$, $d\phi$, dh to the old coordinates of all points in Q (including the geodetic origin of Q) to bring them into sympathy with P. It is, of course, assumed that both surveys have been computed and adjusted in

[7] Three, if the elements of both spheroids are known and the terms in da, df can be evaluated numerically.

accordance with the principles formulated in this paper, so that final discrepancies between common points can be considered local and random and not due to such systematic causes as faulty orientation of a base spheroid, arising from incorrect origin conditions and leading to a wrong use of astronomic observations in adjustment. Few surveys can be said to have avoided such pitfalls altogether and most will need a measure of re-computation before they can be joined with assurance.

No alteration is envisaged in the *orientation* of either geodetic system, although the satisfaction of the conditions (26.1) at the origin of either survey cannot have been error-free. If the procedure recommended in the section on triangulation adjustment has been followed, every astronomic observation, wherever made, will have contributed to setting up and maintaining correct orientation. The only way of further improving the result would be a complete re-adjustment together of the two surveys.

Figure of the Earth

96. The Eqs. (94.1), (94.2) and (94.3) also enable us to determine a geodetic system which departs as little as possible from the astronomic system – a problem which can be considered the three-dimensional extension of the old problem of finding a spheroid which best fits the geoid through minimizing the differences between astronomic and geodetic latitudes and longitudes. There are certain practical advantages in having such a geodetic system, provided it is not changed too often.

An extensive survey on a common geodetic coordinate system is necessary for the purpose, or several surveys which have been brought onto a common system by the method of the last section. Before long, surveys on a common geodetic system spanning the oceans may well have become available, running right round the the Earth and closing on themselves.

97. Wherever astronomical longitude has been measured, we write for $d\omega$ in (94.1) the astronomic minus the geodetic longitude; and similarly for $d\phi$ in (94.2). If astronomic azimuth has been measured, it may be used instead of, but not in addition to, one of these equations by substituting (94.1) and (94.2) in (26.1) and using the result as an azimuth equation in which dx is the astronomic minus the geodetic azimuth. In (94.3), dh becomes the spirit-levelled minus the geodetic height, subject to the considerations discussed in § 90. Spirit-levels are most likely to be available in the flatter areas, where they can probably be used without gravity corrections. It may be argued that geodetic heights should not be included since these are vitiated by atmospheric refraction. If so, Eqs. (94.3) could be given a low weight or ignored. Consistent geodetic heights can nevertheless be obtained, despite atmospheric refraction, over large areas; and it is already becoming clear that the prejudice against "trigonometric heights" derives more from faulty methods of computing them than from any effect of refraction on properly conducted observations.

The resulting observational equations are solved for dx_0, etc., which are finally substituted in similar equations to derive the corrections $d\omega$ etc. to add to old values in order to derive new coordinates of all points on the new geodetic system.

98. The centre of the new spheroid will not necessarily coincide with the centre of mass of the Earth, although it is unlikely to be far away if the survey is sufficiently extensive. We could, however, obtain average values of dx_0, dy_0, dz_0 which would shift the centre of the spheroid to the centre of mass from lunar observations at a number of points (see § 84). These values of dx_0, dy_0, dz_0 would then be substituted in (94.1), (94.2) and (94.3) before solving to obtain values of the remaining two unknowns da and df. In this way we should obtain a new geodetic system which best fits the astronomic system, subject to the condition that the centre of the spheroid coincides with the centre of mass of the Earth, so far as this may be determinable from lunar observations. We should also obtain, by back substitution in (94.1), (94.2) and (94.3) and evaluation of residuals, the deflections and a value of N (see § 90) at every point in the survey in relation to the new geodetic system. These would be directly comparable with the same results obtained by Stokes' integration of gravity anomalies in relation to the same spheroid, apart form the smoothing and first-order approximation inherent in the Stokes's method.

Index of Main Symbols

(Figures in brackets are references to paragraphs where first used).

l^r	A general unit vector (§ 7)
l_r	Covariant components of l^r (§ 8)
l^r, m^r, n^r	A general set of mutually orthogonal unit vectors, right-handed in that order (§ 10)
A^r, B^r, C^r	A set of Cartesian unit vectors, right-handed in that order (§ 12)
λ^r, μ^r, υ^r	A right-handed set of unit vectors defined as follows: (§ 14)
λ^r	Unit "parallel" vector perpendicular to μ^r, υ^r (§ 14)
μ^r	Unit "meridian" vector lying towards the North in the third coordinate surface, coplanar with υ^r, C^r (§ 14)
υ^r	Outward drawn unit vector normal to the third coordinate surface (§ 14)
ω	Longitude of υ^r in relation to A^r, B^r, C^r (§ 15)
ϕ	Latitude of υ^r in relation to A^r, B^r, C^r (§ 15)
h	Geodetic height. This is the third coordinate in geodetic system of coordinates (§ 16)
α	Azimuth measured eastwards from μ^r (§ 18)
β	Zenith distance measured from υ^r (§ 18)
D	Declination of l^r. The angle between l^r and the plane A^r, B^r (§ 19)
H	Origin hour angle of l^r. The angle between the plane A^r, C^r and a plane parallel to C^r and l^r (§ 19)
x^r	A set of general coordinates in the following order; (§ 20)
x^1	Longitude
x^2	Latitude

x^3	Geodetic height or geopotential
x, y, z	Cartesian coordinates in the directions A^r, B^r, C^r (§ 20)
a, b, c	Cartesian components of the unit vector l^r (§ 23)
ϱ, υ	Principal curvatures of the base spheroid (§ 31)
a	Major semi-axis of the base spheroid (§ 31)
e	Eccentricity of the base spheroid (§ 31)
$\bar{e}$	$=\sqrt{1-e^2}$ (§ 31)
f	Flattening of the base spheroid. $e^2 = 2f - f^2 = 1 - \bar{e}^2$ (§ 92)
s	Length of a line (§ 37)
g	Measured gravity (§ 88)
N	Separation of spheroid and geoid (§ 90)
dx_0, dy_0, dz_0	Cartesian coordinates of the centre of mass of the Earth § 84)

(*Note:* ω, ϕ, α, β etc. of the same point or line etc. will not, of course, be equal in different coordinate systems. Whenever two coordinate systems are considered together one is denoted by barred notation, e.g. $\bar{\omega}$, $\bar{\phi}$, $\bar{\alpha}$, $\bar{\beta}$. In some cases, clear from the context, barred notation is also used for a different point in the same coordinate system).

Editorial Commentary

This is unquestionably the most elegant and polished of Hotine's unpublished reports, and it completely deserves the praise that Marussi lavished on it (see our Editorial Introduction and Marussi's tribute reprinted in Whitten's memorial lecture on pp. 171–183).

It encompasses much of the material contained in the two previous reports to the Toronto Assembly in 1957, but curiously *omits* any mention of the differential geometry of the Earth's gravitational field and the Marussi tensor. We believe this was not accidental, but indicative of the fact that first and foremost Hotine was a *practical geodesist*, not merely a desk-bound theorist. This fact is often missed by anyone who casually dips into MG and shrinks away, having been intimidated by its mathematical richness.

Indeed, the perfection of the *Primer* is its only imperfection. As in the writings of Gauss, here we see only the inexorable deduction of the results without any of the scaffolding which led the author to his approach and formulation. It is simply a masterpiece, and if asked which *one* paper of Hotine one should read to gauge the measure of the man and his thought, this would be the unequivocal choice. It is by no means an easy read, but one which repays careful study and consideration.

The content of the report is essentially reproduced with some abridgement in Chapters 25–27 of MG. The *Primer* has an advantage over MG in that it is almost self-contained – even to the extent that index notation of the tensor calculus is re-explained! Actually the use of the tensor calculus per se is very limited and really only index notation and the differentials of the coordinates are required. Roughly speaking, one could almost characterize the mathematical content of the report as three-dimensional analytic geometry as applied to practical geodesy and photogrammetry.

The Venice Symposium was Hotine's *finest hour*, and his report dominated the meeting. Although the symposium had only *19* participants, we are fortunate to have an excellent record of it in Dufour's report (1959a). Subsequent reactions and critiques of the *Primer* were given in Dufour (1959b), Wolf (1963), Näbauer (1965) and Dufour (1968). Chovitz (1974) assessed the relevance of Hotine's methods for the North American Datum.

References to Editorial Commentary

Chovitz B (1974) Three-dimensional model based on Hotine's "Mathematical Geodesy". Can Surv 28:568–573

Dufour H (1959a) Le symposium sur la géodésie à 3 dimensions. (Venice, July 1959) Bull Géod 54:75–92

Dufour H (1959b) Quelques reflexions au sujet du Symposium de Venice. Bull Géod 54:61–64

Dufour H (1968) The whole geodesy without ellipsoid. Bull Géod 88:127–143

Näbauer M (1965) Zu Hotine's "A primer of non-classical geodesy". Deutsche Geodätische Kommission bei der Bayerischen Akademie der Wissenschaften. Reihe A: Theoretische Geodäsie Nr 46, München, 80 p

Wolf H (1963) Die Grundgleichungen der dreidimensionalen Geodäsie in elementarer Darstellung. Z Vermessungswe 6:225–233

6 The Third Dimension in Geodesy[1]

Introduction

In a previous paper (Hotine 1957a, § 40 to § 49) the changes in various functions of position in the gravitational field along lines of finite length have been considered. Some simplification and added precision arise when certain measurements are made at both ends of the line and this case is now considered in particular relation to change in the potential, which is compared with the corresponding change in geodetic height. The results are used to demonstrate the approximations inherent in the method of geoidal sections. Practical tests of the formulae are proposed as a positive item of geodetic research.

A Lemma

1. A scalar function of position F can be expanded along a line of finite length s as follows:

$$(\bar{F}-F) = sF' + \frac{s^2}{2}F'' + \frac{s^3}{6}F''' + \frac{s^4}{24}F^{iv} + \dots$$

in which barred quantities (e.g. $\bar{F}$) refer to values at the far end of the line and unbarred quantities to the near end. Superscripts refer to successive differentiations with respect to s. If these differential coefficients are measured in the same sense at the far end, that is in the direction of the line produced and not in the back direction, then the corresponding expansion from the far end of the line is obtained by interchanging bars and changing the sign of s. Thus:

$$(\bar{F}-F) = s\bar{F}' - \frac{s^2}{6}\bar{F}'' + \frac{s^3}{6}\bar{F}''' - \frac{s^4}{24}\bar{F}^{iv} + \dots .$$

In the mean,

$$(\bar{F}-F) = \frac{s}{2}(F'+\bar{F}') + \frac{s^2}{4}(F''-\bar{F}'') + \frac{s^3}{12}(F'''+\bar{F}''') + \dots . \tag{1.1}$$

Now the differential coefficients can be considered as functions of position, defined at all points along the line, and similarly expanded as:

[1] Report dated 22 June 1960 (Tolworth) and presented to I.A.G. Helsinki Assembly 1960.

$$(\bar{F}'+F') = sF''+\frac{s^2}{2}F'''+\frac{s^3}{6}F^{iv}+\ldots$$

$$= s\bar{F}''-\frac{s^2}{2}\bar{F}'''+\frac{s^3}{6}\bar{F}^{iv}-\ldots$$

so that

$$0 = s(F''-\bar{F}'')+\frac{s^2}{2}(F'''+\bar{F}''')+\frac{s^3}{6}(F^{iv}-\bar{F}^{iv})+\ldots \tag{1.2}$$

and by direct expansion as in (1.1)

$$0 = (F''-\bar{F}'')+\frac{s}{2}(F'''+\bar{F}''')+\frac{s^2}{4}(F^{iv}-\bar{F}^{iv})+\ldots \tag{1.3}$$

with similar equations of higher order starting with the fourth differentials.

We can eliminate the terms containing either the second or the third differentials from (1.1), (1.2) and (1.3) *but not both*; and thereafter one of each succeedding pair of terms. Consequently, we may say that either of the following expansions

$$(\bar{F}-F) = \frac{s}{2}(F'+\bar{F}')+\frac{s^2}{12}(F''-\bar{F}'') \tag{1.4}$$

or

$$(\bar{F}-F) = \frac{s}{2}(F'+\bar{F}')-\frac{s^3}{24}(F'''+\bar{F}''') \tag{1.5}$$

is correct to a fifth order, subject to the usual conditions relating to continuity, differentiability and convergence. We can reasonably be assured from intuitive physical consideration that these conditions are met in the problems we are going to discuss. We may not, however, know how rapidly the original series converges and what will be the effect of neglecting terms of the fifth and higher orders. This will have to be determined experimentally.

When F is the potential, then the second-order terms (F'' etc.) can be measured at the two ends of the line, whereas the third-order terms cannot. Accordingly, we shall develop the terms in (1.4), but to the degree of accuracy at which we are working, the two Eqs. (1.4) and (1.5) are really equivalent and we can say that

$$(F''-\bar{F}'') = -\frac{s}{2}(F'''+\bar{F}''') \ .$$

The result is equally true for vector and tensor functions, provided the coefficients are obtained by successive covariant differentiations along the line.

Potential

2. If F in (1.4) is the potential V, then $\frac{\partial V}{\partial s}$ is the resolved part of the gravitational force in the direction of the line, so that

$$\frac{\partial V}{\partial s} = -g \cos \beta \ . \tag{2.1}$$

β is the zenith distance of the line measured from the astronomical zenith.

3. It has been shown in a previous paper (Hotine 1957a) how various quantities may be expressed in terms of the five parameters of the gravitational field, χ_1, χ_2, τ, γ_1, γ_2 (see list of symbols at end of this paper), and how these parameters may be measured (Hotine 1957a, §§ 32, 33). The variation of certain quantities along an atmospherically refracted ray have also been investigated (Hotine 1957a, § 42 to § 49) on the sole assumption that the gradient of atmospheric density is everywhere in the direction of the plumb line or astronomical zenith. This is the least mathematical assumption which can be made about atmospheric refraction and is equivalent to assuming that there is no lateral refraction in the model atmosphere. In particular, we have, after correcting a slight misprint in Hotine (1957a, § 48):

$$\begin{aligned} \frac{\partial \beta}{\partial s} &= \chi_0 - (\chi_1 \sin^2 \alpha + 2\tau \sin \alpha \cos \alpha + \chi_2 \cos^2 \alpha) \sin \beta \\ &\quad - (\gamma_1 \sin \alpha + \gamma_2 \cos \alpha) \cos \beta \\ &= \chi_0 - \chi_\alpha \sin \beta - (\gamma_1 \sin \alpha + \gamma_2 \cos \alpha) \cos \beta \ . \end{aligned} \tag{3.1}$$

In this expression α is the azimuth of the line; χ_0 is its curvature and χ_α is the normal curvature of the equipotential surface in azimuth α.

By differentiating (2.1) we have

$$\frac{\partial^2 V}{\partial s^2} = -g \cos \beta \left(\frac{\partial \log g}{\partial s} \right) + g \sin \beta \frac{\partial \beta}{\partial s} \ . \tag{3.2}$$

But from Eqs. (16.4) and § 43 (in Hotine 1957a)

$$\frac{\partial \log g}{\partial s} = \gamma_1 \sin \alpha \sin \beta + \gamma_2 \cos \alpha \sin \beta - \left(\chi_1 + \chi_2 + \frac{2\tilde{\omega}^2}{g} \right) \cos \beta \ , \tag{3.3}$$

in which $\tilde{\omega}$ is the angular velocity of the Earth's rotation and all points of the line are assumed to be free air.

Combining this last equation with (3.2), we have

$$\begin{aligned} \frac{\partial^2 V}{\partial s^2} &= (g \sin \beta) \chi_0 - (g \sin^2 \beta) \chi_\alpha - 2 g \sin \beta \cos \beta (\gamma_1 \sin \alpha + \gamma_2 \cos \alpha) \\ &\quad + g \cos^2 \beta \left(\chi_1 + \chi_2 + \frac{2\tilde{\omega}^2}{g} \right) \ . \end{aligned} \tag{3.4}$$

Accordingly, Eq. (1.4) gives us the difference of potential along the line as

$$\begin{aligned} (\bar{V} - V) &= -\frac{s}{2} (g \cos \beta + \bar{g} \cos \bar{\beta}) \\ &\quad + \frac{s^2}{12} [\{(g \sin \beta) \chi_0 - (\bar{g} \sin \bar{\beta}) \bar{\chi}_0\} \end{aligned}$$

$$-\{(g\sin^2\beta)\chi_\alpha-(\bar{g}\sin^2\bar{\beta})\bar{\chi}_\alpha\}$$
$$-\{2g\sin\beta\cos\beta(\gamma_1\sin\alpha+\gamma_2\cos\alpha)$$
$$-2\bar{g}\sin\bar{\beta}\cos\bar{\beta}(\bar{\gamma}_1\sin\bar{\alpha}+\bar{\gamma}_2\cos\bar{\alpha})\}$$
$$+\{g\cos^2\beta(\chi_1+\chi_2)-\bar{g}\cos^2\bar{\beta}(\bar{\chi}_1+\bar{\chi}_2)\}$$
$$+2\bar{\omega}^2(\cos^2\beta-\cos^2\bar{\beta})]\ .$$

Barred quantities refer as usual to their values at the far end of the line. Many of the second-order terms in this expression are likely to be negligible but are given in full here for the sake of completeness.

If we confuse g and $\bar{g}$, and divide (3.5) by either, then *to a first order* the difference spirit levels over the line as ordinarily computed would be $\frac{1}{2}s(\cos\beta+\cos\bar{\beta})$, which is the same as the difference in "trigonometrical heights" as ordinarily computed, using zenith distances measured from the astronomical zenith or plumb line. Refraction does not enter the result computed to this order of accuracy. The second-order term on account of refraction would be very nearly $\frac{1}{12}s^2(\chi_0-\bar{\chi}_0)$, which is zero if the curvature at the two ends is the same, as indeed is usually assumed. The second-order terms containing normal curvatures of the equipotential surfaces (χ_α) and variation in gravity (γ_1, γ_2) may well be more significant, particularly in gravitationally disturbed areas.

Spheroidal Heights

4. The full theory of three-dimensional geodetic coordinate systems has been developed in previous papers (Hotine 1957b, 1959). The recommended system consists of latitude ϕ and longitude ω measured on a base spheroid; and heights h measured along the straight normals to this spheroid, subject to stated conditions for the initial orientation of the spheroid (Hotine 1959, § 24 to § 28).

We can obtain the change in geodetic or spheroidal heights over the line by putting $V=h$; $g=-1$; and $\gamma_1=\gamma_2=0$ in (3.1) and (3.2). (But the Laplacian of V is no longer $2\bar{\omega}^2$.) In the result we have

$$(\bar{h}-h)=\frac{s}{2}(\cos\beta+\cos\bar{\beta})-\frac{s^2}{12}\,[(\chi_0\sin\beta-\bar{\chi}_0\sin\bar{\beta})$$
$$-(\chi_\alpha\sin^2\beta-\bar{\chi}_\alpha\sin^2\bar{\beta})]\ . \qquad (4.1)$$

In this expression, β, $\bar{\beta}$ are now measured from the geodetic zenith, that is the normal to the base spheroid; χ_0 remains the actual curvature of the ray; and χ_α is the normal curvature in azimuth α of the h = const. surface passing through the point:

$$\chi_\alpha=\frac{\cos^2\alpha}{(\varrho+h)}+\frac{\sin^2\alpha}{(\upsilon+h)}\ ,$$

in which ϱ, υ are, as usual, the principal curvatures of the base spheroid. This last expression may usually be approximated as the reciprocal of a mean radius of curvature.

5. Comparison with the exact formulae in previous papers indicates that Eq. (4.1) introduces a negligible computational error (over and above the effect of uncertain refraction curvature). For instance, over a line 50 miles in length in the worst azimuth, the error in height is no more than 3 mm in 2700 meters, that is about one part in a million.

Again note that the only effect of refraction is in the difference of end curvatures in the second-order term.

Potential and Spheroidal Heights

6. We *define* the vector deflection at a point in space as

$$\delta^r = (\upsilon^r) - \upsilon^r \ , \tag{6.1}$$

in which (υ^r) is a unit vector in the direction of the astronomical zenith and υ^r is a unit vector in the direction of the outward-drawn normal to the base spheroid.

7. This rigorous definition accords with the usual first-order theory of deflections. If we bracket astronomical quantities, then the (A minus G) latitude is $\delta\phi = (\phi) - \phi$ and the (A minus G) longitude is $(\omega) - \omega$. From Hotine (1959, § 22), we have

$$\begin{aligned}(\upsilon^r) = {} & \lambda^r \cos(\phi) \sin\delta\omega + \mu^r [\cos\phi \sin(\phi) - \sin\phi \cos(\phi) \cos\delta\omega] \\ & + \upsilon^r [\sin\phi \sin(\phi) + \cos\phi \cos(\phi) \cos\delta\omega] \ .\end{aligned} \tag{7.1}$$

The component of deflection in the direction of the geodetic parallel, λ^r, is accordingly

$$\delta^r \lambda_r = \cos(\phi) \sin\delta\omega = \cos\phi \, \delta\omega \quad \text{to a first order} \ .$$

The component of deflection in the direction of the geodetic meridian μ^r is

$$\begin{aligned}\delta^r \mu_r &= \cos\phi \sin(\phi) - \sin\phi \cos(\phi) \cos\delta\omega \\ &= \delta\phi \quad \text{to a first order} \ .\end{aligned}$$

And the component of deflection in the direction of the geodetic zenith υ^r is

$$\begin{aligned}\delta^r \upsilon_r &= \sin\phi \sin(\phi) + \cos\phi \cos(\phi) \cos\delta\omega - 1 \\ &= 0 \quad \text{to a first order.}\end{aligned}$$

8. Next we find the component of deflection in the direction of a unit vector l^r in astronomical azimuth (α) and zenith distance (β); equivalent to geodetic azimuth α and zenith distance β. The vector can be expressed (see Hotine 1959, § 18) in either of the following ways:

$$\begin{aligned}l^r &= (\lambda^r) \sin(\alpha) \sin(\beta) + (\mu^r) \cos(\alpha) \sin(\beta) + (\upsilon^r) \cos(\beta) \\ &= \lambda^r \sin\alpha \sin\beta + \mu^r \cos\alpha \sin\beta + \upsilon^r \cos\beta \ .\end{aligned} \tag{8.1}$$

Accordingly, the component of deflection in the direction l^r is from (6.1):

$$\Delta = (v^r l_r - v^r l_r = \cos(\beta) - \cos\beta \ . \tag{8.2}$$

This result is rigorously true whatever the magnitude of the deflection, but it accords with the normal first-order theory, in which small deflections, equivalent to small rotations, are compounded like vectors. If we write $\delta\beta = (\beta) - \beta$ and use Hotine (1959) § 26, then (8.2) becomes to a first order

$$\begin{aligned} \Delta &= -\sin\beta \cdot \delta\beta = \sin\alpha \sin\beta\,(\cos\phi \cdot \delta\omega) + \cos\alpha \sin\beta \cdot \delta\phi \\ &= \sin\alpha\,(\cos\phi \cdot \delta\omega) + \cos\alpha \cdot \delta\phi \quad \text{if } \beta \text{ is near } 90^\circ \ , \end{aligned}$$

and this is the usual first-order expression.

9. We denote the component of deflection at the far end of a line of finite length s (not necessarily straight) by

$$\bar{\Delta} = (\bar{v}^r)\bar{l}_r - \bar{v}^r \bar{l}_r = \cos(\bar{\beta}) - \cos\bar{\beta} \ .$$

It is then evident from § 3 and (4.1), to a *first order* and subject to the limitations of § 3, that

$$\frac{s}{2}(\Delta + \bar{\Delta}) = \frac{s}{2}\{\cos(\beta) + \cos(\bar{\beta})\} - \frac{s}{2}\{\cos\beta + \cos\bar{\beta}\} \tag{9.1}$$

= Rise in spirit levels minus the rise in spheroidal heights in proceeding from the unbarred to the barred end of the line .

Geoidal Sections

10. In the normal classical theory $\frac{s}{2}(\Delta + \bar{\Delta})$ is reckoned to be the difference, over the length of the line, in the (underground) separation between geoid and spheroid. This clearly involves the further assumption that there is no change in deflection in proceeding from a point on the topographic surface to a point "vertically" below on the geoid (or spheroid). There is certainly no physical justification for this assumption. Geometrically, it would require the plumb line to have zero curvature underground, and the extent to which this is not so would introduce a direct first-order error in the first-order theory. It might be so in practice for points not much above the geoid (or spheroid) in flat country, but this does not seem to have been satisfactorily demonstrated.

Conclusion

11. Whether or not the less drastic assumptions in Eq. (9.1) ar justified depends on whether the first-order assumption, particularly as applied to Eq. (3.5), is justified. This can only be settled by experiments, which so far have not been done, by comparison between results from (3.5), with and without the second-order terms, and a direct measure of potential by means of spirit levels.

To run lines of spirit levels over mountains is a difficult and costly operation but it should be done to obtain essential topographic information. The use of (3.5) instead of spirit-levelling, if it leads to results of sufficient accuracy, might well have far-reaching practical and economic consequences.

To test Eq. (3.5), we need to measure the end-curvatures of the refracted ray, gravity, and the five parameters of the gravitational field at, say, every main triangulation station. Theoretically, all this is possible, but much research work would have to be done first on instrumentation. In addition to the economic results, such a programme of research would probably contribute far more to our knowledge of the gravitational field than the present collection of random data.

Index of Main Symbols

ϕ Latitude
ω Longitude
h Geodetic of spheroidal height
V Potential
g Gravity
$\tilde{\omega}$ Angular velocity of the Earth's rotation
χ Curvature
χ_0 Curvature of a refracted ray
χ_1 Normal curvature of the equipotential surface in the direction of the astronomical, parallel
χ_2 Normal curvature of the equipotential surface in the direction of the astronomical, meridian
τ Geodesic torsion in the direction of the astronomical, meridian
χ_α Normal curvature in azimuth α.

$$= \chi_1 \sin^2 \alpha + 2\tau \sin \alpha \cos \alpha + \chi_2 \cos^2 \alpha \quad \text{(see Hotine 1957a, § 22)}$$

γ_1 Arc rate of change of (log g) in the direction of the astronomical, parallel
γ_2 Arc rate of change of (log g) in the direction of the astronomical, meridian
α Azimuth
β Zenith distance
λ^r Unit vector in the parallel direction
μ^r Unit vector in the meridian direction
ν^r Unit vector in the zenith direction
l^r A general unit vector

Editorial Commentary

As Hotine said in his abstract, this report was a continuation and refinement of his two Toronto reports *Metrical Properties of the Earth's Gravitational Field, Geodetic Coordinate Systems*, and his *A Primer on Non-Classical Geodesy* (Venice, 1959). In particular, his concern was with the additional precision and simplification which occurs when certain measurements are made at both ends of a line of observation. Hence, in effect, this report dealt with the observational aspects of certain situations considered in these three reports. In particular, special attention is paid to the gravitational potential and spheroidal heights.

The material in this report is developed in somewhat greater detail in Chapter 25, §§ 16–25 of MG.

References to Paper 6

Hotine M (1957a) Metrical properties of the Earth's gravitational field. I.A.G. Toronto
Hotine M (1957b) Geodetic coordinate systems. I.A.G. Toronto
Hotine M (1959) A primer of non-classical geodesy. I.A.G. Venice

7 Harmonic Functions[1]

1. The only harmonic functions so far used in geodesy seem to be spherical and spheroidal harmonics, and very occasionally ellipsoidal harmonics. All these are special cases of a much more general class of harmonic functions, which may well find applications in modern geodesy, although I have not yet attempted to work out the applications in detail.

2. Suppose first that H is any scalar function of position, and that the $\mathbf{n}^{\text{th}}$-order tensor

$$H_{rst\ldots n} \tag{1}$$

is formed by $\mathbf{n}$ successive covariant differentiations of H.

3. The following tensor equation

$$H_{rts\ldots n} = H_{rst\ldots n} \,,$$

in which any two indices have been interchanged, is clearly true in Cartesian coordinates, when the covariant derivatives become ordinary commutable derivatives, and is therefore true in any coordinates. The tensor (1) is accordingly symmetrical in any two indices and therefore has $\frac{1}{2}(\mathbf{n}+1)(\mathbf{n}+2)$ distinct components at most.

4. Now suppose that H is a harmonic function, and that g^{jk} is the metric tensor. Then the Laplacian of the tensor (1) is

$$g^{jk} H_{rst\ldots njk} = (g^{jk} H_{jk})_{rst\ldots n} = 0$$

so that all components of the tensor (1) are harmonic functions.

5. We may similarly write

$$g^{rs} H_{rst\ldots n} = 0 \,. \tag{2}$$

The tensor on the left-hand side is of order $(\mathbf{n}-2)$ with at most $\frac{1}{2}[(\mathbf{n}-2)+1][(\mathbf{n}-2] = \frac{1}{2}\mathbf{n}(\mathbf{n}-1)$ distinct components. If H is a harmonic function, there are then $\frac{1}{2}\mathbf{n}(\mathbf{n}-1)$ relations (2) between the components of the original tensor (1), which can accordingly have

[1] Undated report (Tolworth) presented at the Second Symposium on Three-Dimensional Geodesy (Cortina d'Ampezzo, 1962).

$$\tfrac{1}{2}(\mathbf{n}+1)(\mathbf{n}+2)-\tfrac{1}{2}\,\mathbf{n}(\mathbf{n}-1)=2\mathbf{n}+1$$

independent components at most.

6. Next, for the invariant

$$A^{rst\ldots n}H_{rst\ldots n} \tag{3}$$

in which the multiplying tensor $A^{rst\ldots n}$ is constant under covariant differentiation; that is, all its components are absolute constants in Cartesian coordinates, and are the transforms of Cartesian constants in other coordinates. Then each of the components to be summed in Eq. (3) is a harmonic function and there will be at most $(2\mathbf{n}+1)$ such independent functions. The constant $A^{rst\ldots n}$ should properly by symmetrical in any two indices and have its components further restricted to $(2\mathbf{n}+1)$ by some such relation as (2), but this is not essential. If these conditions are not fulfilled, then there will be more than $(2\mathbf{n}+1)$ terms in the summation (3), but some of them will not be independent.

7. Now consider the summation in series of all invariants such as (3):

$$K=\sum_{n=0}^{\infty} A^{rst\ldots n}H_{rst\ldots n}\ . \tag{4}$$

We assume, subject to test in individual cases, that the successive derivatives of H exist, and together with the constants $A^{rst\ldots n}$, form a convergent series having a finite sum K. It is evident by taking the Laplacian of both sides of Eq. (4) that K is a harmonic function; and considering there are $(2\mathbf{n}+1)$ degrees of freedom in each term of (4), and the function H is at choice, it is evident that K is a very general harmonic function.

8. It can be shown without difficulty that Eq. (3) represents all the *spherical* harmonics of $\mathbf{n}^{th}$-order if $H = 1/r$, the reciprocal of the radius vector, if the differentiations are performed in Cartesian coordinates, and if

$$g_{rs}A^{rst\ldots n}=0\ .$$

But it is well known that *any* harmonic function K can be expressed in a series of spherical harmonics, even though subject to these arbitrary conditions, in such a form as Eq. (4). It is reasonable to assume, therefore, that any harmonic function K can be expressed by Eq. (4), without the restrictions of spherical harmonics, provided the derivatives of H exist and provided Eq. (4) is a convergent series.

9. In geodetic problems, it might be more convenient to make $H = \tan^{-1} u$, where u is that parameter in spheroidal coordinates which is constant over any one of the spheroidal coordinate surfaces. But the main merit of the more generalized expansion (4) may be to make H the actual external attraction potential of the Earth, in which case K could be $1/r$ or any other desired material harmonic function. We can further introduce the geopotential W in a field rotating with angular velocity $\tilde{\omega}$ by making

$$H = W - \tfrac{1}{2}\,\tilde{\omega}^2(x^2+y^2)\ .$$

The second term on the right would vanish in the third order and subsequent terms of Eq. (4) in Cartesian, and therefore in any coordinates.

10. We could re-write the constant tensor in (3) as

$$A^{rst\ldots n} = A\lambda^r\mu^s\upsilon^t\ldots\tau^n\ ,$$

in which λ^r, μ^s, etc. are *fixed* arbitrary unit vectors and A is a Cartesian constant, so that $(2\mathbf{n}+1)$ independent components are still preserved. If all the fixed vectors are the same, say λ^r, and $A = s^n/n!$, then the expansion (4) becomes the tensor form of Taylor's series expressing the potential in the same field at another point distant s in the fixed direction λ^r. This may suggest that the expansion (4) is unnecessarily general in practice. Suppose, for example, that H and K are respectively the "normal" (or "geodetic") and the actual potentials at a point in space. It should always be possible to find another point in the "normal" field where the "normal" potential is K and so relate the two potentials by a Taylor series.

11. It should also be possible to find another point in the "normal" field where the "normal" potential is the actual potential at the original point. If the astronomical latitude and longitude at the original point are (ϕ, ω), then the locus of points in the "normal" field having (ϕ, ω) as the direction of the "normal" plumb line is known. At some point on this locus, the "normal" potential will be the actual potential at the original point. The whole procedure would be simplified by tables for the "normal" field.

12. The "normal" value of gravity at the second point would not, however, be the actual gravity at the first point, although the two might not differ very much in practice. If it were, then tables for the "normal" field would enable us to determine the actual potential at any point from a measure of gravity at that point without any process of integration, except what is implicit in the use of the "normal" field tables. The possibility, however, needs investigation, since this method might well give as accurate results as the first-order Stokes' theory.

13. The new theory of harmonic functions given in this paper does not stand or fall by this one possible application, which is given solely as an example to show the power of the new theory as a weapon of geodetic research. We are perhaps too much wedded to Green and Stokes, and to spherical harmonics, and may need to broaden our horizons.

Editorial Commentary

This brief paper was published as an appendix to Baetsle's comprehensive report (1963) of the Second Symposium on Three-Dimensional Geodesy (Cortina d'Ampezzo, 1962). Actually, Hotine's name does not appear on the paper, but his authorship was indicated in the text of Baetsle's report. See also the discussion of this paper in Baetsle (1963, pp. 43–44).

Essentially this paper was a preliminary version of the material in the first eight sections of Chapter 21 of MG. The treatment was somewhat less concise than that given in MG, and the content of § 9–§ 13 in it are not precisely duplicated in MG. Thus this paper furnishes a slight variant of the elaborate theory of spherical harmonics in Chapter 21 of MG. See also the comments of Chovitz in the second section of his preview article *Hotine's Mathematical Geodesy* re-printed in this monograph. As he noted, although the topic has been exhaustively discussed in the mathematical literature, see in particular Hobson (1911, 1955–65), much of Hotine's treatment was original and illustrated his personal touch. In MG (see § 7 p. 154), Hotine commented that his approach was based on the work of J.C. Maxwell, i.e. see Eq. (7.1) or MG, Chapter 21, Eq. (006). Maxwell's original exposition (1891) (MG Chap. IX, pp. 194–231), is difficult; however, a lucid elementary introduction to it can be found in MacRobert (1927) (MG Chap. XIII, pp. 231–240). In reading Hotine's discussion today, one feels its power, and senses that he has broken new ground, yet one does not quite know how to proceed further. Surely the last word has not been written on the subject.

References to Editorial Commentary

Baetsle PL (1963) Le deuxième symposium de géodésie à trois dimensions. (Cortina d'Ampezzo, 1962) Bull Géod 67:27–62

Hobson EW (1911) Spherical harmonics. The Encyclopaedia Britannica, 11th edn 25:649–661

Hobson EW (1931) The theory of spherical and ellipsoidal harmonics. Cambridge Univ Press, Cambridge. Re-printed by Chelsea Publ Co, New York (1955 and 1965)

MacRobert TM (1927) Spherical harmonics, an elementary treatise on harmonic functions with applications. Methuen, London

Maxwell JC (1891) A treatise on electricity and magnetism. Vol I, 3rd edn. Oxford Univ Press, Oxford. Re-printed by Dover Publ, New York (1954)

8 Downward Continuation of the Gravitational Potential[1]

1. Clerk Maxwell's form of the attraction potential at an external point, distant r from an internal origin, can be written in the tensor form

$$-\frac{V}{G} = \sum_{n=0}^{\infty} \frac{(-1)^n}{n!} I^{stu\ldots(n)} \left(\frac{1}{r}\right)_{stu\ldots(n)} , \tag{1}$$

in which the lower indices imply successive covariant derivatives and the I's are constants, symmetrical in any two indices. The negative sign of the potential used in physics is adopted and G is the gravitational constant. This form of the potential is of great value in theoretical investigations; for example it shows at once the invariance of the potential under rotations of the coordinate axes about a fixed origin.

2. By working straight from the mass distribution, it can be shown that Eq. (1) is equivalent, not by individual terms but by partial sums of terms of the same order n, to the usual expansion in spherical harmonics:

$$-\frac{V}{G} = \sum_{n=0}^{\infty} \sum_{m=0}^{n} \frac{P_n^m(\sin\phi)}{r^{n+1}} \{C_{nm} \cos m\omega + S_{nm} \sin m\omega\} , \tag{2}$$

in which (ϕ, ω) are spherical polar latitude and longitude from the same internal origin. At the same time, it can be shown that the I's in Eq. (1) – the constant inertia tensors – are given by

$$I^{stu\ldots(n)} = \sum \bar{m}\bar{x}^s\bar{x}^t\bar{x}^u\ldots\bar{x}^{(n)} , \tag{3}$$

in which the summation is taken over all particles of mass $\bar{m}$ located at the point whose coordinates are $\bar{x}^q$. An alternative expression for a continuous distribution of density ϱ is

$$I^{stu\ldots(n)} = \int \varrho\varrho^s\varrho^t\varrho^u\ldots\varrho^{(n)} dv , \tag{4}$$

in which ϱ^q is the position vector of the volume element dv and the integral is taken over the entire volume of the attracting body.

3. Again working straight from the mass distribution, it can be shown that a sufficient – but perhaps not necessary – condition for the convergence of Eqs. (1) or (2) is that the point where the potential is sought should lie outside a sphere,

[1] Undated report (Boulder) presented at the I.A.G. Lucerne Assembly 1967.

centered on the origin, which just contains all the attracting matter. The expression of the potental in spherical harmonics, obtained from satellites outside this sphere of convergence, is not necessarily valid at points on the Earth's surface inside this sphere. This question has been extensively and inconclusively argued.

4. We shall now consider expressions for the potential at external points when the origin is also external. In this case the potential in spherical harmonics is well known to be

$$-\frac{V}{G}=\sum_{n=0}^{\infty}\sum_{m=0}^{n} r^n P_n^m(\sin\phi)\{[C_{nm}]\cos m\omega+[S_{nm}]\sin m\omega\} \tag{5}$$

and a sufficient condition for convergence is that the point where the potential is sought shall lie inside a sphere, centered on the origin, which just touches the attracting body. However, by working straight from the mass distribution and using Eq. (1), we find that the potential can also be expressed as

$$V=(V_0)+(V_s)_0\varrho^s+\ldots+\frac{1}{n!}(V_{stu\ldots(n)})_0\varrho^s\varrho^t\varrho^n\ldots\varrho^{(n)}+\ldots\,, \tag{6}$$

in which ϱ^q is the position vector from the external origin to the point where the potential is sought and the quantities within brackets are values of successive derivatives of the potential at the external origin. Moreover, Eqs. (5) and (6) are equivalent by partial sums of terms of the same order. But Eq. (6) is the same as the Taylor expansion of the potential from the external origin. We conclude that Eq. (6) is convergent within the sphere of convergence of Eq. (5).

5. Next, we propose to continue the potential analytically by means of a Taylor series from a Point P in the diagram (Fig. 1), where Eqs. (1) or (2) are certainly valid, to a Point T where Eqs. (5) or (6) are certainly valid. This operation gives the potential at T as

$$-\frac{V_T}{G}=\sum_{m=0}^{\infty}\sum_{n=0}^{\infty}\frac{1}{m!}\frac{(-1)^n}{n!}I^{stu\ldots(n)}\left(\frac{1}{r}\right)_{stu\ldots(n)pqr..(m)}\varrho^p\varrho^q\varrho^r\ldots\varrho^{(m)}\,. \tag{7}$$

The order of summation is important. To reflect the continuation process correctly we must first sum over n in order to derive the potential at P before differentiation and substitution in the Taylor series.

6. Equation (7) can be written as an infinite matrix in the form

$$\begin{matrix} M(1/r)\,, & -I^s(1/r)_s\,, & +\frac{1}{2}I^{st}(1/r)_{st}\,,\ \ldots \\ M(1/r)_p\varrho^p\,, & -I^s(1/r)_{sp}\varrho^p\,, & +\frac{1}{2}I^{st}(1/r)_{stp}\varrho^p\,,\ \ldots \\ \frac{1}{2}M(1/r)_{pq}\varrho^p\varrho^q\,, & \frac{1}{2}I^s(1/r)_{spq}\varrho^p\varrho^q\,, & +\frac{1}{4}I^{st}(1/r)_{stpq}\varrho^p\varrho^q\,,\ \ldots \end{matrix} \tag{8}$$

in which the inertia tensor of zero order is the total mass M. The first row summed represents the potential at P. The second row summed is the first derivative of the potential at P contracted with the fixed bounded vector ϱ^p, and so on.

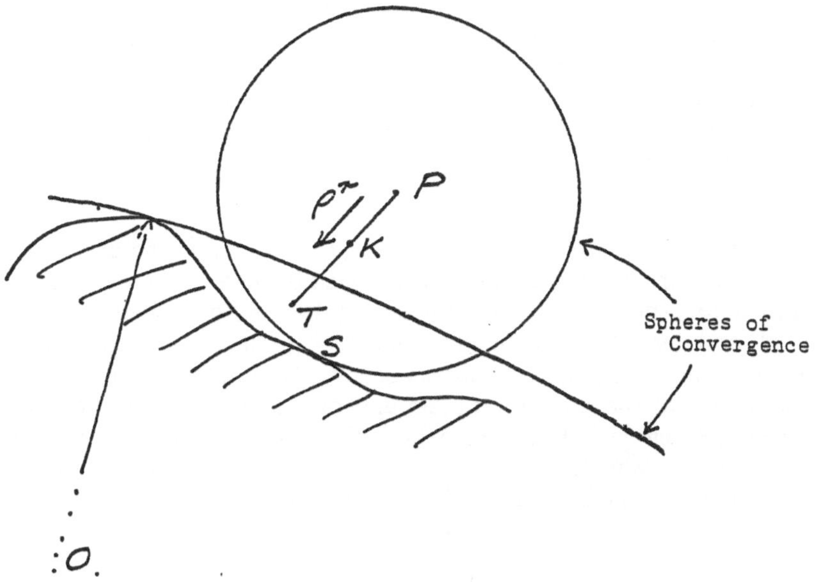

Fig. 1.

The fact that Eq. (7) is convergent implies that the matrix is convergent if the rows are summed first. If, on the other hand, we sum the columns first, and this process is not necessarily valid, then we find after some manipulation that the result would be

$$-\frac{V_T}{G} = \sum_{n=0}^{\infty} \frac{(-1)^n}{n!} I^{stu\ldots(n)} \left(\frac{1}{OT}\right)_{stu\ldots(n)}, \qquad (9)$$

in which the derivatives are now evaluated at T. But this is the same as Eq. (1) evaluated at T, and if Eq. (9) correctly represents the potential at T, then Eq. (1) – or (2) – must be convergent at T, even though T lies inside the sphere of convergence of Eq. (1) – or (2) – and so does not satisfy the sufficient condition for convergence. The convergence of Eq. (1) – or (2) at T – accordingly depends on whether interchange of the order of summation in Eqs. (7) or (8) is valid. Sufficient conditions for this interchange have been worked out in most standard texts, and would not usually be satisfied, but necessary conditions do not seem to have been obtained.

7. Some light may be thrown on this question by considering a point K on PT which lies outside the sphere of convergence of Eq. (1) – or (2). In that case, Eq. (1) – or (2) – certainly represents the potential at K and the summation interchange in Eq. (7), leading to Eq. (9), is certainly valid at K. But the two continuation series, Eq. (7), for K and for T, both of which are convergent, must also have the same properties of absolute and uniform convergence because the coefficients of the vectors are the potential at P, and derivatives of the potential at P, and are

therefore the same for both series. The only difference between the two series is the magnitude, but not the direction, of the contracting vector and this does not affect the convergence of either series. Accordingly, if the necessary and sufficient conditions for the summation interchange depend solely on convergence properties, then these conditions would seem to be satisfied at T as well as at K. By suitable choice of the point P we could reach any point on the Earth's surface by this process.

8. Nevertheless, the proof of convergence of Eq. (7) depends on the absence of matter within the sphere PS, and could be invalidated if there were an alternative distribution of matter, nearer to P than the actual distribution, which gives the same potential, and derivatives of the potential, at P as the actual distribution. For example, if the attracting matter consists solely of a point mass at 0, then a homogeneous sphere of the same total mass and of radius less then OP would give rise to the same field at P. In this case, analytical continuation from P could be completely blocked. Whether or not there is such an alternative distribution in the case of a highly irregular body, such as the Earth, does not appear to have been investigated.

9. Suggestions have been made to overcome the difficulty by expressing the potential in spheroidal harmonics as

$$-\frac{V}{G} = \sum_{n=0}^{\infty} \sum_{m=0}^{\infty} Q_n^m(i \cot \alpha) P_n^m(\sin u)(A_{nm} \cos m\omega + B_{nm} \sin m\omega) \; , \tag{10}$$

in which the spheroidal coordinates (ω, u, α) are given by

$$\begin{aligned} x &= (ae) \csc \alpha \cos u \cos \omega \\ y &= (ae) \csc \alpha \cos u \sin \omega \\ z &= (ae) \cot \alpha \sin u \; , \end{aligned} \tag{11}$$

with (ae) an absolute constant for the (confocal) coordinate spheroids. This constant can be chosen to make one coordinate spheroid approximate closely to the actual surface of the Earth, while enclosing all the attracting matter, in which case Eq. (10) would converge outside this spheroid. However, direct determination of the A_{nm} and B_{nm} from current geodetic measurements is less simple than determination of the spherical harmonic coefficients C_{nm} and S_{nm}. Nevertheless, the two sets of coefficients must be related *for the same mass distribution* in a domain where both series are convergent. In fact, the relationship is linear and is given by

$$\begin{aligned} \begin{bmatrix} A_{nm} \\ B_{nm} \end{bmatrix} = \frac{1\cdot 3\cdot 5\,..(2n+1)}{(n+m)!}\, i^{m+n+1} \Bigg[& \frac{1}{(ae)^{n+1}} \begin{pmatrix} C_{nm} \\ S_{nm} \end{pmatrix} \\ & + \frac{(n-m)(n-m-1)}{2\cdot(2n-1)} \frac{1}{(ae)^{n-1}} \begin{pmatrix} C_{(n-2),m} \\ S_{(n-2),m} \end{pmatrix} \\ & + \frac{(n-m)(n-m-1)(n-m-2)(n-m-3)}{2\cdot 4\cdot(2n-1)(2n-3)} \frac{1}{(ae)^{n-3}} \begin{pmatrix} C_{(n-1),m} \\ S_{(n-4),m} \end{pmatrix} + \dots \Bigg] \end{aligned} \tag{12}$$

together with the inverse equations

$$i^{(m+n+1)}\begin{pmatrix} C_{nm} \\ S_{nm} \end{pmatrix} = \frac{(ae)^{n+1}(n-m)!}{1\cdot 3\cdot 5\ldots(2n+1)}\left[\frac{(n+m)!}{(n-m)!}\begin{pmatrix} A_{nm} \\ B_{nm} \end{pmatrix} + \frac{2n+1}{2}\,\frac{(n+m-2)!}{(n-m-2)!}\begin{pmatrix} A_{(n-2),m} \\ B_{(n-2),m} \end{pmatrix} + \frac{(2n+1)(2n-1)}{2\cdot 4}\,\frac{(n+m-4)!}{(n-m-4)!}\begin{pmatrix} A_{(n-4),m} \\ B_{(n-4),m} \end{pmatrix} + \frac{(2n+1)(2n-1)(2n-3)}{2\cdot 4\cdot 6}\,\frac{(n+m-6)!}{(n-m-6)!}\begin{pmatrix} A_{(n-6),m} \\ B_{(n-6),m} \end{pmatrix} + \ldots\right]. \quad (13)$$

The same formulae give the zonal coefficients for $m = 0$. It is assumed that the spherical and spheroidal harmonics are both related to the same Cartesian system.

10. Equations (12) and (13) enable us easily to transform a potential, validly expressed by Eq. (2), to Eq. (10) and to continue the potential by means of Eq. (10) almost down to the Earth's surface. Even if Eq. (2) is finally agreed to be convergent down to the surface, it will not converge as rapidly as Eq. (10), which may accordingly be a better method of computation. The operation of transforming to spheroidal harmonics may also be a useful tool in further theoretical investigations.

Editorial Commentary

This report considered several of the standard mathematical questions in classical Newtonian potential theory, viz. analytical continuation and convergence. Both continue to be of contemporary interest, and relative to the latter, as Hotine noted at the end of § 3, "This question has been extensively and inconclusively argued". Over the last 25 years since he wrote those words, nothing but the number of memoirs seems to have significantly changed. An excellent survey of more recent work is given in Moritz (1980).

Following the spirit of his report *Harmonic Functions* (Cortina d'Ampezzo, 1962), Hotine attacked the issue using tensor-theoretic methods. This involved reformulating Maxwell's theory of spherical harmonics in tensor form (for references to the classical theory, see our Editorial Commentary on *Harmonic Functions* paper 7, this Vol.), and giving expressions for the inertia tensor of the attractive mass distributions. As no doubt he would have agreed, he solved neither question but merely displayed the difficulties in a new and tantalizing form. However, he would have been the first to maintain that nothing is lost – and something may well be gained – by looking at an old question in a new form. As Georg Christoph Lichtenberg (1742–1799) said

"Neue Blicke durch die alten Löcher".

A more exhaustive treatment of this material may be found in Chapter 21 of MG, beginning with § 12 and continuing through § 107. Note the footnotes on p. 165 of MG, which address different definitions of the inertia tensor. The definition of the inertia tensor employed by Hotine and McConnell (1931) (see § 1 of the latter's Chap. 18) still differ from those employed in standard physics texts on dynamics, e.g. Landau and Lifshitz (1960) (see § 32 of their Chap. 6), and does not seem as unusual as Hotine believed. It would appear that for potential-theoretic purposes (as Hotine observed) his approach is simply different – and more natural – than the definition employed in dynamical considerations. The important fact is to note this discrepancy.

References to Editorial Commentary

Landau LD, Lifshitz EM (1960) Mechanics, Volume 1 of course of theoretical physics. 1st English edn Pergamon, Oxford. Currently in its 4th Russian edn Nauka, Moscow (1988)

McConnell AJ (1931) Applications of the absolute differential calculus. Blackie, London. Corrected printing (1936). Reprinted as Applications of tensor analysis. Dover Publ, New York (1957)

Moritz H (1980) Advanced physical geodesy. Hermann Wichmann, Karlsruhe (see Chaps. 6–8)

9 Curvature Corrections in Electronic Distance Measurements[1]

1. It can be shown (Hotine 1960) that if F is a continuous, differentiable scalar, the expansion

$$(\bar{F}-F) = \frac{1}{2}s(F'+\bar{F}') + \frac{1}{12}s^2(F''-\bar{F}'') \tag{1}$$

along a line of arc length s, is correct to a fourth order. The superscripts refer to successive derivatives of F with respect to s and over-bars indicate values at the far end of the line, while the absence of over-bars indicates values at the near end of the line. It is further assumed that the ordinary Taylor expansion of F along the line exists and is convergent, and that this condition is met, or justified by results, in practical cases.

2. The time (t) of propagation of light or other electromagnetic waves along a line in a medium of refractive index μ is given by the usual formula

$$ct = \int_P^{\bar{P}} \mu \, ds \ , \tag{2}$$

in which c is the (constant) velocity of propagation in vacuo and the integral is taken over the curved line from the initial point P to the far point $\bar{P}$. We take the basic scalar of Eq. (1) to be the *indefinite* integral

$$\int \mu \, ds$$

so that

$$ct = \int_P^{\bar{P}} \mu \, ds = \bar{F} - F \ . \tag{3}$$

We have also

$$F' = \mu$$
$$\bar{F}' = \bar{\mu} \tag{4}$$

retaining the contention that the unbarred quantities refer to values at the initial point P and over-barred quantities refer to the far point $\bar{P}$.

[1] Undated report (Boulder) presented to I.A.G. Lucerne Assembly 1967.

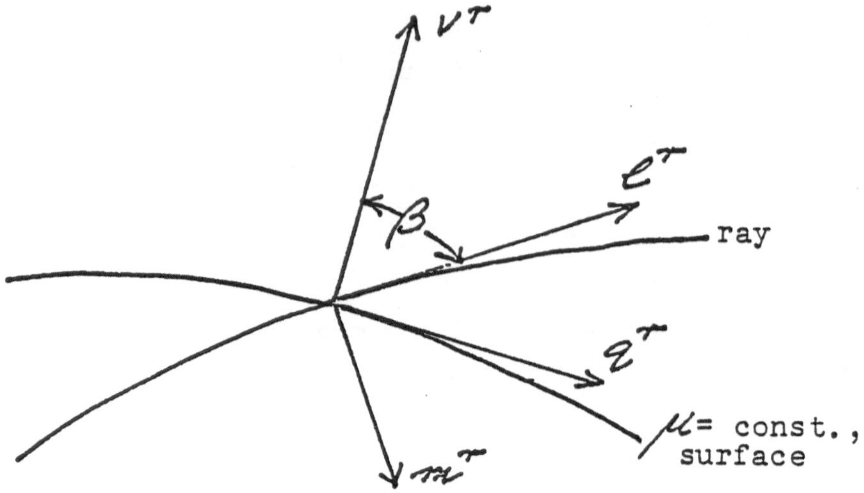

Fig. 1

3. To express the second derivatives we shall have to introduce the laws of refraction and the curvature of the line. In the diagram (Fig. 1) the unit tangent vector of the line in index notation is l^r and the unit normal of the line is m^r. The unit vector υ^r, usually assumed to be in the direction of the geodetic zenith, is normal to the surface of constant refractive index (μ) passing through the point under consideration, and therefore has the same direction as the gradient of μ. The laws of refraction, which can be directly derived from Fermat's variational principle, then require l^r, m^r, υ^r to be coplanar and

$$\chi = (\log n)_r m^r \ , \tag{5}$$

in which χ is the curvature of the line or ray. In this formula, (log n) is the natural logarithm of μ and $(\log n)_r$ in index notation is the gradient of (log n). From the geometry of the figure we have also

$$\begin{aligned} l^r &= \upsilon^r \cos\beta + q^r \sin\beta \\ m^r &= q^r \cos\beta - \upsilon^r \sin\beta \\ &(\log n)_r q^r = 0 \ , \end{aligned} \tag{6}$$

in which q^r is a unit vector tangential to the μ = constant surface in the plane of l^r, m^r, υ^r. Finally we have, for substitution in Eq. (1),

$$\begin{aligned} F'' = \frac{\partial\mu}{\partial s} &= \mu(\log n)_r l^r \\ &= \mu(\log n)_r \upsilon^r \cos\beta \\ &= -\mu(\log n)_r m^r \cot\beta \\ &= -\mu\chi\cot\beta \end{aligned}$$

using in succession Eqs. (4), (6) and (5). Equation (1) becomes finally

$$\mathrm{ct} = \tfrac{1}{2}\,\mathrm{s}(\bar{\mu}+\mu)+\tfrac{1}{12}\,\mathrm{s}^2(\bar{\mu}\bar{\chi}\cot\bar{\beta}-\mu\chi\cot\beta)\ . \tag{7}$$

4. For measured t (half the time for the double journey along the line and return), Eq. (7) can easily be solved for s by iteration in the usual way. The terminal refractive indices are obtained from the standard Barrell and Sears (or Essen and Froome) formula, using measured temperature, pressure and humidity. Ignoring the second-order term, a preliminary value of s is obtained by dividing (ct) by the mean refractive index and this value of s is used to evaluate the second-order term. To complete the evaluation of the second-order term, β and $\bar{\beta}$ can be taken as the observed astronomical or geodetic zenith distance of the curved line, noting that $\bar{\beta}$ refers to the line produced through the far end and not to the reverse zenith distance. To determine the terminal curvatures, we can differentiate the Barrell and Sears (or Essen and Froome) formulae and use Eq. (5) or its equivalent

$$\begin{aligned}\chi &= (\log\mu)_{\mathrm{r}}\mathrm{m}^{\mathrm{r}} = -(\log\mu)_{\mathrm{r}}\upsilon^{\mathrm{r}}\sin\beta\\ &= -\frac{1}{\mu}\,\frac{\mathrm{d}\mu}{\mathrm{d}h}\sin\beta\ ,\end{aligned} \tag{8}$$

in which h refers to geodetic height. In evaluating $\mathrm{d}\mu/\mathrm{d}h$ lapse rates would have to be assumed or obtained from local meteorological services.

5. The second-order term in Eq. (7) in various approximate forms is known in the literature as the "second velocity correction" or the "velocity component of the curvature correction" and it is of some interest to consider what assumptions are made in approximating this correction. If we differentiate the equation

$$\cos\beta = \upsilon_{\mathrm{r}}\mathrm{l}^{\mathrm{r}}$$

covariantly along the line, we have

$$-\sin\beta(\mathrm{d}\beta/\mathrm{ds}) = \upsilon_{\mathrm{rs}}\mathrm{l}^{\mathrm{r}}\mathrm{l}^{\mathrm{s}}+\upsilon_{\mathrm{r}}\mathrm{l}^{\mathrm{r}}_{\mathrm{s}}\mathrm{l}^{\mathrm{s}} = -k\sin^2\beta-\chi\sin\beta\ ,$$

in which k is the normal curvature of the $\mu=$ constant surface. To a first order we have also

$$\begin{aligned}(\cot\bar{\beta}-\cot\beta) &= \mathrm{d}(\cot\beta) = -\mathrm{s}\csc^2\beta(\mathrm{d}\beta/\mathrm{ds})\\ &= -\mathrm{s}(\csc\beta+\chi\csc^2\beta)\ .\end{aligned}$$

If, in evaluating the small second-order correction, we consider that $\mu=\bar{\mu}=1$ and χ is constant along the line, then the second-order term, considered as an additive correction to the preliminary value of s, is

$$\frac{1}{12}\,\mathrm{s}^3\chi(k\csc\beta+\chi\csc^2\beta)\ .$$

If there is no considerable difference in height over the line, $\csc\beta$ is nearly unity; and if we confuse $-1/k$ with a mean radius R of the Earth and write $\mathrm{f}=\chi\mathrm{R}$ for the coefficient of refraction, the correction finally becomes

$$-\frac{1}{12}\frac{s^3}{R^2}f(1-f) \;, \tag{9}$$

which is the form given by Saastamoinen (1964) or Höpcke (1964). In micro-wave measurements, it is usual to assume $f = 0.25$. For precise geodimeter measurements, Saastamoinen recommends evaluating the coefficient of refraction from reciprocal zenith distances, or vertical angles, measured at the same time as the geodimeter observations. But if any such special measurements are to be made, the reciprocal zenith distances β, $180° - \bar{\beta}$ can just as easily enter the precise Eq. (7) together with a mean curvature derived also from the reciprocal zenith distance measurements, without making most of the assumptions implicit in Eq. (9).

6. It should be noted that Eq. (8) gives the actual length S of the curved path. A geometric arc-to-chord correction is necessary in addition to provide the length of the straight line joining the end points, and would also be necessary even if the effect of refraction on the velocity of propagation were removed altogether by a two-wavelength method. By expansion from one end of the line, the magnitude of this arc-to-chord correction is easily found to be

$$-\frac{1}{24}s^3\chi^2 \tag{10}$$

to a high degree of accuracy, whether the curve is plane or twisted. The curvature χ can be a mean curvature $\frac{1}{2}(\chi + \bar{\chi})$, or the correction can be evaluated at the two ends and meaned. Introducing the coefficient of refraction $f = \chi R$, the correction is

$$-\frac{1}{24}\frac{s^3}{R^2}f^2 \;,$$

which can be combined with Eq. (9) in cases where the approximate formula (9) is sufficient.

7. For the adjustment of networks in three dimensions, the chord length is used and no other corrections are required.

Editorial Commentary

The title of this report is somewhat obscure as the "curvature" in question refers not to that of an equipotential surface, but of a refracted ray (a curve) in the Earth's atmosphere. In his characteristic manner of thinking in terms of geometry and basic physical principles, Hotine starts from Fermat's Principle in geometrical optics and applies the Frenet equations to the ray's trajectory. The result is expressed in Eq. (7) which is an elegant blend of geometry and optics. This result is approximate and the second-order term in it is known as the "second velocity correction" or more precisely as the "velocity component of the curvature correction." It is given in several forms and compared with previously known versions which had appeared in the literature.

In MG this material is developed in a more expansive and luxuriant manner in Chapter 24, and the contents of the report correspond to §§ 1–32 of that chapter. The remaining roughly two-thirds of Chapter 24 is concerned with measuring the refractive index and other basic physical properties, e.g. the pressure, and how they affect the problem.

Since this report was written the subject has continued to be of interest, especially due to the more recent use of satellite-to-ground applications like Doppler, GPS, satellite laser ranging, and VLBI. See for example the colloquium proceedings, (Brunner 1984), and the monograph by Iribane and Goodson (1981), which has some geodetic applications in Chapter 8.

References to Paper 9

Höpcke W (1964) On the curvature of electromagnetic waves and its effect on measurement of distance. Vermessungswesen 89:183–200. Translated in Survey Review 141:298–312

Hotine M (1960) The third dimension in geodesy. I.A.G. Helsinki

Saastamoinen J (1964) Curvature correction in electronic distance measurement. Bull Géod 73:265–269

References to Editorial Commentary

Brunner FK (ed) (1984) Geodetic refraction – effects of electromagnetic wave propagation through the atmosphere. Springer, Berlin Heidelberg New York Tokyo

Iribane JV, Goodson WL (1981) Atmospheric thermodynamics. 2nd edn. Reidel, Dordrecht

Bibliography of Martin Hotine

John Nolton

Professional Papers of the Air Survey Survey Committee

(a series of papers published by H. M. Stationery Office, London, for the War Office)

1. Simple methods of surveying from air photographs. No. 3 (1927), 71 pp.
2. The stereoscopic examination of air photographs. No. 4 (1927), 84 pp.
3. Calibration of survey cameras. No. 5 (1929), 81 pp.
4. Extensions of the Arundel method. No. 6 (1929), 115 pp.
5. The Fourcade steriogonimeter. No. 7 (1931), 159 pp..

Book

Surveying from air photographs. Constable and Company Limited, London (1931), 250 pp.

The Empire Survey Review

1. Laplace azimuths I. Vol. I, No. 1 (July, 1931), 24–31.
2. Laplace azimuths II. Vol. 2 (October, 1931), 66–71.
3. An aspect of attraction. Vol. II, No. 7 (January, 1933), 24–28.
4. Geodetic beacons. Vol. II, No. 9 (July, 1933), 151–156.
5. Figures and fancies. Vol. II, No. 11 (January, 1934), 264–268.
6. The East African Arc I: the layout. Vol. II, No. 12 (April, 1934), 357–367.
7. The East African Arc II: marks and beacons. Vol. II, No. 14 (October, 1934), 472–484.
8. The East African Arc III: observations. Vol. III, No. 16 (April, 1935), 72–80.
9. The East African Arc IV: base measurement. Vol. III, No. 18 (October, 1935), 203–218.
10. The re-triangulation of Great Britain I. Vol. IV, No. 25 (July, 1937), 130–136.
11. The re-triangulation of Great Britain II. Vol. IV, No. 26 (October, 1937), 194–206.
12. The re-triangulation of Great Britain III. Vol. IV, No. 29 (July, 1938), 386–405.
13. The general theory of tape suspension in base measurement. Vol. V, No. 31 (January, 1939), 2–36.
14. The re-triangulation of Great Britain IV. Vol. V, No. 34 (October, 1939), 211–255.
15. The orthomorphic projection of the spheroid I. Vol. VIII, No. 62 (October, 1946), 300–311.

16. The orthomorphic projection of the spheroid II. Vol. IX, No. 63 (January, 1947), 25–35.
17. The orthomorphic projection of the spheroid III. Vol. IX, No. 64 (April, 1947), 52–70.
18. The orthomorphic projection of the spheroid IV. Vol. IX, No. 65 (July, 1947), 112–123.
19. The orthomorphic projection of the spheroid V. Vol. IX, No. 66 (October, 1947), 157–166.
20. Professional organization. Vol. IX, No. 67 (January, 1948), 195–205.
21. Survey for colonial development. Vol. X, No. 77 (July, 1950), 290–301.
22. Jottings: battle song of the far flung surveyors. Vol. XI, No. 85 (July, 1952), 326–327.

The Geographical Journal

The East African Arc of Meridian. Vol. 84, No. 3 (September, 1934), 224–235.

Journal of the Institution of Civil Engineers

Surveying from air photographs. Vol. 1 (November-December, 1935 and January 1936), 140–147.

Geographical Magazine

1. Tales of a surveyor I: Ndege Ya Asali. Vol. **25**, No. 4 (August, 1952), 198–200.
2. Tales of a surveyor II: light of the world. Vol. **25**, No. 5 (September, 1952), 248–250.
3. Tales of a surveyor III: the piper's lament, Vol. **25**, No. 6 (October, 1952), 307–309.
4. Tales of a surveyor IV: possession and the law. Vol. **25**, No. 7 (November, 1952), 331–333.
5. Tales of a surveyor V: and the cock crew. Vol. **25**, No. 9 (January, 1953), 480–482.

Survey and Mapping

Rapid topographical surveys of new countries. Vol. XXV, No. 4 (December, 1965), 557–559.

Bollettino di Geodesia e Scienze Affini

1. Trends in mathematical geodesy. XXIV 4 (Ottobre-Novembre-Dicembre, 1965):607–622.
2. Orientamenti nella geodesia matematica. XXIV, 4 (Ottobre-Novembre-Dicembre, 1965):623–638 (an Italian translation of preceding paper).

Bulletin Géodésique

1. Geodetic applications of conformal transformations. 80 (Juin, 1966): 123–140. [See Errata in 81 (Septembre, 1966):287].
2. Triply orthogonal coordinate systems. 81 (Septembre, 1966):195–222.

3. Note by the writer of the paper on "triply orthogonal coordinate systems" 81 (Septembre, 1966):223–224.
4. (with F Morrison) First integrals of the equations of satellite motion. 91 (Mars, 1969):41–45.

Book

Mathematical geodesy. U.S. Department of Commerce, Washington, D.C. (1949), 416 pp.

Editorial Commentary

The above Bibliography is an attempt to compile a list of the books and papers written by Martin Hotine. It makes no pretense at completeness, and is intended to highlight Hotine's activities and interests primarily *before* he became seriously involved in differential geodesy. We know of no previous effort to collect such a bibliography, and if Hotine made up a compilation for his own personal use or reference, it has not been found. Probably during his 40 years' service for the British government he contributed other papers and reports which appeared in various official documents, both unter his name and anonymously, especially during the war years. It would be virtually impossible now to locate such material, but we are confident that none of this material is likely to be relevant to his subsequent work in differential geodesy.

In addition to the listed items, Hotine also wrote about a dozen letters which appeared in *The Empire Survey Review* from 1937–1951. We have omitted these items from the Bibliography, since usually they referred to other letters and papers (not necessarily his own) which appeared in this journal, and, in retrospect, they are of limited interest.

The Bibliography was compiled by John Nolton based on a xerox copy of a card index of the Hotine items from the library of the Directory of Overseas Surveys furnished to Charles Whitten (in preparation for his memorial paper), and a list of Hotine papers in *The Empire Survey Review* provided by Herbert W. Stoughton. It is a pleasure for us to acknowledge the valuable assistance provided by these individuals. A special note of thanks is also due to Grace Sollers of the Technical Information Services of the National Geodetic Survey (Rockville, MD) for her help in obtaining copies of various Hotine papers. The collection of geodetic papers in her care is unique, and her skill in locating an obscure reference is matched only by her willingness to help a geodesist in need.

Finally, my thanks to John Nolton for compiling the final bibliography. His help has been invaluable.

Hotine's *Mathematical Geodesy*[1]

Bernard H. Chovitz

1. Martin Hotine's *Mathematical Geodesy* will be published by the Superintendent of Documents, U.S. Government Printing Office, Washington, D.C. in October 1969.

This symposium has been dedicated to Martin Hotine. In turn, one might almost say that his book is dedicated to this series of symposia. Quoting from the Preface, "The author's main source of inspiration in the subject of this book has been Professor Antonio Marussi of the University of Trieste". But I am sure that Professor Marussi and all of you will agree as soon as you have the opportunity to become acquainted with this book, that it is the finest fruit of these Symposia. Each lends lustre to the other.

It is my intention to describe this work, and then to concentrate on specific instances in the volume which exemplify Hotine's genius. If Martin Hotine were with us now, I would have to extend compliments guardedly in deference to his personal sensibilities. But any words of praise that I put forth today are not meant to be lavish in the sense of speaking conventional phrases over the departed; they are expressions of my evaluation of Hotine's originality and his impact on the geodetic community.

Let me first make some general statements about the book, which I hope will serve to whet your appetite for it. It is comprehensive in that it covers everything substantive in theoretical geodesy today. However, it is completely unlike any other book on geodesy past or present, in English or any other language. Those of you who are familiar with Hotine's *A Primer of Non-Classical Geodesy* should understand what I mean. All of the basic principles of geodesy are completely derived directly from general mathematical considerations. In fact, one of Hotine's principal aims in writing this book was to put geodesy on a firm mathematical and physical foundation. Only in that way, he believed, could geodesy continue forward as a science in the spirit of previous great geodesists like Cassini, Gauss, Stokes and Helmert; otherwise it would deteriorate into technology, providing a means of sustenance for many able minds, but without the original spark that pure science provides. Hotine's book contains nothing concerning instrumentation or field procedure or, in the same sense, computational techniques or methods for the actual reduction of data. The title might well be "*Pure Geodesy*", although that would be much more subject to misinterpretation than its actual name. Indeed, I proposed the title "*Principia Geodesia*" to

[1] Reprinted with minor revisions from Proceedings of the IV Symposium on Mathematical Geodesy (Trieste, 1969), 11–24.

Hotine, half seriously because I thought that it was worthy of the name in the same sense that Newton's and Whitehead and Russell's masterpieces bore to physics and mathematics, and half in jest because I knew Hotine would never allow it.

Let me reiterate again the two points I have tried to make about the book. First, it is comprehensive in the broad sense, that is, in its scope of subject matter. A glance at the table of contents will verify this. Second, it is comprehensive in the deep sense, that is, these subjects are treated in depth; there are no equivocal derivations, and no concession is made anywhere, as is done so often (and in fact almost universally in other geodesy source books) to defer a proof to more advanced (and usually much more obscure) texts. Proofs are complete, direct, and without compromise.

If all this is so, how, you may ask, can so much be contained in a single volume of about 400 pages, even if the subject matter is restricted to what I have termed above "pure geodesy?" In the answer lies the essential strength of the book and the real inspiration of Martin Hotine. The book is developed from a basic single theme and in the most rigorous and economical fashion. The theme is that all geodetic measurements and concepts can be expressed as geometric properties of a three-dimensional manifold. The fashion is the employment from the beginning and throughout of the methods and notation of the tensor calculus.

Martin Hotine, whose sense of humour was well known to all, would anticipate now the immediate rejoinder of some wit "Well, of course, anything, including the collected works of Euler, could be written in 400 pages if converted into the tensor calculus". There is a measure of truth in this. However, this does not imply that if the book were translated into more elementary notation, the equivalent material would re-emerge in, say, 800 pages. The tensor calculus is not merely a notation. I learned this long ago myself from Hotine's old friend, Antonio Marussi. It may be possible to develop equivalent methods (like that of differential forms) which are as good or better; any such statements have yet to be proved for geodetic application. But it is not possible to obtain the same richness and elegance and completeness of results by methods which do not generalize coordinate systems and spatial curvature properties.

These, then, were the basic aims of Hotine: to liberate geodesy from its traditional bifurcation into horizontal and vertical, and to provide adequate tools for advancing geodesy as a respected science. Let us now proceed to particulars of the book itself, in order to see how these aims are fulfilled.

2. The word "tensor" has a forbidding connotation to many people, mainly because of unfamiliarity with its notational aspects. In this respect it is no better or worse than any language; facility comes with practice and application. Part I, the first 66 pages of the book, is devoted to an exposition of the basic principles of the tensor calculus and its application to the differential geometry of 2- and 3-space. On the one hand, Hotine does not pretend that this portion will suffice to make a novice self-sufficient in the tensor calculus; he cannot possibly contain the expository material of an entire textbook in 66 pages. On the other hand, he has managed ingeniously to insert original results, which belong in this section because of their general and abstract nature, as, for example, in his comprehensive discussion of conformal transformations of 3-space. The purpose, then, in

preparing this part was, first, to make the book self-contained, so that the person familiar with tensor calculus would have sufficient material with which to refresh his memory, and all the basic theorems needed for further reference; and, second, to present new or unconventional results not found in standard texts, which bear on the further geodetic development of the subject.

Understanding will not come easily. This book is not an elementary text. Those with no experience in the tensor calculus, who attempt to learn the discipline ab initio here, will find it rough, although not insuperable, going. In the Preface, Hotine himself recommends specific introductory texts for the beginner. If one responds "But I am now too old to learn tensor calculus," Hotine would answer unequivocally "then this book is not for you." He realized, and with full sympathy and understanding, that the older generation is usually well-set in its ways. The book was thus written with the next generation of geodesists in mind. If there was ever a book on geodesy that could be said to bubble with youthful vigour, it is this one. A not inapt parallel here would be Verdi's *Falstaff*.

The subject matter is intrinsically not as difficult as more general texts covering differential geometry, because Euclidean 3-space usually suffices as the arena of action. Since this space has zero Riemannian curvature, many of the general principles of Riemannian or non-Riemannian space are thereby simplified. (But one must remember that the 2-surfaces embedded in 3-space are usually not Euclidean, i.e. plane; otherwise geodesy would become too trivial an exercise). However, in Part I, Hotine makes occasional excursions into curved 3-space, notably in the chapter on conformal transformations. This, and the succeeding chapter on spherical representation, are probably the sections in Part I of most interest to the expert. Neither subject, to my knowledge, is documented to this extent in the literature, although, of course, the basic principles of conformal transformation can be found in Levi-Civita, and Marussi has done extensive research on the subject. Spherical representation – the mapping of a given surface onto a unit sphere – goes back to Gauss, and is discussed in many general texts. However, Hotine, in particular, sets up a train of theorems which prove valuable in applications later in the book. Indeed, the motivation for many apparently broad and abstract derivations is that they are subsequently put to good use in physical situations.

Part II covers about 70 pages. We have not arrived at geodesy in the accepted sense yet, but we are moving in that direction. Whereas Part I can be considered as a set of general instructions for building some useful instruments, in Part II such instruments are actually put together. (Then in Part III, which occupies the bulk of the book, they will be properly applied). The instruments I am referring to are completely delineated specific coordinate systems by which Euclidean 3-space can be described adequately for subsequent purpose. Hotine begins by constructing a specific, but very general, coordinate system, generated in the following fashion. Consider a scalar function of position, N, in 3-space, limited only by the usual continuity and differentiability considerations. The surface $N = \text{constant}$ can be described by two coordinates, say ω and ϕ, defined by the gradient of N, which are subject to no further restrictions than N. Families of $N = \text{constant}$, $\omega = \text{constant}$, and $\phi = \text{constant}$ surfaces thus provide a unique representation of any point in space as the intersection of three surfaces, one from each family. One might feel that such a specification is so general that little of use can be derived. It is true that just about any coordinate system one considers is subsumed under this description. For example, if we specify all the coordinate surfaces to be planes, we get Cartesian (although not necessarily rectangular) coordinates. But Hotine obtains a rich variety of results with definite geodetic application without further restriction of this (ω, ϕ, N) coordinate system. Such familiar concepts as azimuth, zenith distance, parallel, meridian, vertical, the various curvatures of arcs and surfaces, are precisely defined and specific formulae derived. Once such general results are available, particularization to specific coordinates like classical geodetic coordinates (geodetic latitude, longitude, and height) are straightforward, and in fact trivial.

Hotine's general (ω, ϕ, N) system should not be considered, however, as a mathematical playground in which he could display his skills at abstract games before settling down to more mundane tasks. If we consider the actual world in which we take our measurements, we must realize that to tie our mathematical models to reality, our coordinates should directly relate to our measured quantities. Astronomical latitude and longitude and spirit-levelling are examples. Such measurements can be taken as general (ω, ϕ, N) coordinates, but can hardly be specified further without intentional simplification. For example, there is no analytic general formula for the equation of a plumb line which expresses its true curvature in the entire region of interest. But Hotine's (ω, ϕ, N) system is broad enough to encompass this set, and all the results for the (ω, ϕ, N) system hold for the physically measured set. The basic ideas here were first put forth by Marussi, who derived analytic relations for the aforementioned physical set. What Hotine has done is to demonstrate how much can be derived with a bare minimum of particular restrictions.

In order to accomplish his aims, Hotine takes advantage of the fact that we are interested only in Euclidean 3-space. Thus a reference rectangular Cartesian system may be admitted to which the (ω, ϕ, N) system can be related. If this reference system is physically oriented, with, say, one axis parallel to the Earth's axis of rotation, then the subsequent definition of meridian, for example, becomes physically meaningful. But it is important to realize that the specification of a physically oriented Cartesian reference system at this stage is not a restriction; it simply provides a physical reference frame as a model for the results.

Continuing in Part II, Hotine now proceeds to provide more regular systems with which to work. Hotine terms a curve defined by the intersection of two surfaces $(\omega = C_1, \phi = C_2)$ an isozenithal line, that is, one on which only N varies. If all such lines are restricted to be normal to the $N = \text{constant}$ surfaces, then the resulting (ω, ϕ, N) system is called *normal*. The next step in simplification is to make the isozenithal lines straight. Following this, the ω, ϕ coordinates lines are set orthogonal to each other. In each of these three successive systems a series of formulae is generated which specify in turn more simply the geometric properties of the curves and surfaces relating to that particular coordinate system. If, in the last system mentioned, the $N = 0$ surface is taken to be an ellipsoid of revolution, we at last arrive at the familiar set of geodetic reference coordinates. Finally, in this part, Hotine takes up a topic which is essential for the geodetic applications:

means of transforming between different (ω, ϕ, N) systems. General conditions of orientation including a complete and rigorous discussion of the Laplace condition are covered here.

These are the basic essentials of Part II. However, Hotine has discussed other topics, which in a sense are digressions from the main theme, but are interesting in themselves. The application of generalized spherical representation to (ω, ϕ, N) coordinates occupies a chapter, as does the problem of variation of properties along the isozenithals. I call these disgressions because they emphasize projection onto two-dimensional surfaces, which runs counter to Hotine's general insistence on three-dimensional methods. Also, there is a chapter which explores the problem of constructing specific triply orthogonal systems. This is the final form of a paper presented by Hotine at the previous symposium at Turin in 1965.

Part III is the heart of the book – 200 pages which apply the machinery of Parts I and II to the field of geodesy. The subject matter of this part is more conventional than that of the preceding parts (although its treatment may not be!).

If the most general (ω, ϕ, N) system of Part II is restricted by making the N coordinate satisfy the Poisson equation of potential theory, then a unique representation of the Newtonian gravitational field is obtained. In Chapter 20, Hotine proves that this simple condition is necessary and sufficient for this purpose. We immediately begin to appreciate the value of the general resuls of Part II. They can be directly applied now with an insignificant amount of effort to obtain all the properties of the Newtonian field. The many-faceted aspects of Hotine's genius shine brightly here, as he discusses clearly and succinctly the relationship between the theoretical geometric properties of the field and means of actually measuring them by various instruments, especially the Eötvös torsion balance.

Chapter 21, on the presentation of the potential in spherical harmonics, is practically a treatise in itself. Although Hotine relies heavily on results from classical sources like Hobson, there is an extraordinary amount of material here which is original, or not readily available elsewhere. This chapter and the succeeding one on development into spheroidal harmonics comprise truly a source book on the subject. In addition to the usual exposition of the potential in terms of mass-functions and spherical harmonics, there is an equally complete development in terms of inertia tensors. This provides quick and elegant proofs of many results, such as MacCullagh's formula. Expansions for the internal as well as the external potential are developed. But since the latter is of most interest to geodesists, Hotine concentrates on this topic, considering origins both within and outside matter, and questions of convergence. Hobson's form of the external potential, using a modified cylindrical-type coordinate system, and Maxwell's theory of poles are covered. Gravity is represented in spherical harmonics by noting that although gravity is not a harmonic function, the derivatives of the potential with respect to Cartesian coordinates are; thus, for example, gravity times the sine of the astronomical latitude can be expanded in spherical harmonics. Hotine goes one step further and, by obtaining the second Cartesian derivatives of the potential, derives and exhibits rather involved formulae for the curvatures of the field.

Chapter 22 begins by developing, based on the formulae of Part II, the geometric properties of a spheroidal coordinate system: longitude, parametric latitude, and an eccentricity function (which is constant for a given ellipsoid). The standard solution for the potential in terms of Legendre functions of the first and second kinds is derived; the sufficient condition of convergence at the ellipsoidal surface depending on the ellipsoidal flattening is thoroughly discussed. Finally, the relationship between the spherical and spheroidal mass coefficients and the inertia tensor are explicitly exhibited, something I believe no other book on geodesy contains to this degree of completeness.

Chapter 23, the standard gravity field, handles the developments of Pizzetti and Somigliana by deriving expressions for the components of potential and gravity in terms of the geodetic coordinate system of Part II, the base ellipsoid of which is specified to be an equipotential surface. The development is, in my opinion, much more elegant and comprehensive than corresponding recent derivations to higher orders of the flattening by Lambert and Cook. Hotine's original derivation, incidentally, antedates these others; it can be found in his paper *Geodetic Coordinate Systems* presented at the I.A.G. assembly in Toronto in 1957. Also, the curvatures of the standard field are easily obtained from the general formulae of Chapter 21.

In the next four chapters the theme shifts from physical to geometric geodesy. Atmospheric refraction, contained in Chapter 24, was a topic of special interest to Hotine. His last oral presentation at the 1968 I.A.G. assembly in Lucerne was on a new formula for propagation time in a refracting medium. The treatment in this book is notable for its completeness and rigour, beginning from Fermat's principle. To explain his emphasis on this subject I quote: "Atmospheric refraction is particularly important in the three-dimensional methods used throughout this book, although no method of reducing the observations can overcome uncertainty in the refraction; three-dimensional methods are no better and no worse in this respect than any other. Accordingly, we shall treat the subject fully . . . ." After the theoretical development, involving analytic models for the ray and the refracting medium, all the explicit empirical formulae for index of refraction, curvature, and lapse rate are dealt with. Current formulae for astronomical refraction, which have become extremely important now for geodetic satellite work, are also discussed.

In the next chapter we enter the classic heart of geodesy – the adjustment of triangulation – and this is the area where Hotine's three-dimensional methods have been most publicized. It surprised and distressed Hotine that these methods remained misunderstood by many otherwise reputable geodesists. Because the measurement of low vertical angles is vitiated by atmospheric refraction, classical methods have been devised so as to bypass such data. In presenting general three-dimensional methods, Hotine did not compound this error. Although low vertical angles, if used, are very weak, it is no more accurate to ignore them, as is done in classical work. However, it is true that omitting them may be less burdensome. In these days of electronic computing, the extra computation involved in three-dimensional methods is slight compared to the benefit of generality and accuracy gained. And, of course, the question becomes academic in the case of satellite geodesy, where three-dimensional methods really come into their own. The point to be made is that the classical method of horizontal adjustment on a curved surface, treating the vertical dimension as a separate problem, is just a special case of the general three-dimensional method. The former may be simpler in certain cases, but it can never be more accurate, and for modern global geodesy it becomes inadequate.

Chapter 25 begins with the three-dimensional formulation and solution of the so-called direct and inverse problems of geodesy, that is, determining the relationship between two points, given either the coordinates of one and the distance and direction to the other, or the coordinates of both. Since the line joining the two points is a straight line in space, it is not hard to see that the solutions are simple and straightforward. It is a trivial application of the properties of the coordinate systems developed in Part II. A less trivial application arises next in discussing the difference in potential along a line, which is directly related to the determination of heights by levelling. The main problem is relating the general (ω, ϕ, N) system, which might be called the "astronomical" coordinate system, to the mathematically more regular geodetic coordinate system. Typically, Hotine devotes a large measure of this chapter to a little-used method for this purpose – that of employing the Eötvös torsion balance. A rigorous derivation is given and the practical advantages and limitations of the method are discussed.

I would like now to interpolate some remarks concerning Hotine's didactic aims, because they will apply to the remainder of the book. These chapters cover those sections of geodesy which are undergoing the most intense research and application today. It was Hotine's basic purpose to show that his methods provided a sound underpinning for these subject areas, and to open up vistas for further research. In this context, the actual solution of problems, involving standard mathematical techniques, would be out of place. Hotine believed that, in exposition, understanding precisely what we are trying to do and solve is more important than the solution itself. Thus his method is to begin with the most basic principles, either from elementary physical laws or the body of theorems developed in Parts I and II, and to proceed to develop carefully and in detail the exact formulation of the problem. The end result is usually a set of observation (or condition) equations. The formation and solution of normal equations are not dealt with; as Hotine remarks: "These matters are not peculiar to geodesy and are best studied in the standard literature".

In the internal adjustment of networks, the subject of Chapter 26, the basic method is that of variation of coordinates. This follows naturally from Hotine's development. The general coordinates of a spatial triangle, whose sides are straight lines, are specified. Variations in position or length are now easily formulated in whatever coordinate system desired. Specifically, using geodetic coordinates, the complete three-dimensional observation equations in terms of the observed angles (azimuth and zenith distance), observed lengths, and spirit levelling are derived. Observational problems which bear on the accuracy of the solution are not ignored. Corresponding equations in Cartesian coordinates are also presented. The equations thus formulated up to now apply principally to intervisible ground stations, measuring direction, distance and height directly between two points. The great trend in network adjustment over the past 20 years has been to expand geodetic horizons to long lines undreamed of before. It is only natural that Hotine, with his characteristic youthful outlook, should concentrate on such methods. The first step is flare triangulation; next, stellar triangulation which refers the flare to a special (i.e. the stellar) coordinate system; and then satellite triangulation in which the flare becomes a satellite. In discussing the case in which the observations are directions to a satellite, Hotine follows the photogrammetric procedure currently employed at the U.S. Coast & Geodetic Survey. I believe that this exposition is the clearest elementary one of this method available today. But the explanation is general enough to include an understanding of all variations of this method. Hotine next covers the case of distance being the measured quantity, and then the techniques employing lunar observations. In each case the specific observation equations for the method are carefully derived and precisely stated. Finally, equations for systems, such as hiran, which employ line-crossing techniques, are similarly set up.

It is no exaggeration to state that Chapters 25 and 26 provide a Bible – the ultimate sourcebook – for three-dimensional network adjustment techniques. It would be inconceivable for anyone who wishes to work in this field not to have this reference close at hand.

A short Chapter 27 on external adjustment of networks follows. By this is meant the placement of a local network within a world-wide system. Problems like change of shape, orientation, or position of the reference system, which have furiously occupied the energies of geodesists in the past, become almost trivial by Hotine's methods, and are dealt with completely in the space of about six pages. Hotine remarks near the end of this chapter: "In modern language, the old problem of determining a 'figure of the Earth' becomes the problem of finding a geodetic coordinate system which best fits the astronomic system". This statement succinctly exhibits the nub of his method.

The longest chapter of the book, and probably the one on which Hotine laboured hardest, is Chapter 28, on dynamic satellite geodesy. The youthful vigour to which I have already often referred is evident again here. Consider a person almost 60 years old when the first artificial satellite was launched, with no professional background in dynamical astronomy, and who, I personally know, did not become really interested in the subject until 1964. In a very short space of time he turned out a treatise which, first of all, carried out successfully the didactic aims I referred to previously, second, in no sense duplicates other texts as it consistently follows the theme and notational methods of the book as a whole, and third, actually, in my opinion, surpasses other books on the subject in its clear and straightforward treatment of some important but difficult topics. To be specific in this last point, let me mention the exposition of the integration of the canonical equation of motion employing the Hamilton-Jacobi method. One can compare this (paragraphs 103–114) with comparable discussions in Goldstein's *Classical Mechanics* or Plummer's *Dynamical Astronomy*, and then draw one's conclusion on elucidative superiority.

As in his practice, Hotine is very thorough in discussing the basic fundamentals of the subject. Such topics as the meaning of an inertial frame, the formulation of Newton's equations of motion in general coordinate systems and the relation between fixed and moving axes are handled with care. In developing the geometry and dynamics associated with the Kepler ellipse, he gives unusually complete treatment to the parameters α and β, quantities referred to the instantaneous position of the satellite which are analogous to azimuth and zenith distance, the latter being measured in the orbital plane, and the former in a plane perpendicular to the orbital plane. By this means, an astonishing amount of material from Part II can be applied. The discussion of perturbed orbits employs the standard method of variation of parameters, but it is refreshing to see it all translated into the general notation of the tensor calculus. Some might smile at this last remark. However, if one has noticed the difference between Danby's book on celestial mechanics, which uses vector notation consistently, to other standard works, then one can realize how much more is the distinction in passing to general tensor methods. I do not claim it is better, just refreshingly different. In essence, what Hotine has done is to derive generalized versions of the Gauss and Lagrange planetary equations, using elementary geometric methods. If one is concerned that tensor methods complicate matters, let me reassure him that Lagrange brackets never arise. Which would you rather undergo? The flavour of Hotine's treatment is aptly evident in a section which derives the curvature and torsion of the perturbed orbit.

Next the integration of the differential equations of motion is given similar treatment: complete, but novel. The first integral of energy is expounded in its general form, taking into account whether the coordinate system is inertial or rotating. Hotine does not flinch from the task of deriving and computing second-order perturbations. Other topics covered are the by-now classic technique of Kaula for integration of the Lagrange planetary equations to first order, problems of resonance, the aforementioned application of Hamilton-Jacobi theory to the canonical equation, the Vinti potential and von Zeipel's transformation.

There remain discussions of the various physical causes for orbital perturbations: drag, radiation pressure, etc. and a brief but explicit outline for setting up observation equations according to the different methods of measurement. Finally, the chapter is concluded in distinctive Hotine style by considering families of orbits under the conformal transformations properties discussed in Chapter 10. From basic geometric considerations, the familiar physical laws can be freshly derived and perhaps some insight gained from this different approach.

The last two chapters of the book – on the integration of gravity anomalies – continue in form and spirit the preceding portions. I especially stress this because these last chapters were written in 1968 under difficult circumstances. Incidentally, these chapters were circulated to a number of distinguished members of the geodetic community for comments and revisions. Hotine has been generous in his credits in the Preface, most of which follow from comments on these particular sections; but I think it is fair to say that the chapters remains essentially the same as his original version – nothing of substance has been changed.

These chapters overlap to a great extent material found in the excellent volume of Heiskanen and Moritz on *Physical Geodesy.* Hotine's treatment, of course, is more brief – the two chapters occupy less than 40 pages – but I doubt if there is a better and more complete derivation of the fundamental theorems of physical geodesy from basic principles. After one reads this, one really understands Stokes' formula. Beginning with elementary facts about spherical harmonics, and developing these on a purely mathematical basis, Pizzetti's extension of Stokes' function is derived, of which the latter is a special case.

The development of the conception and application of the term "gravity anomaly" is shown to arise from comparison of three coordinate systems; a general (ω, ϕ, N) system, the system corresponding to the standard gravitational field of Chapter 23 in which one coordinate surface is a level ellipsoid, and the geodetic coordinate system. Hotine precisely distinguishes between the "gravity disturbance" defined as the difference between gravity computed at the same point in the first two systems, and the "gravity anomaly" defined as the difference between gravity at a point according to the first system, and gravity in the second system computed at another point at which the potentials and direction of the geodetic normal of the two systems correspond. From these exact definitions, all of the well-known formulae are derived by stating exactly what approximations are involved. There is also a discussion, probably on the controversial side, on the relative value of the gravity disturbance versus the gravity anomaly.

Next, Poisson's integral, providing a means of upward continuation of potential from a reference surface, is derived in full generality, and its use illustrated by a number of applications. A similar treatment is provided by Stokes' integral along with Vening-Meinesz's adaptation to deflections. The limitations involved in applying these formulae are carefully explained. The names most closely associated with modern developments based on Stokes' and Poisson's integrals are Bjerhammer and Moritz. I am sure they will agree that Hotine does full justice to their methods in this chapter. Finally, an introduction to the theory of density layers is provided by considering a simple layer on a spherical surface. This topic will be developed further in the final chapter.

In the last chapter, developments associated with Molodensky and his school are analyzed. Whereas in the previous chapter, the reference surface was no more complicated than an ellipsoid, we now pass to surfaces which are more physically realizable. We begin with the concept of an S-surface, which can be considered to be the actual surface on the Earth, or some smoother version, like de Graaf Hunter's Model Earth. The machinery Hotine has developed in Part II proves invaluable here. All the geometric properties of these surfaces in terms of geodetic coordinates are developed in tensor form. An interesting digression is the discussion of the generation of a family of surfaces by progressive deformation of the S-surface, according to a specific, but very general, law. Green's theorem is applied to derive the basic integral formula for the relation of the S-surface to the potential. Hotine discusses this derivation in more generality and detail than even Heiskanen and Moritz. He is concerned, for example, with the continuity considerations involved if S is taken so that there is matter both internal and external to it. On the other hand, one must turn to Heiskanen and Moritz in order to obtain details on the solution of this integral equation, although Hotine indicates and discusses the work of Molodensky and others. Finally, Hotine tackles the problem of single and double layers on a surface, this time generalized to any S-surface.

That is a rough glance at *Mathematical Geodesy*. What I have tried to do is not so much to inform you on what the book contains, or to analyze it critically, but to try to build up your interest in it. What I fear is that, although you all undoubtedly have the greatest respect for Hotine and his work, you may be in such awe of the abstract-type approach that you may treat the book like a beautiful virgin – fascinating but untouchable. This would be a pity. The rewards to be gained are analogous. It is for this reason that I have interpolated deliberately controversial remarks. If you are curious enough to see if they have any substance, you must go to the book. Then so much the better.

3. What I have talked about up to now has been mostly generalities. I have given a narrative outline of the book and have heaped compliments on it here and there. For those of you who were familiar with Martin Hotine and his work, such praise, I am sure, did not surprise you. But actions speak louder than words, and I would now like to present two specific examples of Hotine's thought and techniques. It is not practical here to go into the development of a complicated or obscure argument. The examples I have chosen were selected primarily because of their simplicity (they can be laid out easily without recourse to tensor notation) and because of their familiarity; they should already be well-known to you. What I hope to convey is a measure of the flavour of his work, which bore unmistakeably his own personal stamp.

(1) The first example concerns the specification of the standard field, that is, the gravitational field at the surface of an ellipsoid of revolution which encloses all matter and is level.

Starting with Pizzetti's and Somigliana's treatment, a closed form for the gravitational potential V is obtained:

$$V = V_0 Q_0(iu) + A_2 Q_2(iu) P_2(\cos U) \ ,$$

where U, u are ellipsoidal coordinates. Transforming to Cartesian coordinates and computing V along the z-axis yields

$$V_2 = A_0 Q_0(iz/c) + A_2 Q_2(iz/c) \ ,$$

where c is an ellipsoidal parameter.

Expanding the Legendre polynomials above results in an expression in terms of $1/z^{2n-1}$; to find the external attraction off the axis of symmetry all that is needed is to replace $1/z^{2n-1}$ by $(1/r^{2n-1})P_{2n-2}(\sin\phi)$, and to substitute known expressions for A_0 and A_2. He thus obtains

$$V = \sum_{n=1}^{\infty} \left(\frac{-GM}{ae(2n-1)} + \frac{4(n-1)\omega^2 a^2}{3(2n-1)(2n+1)F(\alpha)} \right) (-1)^n \left(\frac{ae}{r}\right)^{2n-1} P_{2n-2}(\sin\phi) \tag{1}$$

for the gravitational potential V (*not* the gravity potential) in terms of spherical Earth-fixed coordinates r, ϕ, where e is the ellipsoidal eccentricity (not the flattening) and GM, a, and ω have their usual meaning. If we set $\alpha = \sin^{-1}(e)$, then $F(\alpha)$ is defined by

$$(1 + 3\cot^2\alpha)\alpha - 3\cot\alpha \ .$$

Equation (1) is compared directly with

$$V = \frac{GM}{r}\left[P_0 - \sum_{n=2}^{\infty} J_n \left(\frac{a}{r}\right)^n P_n(\sin\phi)\right] \tag{2}$$

in order to obtain the correspondence between the mass-functions J_n and the standard field given by Eq. (1).

For example $n = 1$ in Eq. (1) yields $\frac{GM}{r}P_0$, the same as the corresponding term in Eq. (2). For $n = 2$ in Eq. (1)

$$V_2 = \left(-\frac{GM}{3ae} + \frac{4\omega^2a^2}{45F(\alpha)}\right)\left(\frac{ae}{r}\right)^3 P_2(\sin\phi)\ ,$$

which corresponds to

$$V_2 = -\frac{GM}{r} J_2 \left(\frac{a}{r}\right)^2 P_2(\sin\phi)\ .$$

Thus the exact relation between J_2 and the standard field is

$$J_2 = \frac{e^2}{3} - \frac{4\omega^2a^3e^3}{45GMF(\alpha)}\ .$$

It then becomes simply a matter of series expansion of $F(\alpha)$ to obtain a polynomial type formula for J_2 in powers of e. The same procedure holds for any mass-function (the odd ones are, of course, zero in the standard field).

Incidentally, matters of priority here are rather touchy. I previously mentioned analogous developments by Cook and Lambert which date from 1959 and 1960 respectively. Of course, the original solution in closed form in ellipsoid coordinates is due to Pizzetti. A general formula equivalent to Hotine's was derived by Caputo in 1963 and presented by him at the Turin symposium in 1965. The exhibited relation between J_2 and the parameters of the standard field can be found in the book of Molodensky, Eremeev and Yurkina. I only suggest that you look at Hotine's derivation, of which I have given just a brief outline, in the paper *Geodetic Coordinate Systems* or in *Mathematical Geodesy* in order to see that it has its own distinctive flavor.

(2) The next example is one in which Hotine takes the mystery out of the meaning of the Laplace condition. In the Preface he noted that his interest in the subject "was aroused some years ago by an argument in print between two leading European geodesists on the correct application of Laplace azimuth adjustment, between points not located on the reference surface, which showed that neither geodesists had clearly defined what he meant by a geodetic azimuth at points in space".

In 3-space, the Laplace conditions are simply a direct elementary consequence of the relation between the precisely defined quantities latitude, longitude, zenith distance, and azimuth as specified in the standard polar triangle. Consider two points P and $\bar{P}$ on a unit sphere with spherical coordinates (ϕ,ω) and $(\bar{\phi},\bar{\omega})$ respectively. Let the azimuth and zenith distance to some zenith be (α,β) and $(\bar{\alpha},\bar{\beta})$ respectively.

Then it is straightforward by the formulae of spherical trigonometry to derive the following exact relation:

$$\begin{pmatrix} \sin\bar{\alpha}\sin\bar{\beta} \\ \cos\bar{\alpha}\sin\bar{\beta} \\ \cos\bar{\beta} \end{pmatrix} = ||a_{ij}|| \begin{pmatrix} \sin\alpha\sin\beta \\ \cos\alpha\sin\beta \\ \cos\beta \end{pmatrix} ,$$

where

$$\Delta\omega = \bar{\omega} - \omega$$

$$\alpha_{11} = \cos\omega$$

$$\alpha_{21} = \sin\bar{\phi}\sin\Delta\omega$$

$$\alpha_{31} = \cos\bar{\phi}\sin\Delta\omega$$

$$\alpha_{12} = \sin\phi\sin\Delta\omega$$

$$\alpha_{22} = \cos\phi\cos\bar{\phi} + \sin\phi\sin\bar{\phi}\cos\Delta\omega$$

$$\alpha_{32} = \cos\phi\sin\bar{\phi} - \sin\phi\cos\bar{\phi}\cos\Delta\omega$$

$$\alpha_{13} = -\cos\phi\sin\delta\omega$$

$$\alpha_{23} = \sin\phi\cos\bar{\phi} - \cos\phi\sin\bar{\phi}\cos\Delta\omega$$

$$\alpha_{33} = \sin\phi\sin\bar{\phi} + \cos\phi\cos\bar{\phi}\cos\Delta\omega \ .$$

There are only two independent relations in the three-part vector equation above, because each term is equivalent to the component of a unit vector.

Now suppose that the change from (ϕ, ω) to $(\bar{\phi}, \bar{\omega})$ is small, so that we can write

$$\delta\phi = \bar{\phi} - \phi \ , \quad \delta\omega = \Delta\omega \ .$$

Then the above exact relations can be approximated by

$$\delta\alpha = \sin\phi\,\delta\omega + \cot\beta(\sin\alpha\,\delta\phi - \cos\alpha\cos\phi\,\delta\omega)$$

$$\delta\beta = -\cos\phi\sin\alpha\,\delta\omega - \cos\alpha\,\delta\omega \ .$$

If the zenith distance, β, is close to 90°, then the equation for $\delta\alpha$ can be further approximated by

$$\delta\alpha = \sin\phi\,\delta\omega \ ,$$

the usual Laplace condition of classical geodesy. But this does not affect the equation for $\delta\beta$ at all, so its neglect must be considered as a defect in the classical system. (It is covered by repeated application of the $\delta\alpha$ equation). The whole operation and the approximations involved become apparent in the three-dimensional approach. A clear account of this problem may also be found in Heiskanen and Moritz's *Physical Geodesy*, where Hotine is credited with first explicitly drawing up the above equations.

4. I should like to conclude with a few words on the impact of *Mathematical Geodesy.* I cannot foretell to what extent this book will influence trends in geodetic research and operations, but if this influence turns out to be insignificant, to me this will reflect more on the current state of geodesy and geodesists than on Hotine. He has provided new, sharper, more efficient tools to work with. In order to take advantage of these tools, as in any other technology, one must learn how to use them. The road to achievement is always uphill. A university provides the best place for systematic acquisition of knowledge in any discipline. It is my hope that this book will be considered seriously as the basis for university courses and seminars by those of you who are in a position to make such policy. My agency, the U.S. National Oceanic and Atmospheric Administration (NOAA), is proud that this book could be printed under its patronage.

Editorial Commentary

This valuable preview of MG appeared on the eve of its publication and was written by one of Hotine's closest associates. Indeed, although it is not mentioned in the paper, the very existence of MG is due in no small measure to the untiring efforts of Bernard Chovitz. He served, together with Ivan Mueller, as one of the official reviewers of the manuscript, and after Hotine's death he was responsible for seeing it through the press.

In addition to this preview, two other papers are especially noteworthy and deserve special notice for anyone interested in MG and Hotine's work. The first of these is the comprehensive review (Thomas 1975) which was written by another of Hotine's colleagues. The other is Chovitz (1982), which assesses the impact of MG during the dozen years following its publication.

MG was issued by the U.S. Government Printing Office in two printings in 1970 and 1971. The only difference between the first and second printing was the insertion of an Errata page in the latter. Currently the book is out of print.

References

Chovitz BH (1982) The influence of Hotine's mathematical geodesy. Boll Geod Sci Aff anno XLI:57–64

Thomas PD (1971) Book review "Mathematical geodesy". Int Hydrogr Rev 48:209–219

Martin Hotine: Friend and Pro[1]

Charles A. Whitten

> *"I hould every man a debtor to his profession; from the which as men of course do seek to receive countenance and profit so ought they of duty to endeavour themselves by way of amends to be a help and ornament there unto". Bacon*

Martin Hotine used this introductory paragraph from Francis Bacon's pen to preface a paper on *Professional Organization* which he presented in London at the 1947 Conference of the Commonwealth Survey Officers. I selected the same paragraph for this memorial lecture because of the insight it provides of Hotine's philosophy on professionalism – whether it be surveyor, photogrammetrist, geodesist, mathematician, administrator or statesman. His talents were great and he shared his tremendous enthusiasm with friends and colleagues throughout his entire life wherever he might be and whatever the responsibility he faced. All of us who were privileged to be associated with him in any way can recall many exciting incidents – some of action, some of discussion, some of debate – each in a very special way identifying this outstanding man as a true professional.

In his associations with friends and coworkers, he requested that titles of rank or distinction not be used; he much preferred the use of surnames for professional colleagues and reserved the first name or familiar nickname for his closer friends. Therefore, whenever I mention any of his colleagues, I will endeavour to follow this informal yet fully respectful pattern.

A playwright might describe Hotine's life as a continuous and logically planned series of episodes. Whether Hotine was preparing the script, setting the stage, leading the action, or greeting the guests after the performance, his keen sense of timing and his natural appreciation of the dramatic have made the story of his life unique for the international surveying and mapping fellowship.

Prologue: Early Life – First World War – Education – Family

We may consider the first third of his life as a fitting prologue to a full career of service to his government plus all the other governments associated with Great Britain. Born June 17, 1898, he received his early education at Southend High School in Essex and was graduated from the Royal Military Academy at Woolwich on June 6, 1917, being commissioned in the Corps of Royal Engineers, the "top cadet of his batch". In the First World War, following additional training at the School of Military Engineering at Chatham, he was sent to India for service in the Queen Victoria's Own Bombay Sappers and Miners. His company moved up

[1] A memorial lecture presented at the Annual Convention of the American Society of Photogrammetry, Washington, D.C., March 1973. Published in Photogrammetric Engineering and Remote Sensing 39 (1973):821–830; and reproduced with permission from the American Society for Photogrammetry and Remote Sensing.

to the Northwest Frontier, where there was incessant guerilla fighting with German-trained hill tribes. Later, his company moved on into Persia at the invitation of the Persian Government to suppress some unruly tribesman. Along with the skirmishing, his company was accomplishing some of the engineering requirements – building roads and railways, placing telegraph lines, and, of course, supplying the supporting surveying and mapping.

His mental challenge during those years was the mastery of unmasterable languages. He learned to read, write, and speak Urdu, the language of the Moslems in India, and also Persian. In 1920, his company was sent into Mesopotamia, now Iraq, where sporadic fighting with Arab guerillas was still going on along the banks of the Euphrates. Peace was slow in coming, but, finally, in 1922 he was ordered back to England for post-war courses at Magdalen College, Cambridge University, and advanced studies at Chatham.

The prologue must also include his marriage to Kate Amelia Pearson on August 9, 1924. I remember sitting under the stars in St. Mark's Square in Venice with Kate and Martin along with other friends listening to them recall their courtship days when they were participating in amateur theatricals. They told us that in 1916 Martin's school had decided to put on Shakespeare's *Merchant of Venice.* Kate had been chosen from a neighbouring girls' school to play Portia and, of course, Martin was playing Bassanio. At the dress rehearsal, before a mixed audience from both schools, Martin went beyond the action indicated in the lines and planted a kiss on Kate's cheek, leaving a tell-tale mark from his, at that time, grease paint mustache, much to her embarrassment, so she said, but, needless to say, to the general amusement of the others.

They could not have selected a better time and place for sharing their memories with friends. Those of you who have known these two, realize that they could have been tremendously successful as professionals in the theatre. However, the story unfolded in the more traditional way: military service, advanced education, home, family, profession. They had been engaged before his long tour of duty in India and Persia, but waited until he had completed his post-war training and education to be married.

Even though the family is basic to the whole drama of life, I should introduce them at this point in the prologue. The Hotines are blessed with three daughters: Margaret has devoted her life to theatrical work, Janet has given her life in service to the church, and Bridget has made a home for a family of her own. Religious faith was a tremendous force in their lives. They were devout Christians and active in their support of the Catholic Church.

Act I: Air Survey Committee – Research Officer – Photogrammetry – Methods – Texts

In 1925 Hotine was appointed to the Air Survey Committee of the British War Office serving as its Research Officer. Here he used the opportunity to devise practical methods of using aerial photographs for topographic mapping. His mathematical ability, his experimental aptitude, his tremendous energy, and his special gift of writing were all combined in producing four professional papers, each of book length, and finally, in 1931, a textbook, *Surveying from Air*

Photographs. The analytical procedures and graphic methods he devised became the basis for all mapping, civilian and military, and as we shall se later, for economic development. The concepts became identified as the *Arundel* Method. Generally, the proper noun identifier relates to a person, but in this case Arundel is the place where the method was first used. Arundel is a small village in Southern England just a few miles north of the Channel. The photogrammetric target would have been Arundel Castle, seat of the Duke of Norfolk, Earl Vice-Marshal of England.

Although this first episode or Act I of the Hotine Drama might be titled Photogrammetry, it was during this period that the prick of his pen began to be noted in the United States. Bowie and Hayford, working through the International Association of Geodesy, had been able to persuade the countries of the world to adopt a new ellipsoid, and, for greater persuasion and geodetic diplomacy, it was identified as the International Ellipsoid. But Hotine was quick to note that Bowie continued to use Clarke's 1866 Spheroid for North America. Also, Hotine expressed his concern about the use of Laplace azimuths in triangulation networks. In the Empire Survey Review, he wrote of the fact that the error of observation for azimuths was considerably larger than that of horizontal directions in triangulation. Therefore, he reasoned that the use of azimuth observations could distort the more accurate triangulation. Quoting in part: "Personally, I know very little about the subject, but I have an open mind and I can appreciate an argument. If the real geodesists can be induced to fight in the same ring, we may get at the truth. Therefore, I shall endeavour to prove that both sides are wrong". I do not know if any others entered the ring, but some months later, Hotine continued his discussion in the Empire Survey Review. He had relaxed somewhat, perhaps realizing that azimuths observed in the latitudes of the United States did not have as large errors as observed in the latitudes of Great Britain.

He also wondered why those responsible for the theoretical applications in the United States seemed to ignore the classical method of defining a datum, in particular, the origin. Bowie has chosen to use the same geographic coordinates for Meade's Ranch, the origin of the then new 1927 Datum, as had been calculated from the original New England Datum and later perpetuated into the U. S. Standard Datum. Quoting in part again from Hotine: "The fact that the constants of the Clarke 1866 Spheroid are just right for this purpose, and that the regional attraction varies with closely approximate uniformity from east to west, is probably the most amazing stroke of luck in the history of Geodesy. The Americans are to be congratulated on this; but their sagacity may be doubted when they suggest in effect that the same chance holds everywhere else". Again later, when writing about azimuths: "After some thought on the question, my own opinion is that the Americans may conceivably be right in using Laplase azimuths, but I am quite sure they are wrong in suggesting that everyone else should use this system of adjustment".

Act II: East Africa – 30th Meridian – Geodetic Engineering – Techniques – Rigours

Hotine's keen knowledge of theoretical geodesy was soon to be applied in the next episode of his career. Late in 1931 he was assigned the task of establishing an arc of triangulation in East Africa along the 30th Meridian from latitude 10° South to 4° South. Now we see his engineering capabilities: the utilization of native manpower, the development of precise base-measuring techniques, the adoption of rigorous observational procedures, and, in fact, the specification of a Laplace azimuth in almost every quadrilateral of the chain of triangulation. The design or strength of the survey was critical. He avoided the use of long lines for the mere purpose of getting from one place to another quickly. I particularly like this sentence of his written in 1933: "The god of least squares, with his unreasoning hatred of small angles and complete ignorance of field conditions, is not a just god and would most likely over-favour the long and possibly inaccurate line".

One can re-read accounts of this African project in the Empire Survey Review, but I must use one more quotation to show the complexity of geodetic problems in considering different continents: "for good or evil, the Clarke 1880 Spheroid has come to stay in Africa and will not be replaced any more than the Clarke 1866 figure has been displaced by Hayford in America".

Extending an arc of triangulation from Rhodesia northward along the east side of Lake Tanganyika through trackless brush and jungle, infested with tsetse flies, was a task that only humans could accomplish. Back-packing food, blankets (for some of the points were at high altitudes), building supplies, surveying equipment, and, of course, rifles, for protection from lions or for acquiring fresh game, required a small army of native porters, sometimes as many as 250. Hotine had the full responsibility. He had one junior officer to assist him with the observing and a few non-commissioned officers to mark stations, post lights, and assist in the training of natives to do some of the work. The task was completed in 2 years, working through two wet seasons and living under canvas the whole time – really a remarkable achievement!

Act III: Great Britain – Retriangulation – Organization – Adjustment – National Grid

The technical aspects of his assignment in East Africa proved to be the basis for the next major role in his service to his country. In the development of urban England after World War I, especially in Northern London, the need for resurvey (as opposed to revision) had become apparent. Any attempt to patch up the existing network to serve as control for the necessary breakdown surveys merely emphasized the inadequacy of the basic framework. In 1935 a decision was made to observe an entirely new primary net and subsequently to reestablish the lower-order networks. Hotine was asked to undertake this project at a time when the resources of the Ordnance Survey had been, to use their words, "pruned to the irreducible minimum". The Great Depression of the 1930's covered the globe, with England affected as much as any country, yet Hotine was called upon to accomplish within a few years a task which, when done the first time, had taken half a century.

The details of how this retriangulation was planned, personnel selected and trained, procedures for reconnaissance and observing established, a special station mark pedestal designed, and the actual work accomplished were well described by Hotine in several issues of the Empire Survey Review. World War II interrupted the work before it was completed. I refer those of you who are interested in reading a vivid account of the total programme to a 1966 publication of the Ordnance Survey, *History of the Retriangulation of Great Britain*, compiled primarily by John Kelsey.

The whole story has, however, never been published. The tales that are told by old-timers, when they sometimes meet, do not become part of an official government publication; but they do become part of geodetic legendry. Fortunately for this occasion, I was provided with a bit of the record which must be passed on to you.

Hotine and the observing party had been bogged down by the weather for weeks on the summit of Ben Mac Dui, a grim Scottish mountain and the second highest in Great Britain. Hotine had gone down to the base camp at Braemar, a distance of some 15 miles, and sent one of his men back to the summit with a supply of the finest Scotch whiskey obtainable. Within a short while the weather cleared, the observations were completed, and a triumphant party staggered down the Luibeg Glen to their trucks parked near a lovely spot, the Linn of Dee. The happy triangulators decided that a more extended celebration was in order, so they went to a nearby hotel. Too bad, though, that it happened to be the Sabbath and the Free Church of Scotland Community objected. A Black Maria provided transportation to the local "jug" for the merrymakers.

Next morning, Hotine, having heard about the incident, hastily made his way to the Court, arriving just at the time when the observer, who had accepted the responsibility for all, was pleading guilty. Hotine asked the Court if the observer was being provided legal counsel. When the Court replied that the defendant was entitled to such, Hotine proposed that inasmuch as he was the defendant's Commanding Officer he was qualified in his military capacity to speak for him. Thereupon, Hotine painted an excellent picture of the hardships of the field men, he emphasized the consequences of delay to the whole operation if the observer should be detained by imprisonment, he stressed the possible loss of rank and even ignominious discharge. Hotine's eloquence was partially effective. The Court ordered a nominal fine of £ 20 to pay his fine. After completing the transaction, Hotine ordered the man to another nearby and difficult station to do penance and work off his "nominal fine".

A few years before Hotine had been assigned this major task of retriangulation, he had expressed some concern, in fact some criticism, of the methods Hayford, Bowie, and others had used in the United States in devising the 1927 Datum for North America. When Hotine was confronted with a similar task in his own country, we must note that he also recognized that expediency sometimes must have primary consideration. Because of the desire not to disturb the graticule of the existing large-scale maps, many at 1/2500, the new triangulation was adjusted to fit the scale and orientation of the old. Thus, Laplace azimuths and precise bases were used for after-the-fact studies or investigations. In a practical sense, the distortions to the new observations were negligible, generally less than one part in 100000, and for the standards of that era, fully acceptable.

In addition to holding overall scale and orientation, the formerly used spheroid of reference, Airy's Figure of the Earth, had to be retained. Airy's parameters had been defined in feet, so there was the complex task of deriving the proper conversion ratios for expressing lengths in meters. There is some satisfaction in noting that we in the U. S. are not the only surveying and mapping group that has been subjected to this numbers game in the effort to achieve uniformity. A unique feature of the new British network is that all coordinates are expressed in metric units on a National Grid based on a single zone of a transverse Mercator projection. The published geographic coordinates were derived from the National Grid coordinates.

Another point of interest to us is that Hotine had acquired some Bilby towers from U. S. He institute a training programme for the erection and dismantling of these towers on the grounds of the Ordnance Survey at Southampton – somewhat different from the on-the-job training we have followed. Because of the area-type network and unusual requirement for tower heights in some of the Eastern counties, Hotine encountered the problem of needing more towers. Rather than face time delays and the cost factors involved when purchasing more towers from the U. S., he solved the problem with the assistance of the Geodetic Survey of Denmark, who generously loaned him two towers of similar design.

Act IV: Dunkirk – East Africa – Greece – Military Survey – Loper-Hotine-Agreement

I mentioned earlier that World War II interrupted the programme of retriangulation. At the outbreak of that war, Hotine was assigned to General Headquarters as the Deputy Director of Survey in the British Expeditionary Force. For those who wish to follow his record from Dunkirk to East Africa to Greece and returning to the War Office to serve the rest of the war as Director of Military Surveys, I suggest reading *Maps and Surveys*, published by the British War Office in 1952. From that historical report you will sense that Hotine and his colleagues could have used 3M as a trademark: not for Minnesota Mining and Manufacturing Company with its various products, but rather for a symbol of military strength – Men, Munitions and Maps – the third element being as essential as the other two.

Hotine was with the British forces in Belgium and France when they were fighting their way back to Dunkirk. He saw the acute problems which could develop if the maps that did exist were not in the hands of the troops needing them. Years later he referred to the episode as "when many of us were seabathing at Dunkirk" and, no doubt, using his keen sense of humour, could reminisce with his colleagues who were with him at the time about many of the little events contributing to this major military miracle.

Hotine was next assigned to survey operations in East Africa and later sent to Greece with a small survey team as a part of a British Expeditionary Force assisting the Greek Army. This mission failed, and many of the personnel were taken prisoner; Hotine was among those who escaped.

He was then ordered back to London to the War Office to be director of Military Surveys. In 1939 he must have been frustrated at times, for in one letter home he wrote: "I'd love to run my own show instead of being an eternally vibrating second string", but now he drew from his experiences of that earlier phase of the war. The requirements for surveys and maps and, in particular, the problems associated with printing and distributing maps were forcefully presented to the General Staff. Hotine came to Washington in May 1942 to discuss the mapping situation with H. B. Loper and members of the Intelligence Branch of the Chief of Engineers, U. S. Army. As a result of this conference, they drew up an agreement, known as the Loper-Hotine Agreement, dealing with the division of responsibility for map production, the exchange of mapping and other survey data, and the selection of military map grids.

Soon afterward, Herb Milwit was ordered to England, to serve as the Chief of the Engineer Intelligence Division, U. S. Army for the European Theatre of Operations. Hotine and Milwit worked closely and effectively throughout the war. Hotine's earlier experiences in aerial mapping, geodetic surveying, and mathematics of projections provided the technical base that, coupled with his tremendous vitality and dynamic and aggressive leadership, gave outstanding and truly professional direction to the Military Survey Office. Recognizing that in earlier years Hotine had been somewhat critical of survey practices in the United States, his ready acceptance of the U. S. competence in mapping with all of the supporting techniques adds to his professional stature. This spirit of cooperation and good will between the various surveying and mapping groups of Great Britain and the United States continues today. The Hotine-Loper Agreement had set the stage for broader international agreements in programme which would follow World War II.

Act V: Overseas Surveys – Statesman – Economist – Mathematician

In 1946, after he had retired from the British Army, Hotine became Survey Advisor to their Secretary of State for the Colonies. He was the first director of the newly formed Colonial Surveys, now known as the Directorate of Overseas Surveys. During Hotine's pre-war assignment in East Africa, he sensed the need for such an organization, and his close contact during the war years with all British territories as well as many others strengthened his position. In this new endeavour, he was to be statesman and economist, using his energy and ability to assist in the development of the Commonwealth. He organized a staff and provided the leadership in applying photogrammetric, geodetic and mathematical techniques to surveying and mapping. Before he retired in 1963, his organization had mapped nearly two million square miles, mostly at a scale of 1/50000.

Even though his administrative responsibilities were great, he continued his mathematical research. He published a series of papers in the Empire Survey Review, giving special treatment to various projections of the spheroid. This work has become a basic reference for all later writings in the field of mathematics dealing with projections. His development of the oblique Mercator projection of the spheroid is particularly unique. Sometimes this projection has been named skew or diagonal. In Hotine's original work, where it was applied to Malaya and Borneo, it is called rectified skew orthomorphic. The same technique was used for the State Plane Coordinate System in Southeast Alaska. Fortunately, the user of the plane coordinates in that part of Alaska has only to know that x and y are in Zone I. More recently, Ralph Berry has proposed the use of these skew projections for each of the Great Lakes, simplifying the cartographic operations where geodetic, photogrammetric and hydrographic surveys must be combined.

Off Stage: Conferences of Commonwealth Survey Officers – Symposia and Assemblies of International Association of Geodesy

During that same time, Hotine was also taking an active part in international organizations, giving unselfishly of his time and talent in planning and conducting conferences, even joining in hearty and friendly debate either in formal session or in the privacy of after-session places. This was the Martin we remember as friend and colleague.

As organizer and leader of several of the Commonwealth Survey Officers Conferences, he set standards which have been followed by many other groups. He enforced discipline on those who prepared papers, he insisted on open discussion, and he maintained a spirit of good will among all participants. At the 1963 Conference, while Sam Gamble was presiding, J.N.C. Rogers from Australia presented a motion "that this Conference of Commonwealth Survey Officers place on record its profound appreciation of the work of Martin Hotine for the distinguished service he has given as President of this Conference and of the previous Conferences of the Commonwealth Survey Officers in Cambridge in 1955 and 1959, and for the stimulation, encouragement, and leadership he has given to surveyors during a long and very distinguished career". The United States has been very fortunate to have been invited to participate in these conferences, even though we withdrew from the Commonwealth almost 200 years ago.

It was through his participation in and contributions to the programmes of the International Association of Geodesy that many of us in the United States learned to know him, grasp some inspiration from him, and join in hearty fellowship with him. I first met him at Oslo in 1948 at the General Assembly of I.A.G. It was at this first post-World War II Assembly that Antonio Marussi presented his classical paper on the differential geometry of the potential field of the earth. This modern treatment of the science of geodesy was appealing to the mathematically attuned Hotine. These two, Hotine and Marussi, initiated a series of symposia on mathematical geodesy – the two of them co-planning with Marussi hosting small groups for intimate discussion at such places as San Georgio in Venice, Cortina d'Ampezzo and Turin, to be followed by others at Trieste and Florence honoring Hotine. Marussi and Hotine seemed to challenge and inspire each other, and in turn do the same for those who had the good fortune to have been participants in these symposia. To show Hotine's personal appreciation for this association, I quote from his book, *Mathematical Geodesy*: "The author's main source of inspiration in the subject of this book has been Professor Antonio Marussi of the University of Trieste, not only for the range and originality of his ideas, but also for continual advice and encouragement". Marussi held the same respect for Hotine. You can grasp the full sincerity of this friendship by reading for yourself Antonio's tribute to Martin reprinted from The Survey Review No. 152, April 1969.

> "I feel inadequate to write the biography of Martin Hotine because I met him late in my life. I know little of his early professional activity, which I am sure was intensive; but in the past two decades through our connection with the International Association of Geodesy I have come to know him as a man, and

as a man of science. On first meeting, I understood many of his thoughts and feelings. Later I studied his work. Then he honoured me with a friendship which I reciprocated with admiration and affection.

"Our friendship was born on the advanced frontier of the discipline dear to us, Geodesy. I remember our first meeting at the General Assembly of the I.U.G.G. in 1948. It was there that the intrinsic and three-dimensional geodesy began to develop, anticipating the oncoming space age which was then knocking at the door. When a report on this subject was read, Martin said that he understood only very little about it, but that it broke with crystallized tradition and that it must therefore be important".

"Later I came to understand that in this thought was all of Martin. He was attracted to the new ideas because they challenged the scholastic framework of the time. Such ideas were perfectly congenial to his non-conformist and politely rebellious temperament.

"I did not meet him again for 4 years. When we did meet it was at the Assembly of the Association in Brussels. In the interim he had mastered tensor calculus, which he later used with rare insight and elegance, finding in it the instrument which permitted him to materialize his intuition of the fundamental problems of modern geodesy, giving life and substance to the creations of his vivid fantasy.

"His first work on the metric properties of the Earth's gravitational field, and on three-dimensional systems of geodetic coordinates appeared at the Toronto Assembly in 1957. Here, through the initiative of President Baeschlin, was born the idea of dedicating a special symposium to three-dimensional geodesy which had already anticipated the use of electromagnetic methods for the measurement of distances in space, and was now anticipating the launching of the first artificial Earth satellites.

"I considered Martin's proposal that this symposium should take place in my country, Italy, to be a token of his great friendship. Italy was very happy to host this symposium at Venice in 1959, and offered Martin its chairmanship. Later, two other symposia were held in Italy, at Cortina in 1963 and in Turin in 1965. The fourth symposium will also be held in Italy at Trieste in 1969, and will be dedicated to his memory.

"At Venice Martin introduced his *Primer of Non-Classical Geodesy*, a work whose title is decorously provocative. It is written in a style which makes it difficult to know which to admire most: its rigour, its conciseness, or its elegance.

"He thought that in the same way as in classical geodesy a clearer vision of many problems could be reached by representing the surface of the ellipsoid on the plane, if three instead of two dimensions were involved one could similarly represent with advantage the three-dimensional space on another three-dimensional space, not necessarily Euclidean, in order to make clearer and more immediate some aspects of the Earth's gravitational field. As in classical geodesy, conformal representations appeared to be the most promising. These ideas were gradually brought to perfection by Martin, and presented by him at Turin in 1965 at the third symposium held on the occasion of the celebration of the 100th anniversary of the Italian Geodetic Commission. In 1967 at Lucerne, Martin introduced new ideas about the downward continuation of the terrestrial potential field. That was the last time we had the pleasure of enjoying his incomparable personality.

"We would have expected that Martin, whose background was in the sphere of practical activity and whose career was anchored in the most deeply rooted traditional habits, would have contributed to geodesy by following the most classical and orthodox trends. However, the truth is exactly the contrary, since he was particularly anti-scholastic in facing the problems of speculative geodesy. This attitude originated in his deep dissatisfaction with compromises by which classical geodesy, burdened by prejudices accumulated during centuries, presented its own problems, managing a way between the rigor of the theoretical approach and the empiricism involved in its practical applications. At the same time, this attitude came from his innate horror for everything that was not simple and rational and, therefore, aesthetically satisfying.

"On this point Martin was intransigent. He detested any compromise whose purpose was to avoid conceptual difficulties if even to the least extent it impaired the rigour of the logical process. He hated approximations, reductions, corrections, or badly defined, unfounded, accommodating hypotheses. He was an aesthete and a purist.

"We must be grateful to Martin for the impetuous vigour with which he always defended his viewpoints, even when they clashed with the deepest-rooted convictions. We must be grateful to him for his courage, since even science sometimes needs courageous men in order to progress.

"Perhaps it is too early to assess how much his work has influenced modern thought concerning our discipline. It is probably too early because we have not had the privilege of reading his book on *Mathematical Geodesy*, which synthesizes the 20 years of thought and creative work with which Martin concluded his earthly spell. This book represents the last and most dignified satisfaction of a man whose work and family were his only reasons for life.

"If I am permitted to express my opinion, perhaps prematurely, I believe his influence will not be found in any particular results of immediate application, but rather in his effectiveness in changing convictions which, because of their venerableness, were removed from critical attack; in clearing away difficulties which hampered the free progress of ideas; and in showing how to get rid of the heavy superstructure heaped up in the long history of our discipline. Martin's spiritual heritage is for this reason embedded in our subconscious by the ways he demonstrated to us for researching the truth through the simplicity of the origin and the elegance of rational thinking".

Of course Hotine's interest extended to all aspects of geodetic work. He naturally had a keen interest in the coordination and improvement of the geodetic networks of Europe. The geographic location of the British Isles and the separation from the continent meant that the geometric impact on the continental networks might not be great, but his contributions to the development of the overall specifications were significant. When he was the leader, he was a strict disciplinarian, as we well know, but I must tell you of an incident I will always remember. Back in 1962 we were meeting in Munich, working on the plans for the readjustment of the European triangulation networks. At the middle of one morning session, a break was taken – not the customary coffee break – but a beer break, Munich's best along with open-faced sandwiches. We had relaxed for at least a half hour when I, who happened to be chairing that particular session, suggested we resume our work. Martin boomed out in his sonorous tone, "Whitten, I thought Abraham Lincoln had freed the slaves!". We did continue our work in good spirit, properly relaxed physically and mentally.

Thus it was that Hotine, through his spirit and vitality, supported with his engineering and scientific ability, enlivened the symposia and assemblies of the Geodetic Association, always supporting and encouraging the officers of that association in their various activities.

Epilogue: U. S. Coast and Geodetic Survey – Research – *Mathematical Geodesy* – Return to Weybridge – Honours

The year was the one that Hotine's own government had established as the point in his life when he should begin to enjoy the rewards of retirement. In that year he was fittingly honoured by his colleagues and countrymen for his long and distinguished service to his country, but he did not seek "the rocking chair". He accepted the invitation of Arnold Karo, the Director of the Coast and Geodetic Survey, to come to Washington and join the ranks of the Survey as a research scientist. I have referred to Hotine's numerous contributions to geodesy. These papers had been in the form of mimeographed reports – somewhat in sequence, all logically interrelated – yet all on the distinctive legal-size paper which would not conform to the standard U. S. notebook, file case or bookshelf. Karo persuaded Hotine to combine, refine and extend his work and produce a hardback monograph that would fit in our bookshelves.

For the next 5 years, Hotine worked toward this goal of producing a text which, in his words, would be "an attempt to free geodesy from its centuries-long bondage in two dimensions." Of course, those same 5 years were those in which geodesists were entering a new era of experimentation and development through the use of artificial satellites. Hotine's interest in and contributions to these developments were just as keen as all of his thinking had been throughout his whole life. Wherever activities of this type were going on, Hotine's advice was sought, so there were excursions from his main thrust. There were even administrative excursions involving agency reorganization. The Hotines accepted with pleasure the physical move to Boulder. He and Kate, with their love of nature and hiking, could explore the trails in the foothills of the Rockies. I was told that the photographic collection they made of Colorado wild flowers is probably equal to that of any ever compiled by the best of American naturalists.

I always noticed that Hotine never hesitated to use a familiar quotation if it helped to emphasize a particular point of interest. I do not think he would have objected to my use of another quote from Francis Bacon: "Reading maketh a full man, conference a ready man, and writing an exact man". All of these phrases apply to the man we honour and, in particular, to his preparation of the manuscript for his text, *Mathematical Geodesy*. Hotine's plan, which he followed explicitly, was to use the methods and notation of tensor calculus for the derivation of theorems and formulae which apply to all of mathematical geodesy. He included geometrical and physical, terrestrial and spheroidal, and, in his words, "internal adjustments" and "external adjustments" of geodetic networks. I urge you to read two outstanding reviews of this text. One is by Bernard Chovitz published in 1970 by the Italian Geodetic Commission in the Proceedings of the Fourth Symposium on Mathematical Geodesy, Trieste, Italy, May 28–30, 1969. The other was written by Paul Thomas and published in the International Hydrographic Review, January 1971. They were close friends and associates of Martin, had worked with him on many sections during the development stages, and Chovitz, in particular, during the final editing and printing.

Early in 1968, while the manuscript was nearing completion, Martin underwent a serious surgical operation. His friends in Boulder told me that he did some of his most effective writing after he returned to his office from the hospital. His friends around the world were concerned and Joe Edge has permitted me to quote from a personally penned letter Martin sent him in July of 1968: "I progress slowly after the extensive revisions, as the tailors might say and do say. It is an up-hill job, but according to the Medicos I have to work at it. If they would only let me off eating and let me concentrate on drinking, I should feel fine, but unfortunately they will do neither".

Martin and Kate Hotine had made their plans to return to England in August of 1968. The manuscript had been sent to the printer and now only the tedious task of checking the galley proof remained. Soon after the Hotines had returned to their apartment in Weybridge in Surrey, Martin wrote in his usual keen manner to Bernie Chovitz: "We have been struggling to get the place straight and in effect start a new life. It would be easier to do this in America; we have forgotten where the ropes are here, even what to do in the absence of yellow pages".

His interest in his work and his love and consideration for others never diminished, but his physical energy and strength, with which he had been blessed in abundance, left him. He died on November 12, 1968. Those who knew him mourned his passing, but gave thanks for the privilege of having known him and having been associated with him. The encouragement and inspiration he gave to others cannot be taken from them.

Governments, national engineering and scientific societies, and other groups recognize the honoured individuals of high distinction. Martin Hotine, modest and unassuming, yet forceful and energetic through his brilliant personality, was an example of true professionalism. Numerous awards were given him at various times throughout his career and significant memorials established after his death.

He was awarded the O. B. E., Commander of the Order of the British Empire in 1945, and the C. M. G., Companion of St. Michael and St. George in 1949.

Also, the Royal Geographical Society awarded him its Founders' Medal in 1947 and in 1955 he was the first recipient of the President's Medal of the British Photogrammetric Society. The Gold Medal of the Institution of Royal Engineers was awarded to him in 1964.

The United States Army made him an officer of the Legion of Merit in 1947 and in 1969, the U. S. Department of Commerce conferred, posthumously, on him its highest honour award, the Gold Medal.

During the 1971 Conference of Commonwealth Survey Officers, a tree, a Norway Maple, was planted on the grounds of the Ordnance Survey at Southampton as a living memorial. Kate Hotine assisted in the ceremony and her daughters and grandchildren were also present. Gifts, which had been received from friends in 31 countries, have been used to create in his honour a Scholarship Fund for graduate study at University College in London. I believe that this method of encouraging and helping young students would have pleased him the most.

Editorial Commentary

This memorial lecture really requires no comment except to say that it is unique and unquestionably the most detailed and authoritative account of Martin Hotine that we will ever have. Its author had the good fortune to be a close personal friend of both Hotine and Marussi, and an eyewitness to their development of mathematical geodesy. Indeed, together with Admiral Karo, Whitten was instrumental in persuading Hotine to come to the United States in 1963, and to write up his ideas in book form.

List of Symbols

Lists of main symbols were given by Hotine for paper 3, p. 61; paper 4, p. 88; paper 5, pp. 128–129; and paper 6, p. 137. His usage of these symbols was essentially the same in all of these papers.

However, some changes of notation, and different choices of signs, occur in MG. A brief indication of the most important of these is given in our Editorial Commentary on paper 3, see pp. 62–63.

In MG, an Index of Symbols was given on pp. 347–351, together with an exhaustive Summary of Formulae on pp. 353–397.

Note that in Chovitz's preview article reprinted in this monograph, see pp. 159–172, he employed the notation Hotine used in MG.

Index of Names

Subject Index

Bonding and Charge Distribution in Polyoxometalates: A Bond Valence Approach

Volume Editor: D.M.P. Mingos

With contributions by
S. Fischer, D. Kurad,
J. Mehmke, K.H. Tytko

Springer

In references Structure and Bonding is abbreviated
Struct.Bond. and is cited as a journal.

Springer WWW home page: hTTP://www.springer.de

ISBN 978-3-540-64934-2 ISBN 978-3-540-68306-3 (eBook)
DOI 10.1007/978-3-540-68306-3

CIP Data applied for

Originally published by Springer-Verlag Berlin Heidelberg New York in 1999

Production editor: Christiane Messerschmidt, Rheinau
Typesetting: Scientific Publishing Services (P) Ltd, Madras
Cover: Medio V. Leins, Berlin
SPIN: 10651510 66/3020 – 5 4 3 2 1 0 – Printed on acid-free paper

Volume Editor

Prof. D. Michael P. Mingos

Department of Chemistry
Imperial College of Science, Technology and Medicine
South Kensington
London SW7 2AY, UK

Editorial Board

Preface

This volume of Structure and Bonding deals with the bond valence method, which came into use some twenty years ago, and its application to the interpretation of the structure and bonding in polyoxometalates of the Group V and VI transition metals and some of their oxides.

The bond valence method, which was developed primarily by I. D. Brown, gives valuable information on the bonding in compounds with a marked difference in the electronegativity of the bond partners. Typical applications involve oxygen atoms which constitute an OH group or H_2O molecule, the oxidation number of the M atoms (the positive bond partners), hydrogen bridge systems, etc. The direct application of this method by previous workers to obtain more ambitious information on polyoxometalates, however, yielded no satisfactory results. Reinvestigation of the methodology has revealed that some of the approximations which were introduced for a more convenient handling of the method are responsible for the failure of the method when applied to polyoxometalate chemistry. (Specifically these approximations involve use of uniform d_0 parameters in the bond length–bond valence function for different oxidation numbers of an element and use of a "universal" B parameter in the function for many elements pairs.) Generally speaking, the method requires *very* exact data (bond lengths) for the reference structures from which the bond length–bond valence parameters have to be derived; the rejection (or at least a critical evaluation) of *universal* parameters, applicable to many element pairs; and an *individual* fitting of the d_0 parameter to each structure determination for the compensation of inevitable (small) systematic bond lengths errors. The necessity for an individual fitting of the d_0 parameter and rejection of the use of uniform d_0 parameters for different oxidation numbers of an element are not contradictory! Under these conditions the method operates very well. The calculated standard deviations for the bond valences arise only from the errors in the bond lengths.

Consideration of the polyoxometalate structures (with given M–O frameworks) in the context of their formation and adequate and consequent application of 'classic' chemical concepts to them explain the distribution of the bond valences (and hence of the bond lengths) over the M–O bonds and of the ionic charge (and of additional charge generated by charge separation) over the O atoms. (The classic chemical concepts which may be applied include the development of Lewis-type structures with extension of the octet to the decet and dodecet rule for M^{V} and M^{VI}; the resonance concept and of resonance bond numbers; the concept of the coordinate bond; Brønsted's acid/base concept; the law of mass action; etc). The governing factors are the stabilization of the polymeric character of the polyoxo species by maximizing the inner bond valence and the increased basicity of the polyoxo species by the preferential acceptance of the negative charge by the terminal oxygen atoms. This results in an increased mutual consumption of the reaction partners according to Le Chatelier's principle. In the literature some of these features had been discussed in a contradictory manner.

Application of the bond valence model to the experimentally determined polyoxo structures confirms the above-stated results which were obtained by the utilization of classic chemical concepts to Lewis-type structures of the polyoxo species. A large number of structural details are now well understood, for example, the "trans influence", the extension of the coordination sphere of the central M atoms; the factors determining the basicity and the protonation behavior of the species; etc. The most remarkable result, however, is the finding that the description of many heteropolyanions as "an anion in an anion" cannot possibly be correct. The central "hetero parts" and the surrounding M–O cages are interconnected by normal bonds of the expected strength (bond valence). The charge electrons are clearly located on the exterior and, mainly on the terminal oxygen atoms of the M–O cages.

The results show very clearly the power of the bond valence method when used in combination with the classic theoretical chemistry. The extraordinarily large number of factors that influence the bond lengths directly or via the charge on the oxygen atoms indirectly is remarkable. It is this large number of generally simple factors, which interact in a complex manner, that causes the complexity of the polyoxometalate systems. It is clear, however, that the solution of the multifactorial problems in this domain is decisively solved by the knowledge obtained by the application of classic chemical concepts and by the large number of structures in the polyoxometalate systems. Occasionally only a few factors determine the structural results and thus allow one to delineate the important features. It is desirable that the bond valence model, which is fully compatible with the valence bond model, is rapidly incorporated in textbooks of inorganic chemistry. This of course will also upgrade the current thinking concerning the valence bond method.

K.H. Tytko, Göttingen

Contents

Contents of Volume 86

Atoms and Molecules in Intense Fields

Volume Editors: L.S. Cederbaum, K.C. Kulander, N.H. March

Contents of Volume 87

Structural and Electronic Paradigms in Cluster Chemistry

Volume Editor: D.M.P. Mingos

Bond Length-Bond Valence Relationships, with Particular Reference to Polyoxometalate Chemistry

K.H. Tytko, J. Mehmke, D. Kurad

Institut für Anorganische Chemie, Universität Göttingen,
Tammannstraße 4, D-37077 Göttingen, Germany

Bond length-bond valence relationships have been investigated fundamentally with emphasis on the M–O bonds in polyoxo compounds ($M = Mo^{VI}, W^{VI}, V^{V}, Nb^{V}, Ta^{V}$). A large number of errors of different types has been made in the derivation of practically all of the published functions and/or of the relevant parameters. Considering all sources of errors, bond length-bond valence functions and the relevant parameters have been derived which represent more shallow curves than most of the functions in the literature. The relationships have been applied classically for identifying O atoms of an OH group or a coordinated H_2O molecule, to elucidate hydrogen bridge systems, to determine the oxidation numbers of M atoms (and to distinguish between different elements via the oxidation numbers), and to verify the coordination numbers assigned to the M atoms, etc. The most important application of the relationship, however, is the calculation of accurate bond valences and in particular the determination of the distribution of the charge over the O atoms of the species. These data can be used to elucidate the relationships between structure, bonding, stability and basicity of the species. However, most functions and/or the relevant parameters stated in the literature produce errors which are most evident in the calculated formal ionic charges of the species and can involve several charge units. Even the best functions and parameters give unreliable results. A first important reason for this is the unsatisfactory identification of erroneous structural data with large random and/or systematic errors in the bond lengths and their rejection from the set of reference structures used for the derivation of the bond length-bond valence parameters B and d_0 or N and d_0 of the commonly used exponential or power functions. This makes the correct determination of B or N difficult. A second important reason is connected with the – at present – unfounded practice of using 'universal' B or N parameters which leads to errors for the proportion in the bond valencies of the inner (bridging) relative to those in the outer (terminal) M–O bonds of the species and for the charge separations. These quantities affect the stability of the species. A third significant reason, which is independent of and hence present even for correctly derived bond length-bond valence parameters, is a small (and inevitable) systematic error in the bond lengths of each structure determination which leads to larger errors for the formal ionic charge. This error can be completely compensated by individual fitting of the d_0 bond length-bond valence parameter for each structural determination.

Keywords: Bond length-bond valence relationships, bond length-bond valence parameters, polyoxometalates

1
Introduction

The conceptualization of the numerical strength of a bond in inorganic chemistry was first introduced by Pauling [1] in the second of his five principles determining the structures of complex ionic crystals. The *electrostatic valence principle* is now commonly referred to as Pauling's second rule and is connected with Pauling's 'bond strength'. Its modified form is now widely used as the *valence sum rule* [2–4] (see Eq. (2)). Introduction of an inverse relationship between the strength and the length of a bond by Byström and Wilhelmi [5], Zachariasen [6], and other authors [7–11] (see Eqs. (1a) and (1b)), based on a relationship proposed by Pauling [12], led to a considerable improvement of the concept. The terms 'bond strength', 'bond order', and other terms that were used at that time have been changed to 'bond valence' [3, 8, 10, 13, 14] since bond strength and bond order are terms also used in a different sense. The parameters of the bond length-bond valence functions have been determined for a large number of bond partners [4, 8, 9, 15–20]. Parallel to the above development, the *bond valence model* was proposed [3, 4, 21, 22] as a bond model describing bonding (bond valences) in a molecule or ion in connection with its neighboring molecules and ions (including solvent molecules).

Bond length-bond valence relationships and the valence sum rule have successfully been used for the following purposes

- to identify oxygen atoms which form an OH group or a coordinated H_2O molecule in polyoxometalate ions [23–26] and minerals [4, 10, 11, 14] via a corresponding deviation of the calculated sum of the bond valences from the bond valence of two expected for oxygen;
- to confirm the interpretations of crystal structure investigations [2–4, 8, 11, 14];
- to elucidate hydrogen bridge systems in polyoxometalates [23, 24, 27–30] and other compounds [2–4, 8, 10, 11];
- to determine the "charge" (oxidation number) of the M and other electropositive atoms and to distinguish between different elements E via their oxidation numbers [2–4, 8, 14, 31–33];
- to assign coordination numbers to the central atoms of coordination polyhedra [2, 3, 8, 10, 11, 31–34]; and others [2–4, 8, 14, 33].

In spite of the success quoted above there are other examples which cannot possibly be correct, for example, statements on the distribution of the bond valences in the polyoxoanions and of the ionic charge over the oxygen atoms in them [35, 36a, 37–41]. There must be some seemingly insignificant but decisive errors which become noticeable only in those cases that require exact bond valence calculations. The present study was undertaken to clarify these questions for (poly)oxometalate ions and related oxides. It is, however, also of more general interest and should also be applicable to other types of compounds.

2 Bond Length-Bond Valence Relationships

2.1 Fundamentals

It has been shown by many authors [3–6, 8–11] that the bond valence s of a bond between two atoms depends inversely on the length of the bond (the distance between the centers of the atoms) d, and that the relationship is non-linear. The most frequently used empirical equations for the description of this relationship are

$$\log s = (d_0 - d)/B \quad \text{and} \quad s = (d_0/d)^N. \qquad (1a, 1b)$$

where d_0 and B or d_0 and N, the bond length-bond valence parameters, are characteristic of the bond partners. The mathematical form of Eqs. (1a) and (1b) specifies d_0 as the single-bond length (s = 1 v.u. [valence unit]). The parameters B and N define the slopes of the functions [3, 4a, 15]. For their interpretation see Refs. 2a, 3, 8, 9, 11, 15, and 18.

The bond valences s_i about an atom sum up (U) to give the atom valence, defined by the absolute amount of the oxidation number n of the respective element:

$$\Sigma s_i \equiv \sum_{i=1}^{k} s_i = U = |n| \quad \text{(valence sum rule)} \qquad (2)$$

(*i*: index for the different bonds about an atom; k: coordination number of the central atom) [2–4, 8, 11, 18]. This applies to all atoms of a compound and is called the 'principle of local valence balance' [18], a generalization of Pauling's principle of local charge balance in ionic crystals [1].

A deficit or an excess of the bond valence sum compared to the absolute amount of the oxidation number (|n|) means a negative or a positive charge on the respective atom in a species, depending on the anionic (the preceding sequence refers to this case) or cationic character of the atom and provided that there are no other reasons (e.g., errors) for the difference. Hence the difference between the sum of the bond valences about an oxygen atom and the absolute amount of the oxidation number two of oxygen is, according to the valence sum rule, the charge c on that oxygen atom:

$$c = \left(\sum_{i=1}^{k_O} s_i \right) - 2 \quad \text{(valance sum rule for oxygen, modified for the charge).} \qquad (3)$$

In a compound, negative and positive charge compensate each other. Therefore an excess negative charge defines unbalanced bonding power of the oxygen atoms in question which is compensated by the cations in the compound (salt), often via molecules of water of crystallization. This charge is given in charge units (c.u.) [42].

In the following k and n mean usually k_M and n_M, respectively.

The sum of all bond valences (V) in a polyoxometalate ion $M_q^n O_u^{m-}$ amounts to

$$\Sigma s_o \equiv \sum_{o=1}^{g} s_o = V = qn = 2u - m \qquad (4)$$

(*o*: index for the M–O bonds in the structure; g: number of M–O bonds in the structure).

The relationships described by the above equations can be interpreted to mean that 1 v.u. corresponds to the bonding power of two electrons (one electron pair), i.e., the bond valence defines the (usually non-integral) number of electron pairs that constitute the corresponding bond [3, 4b, 8, 9, 21, 22].

By use of Eqs. (1) and (2), the parameters d_0 and B or d_0 and N are empirically determined using a large number of structures containing, as a whole or in structural parts, only the bond partners. The bond valence sums are not very sensitive to the values of B or N;[1] the values of d_0, however, are very sensitive to the exact choice of B or N [4a, 8, 9, 15, 17, 31]. Therefore d_0 and B and d_0 and N represent in each case inseparable pairs [4a].

Bond valence in inorganic compounds is nearly always a non-integral quantity and frequently smaller than one [2–4, 8, 9, 21, 22, 43]. (The statements in the literature about bond angles and bond valence [8, 43] are not unambiguously formulated; see Sect. 4.6.10 in Ref. 44.)

2.2 Determination of the Bond Length-Bond Valence Parameters

The following discussion refers mainly to M–O bonds ($M = Mo^{VI}, W^{VI}, V^{V}, Nb^{V}, Ta^{V}$) in polyoxometalate ions and the related oxides, however, the statements and results are more generally applicable.

2.2.1 *Published Bond Length-Bond Valence Functions and Parameters for M–O Bonds*

Published parameters for bond length-bond valence functions are given in Table 1. The relevant plots of the functions for Mo^{VI}–O bonds are shown in Fig. 1. For a comparative discussion of the functions and parameters see Sects. 2.2.2.6 and 2.2.4.6.

2.2.2 *Discussion and Criticism of the Published Bond Length-Bond Valence Functions and Parameters*

The determination of the parameters d_0 and B and d_0 and N using Eqs. (1a) and (1b) presents a range of difficulties [4a]. In the older investigations (ca. 1970–1975) these difficulties were addressed [8–10, 13], however, in the

---

[1] This statement involves a high risk of misinterpretation, see Sect. 3.2.

Table 1. Bond length-bond valence functions and relevant (general) parameters for M–O bonds given in the literature

Authors	d_0 and B in Å or d_0 in Å and N[a,b]									
	Mo^{VI}–O		W^{VI}–O		V^{V}–O		Nb^{V}–O		Ta^{V}–O	
Byström et al. [5]					1.77	0.764[c]				
Kihlborg [47]	1.872	0.668								
Evans [7]					1.81	0.78				
Schröder [45]	1.874	0.600								
Allmann [8]	1.90	0.76	1.91	0.82	1.81	0.78				
Waltersson [18a]					1.791	0.722				
Zachariasen [18][d]	1.890	0.723	1.898	0.725	1.790	0.734	1.921	0.735		
Fuchs et al. [46]	1.915	0.80								
Brown et al. [17]	1.907	0.852	1.917	0.852	1.803	0.852	1.911	0.852	1.920	0.852
Brese et al. [16]	1.907	0.852	1.921	0.852	1.803	0.852	1.911	0.852	1.920	0.852
Present authors[e,f]	1.915	0.953[g]	1.916	0.948	1.799	0.851	1.917	0.859[h]	1.921	0.85[i]
Brown et al. [9][d]					1.799	4.483				
Brown et al. [15]	1.882	6.0	1.904	6.0	1.791	5.1	1.907	5.0	1.907	5.0
Bart et al. [49]	1.882	6.0								
Trömel [31]					1.780	5.51				
Fink et al. [19][j]	1.88	6.4	1.90	6.0			1.92	6.2	1.91	5.4
Present authors[k]	1.904	4.581	1.911	4.341	1.789	5.159	1.908	4.721[h]		

[a] Bond length-bond valence functions (1a) and (1b), respectively. Linear functions have also been used ($s = -2.39d + 5.73$ for Mo–O bonds; d in Å [26]).
[b] If necessary, converted by the present authors.
[c] Recalculated by the present authors. The calculations in the literature could not be reproduced.
[d] Eq. (2) is usually fulfilled to within 5%.
[e] Standard deviation for the sum of the bond valences about M (Eq. (2)): 2%.
[f] The parameters B (and hence d_0) are uncorrected for the systematic error discussed in Sect. 2.3.1.3, +0.010 Å for the Mo–O system.
[g] Provisional function: $\log s = (1.910 - d)/0.882$ [64a].
[h] The set of reference structures included intentionally an unreliable structure determination to have not only MO_6 octahedra in the data set.
[i] Since all available oxotantalate structures consisted only of MO_6 octahedra, the B value was chosen in comparison with those for the other M^{V}–O bonds.
[j] Average agreement between bond valence sum about M and its oxidation number according to Ref. 20: ±5%.
[k] The results obtained with Eq. (1a) are somewhat better.

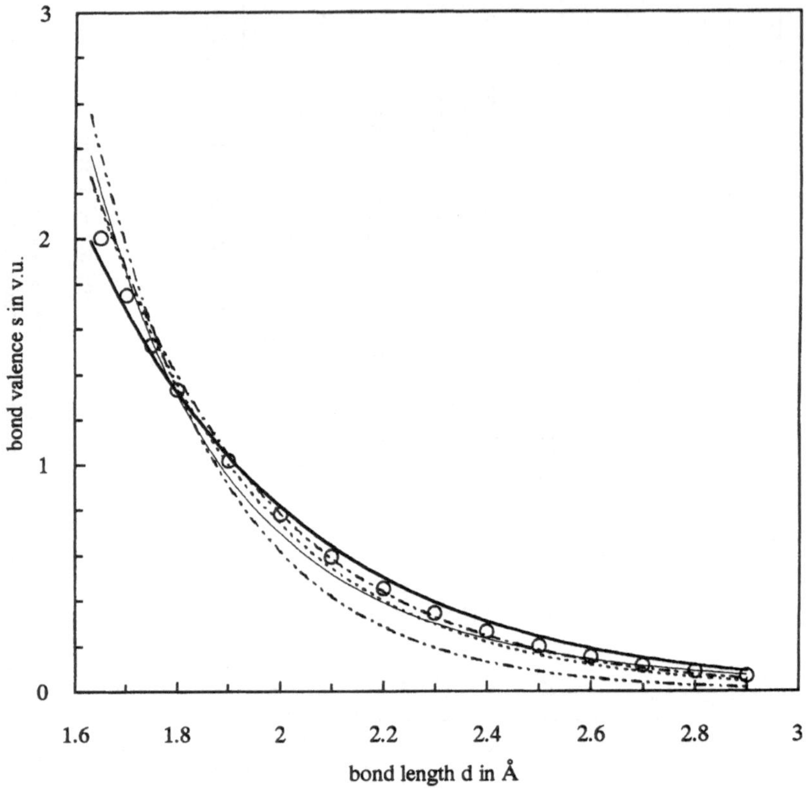

Fig. 1. Bond length-bond valence relationships proposed in the literature for Mo^{VI}–O bonds. — Brown and Wu [15], ········ Allmann [8], — ·· — ·· Schröder [45], — · — · Fuchs and Knöpnadel [46], ▬ present authors (this article). The functions given by Zachariasen [18] and Fink and Trömel [19] are for bond lengths >1.8 Å in the range of those by Brown and Wu, Allmann, and Fuchs and Knöpnadel and for the range <1.8 Å in the range of those by Brown and Wu and Schröder. ○ ○ ○ Function proposed by Brown and Altermatt [17] and Brese and O'Keeffe [16]; this function corresponds nearly perfectly ($|\Delta s| \leqq 0.02$ v.u. in the range >1.67 Å) to a provisional function stated by the present authors [64a].

current literature there appear to be no great problems. Long lists of the bond length-bond valence parameters for element pairs (bond partners) have been published [4, 8, 9, 15–19]. A more critical evaluation of the data suggests that a common B or N (and even d_0 [9a]) has been proposed or even assumed for many element pairs ('universal' B or N parameters) [9, 11, 16, 17, 20] and that the deviations accepted for the valence sum of the central M atoms are rather large (on an average ca. 5% [9, 18, 20, 33], up to ca. 10% [8, 9, 20, 31, 33], or even more [9, 31, 33]). When these parameters are used for more ambitious applications than those quoted above (Sect. 1) they lead to wrong and widely varying results depending on the source (see Fig. 1) of the parameters (see the examples in Tables 4, 5, and 7–9 and the relevant sections in this article and

Table 49, Sect. 4.8.1, in Ref. 44). This has occasionally led to the statement [36a] that the bond valences do not have any precise significance. On the other hand, average deviations for the valence sum ('root-mean-square average R1 of the d_i values') of M exceeding 0.1 v.u. (which would correspond to average deviations of 1.7% for M^{VI} and 2.0% for M^{V}) [21] have been interpreted in terms of "strained bonds".

Standard deviations for the bond valences and for the charges on the oxygen atoms, based on the quoted parameters, are rarely given in the literature.

The large number of proposals for the bond length-bond valence parameters d_0 and B or d_0 and N with little or no discussion of the reasons why the previous literature values are less satisfactory is, in our opinion, an unacceptable situation. Since a comprehensive presentation and discussion of the errors is lacking in the literature, the following section gives a systematic discussion on the types of errors frequently made.

2.2.2.1 *Assumptions for Bond Valences Made in the Literature to Make the Problems Solvable Without a Least-Squares Fitting Procedure*

In the early period (ca. 1950–1970) of establishing relationships between bond length and bond valence and even later (1975 [45], 1982 [46]), the mathematical difficulties associated with handling the structural data resulted in oversimplifications. To make the problem solvable without least-squares fitting procedures, a number of assumptions were made:

- Assignment of a definite bond valence to certain M–O bonds (bond lengths). Examples:
  - A bond valence of 2 v.u. was assigned to the terminal M–O bonds of the neutral (= uncharged) compounds "MoO_3" [47] (and other molybdenum oxides) and "V_2O_5" [5]. In doing so the possibility for negative charge on the terminal O atoms and an equivalent positive charge on bridging O atoms according to resonance structure (17b) in Ref. 48 was neglected. Tables 33–36 in Ref. 44 show that this type of oxygen atom does indeed usually carry a negative charge.
  - A bond valence of 1 v.u. was assigned to the average bond length of two-coordinate O atoms [5, 47, 49]. In this case the possibility for a negative or positive charge on the bridging O atom (compare, e.g., resonance structures (16d), (17b), (17c), and (17e) in Ref. 48) was neglected (or denied). Tables 3–16, 18, 20–23, 25–31, 33, 35, 36, and, summing up, Tables 41 and 42 in Ref. 44 show that this type of oxygen atom does indeed usually carry a negative or positive charge. Additionally, this procedure is inconsistent with the hyperbolic form of the bond length-bond valence function, compare Sect. 4.6.3, in particular Table 44 in Ref. 44.
- Assumptions for the bond valence in certain M–O compounds based on considerations of orbital symmetries [45]:

- A bond valence of 1.67 v.u. (one σ and $\frac{2}{3}$"π_3" bonds) was assigned to the "characteristic" Mo–O distance in Mo(Ot)$_3$ groups (Ot: terminal O atom) of 1.738 Å.
- A bond valence of 1.5 v.u. (one σ and $\frac{1}{2}$"π_3" bonds) was assigned to the "characteristic" Mo–O distance of the MoO_4^{2-} tetrahedra.
- A bond valence of 0.33 v.u. was assigned to the average value of the long Mo–Ob (Ob: bridging O atoms of asymmetric bridges) distances of 2.16 Å (for the arguments supporting this assignment see Ref. 45).

– Assumptions concerning the distribution of the ionic charge over the O atoms and of the bond valence over the M–O bonds of the polyoxometalate ions $M_qO_u^{m-}$ [46, 50]. Assuming
  - an equidistribution of the ionic charge about the O atoms (m/u) and
  - an equidistribution of the bond valence defined by this means ((2u – m)/u = nq/u) about the M–O bonds of the differently coordinated O atoms,

  the average bond valences of the M–O bonds result in $\bar{s}(k_O) = (2u - m)/uk_O = nq/uk_O$ (compare Sect. 4.3.3 in Ref. 44; also given there are the $\bar{s}(k_O)$ values for a large number of polyoxometalate ions) which have been set in relation to the observed bond lengths [46, 50]. However, in this way there are great differences between the average bond valences $\bar{s}(k_O)$ and the bond valences s calculated from the bond lengths via a bond length-bond valence function (e.g., up to 0.66 v.u. for $Mo_7O_{24}^{6-}$ and up to 0.57 v.u. for β-$Mo_8O_{26}^{4-}$), and the bond valences add up to 6.56, 6.27, and 3.79 (!) v.u. for the three independent MoO_6 octahedra in $Mo_7O_{24}^{6-}$ and to 6.83, 6.21, and 4.74 v.u. for those in β-$Mo_8O_{26}^{4-}$). These results clearly indicate the inconsistencies in this approach.

– Assumptions by a more intuitive fitting:
  - use of the average bond lengths $\bar{d}$ of the coordination polyhedra in place of d_0,
  - estimation of N according to $N = \bar{d}/(d_{max} - \bar{d})$ (where d_{max} is the necessarily only roughly definable maximum M–O distance considered as a bond), and
  - introduction of a linear part into the bond length-bond valence curve to reach s = 0 v.u. at finite M–O (or other element–O) bond lengths [8, 10, 13, 51].

The only correct way is the purely mathematical solution, without any assumptions, of a system of equations consisting of equations of types (2) and (1) or another suitable bond length-bond valence function by least-squares fitting procedures [9]. This means simultaneously that diagrams like Fig. 2 in Ref. 45 and Fig. 5 in Ref. 46, showing discrete (d, s) values, are impossible without making some assumptions. Only from regular MO_4 tetrahedra of MO_4^{w-} ions can a single (d, s) pair be derived without a least-squares fitting procedure – provided there are no intermolecular bonds analogous to Scheme (5).

2.2.2.2
Use of Unsuitable Structures for the Derivation
of the Bond Length-Bond Valence Parameters

Some authors used less suitable or unsuitable structures for the derivation of the bond length-bond valence parameters. The following mistakes have been made:

– Use of compounds that contain two (or more) different elements in the coordination sphere of the central atom in question, for example, $MoOF_4$ or $MoO_2Cl_2 \cdot H_2O$ [45]. If there is an additional element in a compound (in the coordination sphere of the central atom), the bond length-bond valence relationship for this element and its bond partner must be known or must be determined in an analogous procedure. It is not allowed, in the above examples, to "cut off" halide ions and subsequently to evaluate the rest molecules as $[Mo{=}O]^{4+}$ and $[MoO_2]^{2+}$, respectively, and to assign 2 v.u. to the Mo–O bond lengths since the following (or similar) types of resonance may exist:

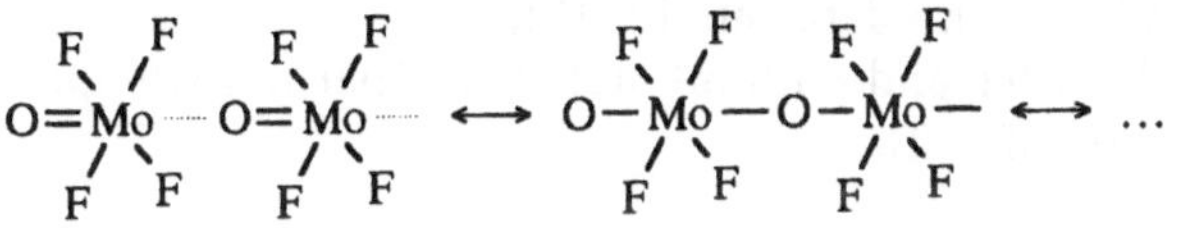

$$
\begin{array}{ccccc}
\text{F} \ \text{F} & \text{F} \ \text{F} & & \text{F} \ \text{F} & \text{F} \ \text{F} \\
O{=}Mo \cdots O{=}Mo & & \longleftrightarrow & O{-}Mo{-}O{-}Mo{-} & \longleftrightarrow \ \ldots \\
\text{F} \ \text{F} & \text{F} \ \text{F} & & \text{F} \ \text{F} & \text{F} \ \text{F}
\end{array}
\tag{5}
$$

$$
O{=}M{-}Y \longleftrightarrow {}^{\ominus}O{-}M{=}Y^{\oplus} \longleftrightarrow \ \ldots
\tag{6}
$$

($=$, $-$, and $\cdots$ mean bonds of bond orders two, one, and zero, respectively, in the individual resonance structures; compare Sect. 2.2 in Ref. 48).

– Use of compounds with M in different oxidation states [18, 19, 31, 45, 49]. Other authors [9, 15–17, 45] report small but clear differences for the bond length-bond valence parameters of the different oxidation states of M; see also Sect. 2.5. It appears, however, to be rather a question of the assessment of the necessary or realizable precision whether or not authors use a uniform d_0 and B (or d_0 and N) for different oxidation states of M.

– Neglect of the requirement of different MO_k coordination polyhedra (*with the correct oxidation number of M*) in the data sets used for the determination of N and d_0; without different values of k the system of equations consisting of equations of type (1) and (2) is unsolvable [9, 13]. According to our investigations the same is true for the derivation of the parameters B and d_0. The reason is that the function $\sigma_\Sigma(B)$ (σ_Σ is the standard deviation for the sum of the bond valences about M, see Sect. 2.2.3.2) approaches – without a minimum at finite B values – the limit $\sigma_\Sigma \to 0$ for large B's if all coordination polyhedra have the same k (compare Sect. 2.3.1.2, in particular Fig. 3a(a1) and (a2)) and hence all M–O bonds attain, e.g., $s = 1$ (if only $M^{VI}O_6$ octahedra are present) or $s = 1.5$ v.u. (if only $M^{VI}O_4$ tetrahedra are present) in the set of reference structures, irrespective of the bond lengths. This item often remains unnoticed due to the inappropriate use of data sets that contain compounds with different oxidation numbers of M or due to the use of

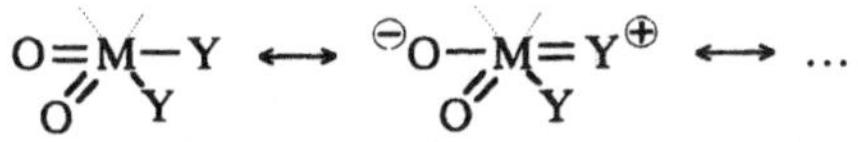

'empirical' values for B or N from other sources. The former case represents an alternative for MO_k polyhedra with different k with the reservations with regard to differences of the bond length-bond valence parameters for different oxidation states; in the latter case there is no independent determination of the bond length-bond valence parameters. (Another possibility would be the simultaneous evaluation of the structural data according to Eqs. (1) and (2) also for oxygen as the central atom. This requires, however, knowledge about the bond length-bond valence function for the bonds between oxygen and the cations of the polyoxometalate salts or their development [9].)

An unequivocal determination of the parameters d_0 and B or N thus requires

- structures containing complete MO_k coordination polyhedra, i.e., there must be no other ligands about M [9, 16, 19];
- structures containing M only in the relevant oxidation state;
- structures with different coordination numbers for M, i.e., different coordination polyhedra in the data sets [9, 13];
- structures without disorder and without statistical (partial) occupancy of atomic sites [16, 17, 19, 32, 52].

2.2.2.3
Use of Incorrect Structural Data

Here we consider only *large* systematic errors (on average large, one-sided deviations from the correct bond lengths in a structure) and/or *large* random errors (large scattering of the values) for the bond lengths of the structures. In the literature one usually finds a selection of the structural data according to the standard deviation for the bond lengths or according to the crystallographic R factor, e.g., rejection of structures for which standard deviations > 0.01 Å for the bond lengths [31–33, 46] or R factors $\geqq 0.12$ [17] have been stated. In our opinion this procedure is neither necessary (compare the commentary in connection with the M···M distances in Sect. 2.2.3.2) nor sufficient (compare, e.g., the results of Fink and Trömel in Sect. 2.2.2.6 and their discussion in Sect. 3.2) as a precautionary measure for polyoxometalates.

The function for the optimization of the bond length-bond valence parameters B (or N) and d_0 and the nature of the structural data of the (poly)oxometalates, i.e., ultimately the form of the $\sigma_\Sigma(B)$ functions (see Figs. 3a and 4 in Sect. 2.3.1.2 and 2.3.1.3, respectively), have the consequence that incorrect structural data in the set of reference structures tend to cause the "emigration" of the B (or N) values in one direction: B becomes greater (compare also Sects. 2.2.4.1 to 2.2.4.4), N smaller. This effect is the more pronounced the more uniform the size of the MO_k polyhedra is. For further discussion of this problem see Sect. 3.2.

For an unequivocal determination of the parameters d_0 and B or d_0 and N it is essential to recognize incorrect structural data by appropriate tests and to eliminate them (see Sect. 2.2.3.2).

2.2.2.4
Use of Improper Mathematical Functions for the Description of the Bond Length-Bond Valence Relationships

Some authors [26, 53] used linear bond length-bond valence functions instead of non-linear ones. The incorrectness of linear functions can be seen directly [3, 8, 9] from the average M–O bond lengths in the MO_6 octahedra which become larger the more distorted the octahedra (the more different the bond lengths) are. As a consequence of the hyperbolic form of the functions weak M–O bonds are longer than strong bonds are shorter with respect to the same alteration of the bond valence (compare Fig. 1).

Occasionally a linear part was appended to a non-linear relationship to attain $s = 0$ v.u. at finite bond lengths [8, 10, 13, 51] (Mo^{VI}–O and W^{VI}–O 2.60, V^{V}–O 2.36, Nb^{V}–O 2.56, and Ta^{V}–O 2.60 Å [8]). This procedure introduces assumptions like those discussed in Sect. 2.2.2.1. Therefore a better way to reach $s = 0$ v.u. at finite bond lengths is the uniform use of a suitable (e.g., three-parametric) bond length-bond valence function for the whole investigation range (see also Sect. 2.2.3.1).

Currently, there is a tendency to accept longer ranges of bonding [9, 11, 20, 22, 33, 43, 54], i.e., ligand atoms of the second coordination sphere of the central atoms are also taken into consideration, and Eq. (1) is used for the whole range of bond lengths. Non-linear functions different from (1a) or (1b) have also been applied for other bond partners [31, 55].

An optimal way to find the optimal mathematical function is to start with a two-parametric non-linear equation (e.g., with Eq. (1a) or (1b)) and to introduce a third parameter to improve the fit between structural data and mathematical function – if the function can be improved at all.

2.2.2.5
Use of Non-Optimal Bond Length-Bond Valence Parameters B or N

It has already been mentioned in Sect. 2.2.2 that there is the tendency in the literature to use a common ('universal') B or N for many element pairs. This procedure is justified [9, 17] by the resemblance of the B or N values for the different M–O pairs in connection with their reported low accuracy (rarely better than 13% [17]; even in the most favorable cases 10% [15]). However, it has neither been convincingly substantiated why the B or N parameters should be so inaccurate, nor recognized that small changes of B or N have a marked influence on the bond valences of the individual M–O bonds and in particular on the charge about the O atoms (see Sect. 2.3.1.1, in particular Table 4). Just the handling and evaluation of these quantities, however, is the real concern when working with bond length-bond valence relationships; the sum of the bond valences about M or its standard deviation, which change only very slightly with a change of B or N, serves, in the absence of a more suitable quantity, merely as a control index ('agreement index' [9]) in the optimization process for determining bond length-bond valence parameters. Hence the use of 'universal' B or N values can only be made at the expense of the accuracy of the method.

If authors [16] accept 'universal' B parameters from the literature [17] they also accept any deficiencies that may exist in the original derivation. Thus, in the case in question, the structural data have not been analyzed and corrected for outliners (outliners in a sense comparable to that used in Sect. 2.2.3.2 and rejection of them), as can be seen from the large standard deviations $\sigma'(\hat{=}\sigma_{d_{0iM}})$ of 0.01–0.05 Å [17]. In our investigations a comparable quantity $(\sigma_{d_{0i}} = \sigma_{d\,syst})$ amounts to 0.005 Å (see Sects. 2.3.2.2 and 3.1). Outliners, however, have a strong influence on the B parameter, see Sects. 2.2.4.1–2.2.4.4, 2.3.1.3, and 3.2. Moreover, it appears from the reports about the accuracy of the established bond length-bond valence parameters and about the lack of sensitivity of the bond valence sum to changes of B [8, 16, 17, 31] that only rather vague ideas about the influence of alterations of the bond length-bond valence parameters on the bond valences and on the charge distribution about the O atoms exist. This topic is treated in Sects. 2.3.1.1, 2.3.2.1, and 3.2.

Hence for an unequivocal determination of the parameters d_0 and B or d_0 and N, an averaging based on the parameters of other bond systems should be strictly avoided.

2.2.2.6
Criticism of the Concrete Bond Length-Bond Valence Parameters for M–O Bonds of the Different Authors

Byström and Wilhelmi's [5] and Kihlborg's [47] investigations are based on only one structure determination each, and some assumptions on bond valences have been made (i.e., a least-squares fitting procedure was not and could not be performed).

Evans [7] accepted the B value derived by Byström and Wilhelmi and determined a new d_0 by means of some new structures, including some with $V^{<V}$.

Schröder's [45] investigations are based on a large number of structure determinations, however, he also made some assumptions on bond valences (and hence a least-squares fitting procedure was not carried out). A number of unsuitable structures were used, i.e., those containing other elements besides oxygen in the coordination sphere of M, for which insufficient corrections were made, and those containing different oxidation states of M for which corrections, adopted from the literature (and hence unchecked by the author), were made.

The investigations by Fuchs and Knöpnadel [46] also include many structural determinations, however, for all bonds assumptions on their bond valence were made (again no least-squares fitting procedure).

Brown and Shannon [9] and Brown and Wu [15] investigated a large number of structures containing only M and O atoms and with the correct oxidation state of M for the first time mathematically correct (i.e., without assumptions on bond valences; see, however, Sect. 3.2) by a least-squares fitting procedure. The deviations between bond valence sum and oxidation number nevertheless amount on an average to ca. 5% (in individual cases even 10–15%). The authors obviously assume that this is due to systematic errors which are greater than the frequently quoted standard errors of 0.005 Å or less for the bond lengths of a structure. However, incorrect structural data have apparently not been identified and rejected.

Bart and Ragaini [49] merely adopted the parameters for Mo^{VI}–O bonds by Brown and Wu [15] on the basis of a selected set of structural data (also containing $Mo^{<VI}$ species) since these parameters gave the best agreement of the bond valence sums with the oxidation number of molybdenum.

Allmann [8] and Zachariasen [18] made only limited statements on the structural data evaluated by them. They apparently used a least-squares fitting procedure. No statements have been made about the identification and elimination of possibly incorrect structural data. The agreement between bond valence sum and oxidation number was usually within 5% [18]. Thus, these investigations are to be viewed similarly as those of Brown et al. [9, 15].

Brown and Altermatt [17] and Brese and O'Keeffe [16] investigated a large number of structures where M with the correct oxidation state was bound only to O atoms. They calculated a separate d_0 for each MO_k polyhedron in each structure on the basis of Eqs. (1a) and (2), assuming a 'universal' B for all element pairs. Rejecting any obvious outliners [16] – justified by the author's belief that, despite claims to the contrary, bond lengths are rarely determined with an accuracy better than 0.01 Å – or rejecting outliners according to the crystallographic R factors ($R \geqq 0.12$) [17], they then averaged the individual d_0 values. The average was considered significant to an accuracy of about ± 0.02 Å in usable cases [16] or when having an estimated standard error of less than 0.01 Å [17]. No statements about the deviation between bond valence sums and oxidation number were made. The more or less critical points of these investigations are obviously an insufficient rejection of outliners (in a sense comparable to that used in Sect. 2.2.3.2) and the assumption of a common, 'universal' B parameter [16, 17] for a large number of element pairs and its acceptance from other sources [16]. For the problem of the use of 'universal' B parameters, see Sects. 2.2.2.5, 2.3.1.1, and 3.2.

The investigations by Fink and Trömel [19, 31] include ca. 20–30 different MO_k polyhedra; however, oxidation states <6 for Mo and W and <5 for V, Nb, and Ta have also been considered. Incorrect structural data were assumed and eliminated when the standard deviation for the bond lengths stated in the structural papers exceeded 0.01 Å. Nevertheless, the average agreement between bond valence sum and oxidation number is only within 5% [20, 33], in individual cases >10% [31, 33]. For the inconsistency of the 0.01 Å value with the 5% or even >10% value, see Sect. 3.2. This might in the first place be due to errors of the bond lengths, as discussed by Brown and Shannon [9] and Brese and O'Keeffe [16], and non-rejection of such (unidentified) structural data.

For another general inconsistency in the literature in connection with the 5% or even 10 and 15% deviations between bond valence sum and oxidation number, namely the statement of (approximately) correct B and N values in spite of the "emigration" effect in cases where B or N have been obtained by a least-squares fitting procedure, see Sects. 2.2.2.3 and 3.2.

2.2.3
Proper Investigation of the Bond Length-Bond Valence Relationships

The mathematical relationships between the accuracy of the bond lengths and that of the bond length-bond valence parameters derived from them or of the bond valence sums about M, of the M–O bond valences, and of the charge on the oxygen atoms calculated via the latter values are very complex. For example, errors of the bond lengths, which are the basis for the determination of the bond length-bond valence parameters, tend to lead to B values that are markedly too large (compare, e.g., Sects. 2.2.4.1–2.2.4.4 and 2.3.1.3) or N values that are too small and hence introduce errors in their use. Also, an agreement between bond valence sum and atom valence (oxidation number) only within 5% [9, 18, 20, 33] or even 10% [8, 9, 20, 31, 33], an accuracy of d_0 of ±0.02 Å [16], ±0.01 Å [17, 31], or a B or N with an error of 10% [15] or 13% [17] are insufficient for adequate applications of the bond length-bond valence functions as, for example, described in Ref. 44. The errors for the charge of the polyoxoanions (and for oxides) are enormous (compare Sects. 2.3.2.1 and 3.2, this article, and Sect. 4.8.1 in Ref. 44). Surprisingly, an agreement between bond valence sum and oxidation number within less than 5% can also be attained by a linear bond length-bond valence function (see Sect. 2.2.4.6, in particular Table 3).

According to the concluding remarks in Sects. 2.2.2.1–2.2.2.5 and to the above notes an unambiguous investigation of bond length-bond valence relationships requires:

- a purely mathematical evaluation of the structural data by a least-squares fitting procedure without any assumptions about bond valences, charges, or bond length-bond valence parameters (e.g., 'universal' B values);
- an optimization of the bond length-bond valence function not only via a minimal standard deviation for the sum of the bond valences about the M atoms, but also for the M–O bond valences and for the charge on the oxygen atoms (so that the standard deviation for these quantities can be obtained);
- a test of the mathematical function itself, expediently by use of a well-fitting function to which an additional third – and possibly fourth – parameter is added (which must give a better fit provided that the function is still capable of improvement);
- consideration of compounds containing only oxygen in the coordination sphere of M and M only in a definite oxidation state; the use of structures with different coordination numbers for M and such without disorder and without statistical occupancy of atomic sites;
- elimination of erroneous structural data by an analysis of the contribution of the individual structures to the error square sum.

2.2.3.1
The Mathematical Evaluation

The (calculated) bond valence sum U for an atom E corresponds precisely to the absolute amount of its oxidation number (its valence) $|n|$ if

- the mathematical form of the bond length-bond valence function represents an excellent approximation to the real bond length-bond valence function,
- the parameters appearing in this function (e.g., d_0 and B in Eq. (1a)) are optimally chosen,
- the structural data (bond lengths) of the set of suitably chosen reference species for the determination of the bond length-bond valence parameters are without error,
- the extent of the coordination sphere of the atoms considered is completely covered, (and if
- there are no 'non-bonded interactions' [14, 56] or 'residual bond strains' [21] or other, still unrecognized effects).

If these items are not fulfilled, each of them makes a contribution to the average error square

$$e_U = \frac{1}{i}\sum_{f=1}^{i}\left(\frac{1}{q}\sum_{y=1}^{q}(U_y - n)^2\right)_f \tag{7}$$

(y: index for the MO_k polyhedra in the structures, q: degree of aggregation, i.e., number of MO_k polyhedra in the individual structures; f: index for the structures in the set of structures chosen for the investigation, i: number of structures in this set of structures) for the sum of the bond valences about the M atoms in the set of the structures chosen for the investigation. In Eq. (7) we assign each structure determination the same weight, independent of the number of MO_k polyhedra in the structures. This is consistent with the fact that there is just one optimization index (e_{Uq}; see Eq. (7a)) for each structure determination (the results for the individual MO_k polyhedra in polynuclear structures depend on each other), otherwise there would be the question of how to take into consideration polymeric structure types with different numbers of crystallographically independent MO_k polyhedra (for example, for $W_{12}O_{38}(OH)_2^{6-}$ there can be twelve different WO_6 octahedra or all of them can be crystallographically equivalent; (β-)$Mo_8O_{26}^{4-}$ is described with eight [57], four [28, 58, 59], or three [46] crystallographically independent MoO_6 octahedra for the different structure determinations).

The optimization of the bond length-bond valence parameters (e.g., d_0 and B or d_0 and N in Eqs. (1a) and (1b), respectively) is accomplished by variation of these parameters until the average error square e_U arrives at a minimum.

The remaining average error square can be reduced further by introduction of an additional third – and possibly fourth – parameter into the two-parametric bond length-bond valence functions (1a) or (1b) (or by suitable choice of another function) and subsequent continuation of the optimization process until the average error square e_U again attains an even lower minimum. (Surprisingly, the sometimes expressed desire for a bond length-bond valence relationship reaching s = 0 v.u. at finite bond lengths [8, 10, 13, 51] can be fulfilled with a single function in this way.)

The average error square thus remaining is merely due to errors in the structural data (the bond lengths) and/or to effects connected with bond lengths (e.g., erroneous assessment of the extent of the coordination sphere, non-bonded interactions).

The extent of the coordination sphere of M is considered in Sect. 2.4 below and particularly in Sects. 3.3, 4.1, and 4.6.2 in Ref. 44. Consideration of the very long M–O distances (Mo–O and W–O distances $\gtrapprox$ 2.6 Å, V–O distances $\gtrapprox$ 2.4 Å, i.e., those beyond the generally accepted coordination polyhedra) would presumably lead to a small decrease of the average error square for the sum of the bond valences about M. Because of the large arithmetical expenditure this item has only been investigated for a relevant collection of structures and for the structures with MO_k polyhedra smaller than MO_6 octahedra. An improvement of the bond valence sums about M has been observed in those cases where the deficit was large (e.g., about the tetrahedrally coordinated M atoms in $Mo_{10}O_{34}^{8-}$ (I) from 5.85 to 5.92 v.u. and in α-$Mo_8O_{26}^{4-}$ from 5.82 to 5.93 v.u. [44]).

Effects that could be assigned to 'non-bonded interactions' [14, 60], 'residual bond strains' [21], or other still unidentified effects have not been observed in polyoxometalate chemistry [44] under the procedure described.

2.2.3.2
The Structural Data: Test for Outliners. The Standard Deviation for the Sum of the Bond Valences About the M Atoms

According to the preceding section, the process of finding the optimal bond length-bond valence function together with the appropriate parameters can be conducted in such a way that the average error square for the sum of the bond valences about the M atoms is solely caused by the quality of the structural data (errors of the bond lengths).

Compounds of the following families can be used for the derivation of the M–O bond length-bond valence functions and appropriate parameters:

- oxides,
- mono- and polyoxometalate ions,
- protonated polyoxometalate ions and oxide hydrates, heteropolyoxometalate ions (presence of other electropositive elements bound to oxygen besides M), and
- other compounds with MO_k polyhedra (e.g., suitable chelate complexes).

In these compounds complete MO_k polyhedra exist (or must exist), i.e., there are no other elements bound to M. The M–O frameworks of the mono- and polyoxometalate ions bear a charge corresponding to the number of cations in the salt. The protonated polyoxometalate ions, species with coordinated H_2O molecules, and heteropolyoxometalate ions with *central* heteroatoms can be included after separation of H^+ and X^{n+} (X: heteroelement) as hypothetical isopolyoxometalate ions. Other compounds (chelate complexes etc.) have not been used.

Since the quality of the mathematical function for the bond length-bond valence relationship and the associated parameters depend on the quality of the structural investigations, usually a selection is made in advance according to the published standard deviations for the bond lengths or the R factor (for example, the rejection of structures for which the standard deviation of the bond lengths exceeds 0.01 Å [31–33, 46] or for which $R \geqq 0.12$ [17]). In other cases the standard deviations are used to statistically weight structural data [9]. We discarded this procedure for the following reason: considering the fact that the M···M distances are usually rather accurately defined (by 0.001 Å [23, 24, 27, 61, 62]), an erroneous position of a bridging O atom between two to six M atoms leads to errors for the respective M–O bond lengths which strongly depend on each other. In this way the effects of the bond lengths errors in question are largely compensated.

Unsuitable structural data have been identified (and rejected) by the large contribution of the respective structures to the average error square for the sum of the bond valences about the M atoms (with use of optimally chosen B and d_0 values, resulting in a minimal e_U according to Eq. (8)), which amounts to

$$e_{Uq} = \frac{1}{q}\sum_{y=1}^{q} (U_y - n)^2 \tag{7a}$$

(for y and q see Eq. (7)). For this purpose the number of structures in the data set with a certain range for the average error square e_{Uq} has been plotted against the range for the average error square e_{Uq} itself. This plot (e.g., Fig. 2a for a set of oxomolybdate structures) should correspond to a normal Gaussian distribution, i.e., of a set of structures ($2\Phi =$) 68.3% of them should lie within the ($\lambda =$) onefold, 95.4% within the twofold, 99.7% within the threefold etc. of the square root of the average error square for the whole of the structures, i.e., within $e_U = (1\sigma_\Sigma)^2, 4e_U = (2\sigma_\Sigma)^2, 9e_U = (3\sigma_\Sigma)^2$, respectively ($\Phi(\lambda)$: Gaussian error integral function). Structures outside a normal distribution (those with a large e_{Uq}) lead to deviations of the percentile parts of the structures from $2\Phi(\lambda)$ for $e_U, 4e_U$, etc. and have to be rejected (compare Figs. 2a and 2b). The bond length-bond valence parameters are reoptimized with the reduced data set and subsequently treated in the same way as described before. The result is a smaller average error square e_U for the new set of structures. The above procedure is repeated (altogether about two to four times) until the deviations of the percentile parts of the structures no longer show deviations from $2\Phi(\lambda)$ for $e_U, 4e_U$, etc. and thus indicate a normal distribution for the set of remaining structures. Structures lying only slightly outside the normal distribution should temporarily be left in the data sets in the first refinement cycles since, after the new optimization of the bond length-bond valence function, these structures

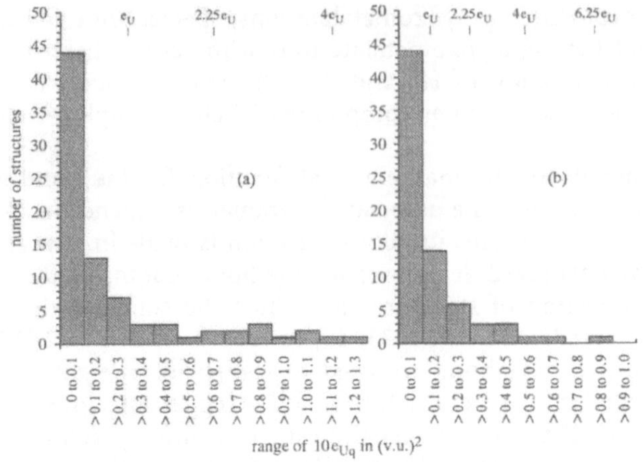

Fig. 2a. Reliability (outliner) test of the structural data (bond lengths) of 66 isopoly-oxomolybdates and 21 monomolybdates by the size of their contribution e_{Uq} to the error square sum $i \cdot e_U$ with respect to the sum of the bond valences about the M atoms on the basis of deviations from a Gaussian normal distribution as example. Ordinate: number of oxomolybdate structures; abscissa: range of the average error square for the sum of the bond valences about the M atoms for the individual structures (e_{Uq}). e_U corresponds to σ_Σ (68.3% range), 2.25 e_U corresponds to 1.5 σ_Σ (86.6% range), 4 e_U to 2 σ_Σ (95.4% range), 6.25 e_U to 2.5 σ_Σ (98.8% range) if there is a normal distribution of the errors. (a) First calculation cycle (all available structures; four structures with the center of gravity at 0.15 (v.u.)^2 lie outside the range depicted); (b) last calculation cycle (for the set of reference structures the three oxomolybdates with $e_{Uq} > 0.05$ (v.u.)^2 have additionally been rejected).

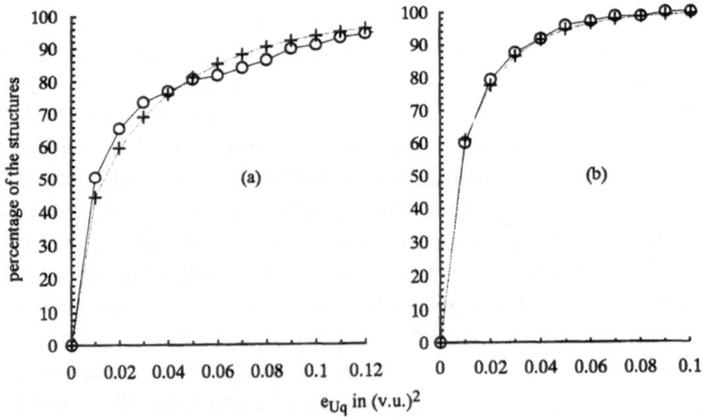

Fig. 2b. Reliability (outliner) test of the structural data of 66 isopolyoxomolybdates and 21 monomolybdates, continued. The cumulative percentage of the structures with a certain average error square for the sum of the bond valences about the M atoms of the structures ——o—— is compared with that for the case of a normal distribution, based on e_U (········+········). (a) The first calculation cycle shows strong deviations. (b) After gradual rejection of the structures showing a large e_{Uq}, there is a good agreement, indicating a normal distribution of the remaining structures.

may lie within the normal distribution. In this way the average error square after rejection of the outliners has been found as $e_U = 0.013\,(\text{v.u.})^2$ and hence $\sigma_\Sigma = 0.12$ v.u. for the set of oxomolybdate reference structures. These numerical relationships have been observed approximately for all M–O systems. In addition to the outliners characterized above, we have rejected on principle all structural determinations resulting in $e_{Uq} > 0.05\,(\text{v.u.})^2$ (corresponding to the range $>4e_U$, i.e., to the range $>2\sigma_\Sigma$) because of the disastrous influence of structures with large errors on the determination of the bond length-bond valence parameters B (compare Sects. 2.2.4.1–2.2.4.4, 2.3.1.1 and 2.3.1.3). There are still other ways to identify (and reject) outliner structures. After having recognized the disastrous influence of structures with $e_{Uq} > 4e_U$, we proceeded in the following way. For a set of structures to be tested for its reference qualities, B and d_0 were optimally determined, resulting in a minimal e_U. All structures with $e_{Uq} > 4e_U$ were rejected. For the new set of structures this procedure was repeated until no further structures had to be rejected.

In the above way complete structure determinations are rejected if they do not fulfill certain minimum requirements. The approach in the literature [9, 16, 17, 19, 32] to consider the single MO_k polyhedra separately even if they are part of a polymeric species and in this case dependent on each other does not allow an adequate identification of erroneous structural data. This problem is most obvious in the case of structures containing only few (two) crystallographically independent MO_k polyhedra which are present in widely different numbers. The error (difference of Σs_i as against $\frac{1}{q}\sum_{y=1}^{q}(\Sigma s_i)_y$) of the MO_k polyhedra forming the majority has to be compensated by a correspondingly larger error of the MO_k polyhedra forming the minority. The latter would then possibly be rejected thus losing the counterbalance to the former; the remaining MO_k polyhedra would on the one hand cause a systematic error and on the other hand simulate an accuracy that is not given. See also the remarks on the weighing of the structures with application of Eq. (7) in Sect. 2.2.3.1.

Also, the use of a large number of MO_k polyhedra for the derivation of the bond length-bond valence parameters [15, 17, 19, 20] is not a guarantee of the reliability of the parameters. Important, above all, is the rejection of ill-defined structures (compare also Sect. 2.3.1.3). However, for the reliable identification of the latter structures via a normal distribution of the e_{Uq} values (compare Figs. 2a and 2b) it may be necessary to have a sufficient number of structures. According to our experience it is also not necessary to reject correct structures or to give structures with rare MO_k coordinations a greater weight only for the sake of having a well-balanced mixture of MO_k polyhedra of different sizes in the set of reference structures [9, 19, 20, 31–33]. As can be seen from Figs. 3a and 3b, 13% non-octahedra (tetrahedra) in a set of reference structures otherwise consisting of MO_6 octahedra results in about 50% of the maximally reachable height of the right branch of the graphs of the $\sigma_\Sigma(B)$ functions, which determines mainly the course of the middle line between the branches and hence the magnitude of a systematic error in B (see Sect. 2.3.1.3, Fig. 4). With regard to the bond lengths, it must be repeated that it is important to have only well-defined structures in the set of reference structures used for determining the bond length-bond valence parameters.

The standard deviation for the sum of the M–O bond valences about the M atoms for the final set of structures,

$$\sigma_\Sigma = (e_{Umin})^{1/2}, \tag{8}$$

represents a criterion for the quality of the set of the structural data if the mathematical function has been fully optimized. Otherwise, it is simultaneously a criterion for the quality of the mathematical function and its parameters. The d_0, B, and, if appropriate, further parameters belonging to this σ_Σ value are the final result of the optimization process.

Incorrect structural data in a set of reference structures can usually be recognized by an "emigration" of the B parameter to larger values, i.e., rejection of the corresponding species from the set of reference structures leads to smaller B parameters (compare Sects. 2.2.4.1–2.2.4.4 and in particular Fig. 4 in Sect. 2.3.1.3). Incorrect structural data due to large systematic bond length errors can be identified by individually fitted d_0 values, d_{0i}, and their comparison with (the general) d_0 (see Sect. 2.3.2.2).

2.2.3.3
Other Paths of Evaluation. The Standard Deviations for the M–O Bond Valences and for the Charge on the Oxygen Atoms

In Eqs. (7) and (7a) the average sum of the bond valences about the individual M atoms of a structure is used to derive the parameters of the bond length-bond valence functions by minimization of the average error square for this quantity. The standard deviation σ_Σ has been calculated from the minimal average error square (Eq. (8)).

The sum of the bond valences about the O atoms can also be used to derive the parameters of the bond length-bond valence functions. However, since the individual O atoms bear charges of unknown size (which occur as bond valence deficit or excess, see Eq. (3)), this can only be accomplished for the sum of the bond valences about all u O atoms, V_f, of a polyoxometalate ion as a whole. In analogy to Eq. (7) we can formulate as average error square for the bond valence of the M–O bonds about the O atoms or for the charge on the O atoms for a set of reference structures

$$e_V = \frac{1}{i}\sum_{f=1}^{i}\left(\frac{V_f - (2u_f - m_f)}{ß_f}\right)^2 = \frac{1}{i}\sum_{f=1}^{i}\left(\frac{V_f - q_f \cdot n}{ß_f}\right)^2 \tag{9}$$

(for f and i see Eq. (7)). The quantity ß is the frequency of the structure element considered, e.g., the number of M–O bonds or the number of O atoms in the individual structures[2], and takes into consideration that there

[2] Note that both expressions of Eq. (9) represent the same and do *not* characterize the paths to σ_s and σ_c, respectively. The paths leading to σ_s and σ_c via e_V (Eq. (10)) differ by the different meaning of $ß_f$ in both cases.

are several M–O bonds or O atoms in the individual structures but only one difference value for each structure. This approach to substitute the difference between the unknown sum of the bond valences about the individual O atoms and their valence by the difference between the sum of the bond valences V about all O atoms (which is equal to the sum of all bond valences in the polyoxometalate ion) and the valence sum 2u − m = qn of all O atoms and to divide it equally among the M–O bonds of the polyoxometalate ions leads to a small error for the average error square e_V in question (see below).

The standard deviations for the M–O bond valences and for the charge on the oxygen atoms,

$$\sigma_{s(uncorr)} = (e_{V\min})^{1/2} \quad \text{and} \quad \sigma_{c(uncorr)} = (e_{V\min})^{1/2}, \tag{10}$$

respectively, represent again criteria for the quality of the sets of reference structures. The d_0, B, and further parameters possibly belonging to these $\sigma_{s(uncorr)}$ and $\sigma_{c(uncorr)}$ values are the final results of these paths of the optimization processes. They correspond very closely (within 0.01 Å for B and 0.001 Å for d_0) to those based on σ_Σ (Eq. (8)). This agreement confirms the statement that the formal ionic charge is exclusively spread over oxygen atoms (a view which also follows from a consideration of the electronegativities of M and O or from attempts to formulate Lewis-type structures for the polyoxometalate ions, see Refs. 44 and 48).

The small deviations in B and d_0 between the different optimization bases just mentioned are probably due to the approximation character of Eq. (9). The approximation character of $\sigma_{s(uncorr)}$ and $\sigma_{c(uncorr)}$ can be corrected according to

$$\sigma_s \approx 1.25\sigma_{s(uncorr)} \quad \text{and} \quad \sigma_c \approx 1.25\sigma_{c(uncorr)}, \tag{10a}$$

respectively. The factor 1.25 considers the difference between a case in which all bond valence errors or charge errors, respectively, are equal (as demanded in Eq. (9)) and a case in which they show a normal distribution and thus have different values. This factor can be calculated from the Gaussian error integral function $\Phi(\lambda)$ according to

$$\frac{\sigma_c \sum_{g=1}^{\nu} \lambda_1}{\sigma_c \sum_{g=1}^{\nu} \lambda_{normal}} : \frac{\sigma_c^2 \sum_{g=1}^{\nu} \lambda_1^2}{\sigma_c^2 \sum_{g=1}^{\nu} \lambda_{normal}^2} = \frac{\sum_{g=1}^{\nu} \lambda_{normal}^2}{\sum_{g=1}^{\nu} \lambda_{normal}} \approx 1.25 \tag{10b}$$

(analogously with σ_s) where ν is the sufficiently finely chosen uniform partition of the Φ range 0 to 0.5, λ_1 the λ value for $\Phi = 0.3413$ (i.e., $\lambda_1 = 1$), and λ_{normal} are normally distributed λ values determined by the sequence $\Phi_g = \left(g - \frac{1}{2}\right)\frac{0.5}{\nu}$ with $g = 1$ to ν where ν, e.g., is chosen as 100.

2.2.4
Results

2.2.4.1
Mo^{VI}-O Bonds

There were 66 structure determinations of isopolyoxomolybdates and oxides (containing some MoO_4 tetrahedra, tetragonal MoO_5 pyramids, and pentagonal MoO_7 bipyramids besides MoO_6 octahedra) and 21 of monomolybdates at our disposal (stand: 1986; heteropolyoxometalates have not been used) which gave for Eq. (1a) $d_0 = 1.918$ Å, $B = 1.032$ Å, and $\sigma_\Sigma = 0.17$ v.u. Elimination of eight isopolyoxomolybdate and six monomolybdate structures as outliners (see Figs. 2a and 2b) resulted in a set of 73 reference structures which led to $d_0 = 1.915$ Å, $B = 0.953$ Å (Table 1), $\sigma_\Sigma = 0.12$ v.u. (2.0%) (Table 2); estimated error for B, derived from the optimization course: ±0.02 Å at most. The set of reference structures contained ca. 87% MO_6 octahedra, 9% MO_4 tetrahedra, and 4% tetragonal MO_5 pyramids plus pentagonal MO_7 bipyramids.

Eq. (1b) resulted for the same set of reference structures in $d_0 = 1.905$ Å, $N = 4.58$, and a somewhat greater $\sigma_\Sigma = 0.13$ v.u. Hence Eq. (1a) gives a somewhat better description of the structural data than Eq. (1b).

Introduction of a third adjustable parameter into Eqs. (1a) and (1b), e.g., according to

$$s = 10^{(d_0'-d)/B'} + a/d \quad \text{or} \quad s = [d_0'/(d-y)]^{N'}, \qquad (11a, 11b)$$

respectively, led for the case of Eq. (11a) to no further improvement of σ_Σ and for the case of Eq. (11b) to the adjustment of $\sigma_\Sigma = 0.12$ v.u., the value which was also obtained with Eqs. (1a) and (11a); the agreement of the three σ_Σ's is to within three places of decimals. We interpret this to indicate that no better fitting functions exist and that the remaining standard deviation of $\sigma_\Sigma = 0.12$ v.u. (2.0%) is virtually solely due to bond length errors in the structures (compare Sects. 2.2.3.1 and 2.2.3.2).

Table 2. Standard deviations for the sum of the bond valences about M (σ_Σ), for the bond valence of the M–O bonds (σ_s), and for the charge on the O atoms (σ_c) at application of Eq. (1a) with the (general) d_0 and B parameters of the present authors given in Table 1, substantiated by the sets of reference structures established in this article

	Mo^{VI}–O	W^{VI}–O[a]	V^{V}–O	Nb^{V}–O[a]	Ta^{V}–O
σ_Σ in v.u.	0.12	0.12	0.11	0.10	
σ_s in v.u.	0.020	0.010	0.020	0.019	
σ_c in c.u.	0.034	0.019	0.030	0.038	

[a] The values for the W^{VI}–O and Nb^{V}–O bonds are less well-defined because of the small number of (reliable) structures in the set of reference structures.

The standard deviations for the M–O bond valences and for the charge on the oxygen atoms on use of the exponential function, based on the set of reference structures, are given in Table 2.

2.2.4.2
W^{VI}–O Bonds

Structure determinations of 32 isopolyoxotungstates and oxides (containing few MO_4 tetrahedra besides WO_6 octahedra) and 5 monotungstates have been at our disposal (state: 1986) which gave for Eq. (1a) $d_0 = 1.938$ Å, B = 2.233(!) Å, and $\sigma_\Sigma = 0.26$ v.u. Identification and elimination of 27 isopolyoxotungstate and 4 monotungstate structures as outliners as described in Sect. 2.2.3.2 resulted in a set of only 6 reference structures (containing only one MO_4 tetrahedron) which led to $d_0 = 1.916$, B = 0.948 Å (Table 1), $\sigma_\Sigma = 0.12$ v.u. (2.0%) (Table 2). The identification procedure for the outliners was facilitated by a selection according to the bond lengths of the terminal M–O bonds which must lie in the range 1.5 to 2.0 v.u. [44, 48] (see also Sect. 3.1). The set of reference structures contained only ca. 3% non-octahedral polyhedra.

Eq. (1b) resulted for the same set of reference structures in a standard deviation σ_Σ of similar size. Because of the small size of the set of reference structures a statement about the better function is not meaningful.

Because of the small size of the set of reference structures a test for an improvement of Eqs. (1a) and (1b) by a third parameter is not reasonable.

The standard deviations for the M–O bond valences and for the charge on the oxygen atoms at application of the exponential function, based on the set of reference structures, are given in Table 2.

2.2.4.3
V^{V}–O Bonds

Structure determinations of 53 isopolyoxovanadates and oxides (containing VO_4 tetrahedra, tetragonal VO_5 pyramids, trigonal VO_5 bipyramids, and VO_6 octahedra) and 40 monovanadates have been at our disposal (state: 1986) which gave for Eq. (1a) $d_0 = 1.801$ Å, B = 0.987 Å, and $\sigma_\Sigma = 0.24$ v.u. Elimination of 10 isopolyoxovanadate and 15 monovanadate structures as outliners as described in Sect. 2.2.3.2 resulted in a set of 68 reference structures which led to $d_0 = 1.799$ Å, B = 0.851 Å (Table 1), $\sigma_\Sigma = 0.11$ v.u. (2.2%) (Table 2); estimated error for B, derived from the optimization course: ±0.02 Å at most. The set of reference structures contained ca. 52% MO_4 tetrahedra, 44% MO_6 octahedra, and 4% trigonal MO_5 bipyramids.

Eq. (1b) resulted for the same set of reference structures in $d_0 = 1.789$ Å, N = 5.12, and a somewhat greater $\sigma_\Sigma = 0.12$ v.u. This shows again that Eq. (1a) gives a somewhat better description of the structural data than Eq. (1b).

Introduction of a third adjustable parameter into Eqs. (1a) and (1b), e.g., according to Eqs. (11a) and (11b), respectively, led for the case of Eq. (11a) to no further improvement of σ_Σ and for the case of Eq. (11b) to the adjustment of $\sigma_\Sigma = 0.11$ v.u., the value which was also obtained with Eqs. (1a) and (11a). We interpret this to indicate that there exist no better fitting functions and that the remaining standard deviation of $\sigma_\Sigma = 0.11$ v.u. (2.2%) is merely due to bond length errors (compare Sects. 2.2.3.1 and 2.2.3.2).

The standard deviations for the M–O bond valences and for the charge on the oxygen atoms at application of the exponential function, based on the set of reference structures, are given in Table 2.

2.2.4.4
Nb^{V}–O Bonds

Structure determinations of 18 isopolyoxoniobates (containing NbO_6 octahedra, tetragonal NbO_5 pyramids, and trigonal NbO_5 bipyramids) have been at our disposal (state: 1986) which gave for Eq. (1a) $d_0 = 1.897$ Å, B = 1.145 Å, and $\sigma_\Sigma = 0.23$ v.u. Identification and elimination of 12 structures as outliners as described in Sect. 2.2.3.2 resulted in a set of merely 6 reference structures containing only MO_6 octahedra and thus being unsolvable. Introduction of one unreliable structure determination of an isopolyoxoniobate composed of tetragonal MO_5 pyramids (resulting in ca. 9% non-octahedral polyhedra for the set of reference structures) led to $d_0 = 1.917$ Å, B = 0.859 Å (Table 1), $\sigma_\Sigma = 0.10$ v.u. (2.0%) (Table 2).

Eq. (1b) resulted for the same set of reference structures in a standard deviation σ_Σ of similar size. Because of the small size of the set of reference structures again a statement about the better function is not meaningful.

Because of the small size of the set of reference structures a test for an improvement of Eqs. (1a) and (1b) by a third parameter is not reasonable.

The standard deviations for the M–O bond valences and for the charge on the oxygen atoms at application of the exponential function, based on the set of reference structures, are given in Table 2.

2.2.4.5
Ta^{V}–O Bonds

Since all available oxotantalate structures consisted only of MO_6 octahedra, an independent determination of B (or N) was impossible. By comparison of the B values of the other M^{V}–O bonds we chose B = 0.85 Å and calculated $d_0 = 1.921$ Å.

2.2.4.6
Comparison of the Bond Length-Bond Valence Functions Derived with Those of the Literature

The numerical values of the bond length-bond valence parameters are rather unsuitable for a comparison of the different functions. A better comparison is given by calculation of the values of the functions over their whole range of validity (see Fig. 1 for the Mo^{VI}–O bonds). In a diagram [63a] representing some Mo^{VI}–O bond length-bond valence functions from the literature it appears that a provisional function of the present authors [64a] plays an outsider role. The investigations by Brown and Altermatt [17] and Brese and O'Keeffe [16], which comprise the least possibilities for errors, however, support our function (compare Fig. 1).

The quality of a function and of the relevant parameters can, of course, only be rated by the standard deviation for the sum of the bond valences about the M atoms (or for other related quantities), referred to a set of reference structures which has been tested for the absence of outliners. Thus we have used the above (Sects. 2.2.4.1–2.2.4.4) sets of reference structures to test some frequently quoted bond length-bond valence functions in the literature (state: 1986) (Table 3). As can be seen, the parameters derived in this article give by far the smallest σ_Σ values in all cases tested. (The more recently proposed functions [16, 17] which have not been tested and come near to our functions are expected to give similarly good results.) The reasons for the greater deviations with the functions in the literature have been discussed in Sects. 2.2.2.1–2.2.2.6. In this context it is surprising that optimally fitted *linear* bond length-bond valence functions lead to σ_Σ values for our sets of reference structures that are of the same order of magnitude (or even better) than many of the non-linear functions in the literature (compare Table 3). This is again a consequence of the presence of unsuitable or erroneous structural data in the sets of structures used by the authors for the derivation of their bond length-bond valence parameters as discussed in Sects. 2.2.2.1–2.2.2.6.

Our bond length-bond valence curves (and those of Refs. 16 and 17) are generally more shallow, i.e., the range of the bond valences in the structures is smaller: the short M–O bonds remain more distant to bond valence two and the bond valences of the long M–O bonds are greater than those based on the functions proposed in the literature (compare Fig. 1).

Table 3. Test of the quality of bond length-bond valence functions from the literature by means of the standard deviation σ_Σ for the sum of the bond valences about M for the sets of reference structures established in this article

Authors	σ_Σ in v.u.				
	Mo^{VI}–O	W^{VI}–O	V^{V}–O	Nb^{V}–O	Ta^{V}–O
Schröder [45]	0.33				
Fuchs et al. [46]	0.25				
Allmann [8]	0.15	0.14	0.28		
Brown et al. [15]	0.21	0.19	0.12	0.11	
Trömel [31]			0.16		
Present authors	0.12	0.12	0.11	0.10	
Linear function[a]	0.22	0.20	0.15	0.15	

[a] Mo^{VI}–O: s = −1.778d + 4.518 v.u.; W^{VI}–O: s = −2.448d + 6.119 v.u.; V^{V}–O: s = −2.022d + 4.720 v.u.; Nb^{V}–O: s = −0.786d + 2.428 v.u. (s and d in Å)

2.3 Errors with the Use of the "General" and of "Universal" Bond Length-Bond Valence Parameters

The bond length-bond valence parameters d_0 and B (or d_0 and N), derived in Sects. 2.2.4.1–2.2.4.5 for the different M–O pairs from the structural data of varying numbers of compounds by optimization processes, are to be understand as "general" parameters. This refers also to all published parameters of other authors. The counterparts are, for each structural determination, individually fitted parameters where usually B (or N) is kept constant and d_0 is individually fitted, see Sect. 2.3.2.2. The term "universal" constant or parameter in the literature [9, 16, 17] is used in a different sense, namely to characterize a B (or N) that is assumed as applicable to a large number of element pairs (compare Sect. 2.2.2.5) as opposed to those that are only valid for a particular element pair.

2.3.1 *Errors of the Bond Length-Bond Valence Parameters B and Errors with the Use of "Universal" Bond Length-Bond Valence Parameters B*

In Sect. 2.2.2.5 it was stated that, according to the literature, the accuracy of B or N is even in the most favorable cases only to within 10% [15] or rarely better than 13% [17]. For our investigations we assume an error for B of ±0.02 Å (2%) at most, estimated from the intermediate results for B during the optimization course (identification and rejection of outliners) to find the optimal set of reference structures for the determination of the bond length-bond valence parameters (compare Sects. 2.2.3.2, 2.2.4.1, and 2.2.4.3). In Sect. 2.3.1.1 we show the importance of exact B values, and subsequently, in Sects. 2.3.1.2 and 2.3.1.3, we make an independent assessment of the achievable accuracy of B.

2.3.1.1 *The Necessary Accuracy of B: Effects of Errors of B on the Distribution of the Bond Valence Over the M–O Bonds and of the Charge Over the Oxygen Atoms*

The form of Eqs. (1a) and (1b) suggests a sensitive dependence of the bond valence upon B. This is confirmed by model calculations for a polyoxometalate structure ($Mo_7O_{24}^{6-}$) in which B is changed according to the statements in the literature [15, 17] about its accuracy by 10% from 0.953 to 0.858 Å, d_0 being in each case optimally fitted (Table 4). There are characteristic alterations. The smaller B value

- results in a too small bond valence (Δs up to 0.04 v.u.) for inner (bridging) and a too large bond valence (Δs up to 0.07 v.u.; for $V_{10}O_{28}^{6-}$ up to 0.12 v.u.) for outer (terminal) M–O bonds;
- results in too much (negative) charge on bridging (Δc up to 0.17 c.u.) and too little charge on terminal (Δc up to 0.07 c.u.) O atoms; thus the charge

separation effect [44, 48] is found to be too small. This error is most pronounced with the most highly coordinated O atoms (for the six-coordinate O atoms in $M_6O_{19}^{m-}$ and $M_{10}O_{28}^{6-}$: $\Delta c = 0.25$ and 0.22 c.u., respectively) since the *sum* of the bond valence alterations about the O atoms corresponds to the charge alterations.

Both items lead to an unfavorable picture of the polyoxometalate structures:

- the structure-stabilizing bond valence in bridging M–O bonds (see, e.g., Sect. 4.3.4.1 in Ref. 44 and Sect. 3.6.7 in Ref. 48) appears to be reduced,
- the negative charge on the terminal and pseudoterminal O atoms, which must necessarily be closely located to the salt cations, appears to be reduced whereas that on centrally located, inaccessible bridging O atoms appears to be enhanced; thus, the charge compensation between cations and polyoxometalate ions is complicated (see, e.g., Sect. 4.4.3.2 in Ref. 44 and Sect. 3.6.5 in Ref. 48).

From the above-mentioned Δs and/or Δc values it can easily be seen that an accuracy wanted for s or c of 0.02 v.u./c.u. admits a maximum error in B of 0.01 Å.

The preceding results indicate that the use of 'universal' bond length-bond valence parameters is by no means justified.

2.3.1.2
Possibility for the Derivation of Bond Length-Bond Valence Parameters from Single Structure Determinations. Tests with "Theoretical" Structures

The nature of the structural data of the polyoxometalates does in many cases fulfill the conditions for the derivation of the bond length-bond valence parameters B and d_0 from a single structure determination even if there are no MO_k polyhedra of different size (in apparent contrast to Sect. 2.2.2.2): For polyoxometalates containing at least two crystallographically independent coordination polyhedra; all structures composed of crystallographically identical coordination polyhedra (most of the monomeric and some polymeric ones) fall flat. The general form of the $\sigma_\Sigma(B)$ functions (d_0 in each case optimally fitted) for faultless structures ["theoretical" structures: structures with calculated bond lengths which fulfill exactly Eqs. (1) and (2) for an assumed (B,d_0) pair] is exemplarily shown in Fig. 3a(a1) for a species composed exclusively of MO_k polyhedra of the same size ($Mo_7O_{24}^{6-}$: seven MO_6 octahedra)[3] and in (b1)–(b3) for species composed of MO_k polyhedra of different size (α-$Mo_8O_{26}^{4-}$: six MO_6 octahedra and two MO_4 tetrahedra; $[Mo_4O_{14}^{4-}]_\infty$: two MO_6 octahedra and two MO_4 tetrahedra in the tetrameric building unit; $[Mo_6O_{20}^{4-}]_\infty$ (o,tp): two MO_6 octahedra and four MO_5 tetragonal

[3] (a2), the $\sigma_\Sigma(B)$ function for $Mo_2O_7^{2-}$ (composed of two MO_4 tetrahedra) has the same form and shows the correct minimum, however, all is laid on the abscissa for the scale used.

Table 4. Influence of an error of −10% in the bond length-bond valence parameter B = 0.953 Å (d_0 in each case optimally fitted) on the bond valences s of the different types and sorts of M–O bonds and on the charge c on the different types and sorts of O atoms in $Mo_7O_{24}^{6-}$ as an example

M–O bond type and sort, O atom type and sort[a]	f_n[b]	f_m[b]	d_n[c] in Å	s_n in v.u.	c_m in c.u.	s_n in v.u.	c_m in c.u.	Δs_n in v.u.	Δc_m in c.u.
				for B = 0.953 Å (d_0 = 1.9159 Å)		for B = 0.858 Å (d_0 = 1.9105 Å)		for ΔB = −0.095 Å	
M1–Ot1	2 ×	2 ×	1.708	1.652	−0.348	1.722	−0.278	+0.070	+0.070
M1–Ot2	2 ×	2 ×	1.728	1.574	−0.426	1.632	−0.368	+0.058	+0.058
M2–Ot3	4 ×	4 ×	1.717	1.617	−0.383	1.681	−0.319	+0.064	+0.064
M2–Ot4	4 ×	4 ×	1.726	1.582	−0.418	1.641	−0.359	+0.059	+0.059
M3–Opa[P]	2 ×	2 ×	1.739	1.533	−0.237	1.585	−0.223	+0.051	+0.014
M1–Opa*[P]	2 ×		2.524	0.230		0.193		−0.037	
M1–O2a1	4 ×	4 ×	1.922	0.985	−0.150	0.970	−0.192	−0.016	−0.042
M2–O2a1	4 ×		1.976	0.865		0.839		−0.026	
M2–O2a2	2 ×	2 ×	1.934	0.957	−0.086	0.939	−0.122	−0.018	−0.036
$M2^i$–O2a2	2 ×		1.934	0.957		0.939		−0.018	

M3–O3a	2 ×		1.896	1.049		1.040		−0.009	
M2–O3a*	2 ×	2 ×	2.267	0.428	−0.095	0.384	−0.192	−0.044	−0.098
M2[i]–O3a*	2 ×		2.267	0.428		0.384		−0.044	
M1–O4l*	2 ×		2.159	0.556		0.513		−0.042	
M2–O4l*	2 ×	2 ×	2.162	0.552	+0.092	0.509	−0.079	−0.042	−0.172
M2[ii]–O4l*	2 ×		2.162	0.552		0.509		−0.042	
M3–O4l*	2 ×		2.262	0.433		0.389		−0.044	
Average bond valence sum of M[d]:				6.000 v.u.		6.000 v.u.			
Sum of the charges[e]:					−6.000 c.u.		−6.000 c.u.		

[a] For the M and O atom types and sorts see Fig. 5. First index of O: coordination number of the O atom (special cases: t [terminal]: 1, p [pseudoterminal]: 2); second index of O (except for Ot): a angularly, l (approximately) linearly M–O–M bridging (these indices of O characterize its type). Next index of O: number to distinguish between several O atoms of the same type, characterizing the sort. * O atom in *trans* position to a terminal or pseudoterminal one; for details about the M–O bond types or types of O atoms see Sect. 3.2, in particular Figs. 2 and 1 in Ref. 44.

[b] Frequency of the structural element.

[c] Average of the M–O distances of eight reliable structure determinations stated in the literature (see Ref. 44) including averaging over the symmetry-equivalent M–O bonds.

[d] Optimization basis: $\Sigma s_o = \sum_{n=1}^{j} f_n s_n = qn = 2u - m$; average bond valence sum about M: $n = \frac{1}{q}\Sigma s_o$ (j: number of sorts of Mo–O bonds in the structure).

[e] Alternative (redundant) optimization basis: $\sum_{m=1}^{l} f_m c_m = [(8-n)q - p] = (2q - p) = m$ (l: number of sorts of O atoms in the structure).

[p] The strong bonds are pseudoterminal M–O bonds, the weak bonds are the complementary bonds of the pseudoterminal oxygen atoms.

pyramids in the hexameric building unit; for the structures see Ref. 44). In all cases B and d_0 can exactly be reproduced (see Fig. 3a).

In practice, for real structure determinations with bond lengths errors, there arise severe but well understandable difficulties. For the case that the bond length data of only two independent MO_k polyhedra are given, the optimization process leads inevitably to $\sigma_\Sigma = 0$ v.u. although the real σ_Σ is 0.12 v.u. (2% of n) for M^{VI} species (see Table 2). This works necessarily at the expense of the correctness of B. For structures containing more than two but still only few crystallographically independent MO_k polyhedra smaller σ_Σ values also result for the same reasons, again indicating that the

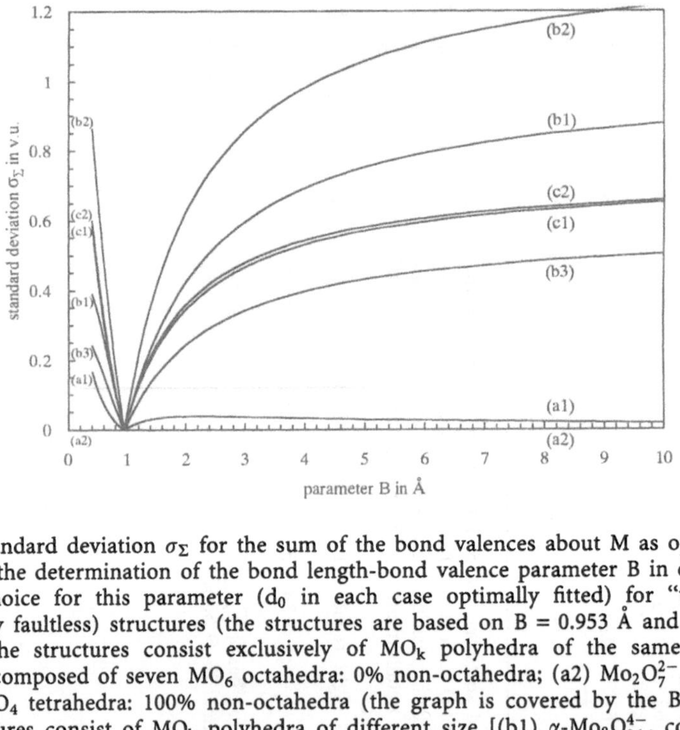

Fig. 3a. Standard deviation σ_Σ for the sum of the bond valences about M as optimization index for the determination of the bond length-bond valence parameter B in dependence on the choice for this parameter (d_0 in each case optimally fitted) for "theoretical" (absolutely faultless) structures (the structures are based on B = 0.953 Å and d_0 = 1.915 Å). (a) The structures consist exclusively of MO_k polyhedra of the same size [(a1) $Mo_7O_{24}^{6-}$, composed of seven MO_6 octahedra: 0% non-octahedra; (a2) $Mo_2O_7^{2-}$, composed of two MO_4 tetrahedra: 100% non-octahedra (the graph is covered by the B axis)]; (b) the structures consist of MO_k polyhedra of different size [(b1) α-$Mo_8O_{26}^{4-}$, composed of six MO_6 octahedra and two MO_4 tetrahedra: 25% non-octahedra; (b2) $[Mo_4O_{14}^{4-}]_\infty$, composed of two MO_6 octahedra and two MO_4 tetrahedra in the tetrameric building unit: 50% non-octahedra; (b3) $[Mo_6O_{20}^{4-}]_\infty$ (o,tp), composed of two MO_6 octahedra and four MO_5 tetragonal pyramids in the hexameric building unit: 66.7% non-octahedra]; (c) mixtures of structures [(c1) 1:1 mixture of $Mo_7O_{24}^{6-}$ and α-$Mo_8O_{26}^{4-}$: 13.3% non-octahedra; (c2) 1:1 mixture of $Mo_7O_{24}^{6-}$ and MoO_4^{2-}: 12.5% non-octahedra]. For the structures of the species see, e.g., Ref. 44. ······Straight line figuring $\sigma_\Sigma = 0.12$ v.u. as the standard deviation observed for M^{VI} species, characterizing by the intersections with the $\sigma_\Sigma(B)$ graphs the magnitude of the deviations from the correct B values in the case of experimentally ascertained structural data (see text). The left-side branches of the graphs show in some instances a different sequence due to a more complicated course of the graphs at small B values (occurrence of a more or less strongly marked shoulder or even an additional minimum).

corresponding B values are incorrect. For a collection of polyoxomolybdate species that have been tested (containing two or more independent MO_k polyhedra), B values in the range 0.5 to 1.3 Å have been observed. In addition, there are many examples (case of species exclusively composed of MO_k polyhedra of the same size) without any minimum for σ_Σ, i.e., the function $\sigma_\Sigma(B)$ approaches steadily the limit $\sigma_\Sigma \rightarrow 0$ v.u. for large B's. These relationships do not need an extradiagrammatic representation, they can also be read from Fig. 3a: The intersections of the straight line $\sigma_\Sigma = 0.12$ v.u. with the $\sigma_\Sigma(B)$ plots characterize the magnitude of the deviations from the correct B values, and the failure to realize an intersection in certain cases explains the examples without any minimum for σ_Σ.

The differences from the correct B values (the values obtained in Sect. 2.2.4) are thus caused by the bond lengths errors of the structures in connection with the small number of crystallographically independent MO_k polyhedra which allow the optimization processes paths to small σ_Σ values at the expense of correct B values. If the number of crystallographically independent MO_k polyhedra is large or if the bond lengths errors are small, the results approach the correct B values. On no account are there different

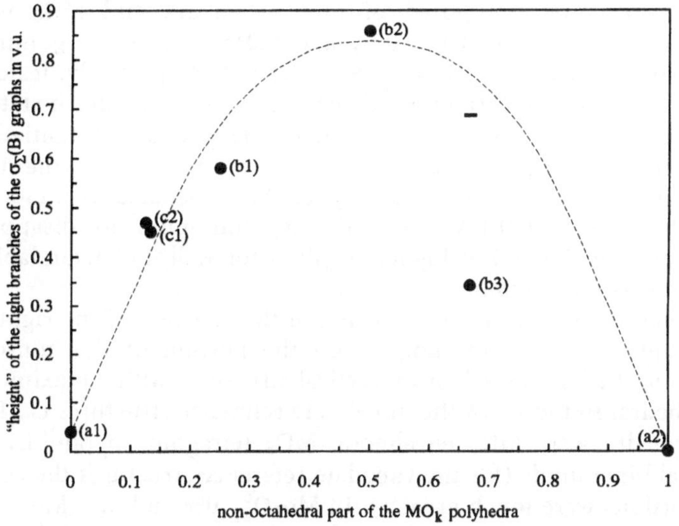

Fig. 3b. The "height" of the right branches of the $\sigma_\Sigma(B)$ graphs of Fig. 3a, characterized by the σ_Σ values at B = 3 Å, as a measure for the sharpness of the σ_Σ minima in dependence on the non-octahedral part of the MO_k polyhedra in the structures or mixtures of structures used for the investigations on the bond length-bond valence parameters. [Deviations from an equalizing curve are to be expected for several reasons. The unusually small value for (b3) is obviously due to the combination of MO_6 octahedra with MO_5 tetragonal pyramids, which do not differ so much, whereas in the other cases there are combinations of MO_6 octahedra with MO_4 tetrahedra. Since MO_5 tetragonal pyramids lie geometrically between MO_4 tetrahedra and MO_6 octahedra, one should in this case perhaps take twice (–) the value of the height.]

(optimal) B values for different structures of a given M (as is the case for d_0 due to systematic bond lengths errors, see Sect. 2.3.2.2). The results for the bond valence distribution over the M–O bonds and in particular for the charge distribution over the O atoms indicate only with constant B values for an M–O system a uniform picture for all structures of the system (see in particular Tables 41 and 42 in Ref. 44. The different types of O atoms show, on the whole, similar charges; differences are in each case well-explained).

2.3.1.3 Tests of Theoretical Structures with Incorporated Well-Defined Errors to Verify the Accuracy of the Obtainable B

The above investigations have been continued with tests using theoretical structures with incorporated well-defined errors. Introduction of systematic errors of +0.005 and −0.005 Å for the bond lengths into a theoretical $Mo_7O_{24}^{6-}$ and α-$Mo_8O_{26}^{4-}$ structure yielded an unchanged B = 0.953 Å and $\sigma_\Sigma = 0$ v.u., but $d_0 = 1.920$ and 1.910 Å, respectively (instead of 1.915 Å); this result was to be expected as can easily be seen from Eq. (1a).[4]

Introduction of random errors for the bond lengths into the theoretical α-$Mo_8O_{26}^{4-}$ structure (composed of MO_6 octahedra and MO_4 tetrahedra) corresponding to $\sigma_d = 0.003$, 0.007, and 0.015 Å gave a σ_Σ minimum at B = 0.953±0.010, 0.956±0.021, and 0.977±0.039 Å, respectively, tested in each case for nine different random distributions of the above-stated errors (σ_d's) (the ± values represent the standard deviations substantiated on the basis of the nine random distributions). In the case of the theoretical $Mo_7O_{24}^{6-}$ structure (composed merely of MO_6 octahedra) random errors greater than $\sigma_d \approx 0.010$ Å cause the σ_Σ minimum to disappear (50% disappearance at this σ_d) at higher B values for well-understandable reasons (see the preceding section).

From Figs. 3a and 3b it can be seen that the "height" of the right branches of the graphs depends non-linearly on the portion of the non-octahedral polyhedra in the "mini sets" of theoretical structures with a maximum at 50% non-octahedra. In the set of the molybdate reference structures ca. 13% of the MO_k polyhedra were MO_4 tetrahedra, MO_5 tetragonal pyramids, and MO_7 pentagonal bipyramids (for the vanadate reference structures the corresponding proportions were much greater). In $Mo_7O_{24}^{6-}$ 0% and in α-$Mo_8O_{26}^{4-}$ 25% of the MO_k polyhedra are not octahedra. We therefore combined the theoretical structures of α-$Mo_8O_{26}^{4-}$ and $Mo_7O_{24}^{6-}$ (the combination contains 13.3% non-octahedra) in a mini set and repeated the above calculations (introduction of random errors for the bond lengths corresponding to $\sigma_d = 0.003$, 0.007, and 0.015 Å, etc.). The result was a σ_Σ minimum at B = 0.953±0.008, 0.960±0.013,

[4] An additive (absolute) systematic error in the bond lengths has no influence on the B parameter of Eq. (1a) but d_0 changes according to the absolute systematic error; a percentage (relative) systematic error in the bond lengths has no influence on the N parameter of Eq. (1b) but d_0 changes according to the relative systematic error.

and 0.988±0.027 Å,[5] respectively, tested again in each case for nine different random distributions of the above-stated errors (σ_d's) (Fig. 4).

From Fig. 4 it can be seen that the B parameters lie within the statistical accuracy of the incorporated random errors for the bond lengths (σ_d) on the middle line between the two branches of the graph for $\sigma_d = 0.000$ Å. This means that the determination of the bond length-bond valence parameter B by localization of the minimum of the σ_Σ(B) curve is not only affected with a statistical (random) error but also with a systematic error. Both errors of B increase with the random error of d. Thus for $\sigma_d = 0.010$ Å, corresponding to $\sigma_\Sigma = 0.07$ v.u. (which equals $\sigma_{\Sigma i} = 0.07$ v.u., the value obtained with individually fitted d_{0i}'s; see Table 6 and Sect. 3.1), and the above case of the combination of the theoretical structures of α-$Mo_8O_{26}^{4-}$ and $Mo_7O_{24}^{6-}$ (containing 13.3% non-octahedra), the systematic error in B amounts to +0.010 Å (taken from Fig. 4 by interpolation). For the same case the random error on the basis of two structure determinations amounts to 0.019 Å (again taken from Fig. 4 by interpolation, i.e., interpolation between $\sigma_\Sigma = 0.013$ and 0.027 Å, see above); hence the 73 structures of the set of molybdate reference compounds create a random error of $((2/73)^{1/2} \cdot 0.019 =)$ 0.003 Å.

2.3.2
Errors of the Bond Length-Bond Valence Parameters d_0 and Systematic Bond Lengths Errors

2.3.2.1
Occurrence of Errors for the Average Bond Valence Sum and for the Formal Ionic Charge

Application of most of the general bond length-bond valence parameters together with the relevant functions stated in the literature [5, 7–9, 15, 18, 19, 31, 45– 47, 49] (Table 1) to (poly)oxometalate ions leads to very poor results for the bond valence sum of the species (i.e., for the *average* sum of the bond valences about the M atoms) and in particular for the formal ionic charge. A typical example of this kind is given in Sect. 4.8.1 (Table 49) in Ref. 44 and is discussed there in more detail; for other examples see Tables 7–9, this article. Some of the reasons for the extremely bad performance of these functions or parameters have been treated in Sects. 2.2.2.1–2.2.2.6: use of unsuitable structures and of inaccurate structural data in the procedure to derive the bond length-bond valence parameters; use of improper mathematical functions; incorrect mathemat-

---

[5] Note that these standard deviations established for the $Mo_7O_{24}^{6-}/\alpha$-$Mo_8O_{26}^{4-}$ mini set (mixture) have to be compared with those established above for α-$Mo_8O_{26}^{4-}$, divided by $\sqrt{2}$, since this mini set contains *two* estimations for random distributions of the errors (for $Mo_7O_{24}^{6-}$ and α-$Mo_8O_{26}^{4-}$) whereas there is only one such estimation for α-$Mo_8O_{26}^{4-}$. The greater deviations of the B's from the nominal value for B are a consequence of the smaller part of non-octahedra in the $Mo_7O_{24}^{6-}/\alpha$-$Mo_8O_{26}^{4-}$ mini set.

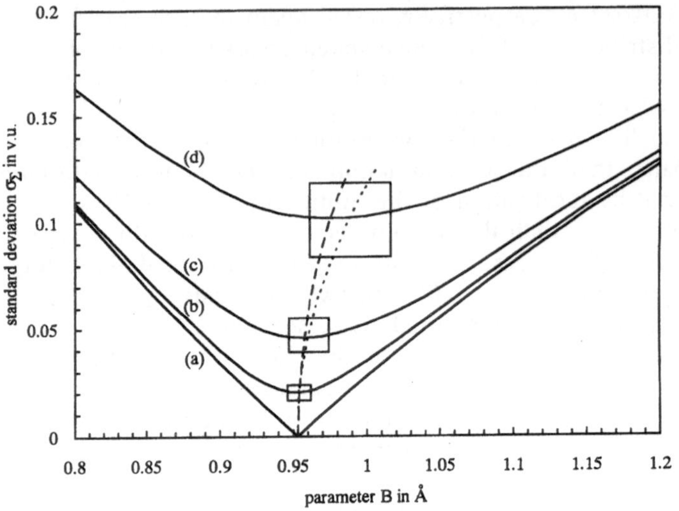

Fig. 4. Standard deviation σ_Σ for the sum of the bond valences about M as optimization index for the determination of the bond length-bond valence parameter B in dependence on the choice for this parameter (d_0 in each case optimally fitted) for a 1:1 mixture of a faultless $Mo_7O_{24}^{6-}$ and a faultless α-$Mo_8O_{26}^{4-}$ structure with different incorporated random errors for the bond lengths d (the structures are based on B = 0.953 Å and d_0 = 1.915 Å). The incorporated random errors correspond to standard deviations for d of (a) 0 Å, (b) 0.003 Å, (c) 0.007 Å, (d) 0.015 Å. The rectangles indicate the standard deviations for B and σ_Σ as found by introduction of nine different random distributions of the above-stated errors for d. In each case the curve for one of the nine incorporated random errors is shown, namely that whose minimum comes next to the center of the rectangles and hence should best describe the average of the nine curves. - - - - Line connecting the centers of the rectangles; - - - middle line between the two branches of graph (a)

ical evaluation of the structural data, including assumptions for quantities to be derived.

Application of the bond length-bond valence functions with the relevant general parameters, derived above (Sects. 2.2.4.1–2.2.4.5) by observance of all possible precautionary measures, and of two more recent, better functions from the literature [16, 17] nevertheless also leads to quite unsatisfactory, although somewhat improved, results for polyoxometalate ions. Table 5 presents a typical example for the results obtained. The bond valences are systematically somewhat too large (if the erroneous d_0 is greater than the correct d_0 – as in the example of Table 5) or somewhat too small (if the erroneous d_0 is smaller than the correct d_0). They are, however, quite acceptable for the identification of O atoms which form an OH group or a coordinated water molecule or for the identification of an M atom by its oxidation number. But for more ambitious considerations of the bond valences and in particular for considerations of the charge as described in Ref. 44 these results are quite unacceptable. An error in d_0 (or a systematic error of the d's) of 0.01 Å causes for the bond valence sum about each M^{VI} atom an error of 0.147 v.u. (2.5%) and for that about each

Table 5. Influence of an error $\Delta d_0 = 0.010$ Å in the bond length-bond valence parameter d_0 (= 1.9159 Å) (or of a systematic error in d of 0.010 Å) on the bond valence s of the different types and sorts of M–O bonds and on the charge c on the different types and sorts of O atoms in $Mo_7O_{24}^{6-}$ (B = 0.953 Å) as an example

M–O bond type and sort, O atom type and sort[a]	f_n[b]	f_m[b]	d_n[c] in Å	s_n in v.u.	c_m in c.u.	s_n in v.u.	c_m in c.u.	Δs_n in v.u.	Δc_m in c.u.
				for $d_0 = 1.9159$ Å		for $d_0 = 1.9259$ Å		for $\Delta d_0 = 0.010$ Å	
M1–Ot1	2 ×	2 ×	1.708	1.652	−0.348	1.693	−0.307	+0.040	+0.040
M1–Ot2	2 ×	2 ×	1.728	1.574	−0.426	1.613	−0.387	+0.039	+0.039
M2–Ot3	4 ×	4 ×	1.717	1.617	−0.383	1.656	−0.344	+0.040	+0.040
M2–Ot4	4 ×	4 ×	1.726	1.582	−0.418	1.621	−0.379	+0.039	+0.039
M3–Opa[P]	2 ×	2 ×	1.739	1.533	−0.237	1.571	−0.194	+0.037	+0.043
M1–Opa*[P]	2 ×		2.524	0.230		0.236		+0.006	
M1–O2a1	4 ×	4 ×	1.922	0.985	−0.150	1.009	−0.105	+0.024	+0.045
M2–O2a1	4 ×		1.976	0.865		0.886		+0.021	
M2–O2a2	2 ×	2 ×	1.934	0.957	−0.086	0.981	−0.039	+0.023	+0.047
M2[i]–O2a2	2 ×		1.934	0.957		0.981		+0.023	
M3–O3a	2 ×	2 ×	1.896	1.049	−0.095	1.075	−0.048	+0.026	+0.047
M2–O3a*	2 ×		2.267	0.428		0.439		+0.010	
M2[i]–O3a*	2 ×		2.267	0.428		0.439		+0.010	
M1–O4l*	2 ×	2 ×	2.159	0.556	+0.092	0.569	+0.144	+0.014	+0.051
M2–O4l*	2 ×		2.162	0.552		0.565		+0.013	
M2[ii]–O4l*	2 ×		2.162	0.552		0.565		+0.013	
M3–O4l*	2 ×		2.262	0.433		0.444		+0.011	
Average bond valence sum of M[d]:				6.000 v.u.		6.147 v.u.			
Sum of the charges[e]:					−6.000 c.u.		−4.973 c.u.		

For the footnotes see Table 4.

M^V atom of 0.123 v.u. (2.5%) and for the formal ionic charge an error of 0.147q c.u. for M^{VI} and 0.123q c.u. for M^V polyoxometalate ions (q: degree of aggregation of the polyoxometalate ion). For the examples of $V_{10}O_{28}^{6-}$ and $Mo_{36}O_{112}(H_2O)_{16}^{8-}$ this results in an error for the ionic charge of 1.2 and 5.3(!) c.u., respectively.

2.3.2.2
Compensation of the Errors for the Bond Valence Sum and for the Formal Ionic Charge by Introduction of Individually Fitted d_o Parameters (d_{oi}); Meaning of Individually Fitted d_o Values

The accuracy of the bond length-bond valence function along with the relevant parameters can be considerably enhanced by individual fitting of one of the parameters for each structure determination, keeping the other parameter(s) constant. Since the measured quantity affected with an error is a length, it is advisable to keep B (or N) constant and to fit d_0 (a length). Moreover, it has been explained (Sect. 2.3.1.2) that the results for the bond valence distribution over the M–O bonds and in particular for the charge distribution over the O atoms indicate a uniform picture for all structures of the system only with constant B values for an M–O system. Finally, the meaning of individually fitted d_0 parameters (see below) demands the improvement of d_0.

The aiming point for the individual fitting (optimization) for charged M–O species (isopolyoxometalate ions, oxides or oxide hydrates as special case with a zero charge; heteropolyoxometalate ions with the central heteroatom cut out) can be the correct formal ionic charge, the correct average sum of the bond valences about M (the correct sum of the bond valences for the whole structure), or the minimum of the sum of the error squares for the sum of the bond valences about the M atoms. The first two possibilities are preferred to find the optimal individual d_0 (termed d_{0i}) because of their simplicity and give identical results, i.e., both charge *and* average sum of the bond valences about the M atoms are correctly found. However, the third possibility can also be applied, in particular to identify outliners in analogy to Eq. (7a) (Sect. 2.2.3.2) by e_{Uqi}, see Sect. 2.3.2.3. The results are very close to those obtained by the first two possibilities. [This outliner test has to be handled with care since in structures containing only few (two) crystallographically independent MO_k polyhedra, which are present in widely differing numbers, the error of the MO_k polyhedra forming the majority has to be compensated by a correspondingly larger error of the MO_k polyhedra forming the minority, see Sect. 2.2.3.2.].

The use of a d_0 fitted to the respective structure determination had already been proposed by Allmann [8] in 1975 with reference to the fact that in this case the [average] bond valence sums about M came out correctly. However, this suggestion was not followed up; on the contrary, all authors attempt to derive bond length-bond valence relationships that are as universally applicable as possible [8] (compare also Sect. 2.2.2). This can necessarily only work at the expense of the accuracy.

An analysis of the d_{0i} values has a remarkable result. For the structure determinations for which $e_{Uq} \leqq 0.05\ (\text{v.u.})^2$ (corresponding to $\leqq 2\sigma_\Sigma$), all d_{0i}

values lie in a small range (±0.012 Å) about the general d_0, which is simultaneously the average of the d_{0i} values with a standard deviation of 0.005 Å. If $e_{U_q} > 0.05$ (v.u.)^2 (case of the outliners), the d_{0i} values vary in a wider range (+0.040 to −0.093(!) Å) and only for statistical reasons do some of them fall in the range $d_0 \pm 0.012$ Å ($= \bar{d}_{0i} \pm 0.012$ Å).

The deviations between d_{0i} and $d_0 = \bar{d}_{0i}$ correspond to *systematic* errors in the determination of the bond lengths of the structures. Systematic errors of bond lengths[6] have also been discussed by other authors [9, 31], albeit without reference to individually fitted d_0 values. In the present case the "good" structures are distinguished by a small standard deviation (caused by random errors) *and* a small systematic error for the bond lengths. The "bad" structures, on the contrary, show *simultaneously* a large standard deviation (expressed by an $e_{Uq} > 0.05$ (v.u.)^2) and a large systematic error for the bond lengths (expressed by a great difference $d_{0i} - d_0$; this difference is for statistical reasons indeed small in some cases). This combination of large random errors *and* large systematic errors underlines the defects of the respective structural data and suggests their rejection. Hence the determination of d_{0i} and its comparison with d_0 is a means to identify systematic bond lengths errors in structure determinations and to gain knowledge of its size.

The use of individually fitted d_0 values, d_{0i}, is a very important measure when working with bond length-bond valence functions and *is absolutely necessary*, for theoretical (compensation of systematic errors in bond lengths measurements) and practical (attainment of reasonable and at least formally correct results for the sum of the bond valences about M and for the formal ionic charge) reasons (see also Sect. 3.2).

Structures with *large* systematic bond lengths errors are disclosed according to Eq. (7a) in Sect. 2.2.3.2 together with those with large random errors in the procedure to derive the bond length-bond valence parameters (and rejected).

2.3.2.3
Standard Deviations for the Sum of the Bond Valences About the M Atoms and in Particular for the M–O Bond Valences and for the Charge on the Oxygen Atoms with the Use of Individually Fitted d_0 Values

When using individually fitted d_0 values, d_{0i}, the average error square for the sum of the bond valences about the M atoms, determined with the set of reference structures obtained according to Sect. 2.2.3.2 and 2.2.3.3,[7] is

---

[6] Note that the systematic error discussed here is not identical with the systematic error arising from thermal motion of the atoms discussed in Refs. 9 and 31.

[7] Note that it is not possible to apply Eqs. (12) and (12a) to a set of structures that has not been tested before by Eqs. (7) and (7a) because structures in which all MO_k polyhedra are symmetry-equivalent (most monometalates; possibly $Mo_6O_{19}^{m-}$, $W_{12}O_{38}(OH)_2^{6-}$) always have $e_{Uqi} = 0$ (v.u.)^2.

$$e_{Ui} = \frac{1}{i}\sum_{f=1}^{i}\left(\frac{1}{q}\sum_{y=1}^{q}(U_{iy} - n)^2\right)_f \tag{12}$$

(the *index* i is exclusively used for 'based on individually fitted d_0 parameters' and must not be confused with the quantity i as the number of structures in a set of reference structures) and the average error square for the sum of the bond valences about the M atoms for the individual structures (elimination of outliners)

$$e_{Uqi} = \frac{1}{q}\sum_{y=1}^{q}(U_{iy} - n)^2. \tag{12a}$$

The standard deviation for the sum of the bond valences about the M atoms in this case is

$$\sigma_{\Sigma i} = (e_{Ui\,min})^{1/2}. \tag{13}$$

The $\sigma_{\Sigma i}$ values amount on an average to 62% of the σ_Σ values and are given in Table 6.

The standard deviations for the M–O bond valences and for the charge on the O atoms cannot be ascertained in an analogous way since for the individual structures there is only *one* optimizable quantity and hence $e_{Vi\,min} = 0$ (v.u.)^2. We made the following estimation to get a surrogate for the standard deviations in question. We assumed that σ_s and σ_c (obtained according to Eqs. (9), (10), and (10a)) are changed to the same extent as σ_Σ (obtained according to Eqs. (7) and (8)) is changed to give $\sigma_{\Sigma i}$ (obtained according to Eqs. (12) and (13)). The resulting σ_{si} and σ_{ci} values are also given in Table 6.

In the absence of systematic bond lengths errors, σ_Σ, σ_s, and σ_c would be identical with $\sigma_{\Sigma i}$, σ_{si}, and σ_{ci}, respectively. Hence the σ_i's state that part of the standard deviation that is due to the random errors whereas the σ's comprise also the part due to the systematic error.

Table 6. Standard deviations for the sum of the bond valences about M ($\sigma_{\Sigma i}$), for the bond valence of the M–O bonds (σ_{si}), and for the charge on the O atoms (σ_{ci}) on application of Eq. (1a) with individually fitted d_0 parameters, d_{0i}, and the general B parameters of the present authors given in Table 1, substantiated by the sets of reference structures established in this article[a]

	Mo^{VI}–O	W^{VI}–O[b]	V^V–O	Nb^V–O[b]	Ta^V–O
$\sigma_{\Sigma i}$ in v.u.	0.07	0.12	0.07	0.04	
σ_{si} in v.u.	0.012	0.010	0.012	0.008	
σ_{ci} in c.u.	0.021	0.019	0.018	0.016	

[a] $\sigma_{\Sigma i}$, σ_{si}, and σ_{ci} define that part of the standard deviation that is due to the random errors.
[b] The values for the W^{VI}–O and Nb^V–O bonds are less well-defined because of the small number of (reliable) structures in the set of reference structures.

2.4 Occurrence of Very Long M···O Bonds and Necessity for Their Consideration in Polyoxometalate Chemistry

The influence of M···O distances $\gtrapprox$ 2.6 Å in Mo–O, W–O, Nb–O, and Ta–O compounds and $\gtrapprox$ 2.4 Å in V–O compounds, which correspond to bond valences $\lessapprox$ 0.2 v.u., has been investigated in Ref. 44 (in the literature M–O distances below the above limits are regarded as bonds in any case). The main results can be summarized as follows:

- In polyoxometalates composed of MO_6 octahedra (and MO_7 polyhedra), bond valences up to 0.03 v.u. – in rare, special cases 0.05 v.u. – are the inevitable consequence of the accumulation of MO_k polyhedra as a building principle of the polyoxometalate ions, i.e., of the connection modes of the MO_k polyhedra [typical M···O distances $\gtrapprox$ 3.2 Å (M = Mo, W, Nb, Ta), $\gtrapprox$ 2.9 Å (M = V)]. Altogether 1.2 to 2.0% of the bond valence of the M atoms are involved in such bonds beyond the octahedral coordination. Neglect of these supernumerary, very weak bonds leads to very small errors. The bond valences come out greater by up to 0.03 v.u. at the expense of the neglected very weak M···O bonds; the charge is changed by up to +0.03 to −0.03 c.u.
- In some (ca. 50%) of the polyoxometalates containing MO_k polyhedra (k taken as usually defined in the literature for the respective species, i.e., only M–O bond valences $\gtrapprox$ 0.2 v.u. are considered [44]) smaller than MO_6 octahedra (trigonal MO_5 bipyramids, tetragonal MO_5 pyramids, MO_4 tetrahedra), bond valences for the very long M···O bonds up to 0.17 v.u. have been observed. Altogether up to 3.9% of the bond valence of the M atoms are involved in bonds beyond the polyhedral coordination defined by k. Neglect of the supernumerary, very weak bonds in this sub-category leads to small or very small errors. The bond valences come out larger by up to 0.05 v.u.; the charge is changed by up to +0.06 to −0.14 c.u.

  In ca. 50% of the polyoxometalates of this category the errors are not greater than for the structures composed of MO_6 octahedra.

For details see Sects. 3.3, 4.1, and 4.6.2 in Ref. 44. (The above study refers, of course, to data obtained with d_{0i} values.)

The above results suggest that M···O distances up to 3.2 Å (M = Mo, W, Nb, Ta) or 2.9 Å (M = V) should still be considered as structure-defining bonds (s > 0.05 v.u.). Those in the range 2.6–3.2 Å or 2.4–2.9 Å, respectively, occur only in some of the structures containing MO_k polyhedra with $k < 6$ and correspond to bond valences of 0.20–0.05 v.u. Bond valences $\leqq$ 0.05 v.u. are an inevitable side-effect of the building principles of the polyoxometalate ions. For other element pairs, consideration of bond valences $\gtrapprox$ 0.08 v.u. [22], 0.08–0.20 v.u. [20], $\leqq$ (?) 0.1 v.u. [33] is regarded as being necessary in the literature.

2.5 Dependence of the Bond Length-Bond Valence Parameters of the M–O Bonds upon the Oxidation Number of M

As has already been mentioned (Sect. 2.2.2.2), some authors [18, 19, 31, 49, 65] assume no dependence of the bond length-bond valence parameters d_0 and B or d_0 and N upon the oxidation number of M, whereas others [15–17, 45] report small but clear differences for the different oxidation states of M. It is obviously a question of the assessment of the necessary or realizable precision of the bond length-bond valence parameters whether or not authors use a uniform d_0 and B or d_0 and N for different oxidation states of M. In view of the high dependence of the M–O bond valences and of the charge on the O atoms upon the bond length-bond valence parameters d_0 and B or d_0 and N (compare Sects. 2.3.1.1 and 2.3.2.1, in particular Tables 4 and 5), there is no basis for a common treatment of different oxidation states of an element.

We have found that d_0 for V^{IV}–O bonds is by 0.020 Å smaller than for V^{V}–O bonds, B assumed to be the same (V^{V}–O: $d_0 = 1.799$ Å, V^{IV}–O: $d_0 = 1.779$ Å; $B = 0.851$ Å). This relationship is also stated in the literature [15–17]. This difference is much greater than the standard deviation of 0.005 Å for d_{0i} which holds true if bad structures ($e_{Uq} > 0.05$ (v.u.)^2) are rejected from the set of reference structures used for the determination of d_0 and B (see Sect. 2.3.2.2).

There are many examples of mixed-valence polyoxometalates. Evaluation of these compounds requires that one of the bond length-bond valence parameters be held constant. Since it is absolutely necessary to correct the bond lengths for systematic errors by individual fitting of d_0 (see Sect. 2.3.2.2), it is indispensable to use a common B for the different oxidation states of M. This requirement appears to be fulfilled at least for $V^{IV,V}$–O compounds, see above and Refs. 15–17.

3 Discussion

3.1 The Quality of the Structure Determinations of the (Poly)oxometalates

In view of the difficulties with the derivation of the bond length-bond valence parameters B and d_0 because of inaccurate bond lengths stated in the literature for the structures, we give in the following a relevant characterization of the published structural data. Of the available data 88% for polymolybdates, 71% for monomolybdates, 81% for polyvanadates, and 63% for monovanadates have been identified as reliably characterized in respect of the bond lengths, i.e., they belong to a normally distributed collective with $\sigma_\Sigma = 0.12$ v.u. (2.0%) for the Mo compounds and 0.11 v.u. (2.2%) for the V compounds. In contrast, only 16% of the polytungstates, and 20% of the monotungstates, and 33% of the polyniobates have been identified as reliably characterized in respect of the bond lengths with $\sigma_\Sigma = 0.12$ v.u. (2.0%) for the W compounds and 0.10 v.u. (2.0%) for the Nb compounds (compare Sects. 2.2.4.1–2.2.4.5 and Table 2).

In the case of the polytungstates the quality of the bond length data was so bad that it was impossible to identify the correct data in the way described in Sect. 2.2.3.2 only. A prior selection of outliners was carried out according to the bond valence of the terminal M–O bonds which must lie in the range 1.5–2.0 v.u. for M^{VI} compounds. This bond valence range is defined by theoretical relationships (see Sect. 3.4 in Ref. 48) and has been experimentally confirmed (see Sect. 4.3.1 in Ref. 44).

We explain the somewhat better definition of the bond lengths of the polymetalates compared to those of the monometalates (molybdates and vanadates) by the presence of bridging O atoms in the polymetalates. Errors of the bond lengths due to erroneous positions of bridging O atoms largely compensate each other (compare Sect. 2.2.3.2) whereas those of terminal O atoms have their full effect. Thus two collectives with somewhat different standard deviations σ_Σ for the polymetalates (containing 14–71% bridging O atoms) and for the monometalates (containing 0% bridging O atoms) can be assumed.

By individual fitting of the d_0 parameters it has been found that the M–O bond lengths of the good structures are affected by a systematic bond length error with a standard deviation of 0.005 Å (Sect. 2.3.2.2). Due to the general rejection of structures which resulted in $e_{Uq} > 0.05$ (v.u.)^2, corresponding to $>2\sigma_\Sigma$ (Sect. 2.2.3.2), all d_{0i} values of the sets of reference structures lie in the range $d_0 \pm 0.012$ Å.

By introduction of well-defined random (statistical) errors into theoretical M^{VI} polyoxo structures it was found that $\sigma_d = 0.010$ Å leads to $\sigma_\Sigma = 0.07$ v.u. (1.2%) (Sect. 2.3.1.3, Fig. 4). Since, on the other hand, the standard deviation for the sum of the bond valences about Mo for the set of reference structures on application of Eq. (1a) with individually fitted d_{0i} parameters comes precisely to $\sigma_{\Sigma i} = 0.07$ v.u. (compare Table 6), the standard deviation for the random (statistical) bond lengths error amounts to merely $\sigma_{di} = 0.010$ Å (for the set of reference structures). (Note that σ_Σ equals $\sigma_{\Sigma i}$ if systematic bond lengths errors are absent.) This value is confirmed by the σ_{si} value (Table 6) which gives with application of Eq. (1a) to the typical bond lengths in the $Mo^{VI}O_6$ octahedra again $\sigma_{di} = 0.010$ Å.

Hence the structures of the sets of reference species are affected with a systematic bond length error of $\sigma_{d\,syst} = 0.005$ Å and a random bond length error of $\sigma_{d\,rand} = \sigma_{di} = 0.010$ Å. The average standard deviations σ_d stated in the literature for the reliably investigated structures amount to 0.009 Å for the Mo compounds, 0.006 Å for the V compounds, 0.025 Å for the few W compounds, and 0.007 Å for the few Nb compounds. This far-reaching correspondence of the σ_d standard deviations in combination with the absence of improvement in the standard deviation σ_Σ by introduction of a third bond length-bond valence parameter (compare Sects. 2.2.3.1, 2.2.4.1, and 2.2.4.3) indicates the high quality of the derived bond length-bond valence functions together with the relevant parameters (at least for Mo–O and V–O bonds).

3.2
Errors of the Bond Length-Bond Valence Parameters and Their Influence on the Individual M–O Bond Valences and on the Distribution of the Ionic Charge Over the O Atoms

The mathematical relationships between the accuracy of the bond lengths, the bond length-bond valence parameters, the bond valence sums about M, the M–O bond valences, and the charge on the O atoms are very complex and not very transparent (Sects. 2.2.3 and 2.3). An agreement between bond valence sum and atom valence within ±5% or even more [8, 9, 18, 20, 31, 33], an accuracy of d_0 of about ±0.02 Å [16], a standard error of d_0 of 0.01 Å [17, 31], or a B or N with an error of 10% [15] or 13% (but with an optimally adjusted d_0) [17] are absolutely insufficient for the determination of the M–O bond valences and particularly for the determination of the distribution of the charge over the oxygen atoms of polyoxometalate ions (Sect. 2.3). *Three* types of errors present even in the best and most recent studies in the literature have been identified to produce incorrect results. They can be divided into two groups: those which refer to the derivation of the bond length-bond valence parameters and those that refer to their application.

As has been assessed from Fig. 4, the standard deviation $\sigma_\Sigma = \sigma_{\Sigma i} =$ 0.07 v.u. (1.2%) for the sum of the bond valences about M^{VI} due to random errors for the set of reference structures corresponds to a standard deviation for the bond lengths of $\sigma_d = \sigma_{di} = 0.010$ Å (compare Table 6 and Sect. 3.1). These values form already the outermost limit to obtain a largely unaffected B (systematic error in B $\leqq$ 0.01 Å). Adequate measures to eliminate the sources for errors of the general bond length-bond valence parameters are *firstly* the outliner test for the identification and elimination of unusable (large random and/or large systematic bond length errors) structure determinations (Sect. 2.2.3.2; note that we reject the structure as a whole and not only single MO_k polyhedra of it) and *secondly* the rejection of 'universal' bond length-bond valence parameters (Sect. 2.3.1.1). The use of the latter in the literature is obviously caused by an insufficient rejection of outliners from the sets of reference structures applied for the determination of the bond length-bond valence parameters. This results in a high standard deviation for the sum of the bond valences about M [$\sigma_\Sigma = 0.3$ v.u. (5%)] as the optimization index (agreement index [9]), lying in a shallow minimum of the σ_Σ(B) function (compare Fig. 4). The latter observations (shallow minimum on a high σ_Σ level) are obviously the reason for the assumption of a large uncertainty for B or N (10% [15], 13% [17]) and of similar 'universal' B or N parameters for many element pairs (provided that d_0 is optimally adjusted for the particular B or N values) in the literature [8, 9, 16, 17] (compare Sect. 2.2.2.5). According to the mode of the investigations (derivation of the bond length-bond valence parameters for a large number of element pairs by a mass handling of structures), we assume that the optimization index was indeed merely a minimal standard deviation for the sum of the bond valences about M and that no investigations about the errors for the M–O bond valences and for the charge on the O atoms with a change of B or N have been carried out. The result of such investigations for (poly)oxometalate ions is (compare Table 4) that an increasing B (or decreasing N) leads to decreasing differences of the bond valences for strong, middle-strong, and weak M–O bonds of the (distorted) MO_6, MO_5, and MO_7 polyhedra and to increasing differences for the charge distributed over the O atoms, and vice versa. In other words: the size of B strongly effects the proportion of the bond valence in inner (bridging) and outer (terminal) M–O bonds and also strongly the results for the charge separation processes (Sect. 2.3.1.1) (the stability of polyoxometalate ions is strongly affected by these factors [44, 48]). In this connection it is surprising that in the literature the low accuracy of B is not seen to bring discredit upon the applicability of such B (or N) values. The obvious reasons are the small changes of the sum of the bond valences about M and also small changes of σ_Σ (provided d_0 is optimally fitted to the particular B) on a high level of σ_Σ (compare Fig. 4) with alterations of B. Summarizing the discussion about 'universal' B parameters we can state that, whereas the sum of the bond valences about M and consequently σ_Σ remain nearly constant and hence suggest negligible alterations for the system considered, the M–O bond valences and the charge on the O atoms actually change considerably even with small changes of B or N. Indeed, an accuracy of B (and presumably N) within at least 1% is necessary and hence desirable to keep the bond valence and charge deviations of the individual M–O bonds or O atoms, respectively, with certainty smaller than 0.02 v.u./c.u. This can hardly be achieved by a 'universal' B parameter. At present, the existence of such a 'universal' B parameter has still to be proved.

An adequate and inevitable measure at application even of properly derived bond length-bond valence parameters is *thirdly* the correction for small systematic bond lengths errors by use of individually fitted d_0 values, d_{0i} (Sect. 2.3.2.2). Evaluation of structures with (even small) systematic bond lengths errors without individual fitting of d_0 results in one-sided bond valence errors and one-sided errors for the charge distribution over the O atoms which sum up in the latter case to great errors for the formal ionic charge and can reach some charge units. Evaluation of outliner structures (with great random and/or great systematic bond lengths errors) results in inconsistent distributions of the bond valences over the M–O bonds and of the charge over the O atoms. Outliner structures with great systematic but small random errors *can* yield quite usable results if d_0 is individually fitted (compare the $V_5O_{14}^{3-}$ example [44]).

Our results about the identification and elimination of outliner structures from the set of reference species used for the derivation of the bond length-bond valence parameters (compare Sects. 2.2.4.1–2.2.4.4) indicate clearly that the presence of structures with great bond lengths errors leads to B parameters that are too great ("migration" of the B values in one direction, see Sect. 2.2.2.3). The results of the introduction of random bond lengths errors into theoretical structures (compare Fig. 4 and the discussion in Sect. 2.3.1.3) show that B becomes indeed greater and explain how this is accomplished. In contrast to these observations we notice that the B parameters published in the literature are often *smaller* (compare Table 1) than ours (for N the reverse is true) despite the presence of great(er) bond lengths errors (resulting in an average error for the bond valence sums about M of 5% = 0.30 v.u. for M^{VI}) in the sets of the reference structures used in the literature. This is an inconceivable result as far as least-squares fitting procedures have been carried out (which is absolutely necessary for an incontestable derivation of the bond length-bond valence parameters, compare Sect. 2.2.2.4). In our opinion the authors must have overlooked the unintentional introduction of assumptions.

In this connection it must also be stated that standard deviations of $\sigma_\Sigma = 5\%$ [20, 31, 33] are inconsistent with the exclusive use of structures with standard deviations for the bond lengths of $\sigma_d \leqq 0.01$ Å [19, 31]. A $\sigma_d = 0.01$ Å causes only a $\sigma_\Sigma = 0.07$ v.u., corresponding to 1.2% for M^{VI} and 1.4% for M^V compounds, see above, this section.

We return to the statement in Sect. 2.1 that the bond valence sums are not very sensitive to the values of B (or N), however, that the values of d_0 are very sensitive to the exact choice of B (or N) [4a, 8, 9, 15, 17, 31]. This statement has to be taken literally, that is, bond valence parameters that have been derived without an exact analysis of the possibilities for errors and their magnitude may be used for the estimation of the bond valence sums *but not for the calculation of bond valences for the individual* M–O *bonds and of charges for the individual O atoms.* They have clearly not been derived for this purpose; however, as a proof of the correctness of interpretations of crystal structure investigations in the usual meaning [2–4, 8, 11, 14] they can be used safely.

3.3 Working with Individually Fitted d_0 Parameters, d_{0i}

In Sect. 2.3.2 we have shown that the simple use of d_0 parameters from the literature (general parameters) is an absolutely inadequate procedure, in particular if considerations about the charge are involved. Only in those cases where a special adjustment of these parameters is impossible, i.e., if no basis for an optimization process for the d_0 parameters of an element pair exists (this applies in particular to the case that this element pair exists only in mixed-ligand coordination polyhedra), is the direct application of literature values justifiable and, in fact, the only adequate way. Incomplete use of the possibilities to fit individually the d_0 bond length-bond valence parameters leads to deviations of the average bond valence sums of the elements from their oxidation number and to deviations of the charge from the formal ionic charge. An individual fitting of the d_0 parameters is possible

- via each kind of complete EL_k coordination polyhedra (e.g., via the MO_k and via the XO_k polyhedra in heteropolyoxometalate ions) and
- via the formal ionic charge

of the species in question. Three examples are discussed to illustrate the procedure.

For the $Mo_7O_{24}^{6-}$ isopolyoxometalate structure [66] (Fig. 5) with seven complete MoO_6 octahedra of, in the ideal case, three different coordination modes we have one possibility to fit individually a d_0 parameter, namely for the Mo–O bonds. This can be done via the correct average sum of the bond valences about the Mo atoms (= 6 v.u.) or via the correct formal ionic charge of the anion (= −6 c.u.) and both lead exactly to the same value. Hence the result of one of the two fitting procedures can be viewed as a redundant control index (Table 7(a)). In Table 7(b) and (c), general bond length-bond valence parameters B and d_0 have been applied to show the necessity of the use of an individually fitted d_0 parameter. [A similar example ($M_6O_{19}^{2-}$), in which results obtained with our B parameter and the corresponding individually fitted d_0 parameter are opposed to results for the same structural data obtained with bond length-bond valence parameters of the literature, is given in Table 49, Sect. 4.8.1 in Ref. 44.]

For the $H_4As_4Mo_{12}O_{50}^{4-}$ heteropolyoxometalate structure [67] (Fig. 6) with twelve complete, in the ideal case symmetry-equivalent, MoO_6 octahedra and four complete, in the ideal case symmetry-equivalent, AsO_4 tetrahedra we have two possibilities to fit individually d_0 parameters, namely for the Mo–O and for the As–O bonds. This can be done via the correct (average) sum of the bond valences about the Mo atoms in the MoO_6 octahedra (= 6 v.u.), via the correct (average) sum of the bond valences about the As atoms in the AsO_4 tetrahedra (= 5 v.u.), and/or via the formal ionic charge of the heteropoly-anion (= −4 c.u.) as the sum of the charges on the O atoms. One of the three possibilities serves as a redundant control index (Table 8(a)). In Table 8(b) and (c), general bond length-bond valence parameters B and d_0 for the Mo–O

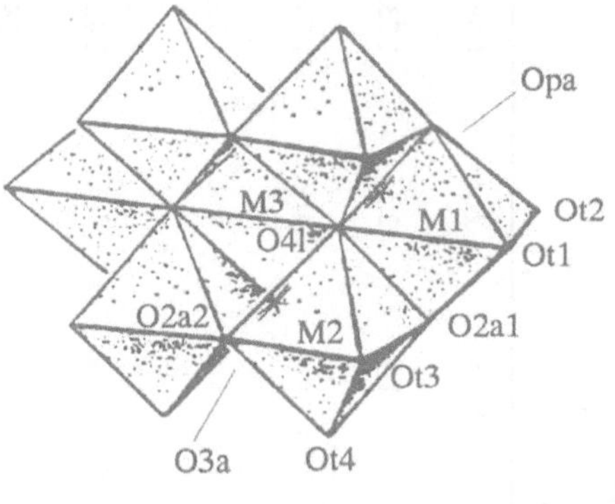

Fig. 5. Coordination polyhedral model of the structure of the $Mo_7O_{24}^{6-}$ isopolyoxometalate ion, composed of seven MoO_6 octahedra, and characterization of the – in the ideal case – crystallographically independent M and O atoms (for the meaning of the indices of O, see footnote a of Table 4)

Table 7. Illustration of the evaluation of the structure determination for the $Mo_7O_{24}^{6-}$ isopolyoxometalate ion in $K_6Mo_7O_{24}{\cdot}4H_2O$ [66] (assessment of the M–O bond valences s and of the charges c on the O atoms) using
(a) our general B = 0.953 Å for Mo-O bonds and an individually fitted d_0, d_{0i} = 1.9202 Å (standard procedure proposed in this article)
(b) our general B = 0.953 Å and general d_0 = 1.915 Å for Mo–O bonds (non-consideration of systematic bond lengths errors)
(c) the general (= 'universal') B = 0.852 Å and general d_0 = 1.907 Å for Mo–O bonds [17] (as the best parameters of the literature)

M–O bond type and sort, O atom type and sort[a]	f_n[b]	f_m[b]	d_n[c] in Å	s_n in v.u. (a)	c_m in c.u. (a)	s_n in v.u. (b)	c_m in c.u. (b)	s_n in v.u. (c)	c_m in c.u. (c)
M1–Ot1	2 ×	2 ×	1.695	1.733	−0.267	1.699	−0.301	1.774	−0.226
M1–Ot2	2 ×	2 ×	1.735	1.573	−0.427	1.543	−0.457	1.592	−0.408
M2–Ot3	4 ×	4 ×	1.745	1.536	−0.464	1.506	−0.494	1.549	−0.451
M2–Ot4	4 ×	4 ×	1.763	1.470	−0.530	1.442	−0.558	1.476	−0.524
M3–Opa M1–Opa*	2 × 2 ×	2 ×	1.744 2.450	1.539 0.280	−0.181	1.509 0.274	−0.216	1.554 0.230	−0.216
M1–O2a1 M2–O2a1	4 × 4 ×	4 ×	1.943 1.938	0.952 0.963	−0.085	0.933 0.945	−0.122	0.907 0.920	−0.173
M2–O2a2 $M2^i$–O2a2	2 × 2 ×	2 ×	1.928 1.928	0.987 0.987	−0.026	0.968 0.968	−0.065	0.945 0.945	−0.110
M3–O3a M2–O3a* $M2^i$–O3a*	2 × 2 × 2 ×	2 ×	1.925 2.220 2.220	0.994 0.487 0.487	−0.031	0.975 0.478 0.478	−0.070	0.953 0.429 0.429	−0.189

M1–O4l*	2 ×		2.155	0.570		0.559		0.512	
M2–O4l*	2 ×	2 ×	2.180	0.537	+0.092	0.526	+0.051	0.478	−0.142
M2^{ii}–O4l*	2 ×		2.180	0.537		0.526		0.478	
M3–O4l*	2 ×		2.255	0.448		0.439		0.390	
Average bond valence sum of M[d]:				6.000 v.u.		5.884 v.u		5.832 v.u	
Sum of the charges[e]:					−6.000 c.u.		−6.815 c.u.		−7.176 c.u.

[c] Average of the M–O distances of the symmetry-equivalent M–O bonds.
For the other footnotes see Table 4.

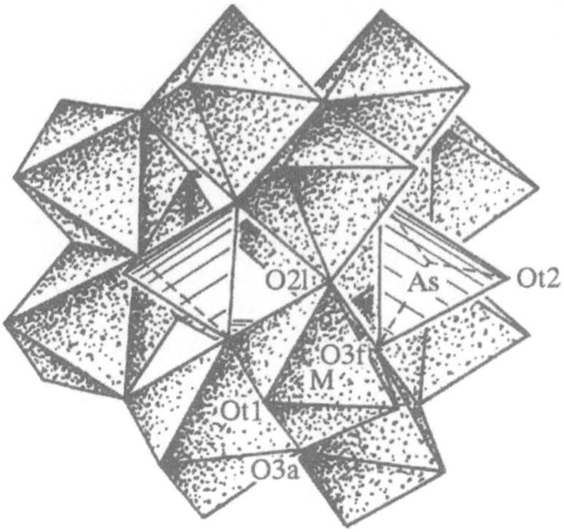

Fig. 6. Coordination polyhedral model of the structure of the $H_4As_4Mo_{12}O_{50}^{4-}$ heteropoly-oxometalate ion, ideally composed of twelve symmetry-equivalent MO_6 octahedra and four symmetry-equivalent AsO_4 tetrahedra (protonated at Ot2), and characterization of the M, As, and O atoms (for the meaning of the indices of O see footnotes a of Tables 4 and 8

and As–O bonds have been applied to show again the necessity of the use of individually fitted d_0 parameters.

For the structure of the $MoO_2(C_5O_5)_2^{2-}$ chelate complex ion [68] with a complete MoO_6 octahedron and two bidentate-chelating $C_5O_5^{2-}$ (croconate) ligands (Fig. 7) we have the possibility to fit individually the d_0 parameter for the Mo–O bonds via the correct sum of the bond valences about the Mo atom in the MoO_6 octahedron (= 6 v.u.). The sum of the charges on the O atoms (= −2 c.u.) is not a redundant control index in this case but serves to fit individually the d_0 parameter for the C–O bonds in which oxygen is the charge-bearing atom. The other bonds in the chelate complex (C–C bonds) act neutrally (Table 9(a)) [from the difference between the sum of the bond valences about the 10 C atoms (= 40 v.u.) and the sum of the 10 C–O bond valences (= 16 v.u.) one can calculate the *average* C–C bond valence which amounts to 1.2 v.u.]. In Table 9(b) and (c) general bond length-bond valence parameters B and d_0 for the Mo–O and C–O bonds have been applied to show again the necessity of the use of individually fitted d_0 parameters.

It has still to be mentioned that the arithmetical accuracy requires consideration of four to five decimals in all calculation steps if two correct decimals are desired in the final results.

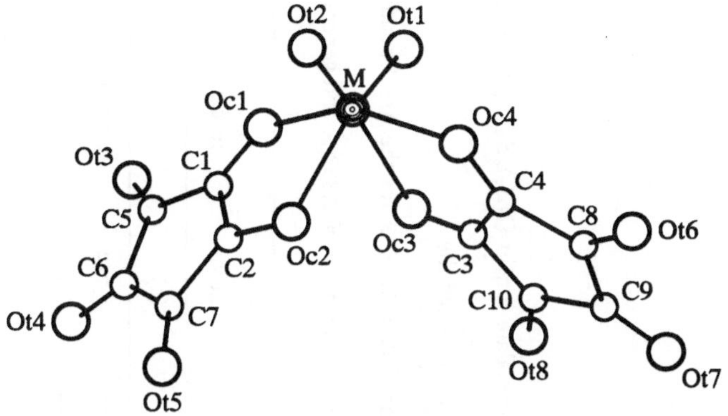

Fig. 7. Structure of the $MoO_2(C_5O_5)_2^{2-}$ chelate complex ion and characterization of the M, C, and O atoms

3.4 Apparent Deviations of Bond Valence Sums for Oxygen (Occurrence of Charge Separation Processes)

Some authors report bond valence sums for O atoms in the range 1.75–2.30 v.u. [9], 1.4–2.3 v.u. [18] (use of general bond length-bond valence parameters), 2.0 ± 0.2 v.u. (use of individually fitted bond length-bond valence parameters) [8] which are seen as real effects [9]. A definite explanation was not given (possible presence of superstructure or disorder [18]). We interpret these effects as the result of charge separation processes (see Sects. 3.2.3, 3.6.7, and others in Ref. 48 and Sects. 4.4.1, 4.4.2.2, and others in Ref. 44). In this case mainly (approximately) linearly bridging O atoms show the positive and terminal O atoms the negative deviations from the nominal valence of O (= 2).

3.5 Apparent Deviations of Bond Valence Sums for M (Effect of "Non-Bonded Interactions", "Residual Bond Strains")?

Some authors report bond valence sums for atoms of electropositive elements E, calculated from the bond lengths according to Eqs. (1) and (2) ("apparent valences", experimental bond valences), deviating by up to 26% from the valences corresponding to the oxidation number of the elements ("actual valences", atomic valences) (use of general bond length-bond valence parameters). The differences have been explained by "non-bonded interactions" ("non-bonded repulsions") [14, 60] or "residual bond strains" [21]. Strained bonds are assumed if R1 [21] [see Sect. 2.2.2; this quantity corresponds to $(e_{Uq})^{1/2}$, related to the general B and d_0 parameters] exceeds 0.1 v.u. This numerical value corresponds to σ_Σ observed in our investigations (compare Table 2).

Table 8. Illustration of the evaluation of the structure determination for the $H_4As_4Mo_{12}O_{50}^{4-}$ heteropolyoxometalate ion in $(NH_4)_4H_4As_4Mo_{12}O_{50} \cdot 4H_2O$ [67] (assessment of the M–O and As–O bond valences s and of the charges c on the O atoms) using
(a) our general B = 0.953 Å for the Mo–O bonds and an individually fitted d_{0i} = 1.9117 Å, the general B = 0.852 Å for the As–O bonds [17] and an individually fitted d_{0i} = 1.7788 Å (standard procedure proposed in this article)
(b) our general B = 0.953 Å and general d_0 = 1.915 Å for the Mo–O bonds, the general B = 0.852 Å and general d_0 = 1.767 Å for the As–O bonds [17] (non-consideration of systematic bond length errors)
(c) the general (= 'universal') B = 0.852 Å and general d_0 = 1.907 Å for the Mo–O bonds [17], the general (= 'universal') B = 0.852 Å and general d_0 = 1.767 Å for the As–O bonds [17] (as the best parameters of the literature)

E–O bond type and sort, O atom type and sort[a]	f_n [b]	f_m [b]	d_n [c] in Å	s_n in v.u. (a)	c_m in c.u. (a)	s_n in v.u. (b)	c_m in c.u. (b)	s_n in v.u. (c)	c_m in c.u (c)
M–Ot1	24 ×	24 ×	1.698	1.677	−0.323	1.687	−0.313	1.759	−0.241
As–Ot2[f]	4 ×	4 ×	1.688	1.278	+0.278	1.238	+0.238	1.238	+0.238
M–O2l	6 ×	6 ×	1.874	1.096	+0.192	1.103	+0.205	1.093	+0.187
M^i–O2l	6 ×		1.874	1.096		1.103		1.093	
M–O3a	4 ×		2.033	0.746		0.751		0.711	
M^i–O3a	4 ×	4 ×	2.033	0.746	+0.239	0.751	+0.253	0.711	+0.134
M^{ii}–O3a	4 ×		2.033	0.746		0.751		0.711	

M–O3f*	12 ×		2.289	0.402		0.405		0.356	
M^{iii}–O3f*	12 ×	12 ×	2.289	0.402	+0.045	0.405	+0.011	0.356	−0.086
As–O3f	12 ×		1.699	1.241		1.202		1.202	
(Average) bond valence sum of M[d]:				6.000 v.u.		6.036 v.u		6.035 v.u.	
(Average) bond valence sum of As[g]:				5.000 v.u.		4.843 v.u.		4.843 v.u.	
Sum of the charges[e]:					−4.000 c.u.		−4.194 c.u.		−4.203 c.u.

[a] For the O atom types and sorts see Figure 6 and footnote a of Table 4; second index of O: f bifurcated bridging.
[d] Optimization basis: $\sum_{n=1}^{j} f_n s_n$ = number of M atoms in the polyanion × oxidation number of M (consideration of the M–O bonds only); (average) bond valence sum about M: $n = \frac{1}{12}\sum_{n=1}^{j} f_n s_n$ (j: number of sorts of Mo–O bonds in the structure).
[e] Alternative (redundant) optimization basis: $\sum_{m=1}^{l} f_m c_m$ = ionic charge of the polyanion (consideration of *all* O atoms) (l: number of sorts of O atoms in the structure).
[f] For the protonated Ot2 atoms formally an O–H bond valence of 1 v.u. has been assumed.
[g] Optimization basis: $\sum_{n=1}^{j} f_n s_n$ = number of As atoms in the polyanion × oxidation number of As (consideration of the As–O bonds only); (average) bond valence sum about As: $n = \frac{1}{4}\sum_{n=1}^{j} f_n s_n$ (j: number of sorts of As–O bonds in the structure).
For the other footnotes see Table 4.

Table 9. Illustration of the evaluation of the structure determination for the $MoO_2(C_5O_5)_2^{2-}$ chelate complex ion in $[(n\text{-}C_4H_9)_4N]_2MoO_2(C_5O_5)_2$ [68] (assessment of the M–O and C–O bond valences s and of the charges c on the O atoms) using
(a) our general B = 0.953 Å for the Mo–O bonds and an individually fitted d_{0i} = 1.9296 Å, the general B = 0.852 Å for the C–O bonds [17] and an individually fitted d_{0i} = 1.4051 Å (standard procedure proposed in this article)
(b) our general B = 0.953 Å and general d_0 = 1.915 Å for the Mo–O bonds, the general B = 0.852 Å and general d_0 = 1.390 Å for the C–O bonds [17] (non-consideration of systematic bond length errors)
(c) the general (= 'universal') B = 0.852 Å and general d_0 = 1.907 Å for the Mo–O bonds [17], the general (= 'universal') B = 0.852 Å and general d_0 = 1.390 Å for the C–O bonds [17] (as the best parameters in the literature)

E–O bond type and sort, O atom type and sort[a]	f_n [b]	f_m [b]	d_n in Å	s_n in v.u. (a)	c_m in c.u. (a)	s_n in v.u. (b)	c_m in c.u. (b)	s_n in v.u. (c)	c_m in c.u. (c)
M–Ot1	1 ×	1 ×	1.684	1.810	−0.190	1.745	−0.255	1.827	−0.173
M–Ot2	1 ×	1 ×	1.710	1.700	−0.300	1.639	−0.361	1.703	−0.297
M–Oc1 C1–Oc1	1 × 1 ×	1 ×	2.019 1.298	0.806 1.336	+0.142	0.777 1.282	+0.059	0.739 1.282	+0.021
M–Oc2* C2–Oc2	1 × 1 ×	1 ×	2.246 1.298	0.466 1.336	−0.199	0.449 1.282	−0.269	0.400 1.282	−0.318
M–Oc3* C3–Oc3	1 × 1 ×	1 ×	2.251 1.275	0.460 1.422	−0.118	0.443 1.365	−0.192	0.395 1.365	−0.241
M–Oc4 C4–Oc4	1 × 1 ×	1 ×	2.044 1.303	0.759 1.318	+0.076	0.731 1.265	−0.004	0.691 1.265	−0.044
C5–Ot3	1 ×	1 ×	1.188	1.798	−0.202	1.726	−0.274	1.726	−0.274
C6–Ot4	1 ×	1 ×	1.194	1.769	−0.231	1.698	−0.302	1.698	−0.302
C7–Ot5	1 ×	1 ×	1.197	1.755	−0.245	1.685	−0.315	1.685	−0.315
C8–Ot6	1 ×	1 ×	1.230	1.605	−0.395	1.541	−0.459	1.541	−0.459
C9–Ot7	1 ×	1 ×	1.184	1.818	−0.182	1.745	−0.255	1.745	−0.255
C10–Ot8	1 ×	1 ×	1.179	1.843	−0.157	1.769	−0.231	1.769	−0.231

Bond valence sum of M[d]:	6.000 v.u.		5.784 v.u.		5.754 v.u.	
Average bond valence in C–O[g]:	1.600 v.u.		1.536 v.u.		1.536 v.u.	
Sum of the charges[e]:		−2.000 c.u.		−2.858 c.u.		−2.887 v.u.

[a] For the O atom types and sorts see Fig. 7 and footnote a of Table 4; special case for the first index of O, indicating its coordination number: c [C-bonded]: 2.
[d] Optimization basis: $\sum_{n=1}^{j} f_n s_n$ = number of M atoms in the polyanion × oxidation number of M (consideration of the M–O bonds only); bond valence sum about M: $n = \frac{1}{1}\sum_{n=1}^{j} f_n s_n$ (j: number of sorts of Mo–O bonds in the structure).
[e] Alternative (redundant) optimization basis: $\sum_{m=1}^{l} f_m c_m$ = ionic charge of the polyanion (consideration of *all* O atoms) (l: number of sorts of O atoms in the structure).
[g] $\frac{1}{10}\sum_{n=1}^{j} f_n s_n$ (consideration of the C–O bonds only); half of the difference as against the oxidation number of C (n = 4) is the average bond valence of the C–C bonds: 1.200 v.u. (a), 1.232 v.u. (b) (c).
For the other footnotes see Table 4.

For our investigations of polyoxometalate ions and relevant oxides (E = M) [44] we disregard these explanations. Structures that produced larger deviations have been identified as incorrect with respect to the bond lengths and been rejected from the set of reference structures used for the derivation of the bond length-bond valence parameters and have usually been excluded completely from an evaluation. The following items indicate incorrect structural data:

- The structures show *simultaneously* large standard deviations for the sum of the bond valences about M (expressed by $e_{U_q} > 0.05$ (v.u.)^2, corresponding to $\sigma_\Sigma > 0.12$ v.u.) *and* large systematic errors for the bond lengths (expressed by great differences $|d_{0i} - d_0| > 0.012$ Å) ("bad" structures in the sense of Sect. 2.3.2.2).
- Incorrect and reliable structures do not show any differences with respect to their structure types.
- The erroneous structures are predominantly characteristic of the heavier metals, M = W, Ta (compare Sects. 2.2.4.1–2.2.4.5), which are more difficult to investigate.

Brese and O'Keeffe [16] mentioned that they rejected any obvious outliners (identified by themselves) from the sets of EO_k polyhedra which served for the derivation of the bond length-bond valence parameters d_0. Hence, in the case of polymeric structures they obviously rejected only the single bad EO_k polyhedra and not the whole structures (for the problems connected with this procedure see Sect. 2.2.3.2). Structures to be investigated in detail later have to be treated in the same way, i.e., they have to be tested for possible outliner characteristics. Therefore it is not clear why the authors could in fact even observe structures with large differences between "apparent" and "actual" valences.

4
Summary

The derivation and application of bond length-bond valence relationships have been investigated with particular reference to the M–O bonds (M = Mo^{VI}, W^{VI}, V^{V}, Nb^{V}, Ta^{V}) in the polyoxometalate ions.

The majority of the published M–O bond length-bond valence functions and the parameters associated with these functions have not been determined unequivocally, even in recent papers. The following types of errors may have been made in the procedures to derive the bond length-bond valence parameters from (a set of) reference structures:

- assumptions concerning bond valences or the distribution of the ionic charge over the oxygen atoms of the reference structures in order to make the problem treatable, i.e., without least-squares fitting procedures (e.g., assignment of a bond valence of 2 v.u. [valence units] to the terminal M–O bonds of neutral [= uncharged] compounds; unintentional and unnoticed erroneous assignment of all bond valences);

- use of unsuitable structures (e.g., use of compounds with different elements in the coordination sphere of M and seemingly faultless correction; use of compounds with M in different oxidation states and sometimes correction; neglect of the requirement of different MO_k coordination polyhedra in the data sets);
- use of incorrect structural data (presence of erroneous bond lengths);
- use of improper mathematical functions for the description of the bond length-bond valence relationships (e.g., taking improper measures to reach a bond valence of 0 v.u. at certain finite bond lengths; no attempts to improve a commonly used function);
- use of non-optimal bond length-bond valence parameters B or N (e.g., unverified use of bond length-bond valence parameters B from other sources).

Additionally the large average deviation between bond valence sum and atom valence (oxidation number of the atom in question) of ca. 5% (or even more) stated by most authors prevents the extension of the functions for more ambitious tasks.

A valid investigation of the bond length-bond valence relationships requires

- a purely mathematical evaluation of the structural data of a set of reference structures by least-squares fitting procedures without any assumptions about bond valences, charges, or bond length-bond valence parameters;
- desirably an optimization of the bond length-bond valence functions not only via a minimal standard deviation for the sum of the bond valences about the M atoms but also for the M–O bond valences and for the charge on the oxygen atoms (via a minimal standard deviation for the sum of the bond valences about the O atoms) of the structures of the set of reference species;
- a test of the mathematical functions (e.g., by use of a well-fitting two-parametric function to which an additional third and possibly fourth parameter to be fitted may be added, which must give a better fit on principle – if the function is capable of further improvement);
- consideration of compounds containing only oxygen in the coordination sphere of M and M only in a definite oxidation state; use of structures with different coordination numbers for M and without disorder as well as without statistical occupancy of atomic sites;
- elimination of erroneous structural data by an analysis of the contribution of the individual structures to the error square sum.

The test of the mathematical functions by introduction of a third parameter to be fitted, carried out for the Mo–O and V–O systems which are characterized by many structure investigations, indicated no further improvement of the standard deviations for the exponential function (standard deviations for the sum of the bond valences about the M atoms on average $\sigma_\Sigma = 2.1\%$; for the bond valence of the individual M–O bonds on average $\sigma_S = 0.020$ v.u.; for the charge on the O atoms on average $\sigma_c = 0.032$ c.u.). For the power function, which gave on average ca. 7% larger standard deviations, the magnitude obtained with the exponential function was again found. Hence the two-parametric exponential functions are the optimal choice for the investigated M–O systems and the respective standard deviations were caused only by errors in the structural data.

The accuracy of the structural data for their use in a set of reference structures for the determination of the bond length-bond valence parameters has not been judged, as is usual, by the crystallographic R factor or by the standard deviation stated in the literature for the bond lengths of the structures since M–O bond length errors due to mislocated bridging O atoms compensate each other to a large extent – provided that the M $\cdots$ M distances are correct. Unsuitable structural data have been identified by the large contribution of the respective structures to the average error squares e_U for the sum of the bond valences about the M atoms and have been rejected if a definite multiple of σ_Σ, i.e., $2\sigma_\Sigma$ (corresponding to $e_{Uq} = 0.05$ (v.u.)^2 for the average error square of the individual species) was exceeded. The most annoying effects of erroneous structures in the sets of reference structures are the production of extremely shallow $\sigma_\Sigma(B)$ optimization curves (an effect apparently well known in the literature) whose minima are extremely shifted to large B (or smaller N) values (a very striking effect, incomprehensively not mentioned in the literature). An optimization procedure via the individual MO_k polyhedra is not permitted since this can give rise to a special error. Interestingly, the great majority of the tested structures of the molybdenum and vanadium oxo compounds have been identified as reliable; by way of contrast, the great majority of the structures of the tungsten oxo compounds have been found to be unreliable (with regard to the bond lengths).

The standard deviations for the bond valence of the M–O bonds and for the charge on the O atoms quoted above have been obtained by a special procedure via an optimization of the bond valence sums about the O atoms.

Our bond length-bond valence functions deviate in general from those of other groups: They are more shallow, i.e., the short M–O bonds remain more distant to the bond valence two and the bond valences of the long M–O bonds are greater than those based on the functions proposed in the literature. Only with the functions (parameters) more recently published by Brown and Altermatt [17] and Brese and O'Keeffe [16] is there a good agreement.

Surprisingly, application of the bond length-bond valence functions and relevant parameters ("general" parameters), derived by the present authors while taking all reasonable precautionary measures (Table 1), leads only to unsatisfactory results for the polyoxometalate species. It is true that the error (standard deviation) for the bond valence sum about M amounts to ca. 2.1% only, but for a systematic error of 0.01 Å in the bond lengths d (or in the bond length-bond valence parameters d_0) the error for the formal ionic charge is 0.147q c.u. for M^{VI} and 0.123q c.u. for M^{V} polyoxometalate ions (q: degree of aggregation of the polyoxometalate ions) which results, e.g., in an error for the ionic charge for $V_{10}O_{28}^{6-}$ of 1.2 c.u. and for $Mo_{36}O_{112}(H_2O)_{16}^{8-}$ of 5.3(!) c.u.

Excellent results are obtained by introduction of individually fitted d_0 values (d_{0i}), B (or N) kept constant. The aim of the optimization is to obtain the correct formal ionic charge and usually simultaneously the correct average sum of the bond valences about the M atoms. This optimization process is mandatory since it corrects a small, inevitable systematic error (standard deviation of this error, established with the sets of reference structures: 0.005 Å) of the bond lengths of the individual structures. This is the most important measure to obtain formally correct values for the formal ionic charge. The standard deviations for the relevant quantities are reduced to approximately 62% by use of the d_{0i} values (sum of the bond valences about M: $\sigma_{\Sigma i} \approx 0.07$ v.u. (1.3%); bond valences of the M–O bonds: $\sigma_{si} \approx 0.012$ v.u.; charge on the O atoms: $\sigma_{ci} \approx 0.020$ c.u.). Both σ_{si} and $\sigma_{\Sigma i}$, obtained via different optimization paths, correspond to a standard deviation (caused by the random errors) for the bond lengths of $\sigma_{di} \approx 0.010$ Å. These results mean that on the basis of a standard deviation (random error) for the bond lengths of 0.010 Å and a systematic bond length error with a standard deviation of 0.005 Å in the set of the reference structures our bond length-bond valence functions of Table 1 are strictly fulfilled.

Application of the bond length-bond valence relationships stated in the literature usually gives poor results for the bond valences (on an average 5% error for the bond valence sum about M) and disastrous results for the charge distribution over the O atoms and for the formal ionic charge (errors up to some charge units) of the polyoxometalate species. Apart from the reasons listed above for the occurrence of erroneous results – in particular the non- or insufficient rejection of structures with inaccurate bond lengths (large random and/or systematic errors) – an additional reason is the effort of the authors to derive 'universal' B (or N) parameters, applicable to many M–O systems, or 'universal' d_0 parameters, applicable to different oxidation numbers of M, with deliberate or unintentional toleration of larger errors. This procedure, obviously based on the insensitivity of the standard deviation for the sum of the bond valences about M on alterations of B (shallow minimum in the $\sigma_\Sigma(B)$ curve on a high σ_Σ level), actually produces considerable alterations of the M–O bond valences and particularly of the charge on the O atoms and of the formal ionic charge with a small change of B. B values which are too small attribute too little bond valence in inner, bridging (except the pseudoterminal) and too much bond valence in outer (terminal) and pseudoterminal M–O bonds and charge separation effects that are too small. All these factors contribute to an unfavorable picture of the polyoxo structures. B should be known with an accuracy of better than 1%. By means of special investigations using "theoretical" structures and theoretical structures with incorporated well-defined errors it was shown that the standard deviations for the sum of the bond valences about M and for the M–O bond lengths in the polyoxometalates, after rejection of outliners according to the above-stated criteria ($e_{Uq} > 0.05$ (v.u.)^2), are sufficiently small to allow the determination of B with the above accuracy: The *systematic* error of B remains below +0.01 Å; the *random* error depends on the number of structures and amounts, e.g., on the basis of the 73 investigated oxomolybdate structures to 0.003 Å.

Thus the common practice of using bond length-bond valence functions from the literature results in three types of severe errors. Two refer to the incorrect derivation of the bond length-bond valence parameters B (and d_0): The unsatisfactory identification and rejection of outliner structures (large random and/or systematic errors) from the sets of reference species used for the derivation of the bond length-bond valence parameters renders the correct determination of B difficult, and the unnecessary and - considering the desirable accuracy of 1% - unfounded use of 'universal' B values makes the situation even more difficult. The third type of error refers to the ill-considered application of the d_0 parameters: Inevitable, small systematic bond length errors (standard deviation of this error 0.005 Å) require an individual fit of d_0 (determination of d_{0i}) to obtain the correct formal ionic charge and bond valence sum for the species.

The general parameters d_0 and B or d_0 and N are necessary if M appears only in a mixed coordination. In this case the parameters for one ligand sort have to be taken from the literature and kept unchanged (general d_0 and B or d_0 and N parameters) and those for the other ligand sort are optimized for d_0 to give d_{0i} (individually fitted d_0 and general B or N parameters). It is advantageous to make the optimization for that element pair that contributes the greater part of the bond valence of M. For isopolyoxometalate ions the optimization of d_0 can be conducted via the correct (average) sum of the bond valences about M or via the correct ionic charge; the result of the other optimization process serves in each case as redundant control index. For heteropolyoxometalate ions two separate optimization processes have to be carried out, namely for the M–O and for the X–O bonds (X = heteroelement) via the correct (average) sum of the bond valences about the M and/or X atoms and/or via the correct ionic charge; one of them serves as redundant control index. For chelate complexes the optimization for the M–O bonds can be put through via the correct (if true: average) sum of the bond valences about the M atom(s) and, in the example cited, for CO groups (bonds) via the correct ionic charge.

Occurrence of very long M···O bonds up to 0.05 v.u. (>3.2 Å for Mo^{VI}, W^{VI}, Nb^{V}, Ta^{V} compounds; >2.9 Å for V^{V} compounds) is the inevitable consequence of the accumulation of MO_k polyhedra as the building principle of polyoxometalate ions, i.e., of the connection mode of the MO_k polyhedra, and hence is structurally irrelevant. Neglect of these very weak bonds leads to very small errors not exceeding 0.03 v.u./c.u. In ca. 50% of the structures containing MO_k polyhedra with $k < 6$ very long M···O bonds in the range 0.05–0.2 v.u. (Mo, W, Nb and Ta compounds: 3.2–2.6 Å; V compounds: 2.9–2.4 Å) have been observed. These bonds should generally be taken into consideration since they are structure-determining and their neglect leads to bond valence errors up to 0.05 v.u. and to errors for the charge on the O atoms of +0.06 to −0.14 c.u. They should, however, not be used for the description of the MO_k polyhedra which would become very difficult.

There is a small but clear dependence of the d_0 parameter upon the oxidation number of M, the B (or N) parameter assumed to be the same (proven for M = V^V: $d_0 = 1.799$ Å, V^{IV}: $d_0 = 1.779$ Å; B = 0.851 Å). In view of the characteristic errors for the bond valence and charge distribution over the structures even with small errors of the bond length-bond valence parameters a use of common parameters for the different oxidation numbers of an element is out of court in any case.

The apparent deviations of the bond valence sums for oxygen, reported in the literature, are obviously due to charge separation processes. Apparent deviations of the bond valence sums for M atoms, also reported in the literature, have never been observed in our investigations in this field and should actually not occur. They are apparently a consequence of inconsistencies in the treatment of structural data.

5 List of Symbols

a	third adjustable parameter of the exponential bond length-bond valence function
B	parameter of the exponential bond length-bond valence function for a definite element pair; in this article with an identical value in certain cases also designated as general B parameter; in the literature in certain cases used as 'universal' B parameter, valid for a large number of element pairs, compared with the (undesignated) B parameter for a definite element pair
B′	B parameter of the exponential bond length-bond valence function in the presence of a third parameter
c	charge (generally)
d	E–O, M–O, or X–O distance (bond length)
d_0	parameter of the (exponential or power) bond length-bond valence function for a definite element pair (equals the single-bond distance); in this article with an identical value also designated as general d_0 parameter, compared with an individually fitted d_0, d_{0i}
d'_0	d_0 parameter of the (exponential or power) bond length-bond valence function in the presence of a third bond length-bond valence parameter (does not equal the single-bond distance!)
d_{0i}	individually (for a given structure determination) fitted d_0 parameter of the (exponential or power) bond length-bond valence function ($d_{0i} - \bar{d}_{0i} = d_{0i} - d_0$ equals the systematic bond lengths error)
$\bar{d}_{0i}$	average of the d_{0i} values of the reference structures (equals d_0)
$\bar{d}$	average bond length in a coordination polyhedron [8]
d_{max}	maximum bond length (maximum M–O distance considered as a bond) [8]

E	electropositive element (M, X, etc.)
e_U	average error square for the sum of the bond valences about the M atoms for a set of i reference structures, each of them assigned the same weight [during the optimization phase of B (or N) and d_0; in the final stage: e_{Umin}]
e_{Uq}	average error square for the sum of the bond valences about the M atoms for a single structure (composed of q MO_k polyhedra)
e_{Ui}	average error square for the sum of the bond valences about the M atoms for a set of i reference structures, each of them assigned the same weight and based on the (general) B value belonging to e_{Umin} and an individually fitted d_0 value (the index i stands always for individual, i.e., related to individually fitted d_0 parameters, see the remark, below)
e_{Uqi}	average error square for the sum of the bond valences about the M atoms for a single structure, composed of q MO_k polyhedra, using the general B value and an individually fitted d_0 value
e_V	(surrogate for the) average error square for the bond valence of the M–O bonds *or* for the charge on the O atoms for a set of i reference structures (containing ß M–O bonds *or* ß O atoms in the individual structures), each of them assigned the same weight [during the optimization phase of B (or N) and d_0; in the final stage: e_{Vmin}]
e_{Vi}	(surrogate for the) average error square for the bond valence of the M–O bonds *or* for the charge on the ß O atoms for a set of i reference structures (containing ß M–O bonds *or* ß O atoms in the individual structures), each of them assigned the same weight and based on the (general) B value belonging to e_{Vmin} and an individually fitted d_0 value [$e_{Vmin} = 0$ (v.u.)^2 !]
f	frequency of a structural element (e.g., number of O atoms or M–O bonds of a certain type or sort)
g	number of M–O bonds in a structure
i	number of structures in a set of reference structures
j	number of sorts of E–O (M–O, X–O, C–O) bonds in a structure
k	coordination number
L	ligand atom or molecule
l	number of sorts of O atoms in a structure
M	group VI or group V transitional metal ("addenda" atom)
M–O	(in textual passages, not in formulas) short, medium-sized, and long M–O bonds in the range 2–0.2 v.u., characterizing the commonly assumed tetrahedral, tetragonal-pyramidal, trigonal-bipyramidal, octahedral, and pentagonal-bipyramidal coordination of the MO_k polyhedra
M···O	(in text passages, not in formulas) additional, very long bonds with <0.2 v.u., extending the commonly assumed coordination sphere of M

m	(positive) number, in combination with the charge symbols (m–; m+) named the ionic charge (formal ionic charge) of the polyoxo species
N	parameter of the bond length-bond valence power function for a definite element pair; for the designations as general and 'universal' parameter see under B
N′	N parameter of the bond length-bond valence power function in the presence of a third bond length-bond valence parameter
n	oxidation number, (atom) valence of the element
Ot	terminal O atom ($M{-}O^-$, M=O)
Ob	bridging (inner) O atom without any distinction; specifically in Ref. 45 (see Sect. 2.2.2.1): bridging O atoms of asymmetric M–O–M bridges
Opa	pseudoterminal (two-coordinate) O atom in an angular M–O–M bridge
O2a	2-coordinate O atom in an angular M–O–M bridge
O3a	3-coordinate O atom in an angular M–O–M bridge
O2l	2-coordinate O atom in an (approximately) linear M–O–M bridge
O4l	4-coordinate O atom in an (approximately) linear M–O–M bridge; the simultaneous presence of angular M–O–M bridges is irrelevant
O3f	3-coordinate O atom in a furcation
Oc	C-bonded (two-coordinate) O atom in an M–O–C bridge
p	stoichiometric coefficient of H^+ in the overall equation for formation of the polyoxo species
q	stoichiometric coefficient of MO_4^{w-} in the overall equation for formation of the polyoxo species; degree of aggregation
R	crystallographic R factor
R1	'root-mean-square average of the d_i values' [21] [corresponds to $(e_{Uq})^{1/2}$, related to the general B and d_0 parameters; approximately equivalent with σ_Σ]
s	bond valence (in the older literature: bond strength, bond order) (generally)
s_i	bond valence of the ith M–O or X–O bond about an M, X, or O atom
s_o	bond valence of the oth M–O bond in a structure
$\bar{s}(k_O)$	average bond valence of the M–O bonds of the O atoms with a definite coordination number against M and a definite protonation state in an oxometalate structure assuming an equidistribution of the bond valence over the M–O bonds and of the ionic charge over the O atoms
U	sum of the bond valences about an atom
U_y	sum of the bond valences about the yth M atom of a polyoxometalate structure (ion)
U_{iy}	sum of the bond valences about the yth M atom of a polyoxometalate structure (ion), based on individually fitted d_0 values

u	number of O atoms in a polyoxometalate ion (species with OH groups and coordinated H_2O molecules require a special approach)
V	sum of all bond valences in a structure
V_f	sum of all bond valences in the fth structure of a set of reference structures
w	charge of the unprotonated monomeric metalate ion
X	heteroelement
y	third adjustable parameter of the power function for the bond length-bond valence relationship
ß	number of M–O bonds *or* number of O atoms in the individual structures
Δ	difference
λ	statistical factor of σ
λ_1	λ for $\Phi = 0.3413$, i.e., $\lambda_1 = 1$
λ_{normal}	normally distributed λ values
ν	partition of the Φ range 0–0.5
Σ	sum
Σs_i	$\equiv \sum_{i=1}^{k} s_i$
σ	standard deviation (generally)
σ_d	standard deviation of the bond lengths (also used for the standard deviation incorporated into "theoretical" structures and for that stated in structural papers)
σ_{di}	estimated standard deviation for the bond lengths, ascertained on the basis of σ_{si} or $\sigma_{\Sigma i}$
$\sigma_{d_{0i}}$	standard deviation of the individually fitted d_0 values, d_{0i}, based on the individual structures
$\sigma_{d_{0iM}}$	standard deviation of the individually fitted d_0 values based on the single MO_k polyhedra of the structures ($\hat{=} \sigma'$ [17])
$\sigma_{d\ rand}$	standard deviation of the random bond lengths error; equals σ_{di}
$\sigma_{d\ syst}$	standard deviation of the systematic bond lengths error; equals $\sigma_{d_{0i}}$
σ_Σ	standard deviation for the sum of the bond valences about M ('agreement index' [9])
$\sigma_{\Sigma i}$	standard deviation for the sum of the bond valences about M, related to individually fitted d_0 values
σ_s	standard deviation for the valence of the M–O bonds
$\sigma_{s(uncorr)}$	standard deviation for the valence of the M–O bonds, uncorrected for the approximation character of Eq. (9)
σ_{si}	estimated standard deviation for the valence of the M–O bonds, related to individually fitted d_0 values
σ_c	standard deviation for the charge on the O atoms
$\sigma_{c(uncorr)}$	standard deviation for the charge on the O atoms, uncorrected for the approximation character of Eq. (9)

σ_{ci}	estimated standard deviation for the charge on the O atoms, related to individually fitted d_0 values
σ'	standard deviation for d_0 on the basis of the individual MO_k polyhedra (defined in Ref. 17)
$\Phi(\lambda)$	Gaussian error integral

Running indices:

f	for the i structures in a set of reference structures
g	for the ν λ values chosen in the range $\Phi = 0$–0.5 with equal distance of the Φ's
i	for the k O atoms about an M atom
m	for the l sorts of O atoms in a structure
n	for the j sorts of E–O (M–O, X–O, C–O) bonds in a structure
o	for the g M–O (and X–O) bonds in a structure
y	for the q M atoms in a structure

Remark: the index i is exclusively used for 'based on individually fitted d_0 parameters'

Acknowledgments. The authors are indebted to the Fonds der Chemie for financial support.

6
References

1. Pauling L (1929) J Am Chem Soc 51: 1010–1026
2. Brown ID (1994) Bond-Length-Bond-Valence Relationships in Inorganic Solids, In: Bürgi HB, Dunitz JD (eds), Structure Correlation, vol 2, VCH, Weinheim, pp 405–429. (a) p 407
3. Brown ID (1978) Chem Soc Rev 7: 359–376
4. Brown ID (1981) The Bond Valence Method: An Empirical Approach to Chemical Structure and Bonding, In: O'Keeffe M, Navrotsky A (eds), Structure and Bonding in Crystals, vol 2, Academic Press, New York, pp 1–30. (a) p 18, (b) pp 2–3
5. Byström A, Wilhelmi KA (1951) Acta Chem Scand 5: 1003–1010
6. Zachariasen WH (1954) Acta Crystallogr 7: 795–799
7. Evans HT Jr (1960) Z Kristallogr 114: 257–277
8. Allmann R (1975) Monatsh Chem 106: 779–793
9. Brown ID, Shannon RD (1973) Acta Crystallogr A 29: 266–282. (a) p 269
10. Donnay G, Allmann R (1970) Am Mineral 55: 1003–1015
11. Pyatenko YuA (1972) Kristallografiya 17: 773–779; Sov Phys-Crystallogr 17: 677–682
12. Pauling L (1947) J Am Chem Soc 69: 542–553
13. Donnay G, Donnay JDH (1973) Acta Crystallogr B 29: 1417–1425
14. O'Keeffe M (1989) Struct Bonding 71: 161–190
15. Brown ID, Wu KK (1976) Acta Crystallogr B 32: 1957–1959
16. Brese NE, O'Keeffe M (1991) Acta Crystallogr B 47: 192–197
17. Brown ID, Altermatt D (1985) Acta Crystallogr B 41: 244–247
18. Zachariasen WH (1978) J Less-Common Metals 62: 1–7. (a) Waltersson K (1976), cited in [18]
19. Fink L, Trömel M (1992) Z Kristallogr 200: 169–175
20. Chladek S, Trömel M (1993) Z Kristallogr 204: 107–113

21. Brown ID (1992) Acta Crystallogr B 48: 553–572
22. Brown ID (1974) J Solid State Chem 11: 214–233
23. D'Amour H, Allmann R (1972) Z Kristallogr 136: 23–47
24. Allmann R (1971) Acta Crystallogr B 27: 1393–1404
25. Krebs B, Paulat-Böschen I (1976) Acta Crystallogr B 32: 1697–1704
26. Perloff, A (1970) Inorg Chem 9: 2228–2239
27. Böschen I, Buss B, Krebs B (1974) Acta Crystallogr B 30: 48–56
28. Vivier H, Bernard J, Djomaa H (1977) Rev Chim Minerale 14: 584–604
29. D'Amour H, Allmann R (1973) Z Kristallogr 138: 5–18
30. Allmann R, D'Amour H (1975) Z Kristallogr 141: 161–173
31. Trömel M (1983) Acta Crystallogr B 39: 664–669
32. Trömel M (1984) Acta Crystallogr B 40: 338–342
33. Trömel M (1986) Acta Crystallogr B 42: 138–141
34. Müller A, Penk M, Krickemeyer E, Bögge H, Walberg HJ (1988) Angew Chem 100: 1787–1789; Angew Chem Int Ed Engl 27: 1719
35. Day VW, Klemperer WG (1985) Science 228: 533–541
36. Pope MT (1983) Heteropoly and Isopoly Oxometalates, Springer, Berlin. (a) pp 20–21
37. Day VW, Fredrich MF, Klemperer WG, Shum W (1977) J Am Chem Soc 99: 952–953
38. Filowitz M, Klemperer WG, Shum W (1978) J Am Chem Soc 100: 2580–2581
39. Klemperer WG (1990) Inorg Synth 27: 71–135; pp 71–74
40. Day VW, Klemperer WG, Yaghi OM (1991) Nature 352: 115–116
41. Klemperer WG, Shum W (1977) J Am Chem Soc 99: 3544–3545
42. Krebs B, Stiller S, Tytko KH, Mehmke J (1991) Eur J Solid State Inorg Chem 28: 883–903
43. Trömel M (1992) Z Kristallogr 200: 177–187
44. Tytko KH, Mehmke J, Fischer S (1999) Struct Bonding 93: 129–321
45. Schröder FA (1975) Acta Crystallogr B 31: 2294–2309
46. Fuchs J, Knöpnadel I (1982) Z Kristallogr 158: 165–179
47. Kihlborg L (1963/64) Arkiv Kemi 21: 471–495
48. Tytko KH (1999) Struct Bonding 93: 67–127
49. Bart JCJ, Ragaini V (1979) Inorg Chim Acta 36: 261–265
50. Fuchs J, Freiwald W, Hartl H (1978) Acta Crystallogr B 34: 1764–1770
51. Zachariasen WH (1963) Acta Crystallogr 16: 385–389
52. Mehmke J (1990) Dissertation, Göttingen; pp 51–52
53. Bauer WH (1972) Am Mineral 57: 709–731
54. Alig H, Lösl J, Trömel M (1994) Z Kristallogr 209: 18–21
55. Clark JR, Appleman D, Papike J (1969) Miner Soc Am Spec Paper 2: 31–50, from [9]
56. Wagner TR, O'Keeffe M (1988) J Solid State Chem 73: 211–216
57. Román P, Gutirrez-Zorilla JM, Martínez-Ripoll M, García-Blanco S (1987) Trans Met Chem 12: 159–167
58. Weakley TJR (1982) Polyhedron 1: 17–19
59. Román P, Martínez-Ripoll M, Jaud J (1982) Z Kristallogr 158: 141–147
60. McGuire NK, O'Keeffe M (1984) J Solid State Chem 54: 49–53
61. Krebs B, Paulat-Böschen I (1982) Acta Crystallogr B 38: 1710–1718
62. Fuchs J, Hartl H (1976) Angew Chem 88: 385–386; Angew Chem Int Ed Engl 15: 375
63. Pope MT (1991) Progr Inorg Chem 39: 181–257. (a) p 183
64. Tytko KH (1989) Reactions of Oxomolybdenum(VI) Species in Aqueous Solution, In: Gmelin Handbook of Inorganic Chemistry, 8th edn, Molybdenum Suppl Vol B 3b, pp 1–207. (a) pp 118–119
65. McCarley RE (1986) Polyhedron 5: 51–61
66. Evans HT Jr, Gatehouse BM, Leverett P (1975) J Chem Soc Dalton Trans 505–514
67. Nishikawa T, Sasaki Y (1975) Chem Lett 1185–1186
68. Qin Chen, Shuncheng Liu, Zubieta J (1990) Inorg Chim Acta 175: 241–245

A Bond Model for Polyoxometalate Ions Composed of MO_6 Octahedra (MO_k Polyhedra with k > 4)

K.H. Tytko

Institut für Anorganische Chemie, Universität Göttingen,
Tammannstraße 4, D-37077 Göttingen, Germany

The literature relating to the bonding in polyoxometalate ions is reviewed. The author's opinions are presented and expanded to give a bond model for polyoxometalate ions of the early transition elements (groups V and VI) composed of MO_6 octahedra (MO_k polyhedra with $k > 4$). This bond model concerns in particular the bond valence (bond lengths) and charge distribution in the polyoxometalate ions and the factors modifying them. It is based on the following commonly used concepts and principles as applied to polyoxometalate ions:

- Lewis's octet rule, extended to the decet and dodecet rule for M^V and M^{VI};
- $p\pi$-$d\pi$ M=O double bond and the coordinate bond (dative bond);
- the resonance concept;
- the resonance bond number (or the bond valence concept and the valence sum rule);
- polydentate ligands and the chelate effect;
- the larger space requirements of unshared electron pairs (cf. the VSEPR model);
- the model of multicenter $p\pi$-$d\pi$ multiple bonds for certain μ-oxo bridges between metal atoms;
- Brønsted's acid/base concept and acid/base equilibria;
- Pauling's rules for the acid/base strength of (monomeric) oxoacids/oxoanions;
- the law of mass action (Le Chatelier's principle), and others.

The result is a set of resonance structures for polyoxometalate species in which three types of resonance can be distinguished. These, in turn, explain:

- the cohesion of the strongly distorted tetrahedral MO_4 building units in the structures by formation of MO_6 octahedra and hence the enhanced stability of the polyoxometalate ions;
- the distribution of the formal ionic charge over (nearly) all types of oxygen atoms, but preferably the terminal ones;
- the occurrence of positively charged oxygen atoms, caused by charge separation processes;
- the enhanced basicity of polyoxometalate ions; and further features.

A "meshing effect" defines the increase of the bond valence in inner (bridging) M–O bonds and hence the stabilization of the structures due to extension of the coordination spheres of MO_4 tetrahedra. Thus, the bond lengths of the different structure types are governed by a maximization of the bond valence of the inner, bridging (or minimization of the bond valence of the outer, terminal, in contrast to frequent statements in the literature) M–O bonds. The limits of the maximization of the inner bond valences of the polyoxometalate ions are determined

- by minimum stoichiometric requirements for corresponding resonance formulae (inevitability of charge on bridging oxygen atoms);
- by the necessity to fulfill simultaneously the geometrical relationships of the M–O bond lengths (as defined by their interdependence in the M–O frameworks) and the valence

sum rule for M and O atoms which are interrelated through the bond length-bond valence function;
- by (for the solid state) the packing of (poly)anions, cations, and molecules of water of crystallization with respect to the requirements of their size, shape, and charge;
- and/or by the capacity of the terminal oxygen atoms for the acceptance of unshared electron pairs in relation to that of the bridging oxygen atoms.

The principle of the stabilization of polymeric species by maximization of their inner bond valence, which is simultaneously connected with an increase of the basicity of the species, is, however, only valid for a certain range of acidification of the oxometalate solution from which or in which the species form. The overlying principle is ultimately the consumption of H^+ ions in the MO_4^{w-}/H^+ systems, thus fulfilling the requirements of Le Chatelier's principle.

Keywords: Polyoxometalates, bond model, bond valence model

1 Introduction

The bonding of polyoxometalate ions of the early transition elements (M^{VI}, M^{V}) composed of MO_6 octahedra has not previously been reviewed comprehensively, most contributions being restricted to small sections of the topic. The bonding arguments are usually directed towards the stabilities and/or on the molecular and electronic structures of the polyoxometalate ions. Thus, metal/terminal-oxygen π-bonding is considered to be important for the stabilization of polyoxometalate structures [1a, 2, 3, 4a]. Accordingly, it has been proposed that a major reason for the existence (stability) of polyoxometalate ions is the ability of the corresponding elements to form M^{n+}–O bonds of bond order two so that the oxygen atom has little or no basic character and that the M^{n+}–O π-bonds are relatively localized [1a, 2, 4a]. From the sum of the bond valences ("bond strengths")[1] about the oxygen atoms in several isopolyoxometalate ions [8–13], the present author [14a, 15a] concluded, contrary to the above proposal, that the ionic charge is mainly distributed over the terminal oxygen atoms which, therefore, are the most basic oxygen atoms of a polyoxometalate structure. This is in line with

- the observation that the cations in polyoxometalates are in contact mainly with the terminal oxygen atoms of the polyanions [8–13, 16–19],
- the assumption that in ion pairs (i.e., in solution) the cations are in contact with terminal oxygen atoms of the polyanions [10, 11, 20], and
- the view that terminal oxygen atoms behave as if they are bigger than the other oxygen atoms [21] or make available more space to the unshared electron pairs in which the charge electrons are merged in [14a, 22a].

Long M-O distances are considered to characterize ionic bonds [23-26]. Thus a number of polyoxometalate ions are described to be composed of an O^{2-} or another anion surrounded by or otherwise combined with an $(MO_3)_\xi$ cage, ring, or something similar, for example, $(O^{2-})(M_6O_{18})$ $(= M_6O_{19}^{2-})$ or $(PO_4^{3-})(M_{12}O_{36})$ $(=PM_{12}O_{40}^{3-})$ [26–31]. For the case of the $XM_{12}O_{40}^{m-}$ heteropolyoxo anions, where an XO_4^{m-} tetrahedron is assumed to be surrounded by an $M_{12}O_{36}$ cage [26, 29–31], this view has been questioned [32, 33a, 34a].

In the following sections, a bond model for polyoxometalate ions $H_{p-2r+2o}M^n_qO^{[(8-n)q-p]-}_{4q-r+o}$ with an extended coordination sphere of the metal M^n, formed by q $MO_4^{(8-n)-}$ and p H^+ ions with elimination of r H_2O molecules as well as, in rare cases, incorporation of o coordinated H_2O molecules is described. Current views on simple non-polymeric compounds and complexes have been applied to polynuclear oxo complexes (the above formula for the polyoxometalate ions can be simplified according to $M^n_qO_{u-h}(OH)_h(H_2O)_o^{m-}$ or, in the absence of H atoms, even to $M^n_qO_u^{m-}$). This model was developed in its simplest form during our first theoretical studies on the mechanisms of formation for polyoxometalate ions and an analysis of the driving and directing factors [14b, 22, 35, 37], see Ref. 15b. Its simplicity and its analogy to bond models for non-polymeric compounds and complexes resulted because the main concepts used were those named above (except for the model of multicenter pπ-dπ multiple bonds for certain μ-oxo bridges between metal atoms) – and therefore a detailed explanation was not necessary. From the papers dealing with bonding in polyoxometalate ions published since then it became clear that there are often contradictory opinions and that our model was only marginally used in these papers. These aspects are discussed below in the light of our bond model.

[1] The terms "bond valence", "bond strength", "bond order", and other terms are not uniformly used in the literature [5–7]. In this article bond order is used to indicate the number of bonds between two atoms in a classical formula (in one of the resonance formulae), whereas bond valence is used to characterize the weighted mean of the bond orders of a bond considered in the different resonance formulae or to characterize a bond according to Eqs. (20) and (21).

2
Relevant Features of Non-Polymeric Analogues Applicable to Polyoxometalate Ions

Most of the following relevant features of non-polymeric species and complexes are generally accepted and have been treated in textbooks and monographs.

2.1
Charge Distribution Over the Oxygen Atoms and π-Bond Delocalization Over the Element-Oxygen Bonds of Small Oxoanions

The most frequent examples of negatively charged oxo species are the anions of the oxo acids of elements (E) such as, for example, NO_3^-, ClO_4^-, SO_4^{2-}, CrO_4^{2-}. The charge and the π-bonds are assumed to be delocalized according to, for example,

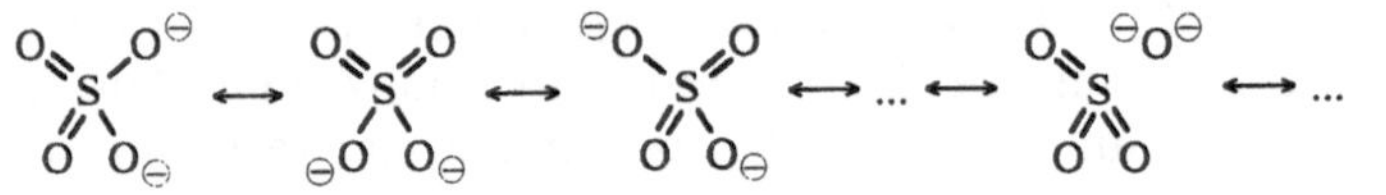
(1)

If one of the oxygen atoms is involved in a bond to an additional atom, for example, as in HSO_4^- or $S_2O_7^{2-}$, the contribution of the corresponding oxygen atom to the charge and π-bond delocalization is strongly reduced:

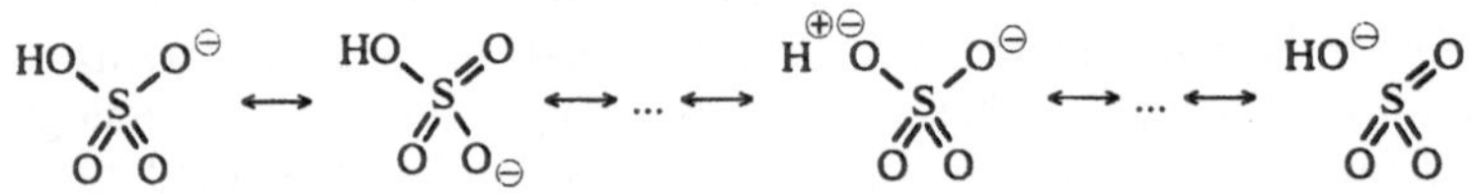
(2)

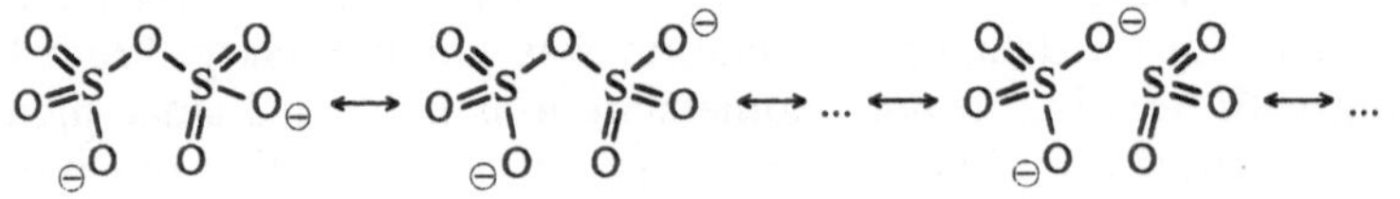
(3a)

In linear or approximately linear E–O–E bridges, multicenter pπ-dπ multiple bonds are assumed to occur, which become stronger the more the E–O–E angles are widened [38a, 39–41, 42a] and lead to positively charged bridging oxygen atoms [43]:

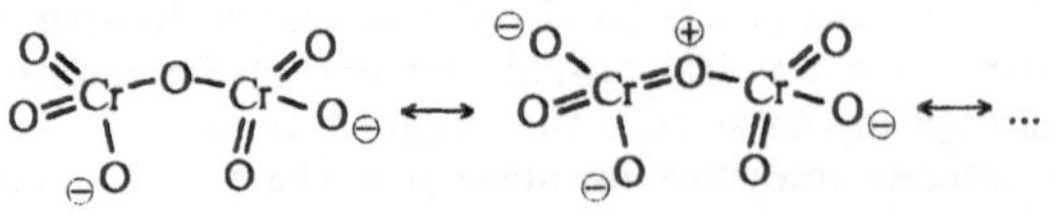
(3b)

2.2
Formation and Formulation of Species with Coordinate Bonds (Dative Bonds)

A simple example is the complex formation between the coordinatively unsaturated SiF_4 molecule and F^- ions. In the SiF_6^{2-} anion the six Si–F bonds cannot be distinguished, which is explained by resonance:

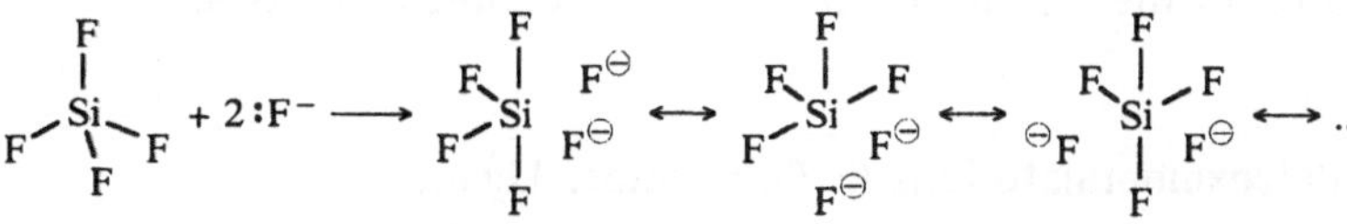
(4)

Many chemists [44a, 45–47] only unwillingly distinguish between covalent and coordinate-covalent (dative-covalent) (in short: coordinate) bond types. However, it is now well established that there is generally a significant difference between the two bond types which can, in the case of small, symmetrical molecules be hidden by resonance [48] according to Scheme (4). In the case of less symmetrical molecules like the chelate complexes based on (at least formally) protonated metalate ions, the difference can easily be recognized (see, e.g., the structure of $MoO_2(CH_3COCHCOCH_3)_2$ [49], formally derived from diprotonated monomolybdate and acetylacetonate under elimination of two H_2O molecules). This must also be the case for the polyoxometalate ions, see below.

Coordinate bonds, which occur also in individual resonance formulae of Schemes (1)–(3), have not expressly been characterized in those schemes. They are sometimes designated by an arrow (I), by a plus and minus sign (II), by a dotted or dashed bond (III), or by a formal charge (IV). In the structures

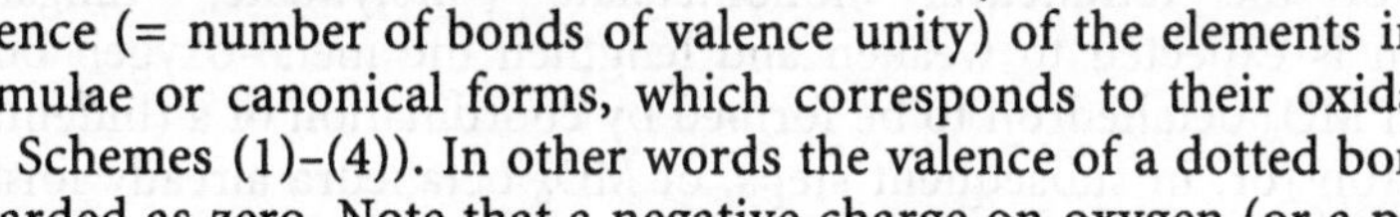

of the polyoxometalate ions given below they are represented by dotted lines to indicate the bond partners in the polyatomic molecules. We avoid drawing full lines for coordinate bonds since in this way we can correctly express the valence (= number of bonds of valence unity) of the elements in the classical formulae or canonical forms, which corresponds to their oxidation number (cf. Schemes (1)–(4)). In other words the valence of a dotted bond is formally regarded as zero. Note that a negative charge on oxygen (or a positive charge on E, cf. Scheme (22)) replaces and a positive charge on oxygen (or a negative charge on E) eliminates formally a covalent bond of valence unity of the respective atom. (The possibilities written in parenthesis, leading to a charge on E, must be avoided if the valence of the element E is to be expressed by the number of bonding dashes, i.e., the number of bonds of valence unity in the individual resonance formulae; see Sect. 4.3).

2.3 Space Requirements of Unshared Electron Pairs

Unshared electron pairs require more space than electron pairs that form a bond between two atoms and therefore arrange themselves in a structure in such a way as to occupy a maximum of space (cf. the Valence Shell Electron Pair Repulsion (VSEPR) model [50]). This view has certain consequences for the distribution of the electronic charge in polyoxometalate ions.

2.4 Mono- and Polyoxometalate Ions As Polydentate Ligands

It is known that the monomolybdate ion, like other oxoanions, can act as a polydentate (bidentate) ligand [4b, 38b, 51, 52].

Polyoxometalate ions can also act as polydentate ligands, as seen from many heteropolyoxometalate ions (e.g., $Mn^{IV}(Nb_6O_{19})_2^{12-}$ [53], $U^{IV}(W_5O_{18})_2^{8-}$ [54]); see also Ref. 3.

2.5 Expansion of the Coordination Sphere of Protonated Monomolybdate, -tungstate, and -vanadate Ions

It is established that protonated monomolybdate, -tungstate, and -vanadate ions expand their coordination sphere from tetrahedral to octahedral on reaction with bidentate ligands such as 8-hydroxyquinoline, *o*-phenols [55–57], and others. Therefore, it can be assumed that metalate ions like mono- and polymolybdates, -tungstates, etc. (see Sect. 2.4) can also act as bidentate ligands towards protonated monometalate ions [15c, 15e, 22a, 35, 36a].

3 The Bond Model

3.1 The Mechanistic Basis of the Model

Protonation of the tetrahedral monometalate (-molybdate, -tungstate, -vanadate) ion is expected to weaken and lengthen the metal-oxygen bond. This allows an MO_6 octahedron to be formed by coordination of a (bidentate) MO_4 tetrahedron (or, in subsequent steps, of MO_6 octahedra already formed according to this mechanism) through simple changes of the O–M–O angles at the protonated MO_4 tetrahedron [35, 36a, 58–60]. In a similar manner, an additional protonated MO_4 tetrahedron can rapidly add to this dimetalate ion to form a trimetalate ion, and so on ("addition mechanism"). The ion can grow rapidly at either end (actually: at several sites since the chain may branch [35, 36a]) in a variety of linear orientations, also with the addition of dimetalate or higher units [35, 36a, 58, 60]. A "limiting trimetalate ion" was claimed to be more stable than the linear polymers [58, 60]. However, it was then shown that the ring-like ("closed") polymers built according to this principle should be the most stable ones and that, for statistical reasons, the smallest rings should be the most favorable, namely tetrameric "rings" existing in two isomeric forms, planar (see Fig. 1(a)) and puckered [35, 36a]. These polymetalate ions have been described as complexes in which monomeric MO_4 units capable of extending their coordination sphere ($HMoO_4^-$, HWO_4^-, $H_2VO_4^-$, and their more highly protonated forms) act as mono- or bidentate as well as even multiply bridging ligands. They also simultaneously make two coordination sites available to other MO_4 units, thus meshing the MO_4 units to give the polymetalate ion (Fig. 1(a)) [15c, 22a, 35, 36a]. An equivalent view is that any one of the MO_4 units (which are all capable of extending their coordination sphere) makes two coordination sites available to the rest of the polymetalate ion acting as a bidentate ligand. The meshing stabilizes the polymeric character of the structures. Hence, the "meshing effect" resembles the chelate effect.

Further aggregation to give higher polymetalate ions is only possible via elimination of oxygen as H_2O. Stepwise construction of new MO_6 octahedra to give "block"-type polymetalate structures ($M_7O_{24}^{6-}$- and (β-)$Mo_8O_{26}^{4-}$-like structures; see, for example, Ref. 61) can be accomplished by an initial addition, followed by the actual elimination step arising from the condensation of an H_2O molecule from two OH groups ("condensation mechanism"). The ion can grow rapidly at several sites [22, 36a, 62]. "Closed" polymers constructed through a final condensation step with elimination of *two* (M_7O_{24}, M_8O_{26}) or *three* H_2O molecules (M_6O_{19}, $M_{10}O_{28}$) have the most favorable structures [14c, 22, 36a, 62–64]. These polymetalate ions have been described as complexes in which the MO_4 units are even more highly meshed [15d, 15e, 22a, 36a, 62]. In addition to the above meshing principle (Fig. 1(a)), some or all of the MO_4 units now appear as oligotetrahedral, chain-like M_2O_7, M_3O_{10}, or higher $M_\xi O_{3\xi+1}$ entities (or even as ring-shaped $M_\xi O_{3\xi}$ entities) consisting of corner-sharing MO_4 tetrahedra – depending on the number of condensed H_2O molecules. Thus, these isopolyoxometalate ions can also be described as an assembly of mono- and/or oligotetrahedral entities capable of extending their tetrahedral coordination spheres to octahedral ones. The latter simultaneously act as mono- and bidentate or even as multiply bridging ligands and make two coordination sites available to MO_4 tetrahedra of their own and/or of other oligotetrahedral entities by appropriate three-dimensional bending of the chains and rings (Fig. 1(b)–(h)). The $M_{10}O_{28}$ structure can be described as an $M_qO_{\leq 3q}$ entity consisting of corner-sharing MO_4 tetrahedra in the form of cyclic, oligocyclic, and occasionally suitably bent, short, chain-like units (Fig. 1(i), (j)).

The "cage-like" polymetalate ions (e.g., $W_{12}O_{40}(OH)_2^{10-}$, $W_{10}O_{32}^{4-}$; structures see, e.g., Ref. 61) can be visualized as being built up of small, block-type polymetalate ions by addition and/or condensation reactions with elimination of oxygen as H_2O formed from two OH groups [14d, 15f]. There are no fundamentally new meshing principles.

Each MO_4 unit forming an MO_6 octahedron chosen from a formula like those given in Fig. 1 can be visualized as shown in (V) [22a]. This involves an

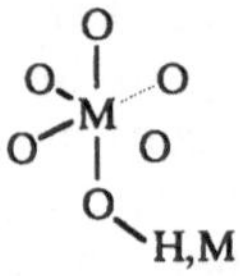

(V)

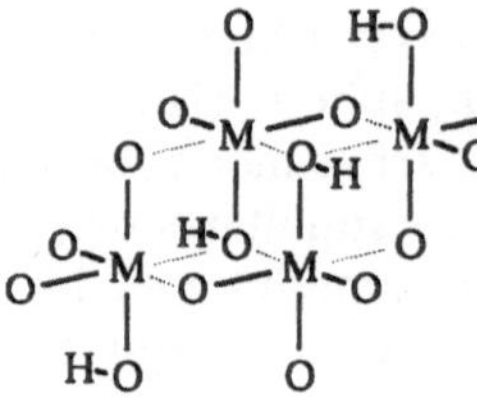

(a)

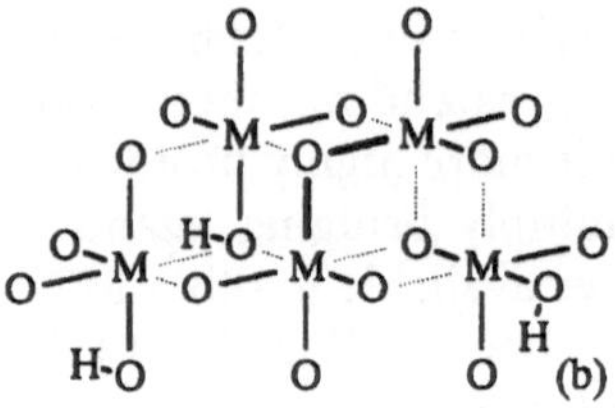

(b)

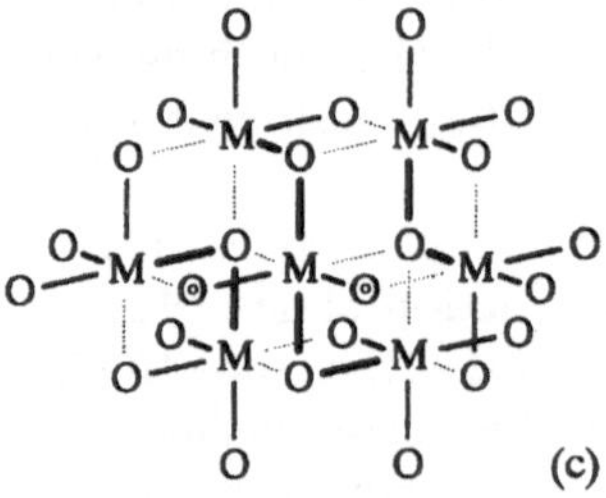

(c)

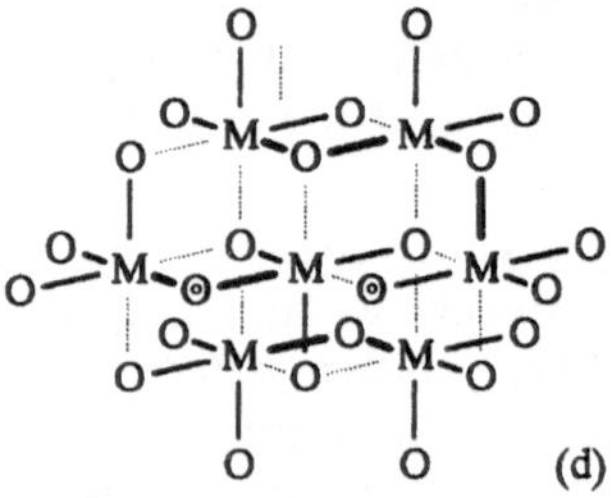

(d)

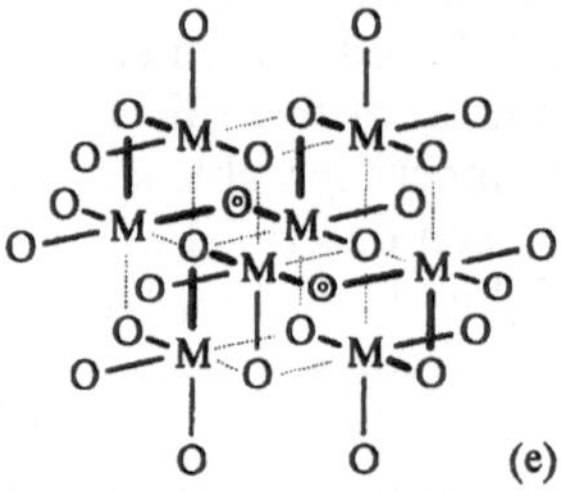

(e)

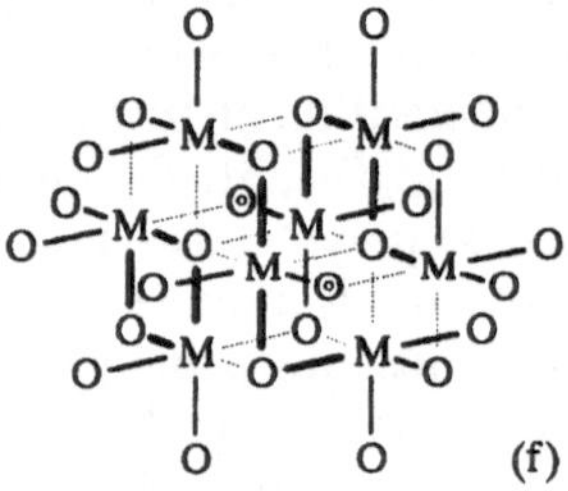

(f)

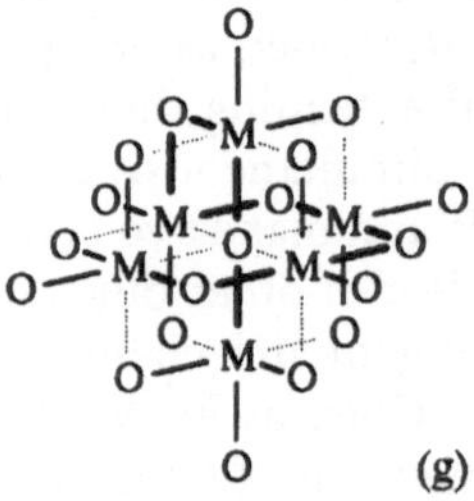

(g)

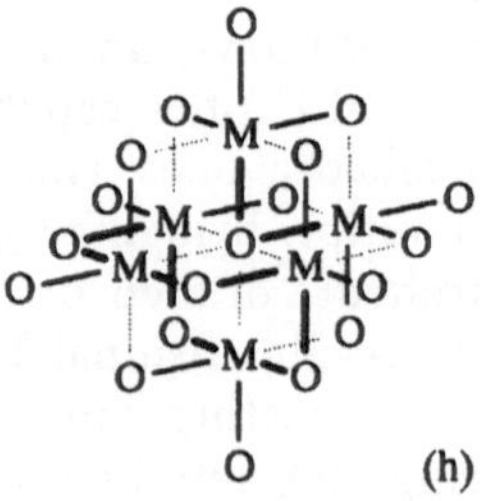

(h)

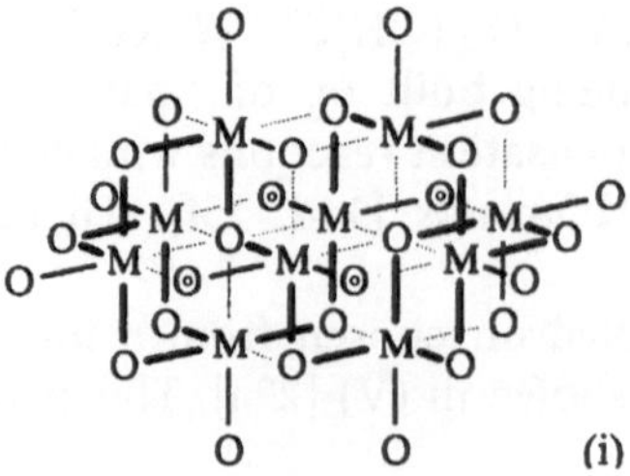

(i)

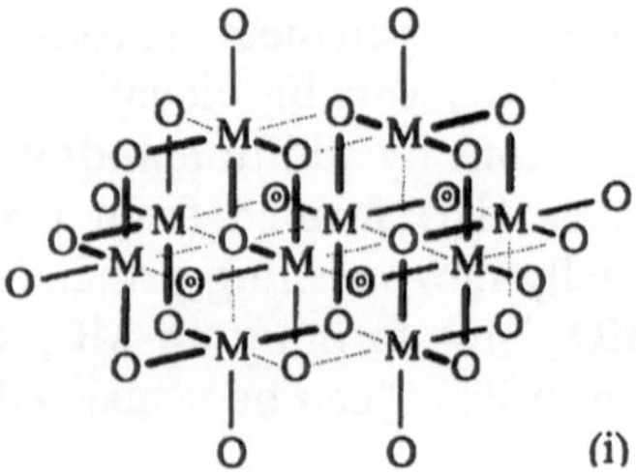

(i)

arrangement of four oxygen atoms and a central M atom in the form of a "seesaw" (as in SF_4) which can be assumed (at least formally) to originate from the original, protonated monometalate ion. One of the oxygen atoms forming the "beam" of the seesaw is connected to an H or M atom. The elongation of the corresponding M–O bond permits the development of an octahedral arrangement about the central M atom by coordination of two additional oxygen atoms ("coordinatively" bonded oxygen atoms) originating from other MO_4 units [22, 35, 36a]. For M = Mo^{VI} and W^{VI}, a second of the four oxygen atoms originating from the original monometalate ion can be connected to an M (or H) atom. For M = V^V it is assumed that at least two of the four oxygen atoms originating from the original monometalate ion must be connected to an M (or H) atom to achieve the octahedral coordination. (This requires at least diprotonation of the monovanadate ion [15c, 62, 65, 66].) For experimental support of this view see Sect. 4.6.1 in Ref. 61. For M = Nb^V and Ta^V, it is claimed [60, 62, 67, 68, 69b] that these elements occur with oxygen always in octahedral coordination, at least in MO_k polyhedra with $k > 4$.

Fig. 1. Resonance formulae for different polyoxometalate ions composed of MO_6 octahedra showing meshing and extension of the MO_4 tetrahedral building units through simultaneous action of the MO_4 tetrahedra as mono- or bidentate and even as multiply bridging ligands and as acceptors of ligands (representation as used in our early papers, i.e., without notation of the charge and double bonds and with one of different possibilities for the distribution of the H atoms (if present). In contrast to our early investigations, the positions of protons in the structures being still influenced by the addition mechanism have now been allocated according to the results of Sect. 3.7). (a) O:M ratio = 4 (structures exclusively built according to the "addition mechanism"): In a tetrametalate structure M_4O_{16} only monomeric MO_4 units occur since the formation of M_4O_{16} frameworks from MO_4 monomers does not require elimination of oxygen. (b) 4 > O:M ratio > 3.5 (structures built according to the "condensation mechanism" in the starting phase where H atoms originating from the addition mechanism are still present): In a pentametalate structure M_5O_{19} a ditetrahedral M_2O_7 unit occurs according to the number of eliminated oxygen atoms in the aggregation process besides three monomeric MO_4 units. (c)–(h) 3.5 ≧ O:M ratio > 3 (structures built according to the "condensation mechanism" or any other mechanism under elimination of at least 0.5 but less than 1 O atom per M atom): Elimination of more oxygen leads to polymetalate structures formed exclusively by oligotetrahedral $M_\xi O_{3\xi+1}$ and occasionally $M_\xi O_{3\xi}$ units, according to the number of eliminated oxygen atoms. The heptametalate structure M_7O_{24} is formed by one M_3O_{10} and two M_2O_7 units; the (β-)Mo_8O_{26} structure by two M_4O_{13} units (as shown), one M_5O_{16} and one M_3O_{10} units, two M_3O_{10} and one M_2O_6 units (as shown), or in yet other ways; the M_6O_{19} structure by one M_6O_{19} unit (as shown), one M_4O_{13} and one M_2O_6 units, one M_2O_7 and one M_4O_{12} units (as shown), or in yet other ways; etc. (i), (j) O:M ratio ≦ 3 (structures built according to the "condensation mechanism" or any other mechanism under elimination of at least 1 O atom per M atom): $M_qO_{\leqq 3q}$ entities composed of rings and short chainlike parts of corner-sharing MO_4 tetrahedra occur. A tricyclic $M_{10}O_{28}$ entity consisting of a decameric and two trimeric rings (as shown), an $M_{10}O_{28}$ entity composed of a tetrameric, a trimeric, and a dimeric ring, connected by short chains (as shown), or yet other $M_{10}O_{28}$ entities form the well-known $M_{10}O_{28}$ structure by appropriate three-dimensional bending of the entities. The $M_\xi O_{3\xi+1}$ ((b)–(h)), $M_\xi O_{3\xi}$ ((f), (h)), and $M_qO_{\leqq 3q}$ ((i), (j)) units are characterized by heavy bonds. Note that in resonance formulae other combinations of the M atoms can form the $M_\xi O_{3\xi+1}$, $M_\xi O_{3\xi}$, and $M_qO_{\leqq 3q}$ units, or other bond patterns for the same combination of the M atoms can occur. Note in particular that in (d) the MO_4 group quite on the right violates coordination mode (V). ⊚: pseudoterminal O atoms

The above-mentioned mechanisms of formation for polyoxometalate ions have been considered to be the simplest of the conceivable mechanisms [15g, 22, 70]. In particular, they represent routes that are geometrically and electronically exactly defined (i.e., all steps occur with a minimum of disruption and can proceed in concerted steps in which bond cleavage and bond linkage occur simultaneously, see, e.g., Fig. 5 in Ref. 22a) [15g]. Furthermore, the individual steps correspond to mechanisms that are currently being discussed for the formation of mononuclear M^{VI} (and M^V) complexes [55–57, 71, 72]. However, it must be emphasized that the above-mentioned connection modes of the MO_4 units in the polyoxometalate ions, although directly derived from our mechanistic ideas, do not depend on any specific concept for the mechanism of formation of the polyoxometalate ions themselves.

These relationships have been described as part of our theoretical studies on the mechanisms of formation of isopolyoxometalate ions. Unless otherwise stated, the specifications in this and the following sections were assumed to be general knowledge and not explained in detail.

Formula (V) is an abridged version of "complete" formulae of the types (VI) and (VII) or (VIII) and (IX), including all their resonance forms

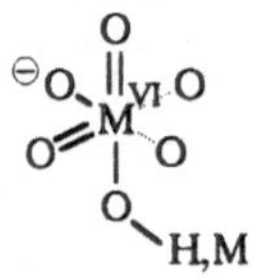

$$
\text{(VI)} \qquad \text{(VII)} \qquad \text{(VIII)} \qquad \text{(IX)}
$$

(see Sect. 3.2), in which M^{VI} or M^V have, according to their position in the periodic table, an electron dodecet and decet, respectively, and form two and one $p\pi$-$d\pi$ double bonds, respectively. Hence formulas (VI) to (IX) take into consideration the hexa- and pentavalency of M^{VI} and M^V, respectively; see also Reference 43. The relationship between (V) and (VI) to (IX) is just the same as between the commonly used formulas (X) and (XI) for the sulfate ion. The

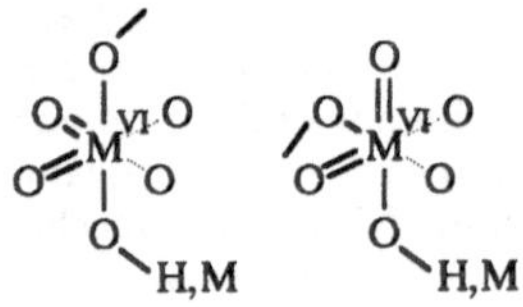

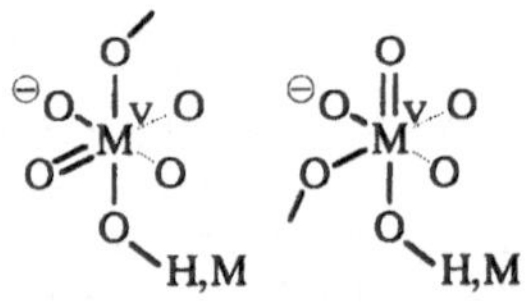

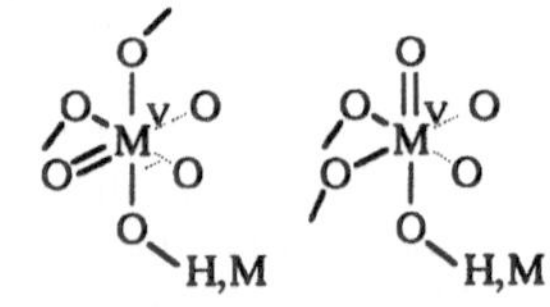

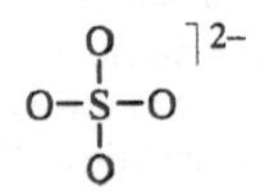

$$
\left[\, O\!-\!\overset{\displaystyle O}{\underset{\displaystyle O}{S}}\!-\!O \,\right]^{2-} \text{(X)} \qquad O\!=\!\overset{\displaystyle O^{\ominus}}{\underset{\displaystyle O_{\ominus}}{S}}\!=\!O \quad \text{(XI)}
$$

same relationship exists between structure (b) of Figure 1 and the structures of Figure 2.

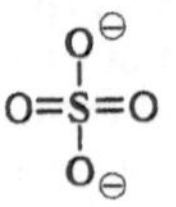

3.2
Occuring Resonance Types

3.2.1
A Resonance Leading to a Certain Equalization of the Inner M-O Bonds and Explaining the Cohesion of the MO_4 Building Units of the Polyoxometalate Ions

As has been stated before, dotted lines are used to indicate the bond partners of a coordinate bond in the resonance formulae of a molecule. Owing to resonance (analogous to resonance scheme (4)), bond compensation occurs, that means, precisely to the degree a coordinate bond (dotted line) develops, to the same degree "bonding power" is delivered of other (covalent) bonds (drawn with full lines) with the consequence that the sum of the bonding power (bond valences) about M remains constant and corresponds to the oxidation number of M. Thus the fundamental resonance scheme

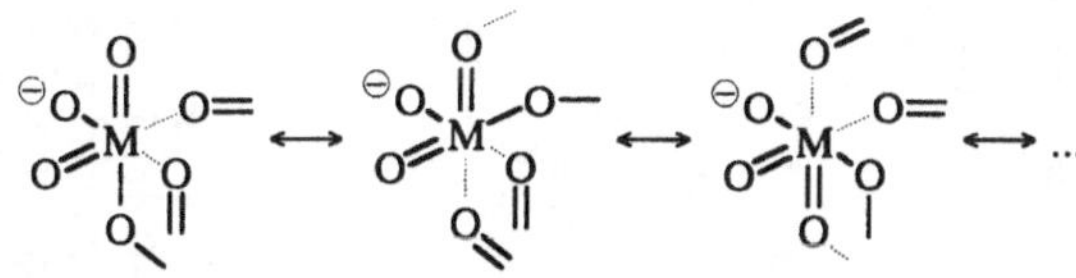
(5)

can be formulated for the example of an $M^{VI}O_6$ octahedron with two terminal oxygen atoms. This type of resonance describes a certain equalization of the inner (bridging) M–O bonds and the cohesion of the MO_4 building units, see Fig. 2(a).

3.2.2
A Resonance Explaining the Distribution of the Negative Ionic Charge Over (Nearly) All Types of Oxygen Atoms, the Equalization of the Outer and of Some Inner M–O Bonds, and the Cohesion of the MO_4 Building Units of the Polyoxometalate Ions

A second fundamental resonance scheme for the MO_6 octahedra describes the acceptance of the negative ionic charge by (nearly) all oxygen atoms:

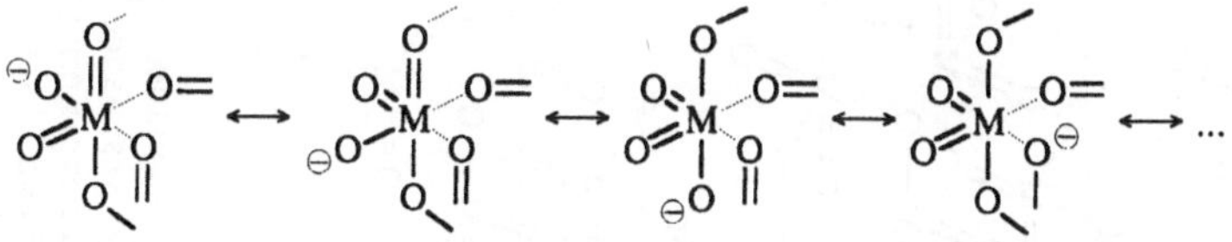
(6)

This type of resonance ensures the distribution of the negative ionic charge over (nearly) all types of oxygen atoms, terminal and bridging. It also contributes to the equalization of the outer and some inner M–O bonds and to the cohesion of the MO_4 building units, see Fig. 2(b).

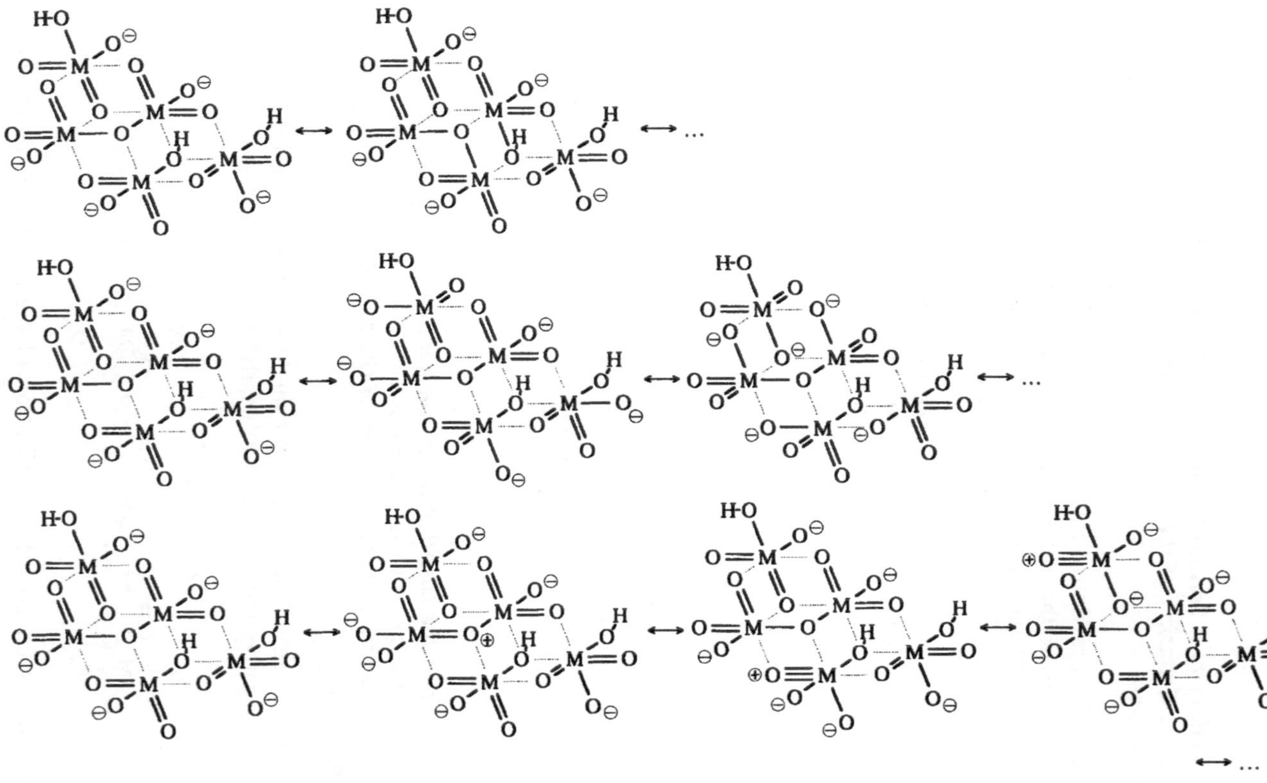

(a)

(b)

(c)

Fig. 2a–c. Resonance types for the M–O frameworks of polyoxometalate ions illustrated for a hypothetical pentametalate(VI) ion (structure (b) of Fig. 1). **(a)** Resonance leading to a certain equalization of the inner M–O bonds explaining the holding together of the MO_4 building units. **(b)** Resonance explaining the distribution of the negative ionic charge over (nearly) all types of oxygen atoms and contributing to an equalization of the outer and of some inner M–O bonds and, hence, to the holding together of the MO_4 building units. **(c)** Resonance explaining the enhanced basicity of the polyoxometalate ions, the occurrence of positively charged oxygen atoms, and the strengthening of some inner M–O bonds and, hence, the holding together of the MO_4 building units

3.2.3
A Resonance Explaining the Enhanced Basicity of the Polyoxometalate Ions, the Occurrence of Positively Charged Oxygen Atoms, the Strengthening of Some Inner M–O Bonds, and the Cohesion of the MO_4 Building Units

A third fundamental type of resonance is the "charge separation" which results from the generation of positive and negative charges on different oxygen atoms of the oxoanion. This has recently been proposed for monomeric oxometalate species and polyoxometalate ions formed by corner-sharing MO_4 tetrahedra to explain their enhanced basicity as compared to that of the monomeric oxo species and polyoxo anions of the corresponding main group elements [43]. Applied to the polyoxometalate ions built up from MO_6 octahedra, this resonance has to be formulated as

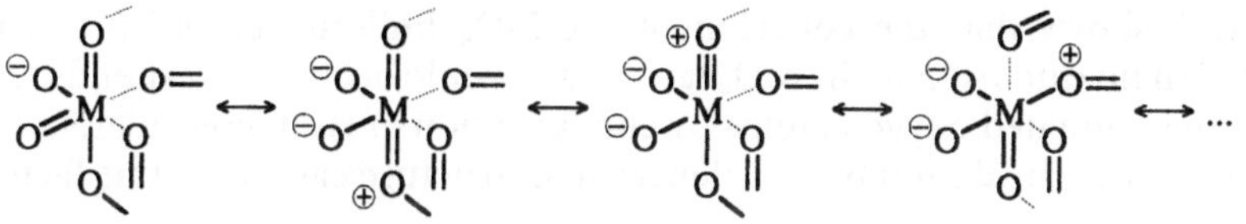

(7)

(see also Ref. 15h). This type of resonance explains the enhanced basicity of the polyoxometalate ions consisting of MO_6 octahedra (and other MO_k polyhedra), the occurrence of positively charged oxygen atoms, and the strengthening of some inner M–O bonds. Hence, it contributes also to the cohesion of the MO_4 building units, see Fig. 2(c). (This type of resonance was not included in our original bond model.)

3.2.4
A Resonance Involving Central Heteroatoms of Heteropolyoxometalate Ions

The involvement of central [34b] heteroatoms generally leads (if the coordination number and oxidation number of X are different) to an additional type of resonance, e.g.,

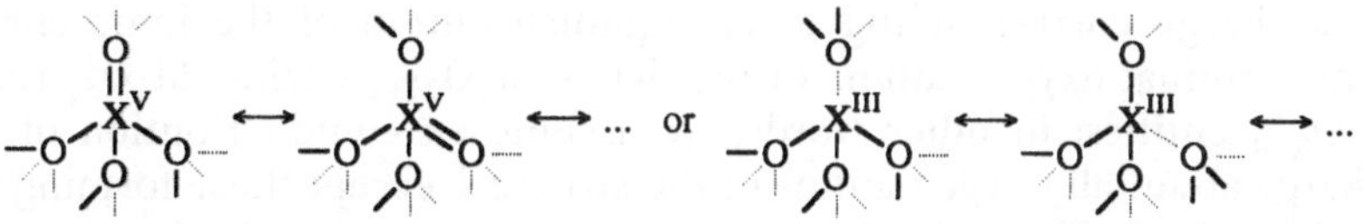

(8)

Because of their presence in M–O frameworks, the O atoms of the heteroatom polyhedra can be negatively or positively charged according to Schemes (6) and (7).

Polyhedra of peripheral [34c] heteroatoms show a similar behavior as the addenda (M) atom polyhedra, i.e., resonance schemes corresponding to Schemes (5)–(7) exist, albeit without coordinate bonds within the (non-extended) heteroatom coordination polyhedra.

3.3 The Meshing Effect

The construction of MO_6 octahedra from MO_4 tetrahedra in the course of the formation of polymetalate ions is an energetically favorable process. This is believed to be due to the larger number of (inner) M–O bonds and a type of entropic chelate effect. The latter is expressed by the simultaneous action of the tetrahedral MO_4 units as mono- and bidentate and even as multiply bridging ligands, as acceptors of ligands, and by condensation of H_2O molecules during the aggregation process [15i, 22a, 36a, 62].

Since the sums of the bond valences about M (the "bonding power" of M) in MO_4 tetrahedra and MO_6 octahedra are the same and correspond to the oxidation number of M, it is not immediately clear why the larger number of M–O bonds about M in MO_6 octahedra is a favorable feature.

Figure 1 shows that the cohesion of the MO_4 building units by "normal bonding" (if no extension of the MO_4 tetrahedra would occur) corresponds to the number of eliminated oxygen atoms by condensation of H_2O molecules, i.e., the sum of the inner (bridging) bond valences in a structure caused by this factor is

$$S_n = 2r = p - h = qZ^+ - h \tag{9}$$

(r, p, q: stoichiometric quantities in the equation of formation of the polyoxo species; h: number of OH groups in the structure, see Sect. 1; $Z^+ = p/q$). In Fig. 1 the corresponding bonds are characterized by thick lines.

The rest of the inner (bridging) M–O bonds, characterized by thin lines, are produced by the meshing effect (note that, when counting these bonds, some of them are double bonds – compare Fig. 2 – which was not noted in Fig. 1). They are (formally) converted from terminal to bridging M–O bonds and thus the cohesion of the MO_4 units as well as the stability of the polyoxometalate ions $M^n_qO_{u-h}(OH)_h(H_2O)^{m-}_o$ is strongly enhanced. This is precisely the effect of the larger number of (inner) M–O bonds about M in MO_6 octahedra. The sum of the bond valences in bridging M–O bonds, generated by the meshing effect, can be calculated (estimated) by subtraction from the overall bond valence in the structure (qn) the bond valence of the unshared O atoms and that of the inner M–O bonds by normal bonding. For the unshared O atoms we assume a charge corresponding to an equidistribution of the ionic charge about the terminal oxygen atoms of the fictive $M_\xi O_{3\xi+1}$ chains, $M_\xi O_{3\xi}$ rings, and $M_qO_{\leqq 3q}$ entities; in other words, we assume an equidistribution of the ionic charge about all oxygen atoms of the structure except those forming the μ-oxo bridges in the fictive $M_\xi O_{3\xi+1}$ chains, $M_\xi O_{3\xi}$ rings, and $M_qO_{\leqq 3q}$ entities (those forming the normal bonding in the polyoxometalate ions). Accordingly

$$S_m = qn - \left(2 - \frac{(8-n)q - p}{4q - r - (p-h)/2}\right)(t + h_u/2) - S_n$$

$$= qn - \left(2 - \frac{m}{u - (p-h)/2}\right)(t + h_u/2) - S_n \tag{10}$$

(t: number of terminal O atoms in the structure; h_u: number of OH groups in the structure as unshared O atoms; for the other quantities see the general formulae for the polyoxometalate ions here and in Sect. 1). The quantitative view of this point, expressed by Eq. (10), was not included in our original bond model.

Note that Eq. (9) requires no knowledge about the structure. Only the stoichiometry of the compound (r) has to be known. Eq. (10) requires additionally some knowledge about the connectivity of the atoms in the polyoxometalate framework (t, h_u), but no knowledge of bond lengths. For numerical values of S_n and S_m see Sect. 4.6.8, in particular Table 46 in Ref. 61.[2] Remarkably, in some structures there is no meshing effect despite the occurrence of extended MO_k polyhedra since bridging (and not terminal) oxygen atoms are used for the extension of the MO_k polyhedra, see Sect. 4.6.8, in particular Fig. 35 in Ref. 61.

3.4 The Valence of the M–O Bonds and the Distribution of the Ionic Charge Over the Oxygen Atoms

The different types of resonance exist simultaneously and thus allow the formulation of a large number of resonance structures for a polyoxometalate ion. The valence of an M–O bond is represented by the statistical mean of the bond orders (cf. footnote 1) of the M–O bond under consideration in the different resonance formulae, provided that no other factors become operative. If necessary the different resonance formulae have to be weighted appropriately. Thus the sum of the bond valences about the M atoms (the sum of the "bonding power" of the M atoms) is constant and equals (exactly) the oxidation number of M. This view corresponds to the definition of the resonance bond numbers [73a, 74, 75a].

Similarly, the distribution of the negative ionic charge over the oxygen atoms is the statistical mean of the charge distribution in the different resonance formulae, provided that no other factors become operative. If necessary the different resonance formulae have to be weighted appropriately, as above. Thus, the sum of the bond valences about each oxygen atom plus the negative charge (or minus a positive charge) on it results (exactly) in the absolute amount of the oxidation number of oxygen, i.e., two.

For a given polyoxometalate ion, characterized merely by the network of bonds connecting neighboring atoms and by the oxidation number of the atoms (this includes the ionic charge), it is, in principle, possible to calculate the valences of the M–O bonds and the charge distribution over the oxygen

[2] They can be summarized as follows: The average (related to 1 M atom) stabilization by normal bonding results for the studied M^{VI} polyoxo species with extended MO_k polyhedra in 0.40 to 2.00 v.u. [valence units] and for the corresponding M^V species in 1.67 to 3.00 v.u. These values correspond to the minimum $Z^+ - h/q$ ($HW_5O_{19}^{7-}$, $Z^+ = 0,60$; $Nb_6O_{19}^{8-}$, $Z^+ = 1.67$) and maximum $Z^+ - h/q$ ("MoO_3", "(α-)$MoO_3 \cdot H_2O$", "$MoO_3 \cdot 2H_2O$", $Z^+ = 2$; "V_2O_5", $Z^+ = 3$). The average stabilization by the meshing effect for the M^{VI} species is up to 2.85 v.u. and for the M^V species is up to 1.90 v.u.

atoms on the basis of Schemes (5) and (6) (and (8)), which are based, in turn, on the building Scheme (V) (or, more detailed, Schemes like (VI) to (IX)) of the MO_6 octahedra (MO_k polyhedra). Unfortunately, such calculations cannot yet be performed. They would, on the one hand, reveal certain principles about the distribution of the valences of the M–O bonds in the M–O frameworks and about the charge distribution over the oxygen atoms. [In the former case, for instance, it is supposed that the existence of the type I MO_6 octahedra – consisting of one strong (short), four middle-strong (medium-sized), and one weak (long) M–O bond in a *trans* position to the strong (short) one – and type II MO_6 octahedra – consisting of two strong (short) *cis*-located, two middle-strong (medium-sized), and two weak (long) M–O bonds in a *trans* position to the strong (short) ones – [76–78] would result.] On the other hand, they would disclose, by comparison with experimental results (cf. Sects. 4.3 and 4.4 in Ref. 61), the influence of charge separation processes (resonance scheme (7)) and of other possibly operative factors (items (iii)–(v), below).

Fortunately, it is possible to make reasonably good estimates or even sometimes exact statements on some special topics:

- The requirements of (resonance) formulae (VI) and (VII) for M^{VI} species, which to a first approximation have two M–O bonds with near double-bond character, two with near single-bond character, and two rather weak M–O bonds, lead to, in the case of the type II MO_6 octahedra[3a] without two terminal oxygen atoms, the definition of "pseudoterminal" oxygen atoms, i.e., two-coordinate oxygen atoms in the function of terminal ones. These disclose their identity by one rather short and a second, usually rather long M–O bond (for the type of the pseudo- or quasiterminal oxygen atom, see also Refs. 1b, 4, 77, and 79).
- The coordination mode (V) (defined by schemes like (VI) to (IX)) requires one of the oxygen atoms, forming the "beam" of the "seesaw," to be connected to an H or M (or X) atom. For this reason there can be no resonance formula with a negative charge on the most highly coordinated oxygen atoms of some polyoxometalate structures, e.g., for $M_7O_{24}^{6-}$, $M_6O_{19}^{2-}$, $W_{12}O_{38}(OH)_2^{6-}$, and $W_{10}O_{32}^{4-}$ (however, for (β-)$Mo_8O_{26}^{4-}$, $Mo_8O_{26}(OH)_2^{6-}$, $Mo_{10}O_{34}^{8-}$ (I), – (II), $[Mo_8O_{27}^{6-}]_\infty$, and $M_{10}O_{28}^{6-}$ resonance formulae of this type can be drawn)[3b]; compare Fig. 1(c)–(j). [In the resonance structure for $M_7O_{24}^{6-}$ of Fig. 1(d) a negative charge on the four-coordinate O atoms is possible only by violation of coordination mode (V) (i.e., Schemes (VI) and (VII)) for the MO_4 unit to the right.]
- In the resonance structure for $M_7O_{24}^{6-}$ of Fig. 1(d) bonds between the pseudoterminal O atoms and the neighboring outer M atoms are only

[3a] Note that the formally identical descriptions of the MO_6 octahedra in the resonance formulae of polyoxometalate ions (above) and of the type II MO_6 octahedra in (real) polyoxometalate structures refer to fundamentally different objects: According to the M–O framework the statistical mean of the resonance formulae can lead to type I or type II MO_6 octahedra, and the latter can appear with two terminal (normal case), one terminal and one two-coordinate, or two two-coordinate O atoms.

[3b] For the structures of these species see Ref. 61.

possible by violation of coordination mode (V) for the MO_4 unit to the right. This may be a reason for the rather large (average) lengths of 2.52 Å (very small bond valence of 0.23 v.u.) [61] for the bonds in question (see Sect. 4.1).

On considering the resonance schemes for the different types of oxygen atoms we can obtain further information. Fig. 3 shows that the importance of resonance formulae with a negative charge on oxygen decreases with increasing coordination number of the oxygen atom for purely statistical reasons. This can be expressed as follows: If the number of M–O bonds originating from an oxygen atom is high, the oxygen atom "mobilizes" its "bonding power" (small negative or even positive charge on the oxygen atom) and distributes it among all bonds (no M–O multiple bonds). The resonance schemes of Fig. 3 indicate:

- Terminal M–O bonds have multiple-bond character (Fig. 3(a) and (b)). In (poly)oxometalate ions of M^{VI} where for structural reasons (necessity of at least monoprotonation of the original monometalate ions; occurrence of polyhedra with at most two terminal O atoms) usually (compare Table 41 in Ref. 61) $m \leqq t/2$, the average multiple bond character can be in the range 1.5–2 v.u. for a terminal M–O bond, depending on the Z^+ value of the polyoxometalate ion and on the weight of the resonance structures. In (poly)oxometalate ions of M^V where for structural reasons (necessity of at least diprotonation of the original monometalate ions; occurrence of polyhedra with at most one terminal O atom) usually $m \leqq t$ (compare Table 42 in Ref. 61), the multiple bond character can be in the range 1–2 v.u. for a terminal M-O bond, again depending on the Z^+ value of the polyoxometalate ion and on the weight of the resonance structures. (See also the discussion of resonance scheme (23) in Sect. 4.6.2.) (Note that we have considered *unprotonated* polyoxometalate *ions* composed of MO_k polyhedra with $k > 4$.)
- Pseudoterminal M–O bonds also have multiple-bond character (somewhat smaller than that of terminal M–O bonds); the other M–O bond is usually a coordinate bond (Fig. 3(c)–(e) with only a small portion of resonance formula (e) and of the resonance formulae not depicted in Fig. 3(c) and (d). The situation thus resembles that for terminal M–O bonds).
- The bonds of two-coordinate oxygen atoms (i.e., M–O–M bridges except those where pseudoterminal oxygen atoms are involved) have (approximately) single-bond character in both bonds (Fig. 3(c)–(e)).
- The bonds of three-coordinate oxygen atoms (M_3O bridges) have partly (approximately) single-bond character and partly coordinate bond character (Fig. 3(f)–(h)). [In extreme cases three M–O bonds of single-bond character ($H_4As_4Mo_{12}O_{50}^{4-}$, compare Sect. 4.3.4.3 in Ref. 61) and no bonds of single-bond character ($W_{12}O_{40}(OH)_2^{10-}$ and $W_{12}O_{38}(OH)_2^{6-}$, compare Sects. 3.17, 3.18, and 4.3.4.4 in Ref. 61) have been observed.]
- The bonds of five- and six-coordinate oxygen atoms have the character of coordinate bonds (Fig. 3(l)–(q)). [The character of the bonds of four-coordinate oxygen atoms (Fig. 3(i)–(k)) lies between that of three- and five-coordinate oxygen atoms.]

	A	B	C	
k = 1:	M−O⁻ (1) (a)	M=O (1) (b)		$\bar{s} = 1.50$ v.u. $\bar{c} = 0.50$ c.u.
k = 2[a]:	M−O⁻···M (2) (c)	M=O···M (2) (d)	M−O−M (1) (e)	$\bar{s} = 0.80$ v.u. $\bar{c} = 0.40$ c.u.
k = 3[b]:	M−O⁻···M (3) │ M (f)	M=O···M (3) │ M (g)	M−O−M (3) │ M (h)	$\bar{s} = 0.56$ v.u. $\bar{c} = 0.33$ c.u.
k = 4[c]:	M−O⁻···M (4) M M (i)	M=O···M (4) M M (j)	M−O−M (6) M M (k)	$\bar{s} = 0.43$ v.u. $\bar{c} = 0.28$ c.u.
k = 5[d]:	M M−O⁻···M (5) M M (l)	M M=O···M (5) M M (m)	M M−O−M (10) M M (n)	$\bar{s} = 0.35$ v.u. $\bar{c} = 0.25$ c.u.
k = 6[e]:	M M M−O⁻···M (6) M M (o)	M M M=O···M (6) M M (p)	M M M−O−M (15) M M (q)	$\bar{s} = 0.30$ v.u. $\bar{c} = 0.22$ c.u.

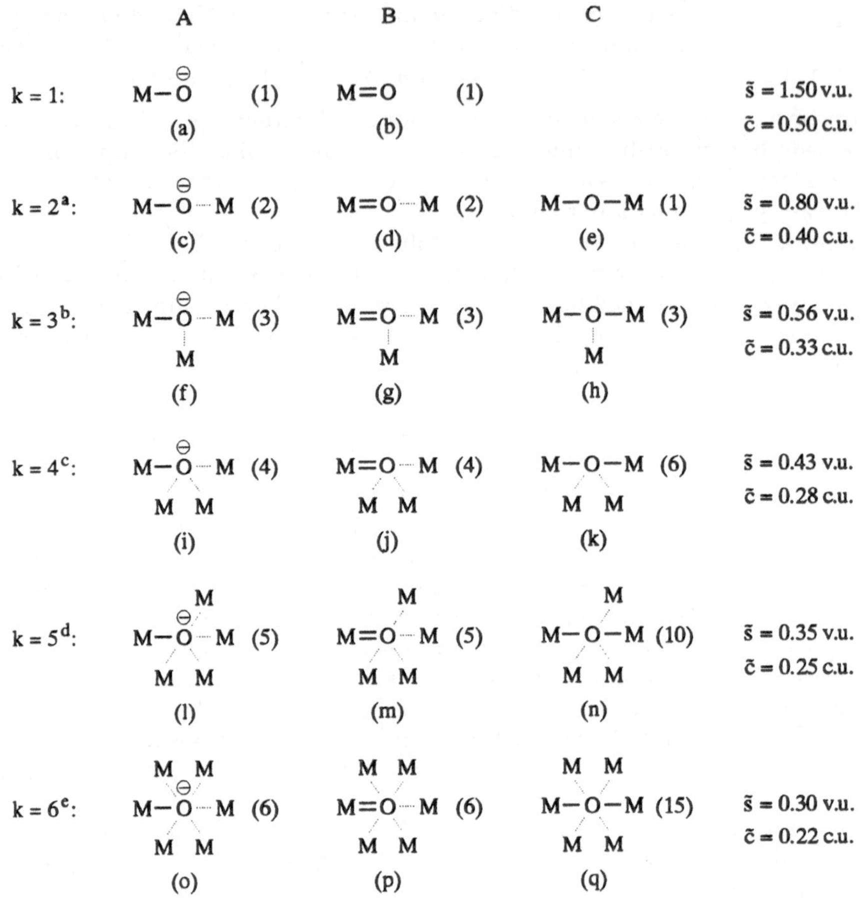

Fig. 3. Resonance schemes for the M–O bonds of the differently coordinated oxygen atoms in polyoxometalate ions. The numbers given in parentheses state the number of resonance structures in which combinations of the bonds of bond valences two, one and zero for the resonance types A, B, and C occur. k: coordination number of oxygen; the coordination geometries about the oxygen atoms are as follows (see Fig. 2 in Ref. 61): [a] angular or – as shown – (approximately) linearly two-coordinate; in the case of pseudoterminal oxygen atoms the figured resonance formulae (c) and (d) make the main contribution, the situation thus resembling that for k = 1, [b] trigonal-pyramidal with oxygen at the apex, T-shaped – as indicated – , or approximately trigonal-planar three-coordinate, [c] seesaw-like – as indicated – or approximately tetrahedral four-coordinate, [d] tetragonal-pyramidal five-coordinate, [e] octahedral six-coordinate. $\bar{s}$, $\bar{c}$: average bond valence of the M–O bonds of the O atoms and average charge on the O atoms with the coordination number k in the collection of resonance formulae specified assuming equal probability for all resonance structures of the $M_k O$ groups ($\bar{s}$ and $\bar{c}$ represent proportional numbers)

Note that other factors which may become operative have been excluded here.

In a first, rough approximation the M–O bond valences and the charge distribution over the oxygen atoms of the polyoxometalate ions can be

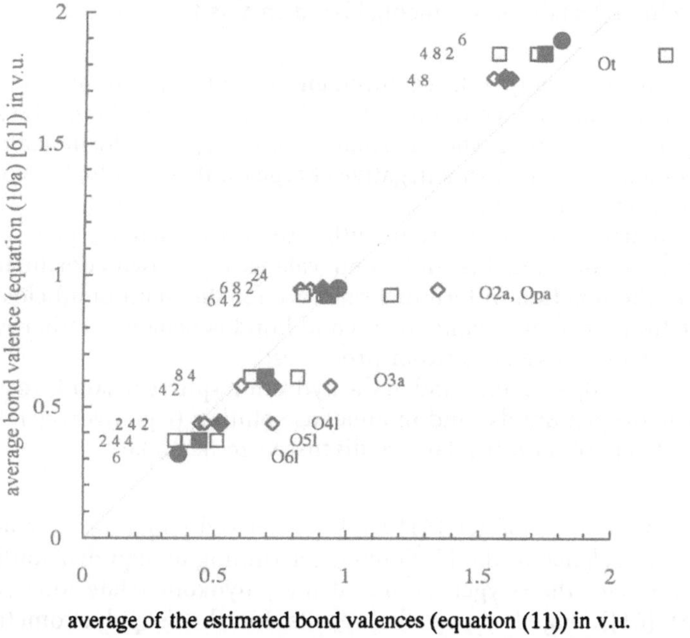

Fig. 4. Average of the estimated M–O bond valences according to Eq. (11) (which take into consideration the high statistical weight of the resonance structures with charged terminal oxygen atoms and the small statistical weight of the resonance structures with charged bridging oxygen atoms) and average bond valences according to Eq. (10a) of Ref. 61 (assuming an equidistribution of the total bond valence about the M–O bonds) for the M–O bonds of the different types of oxygen atoms of the polyoxometalate ions $M_7O_{24}^{6-}$ (◆), (β-)$Mo_8O_{26}^{4-}$ (■), and $M_6O_{19}^{2-}$ (●); Ot: terminal O atoms; O2a and Opa, O3a: two- and three-coordinate, respectively, angularly bridging O atoms; O4l, O5l, O6l: four-, five- and six-coordinate, respectively, (approximately) linearly bridging O atoms (for the structures see Ref. 61). ◇, □, ○ individual values with designation of their frequency (○ coincides with ●)

estimated in the following way. (Approximate) valences s_i of the M–O bonds can be obtained by a normalization of the sum of the "probability quantities" $\tilde{s}_i$ for the different M–O bonds of the individual MO_k polyhedra to the oxidation number n of M according to

$$s_i \approx n \frac{\tilde{s}_i}{\sum_{i=1}^{k_M} \tilde{s}_i} . \tag{11}$$

Similarly [80], (approximate) charges c_j on the oxygen atoms can be obtained by a normalization of the sum of the "probability quantities" $\tilde{c}_j$ for all u oxygen atoms of the polyoxometalate ion $M_q^n O_u^{m-}$ to its charge number m according to

$$c_j \approx m \frac{\tilde{c}_j}{\sum_{j=1}^{u} \tilde{c}_j} . \tag{12}$$

The following effects remain unconsidered in this process:

- (i) peculiarities owing to the requirement of coordination mode (V) that one of the oxygen atoms forming the "beam" of the "seesaw" must be connected to an H or M atom (non-existence of resonance formulae for some polyoxometalate ions with a negative charge on the most highly coordinated oxygen atoms, see above);
- (ii) peculiarities in connection with pseudoterminal oxygen atoms (the properties - bond lengths and bond valences - of two-coordinate oxygen atoms in the function of terminal ones are for the main bond close to those of the latter, see above, while the second bond is usually a rather weak one);
- (iii) effects of charge separation processes;
- (iv) effects of adjacent ions and of the hydration sphere in solids (e.g., packing effects, hydrogen bonds) and in aqueous solution (e.g., hydrogen bonds);
- (v) the effects of all other factors discussed in Sect. 3.6.

In an accompanying chapter [61] we have derived equations to calculate the *average* bond valence of the M–O bonds, assuming an equidistribution of the ionic charge over the oxygen atoms of the polyoxometalate ions (Eqs. (10a) and (10b) [61]). Fig. 4 shows the results for the isopolyoxometalate ions $M_7O_{24}^{6-}$, (β-)$Mo_8O_{26}^{4-}$, and $M_6O_{19}^{2-}$ according to Eq. (11) above and Eq. (10a) in Ref. 61. Eq. (11) above gives somewhat smaller (0.1 to 0.15 v.u.) bond valences for the strong, terminal M–O bonds than Eq. (10a) of Ref. 61. For the weak, usually multiply bridging M–O bonds the reverse situation arises. For the middle-strong, usually low-coordinate bridging M–O bonds there is practically no difference. The differences are the consequence of the high statistical weight of the resonance structures with charged terminal oxygen atoms and consequently smaller bond valences in the terminal M–O bonds and the small statistical weight of the resonance structures with charged bridging oxygen atoms and consequently greater bond valences in the bridging M–O bonds (see Fig. 3). In Fig. 30 of Ref. 61 the individual M–O bond valences estimated according to Eq. (11) are compared with the actual bond valences based on the bond lengths for the species $M_7O_{24}^{6-}$, (β-)$Mo_8O_{26}^{4-}$, $M_6O_{19}^{2-}$, and $[Mo_6O_{20}^{4-}]_\infty$(o).

The above methods for estimating M–O bond valences and charges on oxygen atoms of polyoxometalate ions, may be augmented by the possibility of calculating simply the (exact) sum of normal bonding in the M–O–M bridges from the stoichiometry of the polyoxo species (i.e., the sum of the inner bond valences if no extension of the MO_4 tetrahedra would occur) and the (in relation to a reference state exact) sum of the bond valences produced by the meshing effect from the stoichiometry and the connectivity of the atoms in the polyoxo structures (i.e., the sum of the inner bond valences produced by the extension of the MO_4 tetrahedra), see Sect. 3.3, above, and Sect. 4.6.8, in particular Table 46 in Ref. 61.

3.5
The Valence of the X–O Bonds of Central Heteroatoms

The electronegativity of the common heteroelements X is always by far smaller than that of oxygen. Hence X cannot bear a negative charge. It also cannot bear a positive charge since this has to be compensated by corresponding negative charges on the directly adjacent O atoms, and this situation would be equivalent to normal covalent bonds (compare Sect. 4.3, in particular Scheme (22)). Hence the sum of the bond valences about X corresponds to the oxidation number of X (just as is the case for M). Because of the high symmetry of the heteropolyoxo anions the X–O bond valences of central heteroatoms can exactly be established as

$$s_{X\text{-}O} = n_X/k_X. \tag{13}$$

Due to packing effects in the crystal the X–O bonds may be somewhat different in length and bond valence.

Peripheral heteroatom polyhedra behave like the addenda (M) atom polyhedra, the coordinate bonds excluded (compare Sect. 3.2.4).

3.6
Other Features Exerting an Influence on the M–O Bonds and on the Charge Distribution

The statistical mean of the different resonance formulae as discussed in Sect. 3.4 should be seen as a provisional reference state. It can be influenced by several other factors.

3.6.1
The Geometrical Requirements of Rigid Structures

The interdependence of the M–O and X–O bond lengths and fixed bond lengths (e.g., for central X–O bonds) in a polyoxometalate structure allow only restricted ranges for possible bond lengths of the *inner* M–O bonds. This can lead to geometrically constrained structures since the bond valence sum for all atoms of a compound has to yield their oxidation number and additionally a bond length-bond valence relationship has to be fulfilled (compare Eqs. (20) and (21)). The constraint is the more pronounced the more compact and/or the more symmetrical (e.g., M_6O_{19}, $XM_{12}O_{40}$), i.e., the more rigid a structure is. This favors those resonance formulae which comply with suitable bond lengths and bond valences. (This feature was not included in our original bond model.)

3.6.2
The Symmetry Properties of the Orbitals on the M and on the O Atoms

The symmetry properties of the orbitals on the M and on the O atoms can have an influence on the weight of the different resonance formulae; cf. the tendency for *cis*-location of the π-interaction in each of the type II MO_6

octahedra [1a, 4a, 16, 77, 81, 82] and the formation of multicenter pπ-dπ multiple bonds in (approximately) linear M–O–M and X–O–M bridges [38a, 39–41, 42a, 43, 66, 83–85]. Multiple bonds in M–O–M bridges favor the cohesion of the MO_k polyhedra and hence stabilize the polyoxometalate structures, see resonance scheme (17) in Sect. 3.6.7.

However, based on the great variability of the coordination modes observed for the M and O atoms (see Sects. 3.2 and 4.8.4 in Ref. 61) and on the typical bond valences (bond lengths) of the M–O bonds of the different types of O atoms (see Sect. 3.4), it has been argued [61] that there is at present no possibility for an unequivocal proof of electronically driven asymmetries in the coordination environments of M in (poly)oxometalate ions as discussed, e.g., in Refs. 1a, 3, 75b, 77, 78, and 86. This is because all effects ascribed to electronic asymmetries can conclusively be attributed to other factors.

3.6.3
The Location of Unshared Electron Pairs

Unshared electron pairs, in which the charge electrons are contained, require more space than bonding electron pairs and therefore prefer terminal oxygen atoms [14a, 21, 22a]. Resonance formulae with the charge on terminal oxygen atoms are in general more favorable (cf. the VSEPR model [50]) [14a, 22a].

For the structures with $M_qO_{<3q}$ entities ($Z^+ > 2$) (e.g., $M_{10}O_{28}^{6-}$) or with $m > t$ (e.g., $M_6O_{19}^{2-}$) the choice to place the ionic charge on terminal or bridging O atoms of the resonance structures is restricted. In both examples at least two charge units *must* be accepted by bridging O atoms.

If the negative charge density becomes greater, the terminal O atoms may become overcrowded and then bridging O atoms at the surface of the polyanion may be the better choice for additional charge.

3.6.4
The Coulombic Repulsion Between Like Charges

To keep the charge electrons as far away as possible from each other because of the Coulombic repulsion [22, 35, 36a, 87–89], the ionic charge of the polyanion resides preferably on the surface oxygen atoms; that is, the corresponding resonance formulae are more favorable [22, 35, 36a].

There now appears to be agreement [1c, 15j, 33b] that, concerning the repulsion between the highly charged M^{n+} cations [67, 81, 86, 90], the effect is too weak in comparison to the factors driving polyoxometalate ion formation and directing the course of aggregation [15j, 15k, 22a, 36a, 91]. If repulsion between the M atoms were significant, aggregation reactions would generally be suppressed [15j].

Only the Coulombic repulsion between the negative charges is important because the negative charges are *real* charges whereas the positive charges on the M atoms are not real but fictional charges and are actually compensated by the formation of (covalent) M–O bonds between M^{n+} and O^{2-}; see also Sect. 4.3.

3.6.5
The Compensation of the Ionic Charge of the Polyoxometalate Ions by the Cations

To effectively compensate the ionic charge of the polyanions by the cations and to hold anions and cations together in crystalline solids [8–13, 15a, 92], the ionic charge of the polyanion resides preferably on the surface oxygen atoms; that is, the corresponding resonance formulae are more favorable (see also Sect. 4.4).

3.6.6
The Tendency for the Formation of Closely Packed Structures

For the solid state the formation of closely packed structures is to be expected. The structures of the polyoxoanions composed of MO_6 octahedra consist of closely packed arrays of O atoms [8, 9, 11, 21, 93]. In our opinion the close packing is facilitated by bond valence and bond length differentiation in the MO_6 octahedra due to the ionic character of the structures. In solid polyoxometalates the molecules of water of crystallization and the cations supplement the closely packed arrangements [8, 11, 93].

Whether the tendency to closely packed structures exerts its own influence on bond length differentiation cannot be recognized for the polyoxometalate structures since other factors also lead to bond length differentiation and to closely packed structures. The situation thus resembles that discussed in Sect. 3.6.2 with regard to an influence of the symmetry properties of the orbitals on the M and O atoms. For non-ionic structures (e.g., for oxides) which, in contrast to polyoxo anions, do not necessarily require different bond lengths because of an ionic charge, there is also a differentiation of the bond lengths (formation of chain and layer structures, see Sects. 3.25–3.28 in Ref. 61). We attribute this to the gain in space for the central M atoms of the MO_6 octahedra by the bond valence/bond length differentiation (this effect applies, of course, also to the polyoxo anions); see also Sect. 4.6.1 in Ref. 61.

3.6.7
The Law of Mass Action (Le Chatelier's Principle)

Depending on the conditions in the solution in which the oxometalate ions occur (e.g., metalate concentration, degree of acidification, pH) the law of mass action favors the more proton-consuming species, i.e., the more stable and/or the more basic oxometalate ions (poly- and monometalate ions). (A feature that was not included in our original bond model [15l]). This item now requires discussion in greater detail.

When a solution of a polyvalent base, e.g., a monometalate ion, is acidified, protonation of the monometalate ion takes place since it is a stronger Brønsted base than the H_2O. The strength of the Brønsted bases is determined by Pauling's rules [44b] on the acid/base strength of oxoacids/oxoanions, with modifications for oxoacids/oxoanions of the transition metals [43]:

$$MO_4^{w-} + H_3O^+ \rightleftharpoons MO_3(OH)^{(w-1)-} + H_2O \quad (14a)$$

$$MO_3(OH)^{(w-1)-} + H_3O^+ \rightleftharpoons MO_2(OH)_2^{(w-2)-} + H_2O, \text{ etc.} \quad (14b)$$

(basis reactions).

The basis reactions (14) are influenced in a first stage by consecutive reactions of the protonated monometalate ions in which the primary reaction products are removed from the equilibria[4], for example, by aggregation reactions [15o, 22a, 36a]:

$$2MO_3(OH)^{2-} \rightleftharpoons M_2O_7^{4-} + H_2O \quad (M = V) \quad (15a)$$

$$8MO_3(OH)^{-} \rightleftharpoons M_7O_{24}^{6-} + MO_4^{2-} + 4H_2O \quad (M = Mo, W) \quad (15b)$$

or by reactions with solvent molecules [36a, 43]:

$$MO_2(OH_2) + 2H_2O \rightleftharpoons MO_2(OH)_2(OH_2)_2 \quad (M = Mo, W). \quad (15c)$$

The concentration of the protonated monomers on the right side of Eqs. (14) decreases, whereas the concentration of the unprotonated (or less protonated) monomers on the left side remains - for a given degree of acidification - nearly constant. Hence the pH value (in the sense of the negative decadic logarithm of the H^+ *concentration*) rises[5] according to the Henderson-Hasselbalch equation; that is, the protons introduced are increasingly consumed by the occurrence of the consecutive reactions.

In a second stage, all effects that enhance the stability or basicity of the (mono- or poly)oxometalate species formed in reactions (15) act in a similar manner and lead to a further increased consumption of protons and hence to still more increased pH values. Considering the stability of the oxometalate species, the condensation of H_2O molecules (an entropy effect) [22a, 36a], the meshing of the MO_4 building units (favored by the "meshing effect", see

[4] The H_2O molecules of the basis reactions (14) can also be removed from the equilibria by a type of consecutive reaction. They are, to a certain degree, bound by the medium salt (hydration of its ions) with the consequence that the metalate concentration appears to be increased as indicated by correspondingly enhanced formation constants for the oxometalate species; that is, the consecutive reactions (15) are favored by this effect of the medium salt (decrease of the water activity [15m, 94]) whereas the basis reactions (14) are only slightly affected since simple protonation equilibria are not dependent on the metalate concentration (cf. the Henderson-Hasselbalch equation and the existence of the "mononuclear wall" [15n] in (Z, log[H^+], C_M) equilibrium curves for the MO_4^{w-}/H^+ systems).

[5] For example, an MoO_4^{2-} ($K_p = 10^{3.87}$ [95]) solution (ionic medium: 3 M $Na(ClO_4)$, 25 °C), acidified to P = 0.5 (the degree of acidification P is the molar ratio of H^+ ions introduced to MO_4^{w-} ions introduced), would have a pH of 3.87 if no consecutive reactions occur (Eq. (14a)). Since a marked aggregation to give $Mo_7O_{24}^{6-}$ takes place (Eq. (15b)), an MoO_4^{2-} solution acidified to P = 0.5 has a higher pH, 6.40 for a 0.1 M molybdate solution (compare Fig. 8 in Ref. 95; note that in the acidification range concerned P numerically equals Z, see Refs. 15p and 36b).

Sect. 3.3) in the case of the structures composed of MO_k polyhedra with $k > 4$ [22a, 36a], and, for polyoxometalate ions composed of corner-sharing MO_4 tetrahedra, the formation of multicenter $p\pi$-$d\pi$ multiple bonds in (approximately) linear M–O–M bridges [43, 66] have all been mentioned [15i]. Considering the basicity of the oxometalate species, for example, the following protonation constants of successively occurring oxomolybdate ions have been cited in the literature [15a, 15q]:

	MoO_4^{2-}	$MoO_3(OH)^-$	$Mo_7O_{24}^{6-}$
K_p :	$10^{3.8}$	$10^{3.8}$	$10^{4.5}$

[Notice in respect of $MoO_3(OH)^-$ that – according to the rule that successive acid constants of a polyvalent (monomeric) acid differ by a factor of 10^4 to 10^5 [38c, 42b, 44b, 96] – the expected protonation constant is $K_p(MoO_3(OH)^-) \approx K_p(MoO_4^{2-})/10^{4.5} = 10^{-0.7}$.] The reasons for the enhanced basicity are seen in structural changes [coordination of two H_2O molecules by $MoO_2(OH)_2$ and occurrence of charge separation and charge delocalization processes [43] (protonation of $MoO_3(OH)^-$); decrease of the number of terminal O atoms in the aggregation process [15o, 22a, 36a] by condensation of H_2O molecules and by the meshing effect (formation of $Mo_7O_{24}^{6-}$). For an experimental proof of the dependence of the basicity upon the number of terminal O atoms as the most basic ones see Fig. 33 in Section 4.6.6 of Ref. 61].

Scheme (16), which is a consequence of Scheme (6) for bridged MO_k polyhedra, shows that the negative ionic charge on bridging oxygen atoms reduces the bonding power of the latter, i.e., the bonding power between the MO_k polyhedra, by which the polymeric character of the structure is destabilized ((16d)). Conversely, the polymeric structure is more stable if the negative charge is accepted by the terminal oxygen atoms ((16a)–(16c)):

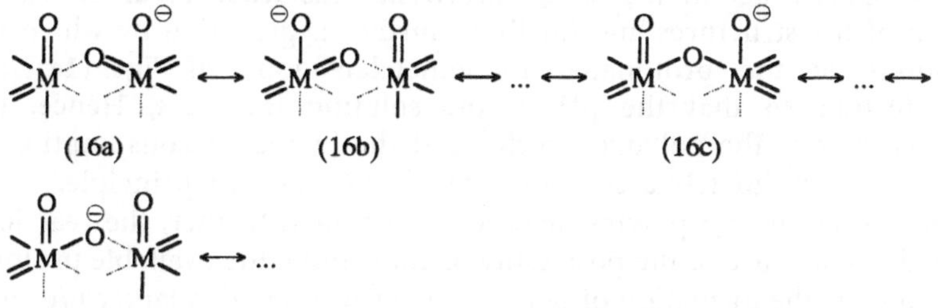

(16a) (16b) (16c) (16d) (16)

Similarly, Scheme (17), which is a consequence of Scheme (7) for bridged MO_k polyhedra, shows that the positive charge (produced by charge separation) on a bridging and the corresponding negative charge on a terminal oxygen atom leads to a stabilization of the polymeric character of the structure ((17b)). The reverse leads to a destabilization of the polymeric structure ((17c)), whereas the positive and the negative charge both on terminal or both on bridging oxygen atoms has virtually no effect ((17d), (17e)):

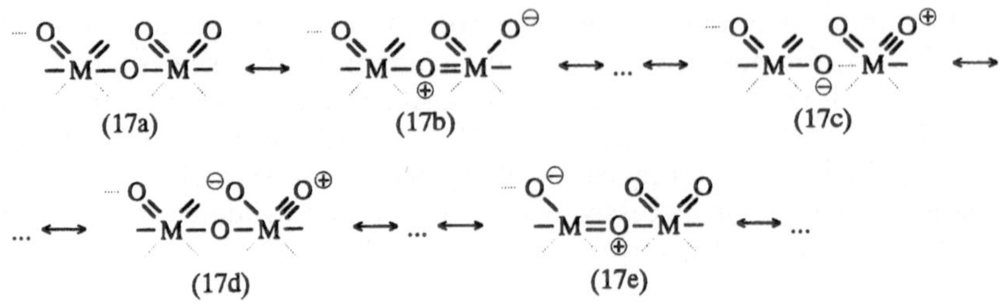

(17)

In all favorable resonance formulae of Schemes (16) and (17) there is (simultaneously) an increase of the inner bond valence[6] and an increase in the basicity of the structures. Formation of pπ-dπ multiple bonds in approximately linear M–O–M bridges is hence not only an important stabilizing factor for the polyoxometalate ions composed of corner-sharing MO_4 tetrahedra [43, 66] but also for the polyoxometalate ions composed of MO_6 octahedra (MO_k polyhedra with $k > 4$) (this feature was not included in our original bond model).

In the above concept the stabilization of the polymeric character of a structure is of no advantage for the system as such, as has often been claimed or tacitly assumed in the literature [1a, 26, 81, 97–99] in other contexts (see also Ref. 15r). The enhanced consumption of protons by the system in the course of the acidification of the metalate solution is the decisive factor. Hence, the optimal stabilization of the polymeric character of the structures caused by the adjustment of the bond lengths is ultimately a consequence of Le Chatelier's principle.

Similarly, the preferred acceptance of the negative charge by the terminal oxygen atoms according to (16a)–(16c) (ionic charge) and (17b) (charge produced by charge separation) and the resultant enhanced basicity of the oxometalate species act in the same direction. This leads to an increased protonation of the structures and finally to further aggregation by which the concentrations of the protonated monometalate ions of Eqs. (14) are decreased further so that the pH of the solution increases. Hence, the increased basicity of the polyoxo species and the increased consumption of protons are again ultimately a consequence of Le Chatelier's principle.

When the solution of a polymetalate ion is acidified further, the reactions that occur also make use of the possibility of consuming the available protons. The main route is the formation of new polyoxo species with a larger p/q ratio (Z^+ value) by protonation of existing polymetalate ions (according to their protonation constants and the pH of the solution) or by formation of new structural types. In this process the factors operating above continue to apply. In the region of high acidification of the oxometalate solution it is necessary that species with high p/q ratios form and this requires the decomposition of the polymeric structures until, in the final stages, highly protonated di- and

[6] These effects can also be seen directly from Fig. 2(b) and (c) by counting the number of the inner M–O bonds in the different resonance formulae.

monometalate ions arise [15o, 43] (the maximum for a *polyoxo anion* of M^{VI} approximates to $Z^+ = 2$; the maximum for a protonated *cationic dimer* is $Z^+ = 3.5$ and for a protonated *cationic monomer* $Z^+ = 4$ [15bb]). (Note that some of the H^+ ions are required in each case to adjust the pH value of the solution in which the reactions take place.)

3.6.8
Choice of a (New) Reference State for an Optimal Distribution of the Bond Valences Over the M–O Bonds and of the Ionic Charge Over the O Atoms; Factors Altering the New Reference State (Hierarchy of the Structural Features)

We have now recognized the generally favorable effects of the acceptance of the ionic charge by the terminal oxygen atoms of polyoxometalate ions, namely

- the stabilization and enhanced basicity of the polymeric structures by acceptance of the ionic charge and the negative charge produced by charge separation through terminal oxygen atoms (while the positive charge is accepted by bridging oxygen atoms) and their consequences according to Le Chatelier's principle,
- the high statistical weight of the corresponding resonance structures,
- the space requirements of lone electron pairs,
- the decrease of the Coulombic repulsion between the negative ionic charges,
- the contacts to the cations and the cohesion of anions and cations in the salt.

Accordingly, we can now choose a new reference state in which the ionic charge is distributed over the terminal oxygen atoms. In this case the sum of the inner bond valences amounts to

$$S_i = qn - h_u - m - 2(t - m) = qn - h_u - 2t + m \tag{18}$$

(h_u: number of "terminal" M–OH bonds; m: number of terminal $M–O^{\ominus}$ bonds; 2(t − m): number of terminal M=O bonds). This state will occur when the requirements of the following features do not become operative.

In compact and/or symmetrical structures (M_6O_{19}, $XM_{12}O_{40}$, and others) the geometrical consequences of the rigidity of the structures and given, fixed (e.g., central X–O [11, 13, 34a]) bond lengths (bond valences) lead to a restricted range for possible lengths and hence bond valences of the inner (bridging) M–O bonds. If these M–O bond valences differ from those for the reference state with the ionic charge on the terminal oxygen atoms, the resonance formulae which describe the bond valences (bond lengths) better become more important; all three resonance types (5)–(7) can serve for this purpose. If the necessary bond lengths (bond valences) cannot be achieved, the structure cannot be realized. The charge resulting on the oxygen atoms is a secondary factor, i.e., the resulting charge distribution may differ from the most favorable distribution. In the less compact and less symmetrical structures of polyoxometalate ions ($M_7O_{24}^{6-}$, $Mo_{10}O_{34}^{8-}$, and others) the M–O distances can be adjusted more independently, and the favorable distribution of the negative charge over the terminal oxygen atoms can be better realized. The interdependence of the lengths (and hence of the bond valences) of certain inner M–O bonds and of the size of the heteroatoms (the X–O bond valences) of the polyoxometalate ions have precedence over the charge distribution according to the type of oxygen atoms (preference of terminal oxygen atoms for accepting the ionic charge). (For the structure types see Ref. 61.)

The requirements for the packing of cations, oxometalate ions, and molecules of water of crystallization in the solid state can similarly be adjusted by bond length (bond valence) variations within the polyoxo species. However, the influence of this factor on the inner M–O bonds should be only small since the flexibility of the terminal M–O bonds for bond length compensations gives the polyoxometalate ions a great versatility for adapting their shape and charge to the conditions required for suitable packing. Furthermore outer (terminal) and inner (bridging) M–O bonds are geometrically largely independent of each other.

The symmetry properties of the orbitals on the M atoms do not appear to be a restricting factor since a multitude of MO_k polyhedra (MO_4 tetrahedra, MO_5 tetragonal pyramids and trigonal bipyramids, MO_6 octahedra, and MO_7 pentagonal bipyramids with different patterns of short, medium-sized, and long M–O bonds) are realized in polyoxometalate structures, see, e.g., Fig. 1 in Ref. 61. Polyhedra not resembling MO_6 octahedra with $k > 4$ can be visualized as being built up in a similar way as the MO_6 octahedra. Likewise, the symmetry properties of the orbitals on the oxygen atoms do not appear to be a restricting factor since a large number of coordination modes of oxygen are realized, see Fig. 2 in Ref. 61. However, suitable oxygen bridges between M atoms (e.g., approximately linear M–O–M bridges) are expected to favor resonance structures with multicenter $p\pi$-$d\pi$ multiple bonds [38a, 39–41, 42a, 43, 85] and a positive charge on the bridging oxygen atom (charge separation) [43]. The compensating negative charge is accepted by a terminal oxygen atom. This feature leads to an additional stabilization of the polymeric structures, as discussed above (Sect. 3.6.7).

When the negative charge density on the terminal O atoms becomes large because of a high ionic charge and a small number of terminal O atoms in the polyanion (as is in particular the case for the highly compact polyanions $M_6O_{19}^{8-}$ and $M_{10}O_{28}^{6-}$ of M^V), further acceptance of negative charge by terminal O atoms can be suppressed. In this case the bridging O atoms (at the surface of the polyanion) must also accept some charge, and the charge separation processes are suppressed. This factor has precedence over the charge distribution according to the type of oxygen atoms.

The acceptance of the ionic charge by bridging O atoms is commanding when corresponding resonance formulae for the polyoxo species cannot be avoided [polyoxometalate structures with $t < m$, e.g., $M^{V}_{6}O^{8-}_{19}$ (compare Figs. 1(g), (h)); polyoxometalate structures with $M_qO_{<3q}$ entities ($Z^+ > 2$), e.g., $M^{V}_{10}O^{6-}_{28}$ (compare Figs. 1(i), (j))]. This factor has again precedence over the charge distribution according to the type of O atom.

It is proposed that the alterations caused by the effects listed in Sect. 3.4 under (i)–(v) might reach some percentage of the M–O bond valences, with the M–O bond character thus not being fundamentally altered. However, bond valence changes of this magnitude have a significant influence on the distribution of the ionic charge over the oxygen atoms (which is the greater the more highly coordinated an oxygen atom is) since the charge results as the small difference between the sum of the bond valences of the oxygen atom under consideration and the absolute amount of the oxidation number, two, of oxygen.

3.7
Protonation of the Polyoxometalate Ions

3.7.1
Protonation of "Normal" Polyoxometalate Ions

"Normal" polyoxometalate ions are considered to be composed of type I and/ or type II octahedra (polyhedra), species with coordinated H_2O molecules and/or those containing MO_6 octahedra with three unshared oxygen atoms are excluded (examples: $M_7O^{6-}_{24}$, (β-)$Mo_8O^{4-}_{26}$, $M_6O^{2-}_{19}$, $V_{10}O^{6-}_{28}$).

Protonation of a terminal O atom causes a decrease in the number of terminal O atoms in the respective MO_6 octahedron and thus changes its type and requires appreciable alterations of the bond valences and bond lengths within the M–O framework to transmit the originally local disturbance in such a way that the interdepending bond lengths (compare Sect. 3.6.1) comply with the bond length-bond valence relationship and the valence sum rule according to Eqs. (20) and (21).

Protonation of a bridging O atom (at the surface of the polyanion) does not cause a change in the type of an MO_6 octahedron and hence requires less marked alterations of inner M–O bond valences and bond lengths to comply with the interdependence of the M–O bond lengths in the polyoxometalate structure, etc. Additionally these alterations can be better transmitted from the protonation site over the structure via two or three paths (two or three neighboring M atoms).

In our original bond model this feature was considered with a different result since we, like other authors (compare Sect. 4.6.7 in Ref. 61), had assumed that protonation occurs at the most basic sites of the structures, which are indeed the terminal O atoms (see Sect. 4.6.6 in Ref. 61).

3.7.2
Protonation of MO_6 Octahedra with Three Unshared Oxygen Atoms

$M^{VI}O_6$ octahedra with three unshared oxygen atoms, protonated at bridging oxygen atoms, have the configuration (19a)/(19b) with three strongly (on an average 1.67–2 v.u.) bonded O atoms (three short M–O bonds), hence their occurrence as type II octahedra is excluded. These MO_6 octahedra are integrated within the M–O framework by 1–0 v.u. only since the maximum charge on the terminal O atoms of an MO_6 octahedron of this type is −1 c.u. (due to its derivation from $HM^{VI}O_4^-$, compare (19a) where H may also be substituted by M)[7]. If one of the outer O atoms is protonated, only two O atoms are strongly (1.5–2 v.u.) bound, and the configuration (19c)/(19d) occurs, allowing a stronger integration of the respective MO_6 octahedron

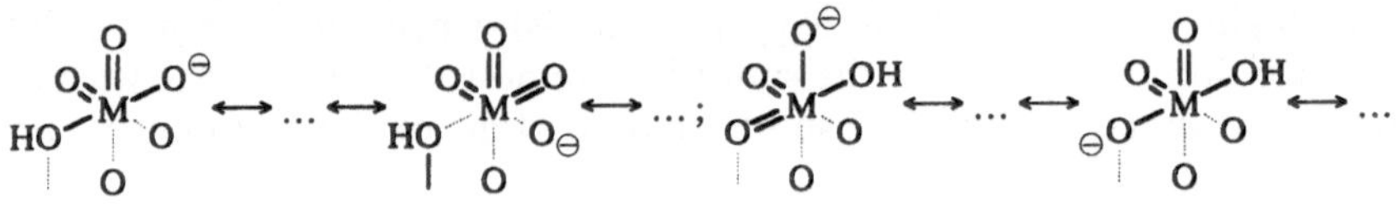

(19a) (19b) (19c) (19d)

(19)

within the M–O framework (by 2–1 v.u.) and its existence as a type II octahedron since the M–OH bond is a medium strength bond. (Corresponding M^VO_6 octahedra do not exist.)

In our original bond model this feature was considered in a different manner: For the case of non-acidic H atoms (H atoms in formula V originating from the addition mechanism in the starting phase of the aggregation processes, i.e., at small Z^+ values) it was assumed [15aa] that their positions are preferably bridging O atoms since in this case the number of terminal O atoms available for the negative charges is larger. In other words, the basicity is smaller and hence there is no need for the use of protons for pure protonation reactions in the case of a shortage of protons (i.e., in the starting phase at small degrees of protonation, $Z \lessapprox 1$), those available can then be used for the construction of additional MO_6 octahedra.

3.7.3
Coordinated Water Molecules

Coordinated water molecules occur usually with unshared but sometimes also with bridging oxygen atoms. They are suited to take the position of a weakly bound O atom in a polyoxo structure and hence to provide weak outer and

[7] In the trimeric chelate complex ion $Mo_3O_7(hmmp)_2^{2-}$ (H_3hmmp: 2-hydroxymethyl-2-methylpropane-1,3-diol) [100] one MO_6 octahedron with three terminal O atoms and two MO_6 octahedra with two terminal O atoms are present. The one with three terminal O atoms occurs in protonated form when the compound is allowed to react with chloroacetic acid (reaction medium not named). This observation was generalized, tacitly ascribed to an aqueous medium, and used to contend that oxygen atoms of MO_6 octahedra with three (*fac-*) terminal oxygens are more basic than such with *cis*-dioxo- and monooxo-terminal oxygens [3, 69a].

accordingly stronger inner M–O bonds, particularly in the case of MO_6 octahedra with a larger number (three) of unshared O atoms. Thus MO_6 octahedra with three unshared O atoms are usually of the type $MO_2(OH_2)$ (or $MO_2(OH)$), concerning the latter. (In our original bond model this case was not considered.)

4 Discussion

4.1 The Basic Features of the Bond Model

As stated above (Sect. 3.1) the bond model, although derived on the basis of (or initiated by) mechanistic studies on the formation of polyoxometalate ions, is completely independent of the mechanisms. Apart from the network of bonds connecting neighboring atoms in the M–O (or X–M–O) framework, one has only to consider

- the valence of the elements, which corresponds to their oxidation numbers (that is, the hexa- or pentavalency of the addenda element M, the corresponding valence of the heteroelement X, and the divalency of oxygen),
- the seesaw-like structure of the MO_4 units with two additional, *cis*-coordinated oxygen atoms (which means the expansion of the tetrahedral to an octahedral coordination: Schemes (VI)–(IX)), and
- to set up the Lewis-type structure of the polyoxo anion including all resonance structures according to Schemes (5)–(8) (cf. Fig. 2).

Meaningful formulae can only be given with the ionic charge on the oxygen atoms (see also Sects. 4.3 and 4.4), and the sum of the charges on all atoms must equal the ionic charge of the polyoxo anion. This procedure is completely analogous to the setting up of formulae for small molecules as shown in Schemes (1)–(4).

The above procedure results in Lewis-type structures for isopolyoxo anions in which imaginary chain-like $M_\xi O_{3\xi+1}$ ($\xi = 1$ to q, depending on the Z^+ value of the species; q: degree of aggregation of the polyoxo anion) and even ring-shaped $M_\xi O_{3\xi}$ building groups composed of strongly distorted MO_4 tetrahedra sharing corners are arranged in such a way that oxygen atoms of the MO_4 units coordinate to other MO_4 units thus meshing them (stabilization by the "meshing effect") and forming MO_6 octahedra (Figs. 1(a)–(h) and 2). (Note that this description refers to single resonance structures and therefore is by no means equivalent to a similar description of type II polyoxometalate structures given by some authors, for example, in Refs. 28, 101, and 102; see also footnote 3) In the case of an O:M ratio $\leqq 3$, e.g., with $V_{10}O_{28}^{6-}$, there are cyclic or oligocyclic $M_qO_{\leqq 3q}$ entities of corner-sharing MO_4 tetrahedra (often containing short, chain-like parts) that form the polymetalate ion in the same way, cf. Fig. 1(i), (j). In rare cases MO_5 tetragonal pyramids, MO_5 trigonal bipyramids, or MO_7 pentagonal bipyramids form according to the same principle, or MO_4 tetrahedra are retained. In some cases H_2O molecules are additionally included. Note that the $M_\xi O_{3\xi+1}$ (and $M_\xi O_{3\xi}$) building groups as well as the cyclic or oligocyclic and relevant chain-like parts of the $M_qO_{\leqq 3q}$ entities in the different resonance structures of a polyoxometalate ion can be formed by different combinations of the M atoms, see Figs. 1(b)–(j) and 2.

The X–O bond lengths of central heteroatoms in highly symmetrical heteropolyoxometalate ions are defined in advance (Scheme (8), Eq. (13)). The inner M–O bond lengths must accommodate to

- the stoichiometric requirements for Lewis-type resonance structures (controls the stoichiometrically necessary minimum of the ionic charge on bridging O atoms, in particular in the case of M^V structures),
- the geometrical requirements of the M–O framework (interdependence of the M–O bond lengths and hence bond valences) of the polymetalate ion and, if present, to the size of the heteroatom,
- (for the solid state) the geometrical and charge requirements of the packing of cations, oxometalate ions, and molecules of water of crystallization (the main effect of the packing of the named units, however, is directed towards the outer, mainly terminal M–O bonds),
- the capacity of the terminal O atoms for the acceptance of unshared electron pairs in relation to that of bridging O atoms (repression of the charge acceptance by terminal O atoms and suppression of charge separation processes in the case of structures with a high density of the ionic charge like some of M^V),
- the preferential acceptance of the negative ionic charge by terminal oxygen atoms (stabilization and enhanced basicity of the polymeric structure and their consequences according to Le Chatelier's principle, cf. Scheme (16); statistical weight of the fundamental resonance structures with negatively charged oxygen atoms in dependence on the coordination number of oxygen, cf. Fig. 3; space requirements of unshared electron pairs; decrease of the Coulombic repulsion between the negative ionic charges; contacts to the cations), and
- to the conditions for bond strengthening by multicenter $p\pi$-$d\pi$ multiple bonds in (approximately) linear M–O–M bridges as a consequence of charge separation processes (additional stabilization and enhanced basicity of the polymeric structure according to resonance formula (17b)).

The first four items take precedence over the others, according to Schemes (5)–(7), with the consequence of a corresponding distribution of the ionic charge over the oxygen atoms. Thus the ability of the oxygen atoms to accept variable partial amounts of the ionic charge or even to accept positive charge is a fundamental quality of oxygen for the realization of the (X–)M–O frameworks of the polyoxometalate ions. It is also decisive for the fact that polyoxometalate types can be realized with single addenda atoms and/or heteroatoms of different size and oxidation number.

The occurrence of a positive charge on oxygen as a consequence of the charge separation processes supports the amphoteric character of the species.

As already stated in Sect. 3.4, a resonance structure with a bond between the pseudoterminal O atoms and the neighboring outer M atoms (i.e., the formulation of a two-coordinate pseudoterminal O atom and hence of MO_6 octahedra for the outer M atoms of the row of three M's) for $M_7O_{24}^{6-}$ can only be formulated by violation of coordination mode (V) for (at least) one of the M atoms. Violation of coordination mode (V) is not necessary for any other polyoxometalate structure containing pseudoterminal O atoms. This is due to their greater Z^+ values, i.e., to the greater proportion of type (VII) compared to type (VI) formulae by which coordination mode (V) is favored. These relationships might be the reason for the conspicuous small bond valence of just 0.23 v.u. [61] for the M–O bond in question.

Polyoxometalate ions formed exclusively by MO_4 tetrahedra sharing corners have been discussed somewhere else [43, 66][8]. For tetrahedral parts of polyoxometalate ions otherwise composed of MO_6 octahedra see Ref. 61.

4.2
Test of the Bond Model

Several authors [5–7, 103–108] have shown that the valence s of the E–O(M–O) bonds is dependent on the bond length d (the dependence on the bond angle is very small [7]; a dependence of the bond length-bond valence relationship on bond angles does not exist and is not meaningful [61]) according to

$$\log s = (d_0 - d)/B \text{ or } s = (d_0/d)^N \tag{20}$$

[8] The main results of these investigations can be summarized as follows: In small MO_4^{w-} ions and their protonated forms (CrO_4^{2-}, $HCrO_4^-$, H_2CrO_4; VO_4^{3-}, HVO_4^{2-}) the forces are seen in the formation of M–O–M bridges ("normal" bonding and bond strengthening by multicenter pπ-dπ multiple bonds) and in the condensation of H_2O molecules (an entropy effect); compare Sect. 3.6.7. If only monoprotonated monometalate ions are available, aggregation stops at the dimetalate ions. If mono- and diprotonated monometalate ions are available, chain-like structures form whose length is determined primarily by the ratio of the diprotonated (forming the inner links) to the monoprotonated (forming the terminal links) monometalate ions, i.e., by the ratio of H^+ to MO_4^{w-} ions in the formation reaction for the polymetalate ion. However, at lower metalate concentrations the formation of long polyanions is suppressed as a result of the law of mass action in favor of shorter, protonated (for reasons of compensation) chain-like polyanions. The formation of ring-shaped structures (only possible for M^V) represents the optimum according to this building principle: there is a maximum condensation of H_2O molecules (1 H_2O per M) and a maximum of normal bonding and bond strengthening by pπ-dπ multiple bonds in M–O–M bridges (1 M–O–M bridge per M). The mathematical expressions of the equilibrium constants favor small rings over large rings at normal concentrations. However, trimeric rings as the geometrically smallest possible rings are excluded because the small M–O–M angles ($\lesssim 130°$) do not allow the formation of multicenter pπ-dπ multiple bonds with their favorable consequences. (The formation of polyoxometalate ions composed of M^VO_6 octahedra based on diprotonated monomers does not come into play because there is no possibility for such a structure with $Z^+ = 2$).

($d_0 \approx 1.90$ Å for Mo^{VI}, W^{VI}, Nb^{V}, and Ta^{V}, 1.80 Å for V^{V}; $N \approx 5$, $B \approx 0.9$ Å) and that the bond valences about the atoms sum up to the absolute amount of their oxidation number, $|n|$ e.g., 6 v.u. for M^{VI} and 5 v.u. for M^{V} (the sum of the "bonding power" of the atoms is constant):

$$\sum_{i=1}^{k} s_i = |n| \quad \text{('valence sum rule')} \tag{21}$$

(i: index for the M–O bonds about an atom, k: coordination number of the atom); the parameters d_0 and B or d_0 and N are characteristic of the bond partners (an element pair). These relationships are the basis of the so-called bond valence model [5, 73, 75, 109]. The equations have been used:

- to identify oxygen atoms which constitute an OH group or H_2O molecule in minerals [6, 105, 106, 110] and polyoxometalate ions [10, 11, 101, 111] via a corresponding deviation of the calculated sum of the bond valences from the bond valence two expected for oxygen;
- to control the validity of interpretations of crystal structure investigations [6, 7, 106, 109, 110];
- to elucidate the hydrogen bridge system of polyoxometalates [8–13] and other compounds [6, 7, 105, 106, 109];
- to determine the charge (oxidation number) of the M atoms [5, 7, 107, 109, 110];
- to control the coordination numbers assigned to the central atoms of coordination polyhedra [5–7, 105–107, 111], and others [5–7, 107, 109, 110].

Equation (21) corresponds precisely to a basic assumption of the bond model for polyoxometalate ions described above. The relationships (20) and (21) can therefore be applied to test this bond model by means of structural data of polyoxometalate ions utilizing improved values [113] for the parameters d_0 and B (or d_0 and N) and, above all, statistical reliability criteria. It was found that the distribution of the ionic charge over the oxygen atoms of iso- and heteropolyoxometalate ions is in full agreement with the distribution expected according to the canonical resonance formulae, as modified by the above-named features exerting an influence on the M–O bonds. The features of rigid structures and of given, fixed bond lengths as well as those of a high charge density do indeed take priority over the other factors as discussed above. These investigations are published in a separate article [61].

4.3 Presence of Partial Charges or Atomic Charges

Structures are often formulated with a partial charge or atomic charge on the component atoms of the compound. Published data are very questionable [114a, 115]. For example, in the ClO_4^- ion the partial/atomic charge on the oxygen atoms ranges from −0.21 to −0.82 and on the chlorine atom from

−0.16 to +2.27 [114b, 115]. A similar situation [116, 117] seems to exist for the oxometalate ions.

If there is a partial/atomic charge on the M atoms of the polyoxometalate ions, this must be compensated by an equal charge of opposite sign on the oxygen atoms (see Scheme (22)). These charges would be present in addition to the ionic charge as indicated by the formulae of the polyoxometalate ions. The bonding power resulting from the presence of partial charges of opposite sign on the two atoms of a bond must be considered as being included in the bond valences which amount to the valence (oxidation number) of the atoms under consideration, compare Scheme (22) (for $M = M^{VI}$)

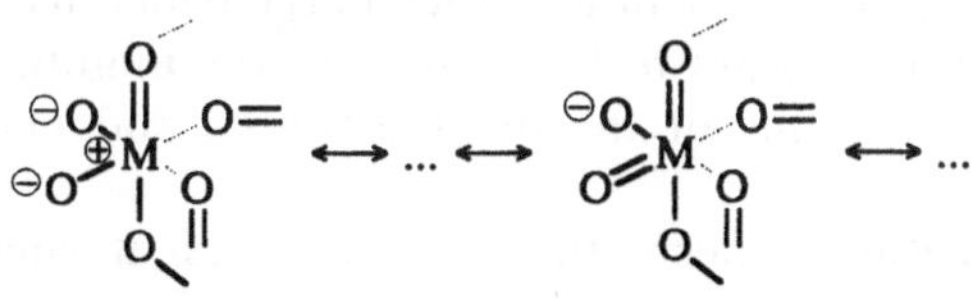

(22a) (22b) (22)

Therefore formulae like (22a) are not considered in this article (see also Sect. 2.2), especially since the bond lengths in polyoxometalate ions show no indications for unbalanced binding forces apart from those defined by Eqs. (20) and (21) [61].

Charges on oxygen atoms considered in this article require a compensation as discussed in the next section.

4.4
Charge Compensation

The net charge [negative charge minus positive charge on the (oxygen) atoms] is the formal ionic charge of the polyoxometalate ion and is ionically balanced, either by a direct contact to the cations (XII) or by an indirect contact to the cations via hydrogen bonds of water molecules of the crystal water (in solid salts) or the solvent (in solution) (XIII).

$A^{\oplus}\cdots{}^{\ominus}O{-}M$ (XII) $A^{\oplus}\cdots O(H){-}H\cdots{}^{\ominus}O{-}M$ (XIII)

This requires that the ionic charge is mainly located on oxygen atoms at the surface of the polyanion (see Sect. 3.6.5).

The negative and positive charges generated by charge separation are balanced in the zwitterion (XIV). In this case the positive charge is located on

$M{-}\overset{\oplus}{O}{=}M{-}\overset{\ominus}{O}$ (XIV)

inner, often inaccessible O atoms. An ionic intra- or intermolecular compensation is possible when the oxygen atom bearing the positive charge is accessible to a negatively charged oxygen atom, i.e., when it is located at the surface or at most in an indentation of the polyoxo anion, either by direct contact with oxygen atoms of opposite charge (XV) or by indirect contact via hydrogen bonds with water molecules of the crystal water (in solid salts) or of the solvent (in solution) (XVI).

$$
\begin{array}{cc}
\text{M} \quad \text{M} \\
\overset{\|}{\underset{\text{M}}{\overset{\oplus}{\text{O}}}}\text{...}\overset{\ominus}{\overset{|}{\text{O}}} \quad \text{(XV)}
\end{array}
\qquad
\begin{array}{cc}
\text{M} \qquad\qquad \text{M} \\
\overset{\|}{\underset{\underset{\text{H}}{|}}{\underset{\text{M}}{\overset{\oplus}{\text{O}}}}}\text{...O--H...}\overset{\ominus}{\overset{|}{\text{O}}} \quad \text{(XVI)}
\end{array}
$$

The ionic intra- and intermolecular compensation of the charge generated by charge separation ((XV) and (XVI)) appears to be a *direct* factor mobilizing additional bonding power between atoms of the polyoxoanions and hence stabilizing the structure.

Note that the above formulations mean that a cation is normally coordinated to a number of oxygen atoms, just as a charged oxygen atom is normally coordinated to several cations, and that the hydrogen-bonded water molecules may be sites of branching (see, for example, Refs. 8–11 and 17). It is well established [9–11] that the valence sum rule and hence Eqs. (20) and (21) have to be fulfilled for $A^{\oplus}\cdots{}^{\ominus}O-$ contacts of ionic bonds, as well as for hydrogen bonds, etc., too (see also comments on the "valence-sum map" in Ref. 73b). This means that the packing of the constituents of a salt requires a very intricate and strict charge balance. In this context the occurrence of hydrogen bonds in the presence of water molecules ((XIII), (XVI)) can also be seen as a means to accommodate spatial distance in cases where the cation-oxygen(polyoxometalate ion) bond valences according to Eqs. (20) and (21) could not otherwise be fulfilled ("indirect charge transfer" [11]). See also Ref. 118.

For charge on oxygen atoms forming an OH group or a coordinated H_2O molecule see Sect. 4.4.2.3 in Ref. 61.

4.5
Stabilization of Reduced Type I Polymetalate Ions

As stated in Sect. 4.7.2, polymetalate ions composed of type I MO_6 octahedra can be reduced with only minor structural changes. The reduction leads to more highly charged and hence more basic and more stable species (cf. Sect. 3.6.7); the greater stability has been proven experimentally [34d]. However, apart from these general statements on the charge, basicity, and stability of the polymetalate species, in the present case of type I structures the stability must also be viewed in terms of the necessary degree of protonation of the reacting species for the last aggregation step with elimination of three O atoms by condensation of three H_2O molecules (compare Sect. 3.1). This requires in particular a triply protonated monometalate ion. The corresponding type I polyoxometalate structures form more easily and are more stable the more readily (high protonation constant) a high degree of protonation (high

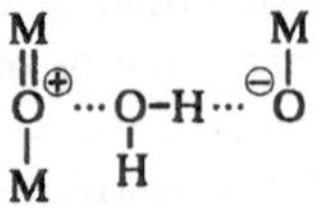

charge numbers w and m), in particular of the monometalate ion, can be reached [34d].

4.6 Aspects of Structure and Bonding of Polyoxometalate Ions Considered in the Literature in the Light of the Bond Model Proposed

Isolated aspects of structure and bonding in polyoxometalate ions have been discussed in the literature. Usually one or two factors are named as being responsible for an observed feature with a clear preference for "electronic" explanations. In contrast, Sects. 3.6, 3.7, 4.1, and 4.4 indicate that a very complex interaction of a large number of factors determines structure and bonding in polyoxometalate ions.

4.6.1 *Application of Valence Bond Formalisms*

A valence bond formalism has also been applied by Linnett [21] in which four- (short bonds), three-, two-, and one-electron bonds (long bonds) are considered in the electronic structures, and the valence shells of M^{VI} and O are formulated with twelve and eight electrons, respectively. However, $X = P^V$ in $PM_{12}O_{40}^{3-}$ is described with an electron octet and Mn^{IV} in $MnMo_9O_{32}^{6-}$ with an electron sextet (?), while merely Te^{VI} in $TeMo_6O_{24}^{6-}$ is formulated with an electron dodecet, in accord with its oxidation number. The formulae usually concentrate positive charge on the (X and O) atoms at the centers of the heteropolyanions and negative charge on the outer oxygen atoms.

In a short paper, Waugh, Shoemaker, and Pauling [119] used an electron octet for Mn^{IV} and a dodecet for Mo^{VI} in $MnMo_9O_{32}^{6-}$ and described a well-balanced X–Mo–O framework (requiring a multitude of resonance structures [21]) with the charge on the terminal oxygen atoms. It is reasonable to assume that these authors considered it unnecessary to explain their procedure in more detail since the results follow directly from a consequent application of commonly accepted concepts. For just this reason we did not describe bond questions in more detail in our investigations on the mechanisms of formation and on the driving forces and directing factors for polyoxometalate ion formation [22, 35, 36a]. However, the contradictory papers published on this subject since then prompted us to extend the formulations and studies as presented in this article (cf. Sect. 1). In terms of our bond model, the authors [119] could have considered the resonance types (5), (6), and (8). However, a more detailed analysis of the polyoxometalate structures as described in Sects. 3 and 4.1–4.5 of the present investigation has now revealed a broad spectrum of the additional factors determining polyoxometalate structures beyond the standard solution presumably reached by Waugh, Shoemaker, and Pauling.

4.6.2
Bonding Power of Bridging Oxygen Atoms, Law of Mass Action, Opposition of Covalent and Coordinate Bonds, and Other Arguments versus (cis-) Terminal pπ-dπ Interactions of Bond Order Two and their Maximization as Stabilizing Factor, Non-Basic Character of Terminal Oxygen Atoms, Trans Influence, and Other Arguments

According to a group of Russian authors [77, 78, 81] the *cis*-arrangement of the terminal (or pseudoterminal), multiple-bonded oxygen atoms in octahedral d^0 complexes of group VI and V transition metals, the strong trans influence of the M–O multiple bond (leading to considerable weakening and lengthening of the bond with its *trans*-partner), as well as the self-consistency of the trans influence of a multiple bond and the donor rigidity of the *trans*-partner lead to the appearance of stable oligomeric species constructed of MO_6 octahedra linked by common edges in such a way that multiple-bonded oxygen atoms are peripheral and internal oxygen atoms having high coordination numbers are in the *trans*-positions to them. The authors concede, however, that it might be a trivial consequence of the block-type polymetalate structures that terminal oxygen atoms are found outside the block, whereas oxygen atoms with maximum bridge multiplicity are found within the block.

If we make allowance for the fact that the MO_4 units in the polyoxometalate ions retain the major characteristics of the monomeric MO_4 species (see Sect. 3.1 and Scheme (23)), it is not appropriate to speak of a trans influence as an effect considerably weakening and lengthening the bond with the trans-partner of an M–O multiple bond. The reference states are the monomeric MO_4 species [15s]. Hence from a mechanistic point of view the "trans influence" consists of a weakening and lengthening of the original M–O bonds of the monomeric MO_4 species (with an average bond valence of 1.5 v.u. ((23a)) in the case of M^{VI}) by coordination of two oxygen atoms according to

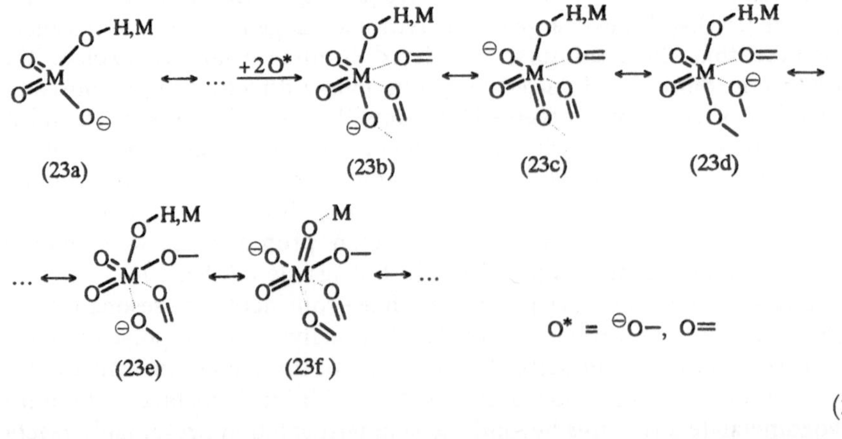

(23)

where, in single resonance structures ((23e), (23f)), a minimum average bond valence of 1.25 v.u. (disregarding resonance type (7)) for the original M^{VI}–O bonds of the monomeric MO_4 species occurs. The maximum average bond valence for the original M^{VI}–O bonds of the monomeric MO_4 species in a resonance structure is, as before, 1.5 v.u. ((23b), (23c), (23d)). From a static point of view the relationships between the bond valences (i.e., the typical distortion patterns of the MO_6 octahedra and other MO_k polyhedra with $k > 4$) within a polyoxometalate ion can also be expressed as follows: The valence of the atoms (O: two; M^{VI}: six; etc.) constituting a polyoxometalate ion with a given (X–)M–O framework composed of MO_k polyhedra ($k > 4$) and the range of charge of the polyoxometalate ions (average negative charge per $M^{VI}O_6$ octahedron: $1 \geqq z > 0$[9]), i.e., their Z^+ values determine sufficiently and decisively that the bond valences of the outer, terminal (or pseudoterminal) M^{VI}–O bonds lie in the range 1.5 v.u. (e.g., if all MO_6 octahedra were of the types (23c) or (23f)) to 2 v.u. (e.g., if all MO_6 octahedra were of the types (23b), (23d), or (23e)). The exact values then depend on the weight of the resonance structures (see Sect. 3.4). The bond valences of the inner, bridging M^{VI}–O bonds lie, according to the coordination number of the oxygen atoms (two to six), in the range 1–0.33 v.u., disregarding the factors listed in Sect. 3.4 under (i)–(v)[10]. Analogous considerations can be made for the M^V polymetalate ions.

Pope and Baker assumed that the major reasons for the formation of polyoxo anions consisting of MO_6 octahedra are

- the ability of the early transition elements to form M–O bonds of bond order two arising from $p\pi$-$d\pi$ overlap,
- the size of the M cation, and
- its ability to change the coordination number with oxygen.

Thus, short M–Ot bonds (Ot: terminal oxygen atom) are produced by the displacement of the metal atoms from the centers of the MO_6 octahedra towards a corner or an edge. This feature is seen as the key to the stabilities of all polyoxo anion structures. If there is a second (or possibly third) terminal oxygen atom on the same metal atom these oxygen atoms occupy *cis*-(*fac*-) rather than *trans*-(*mer*-)related positions to avoid competition with the same (vacant) t_{2g} orbital on the metal. The *trans*-related bonds are considerably weakened. Since the short M–O bonds are directed towards the exterior of the polyanion they form a layer of surface oxygen atoms that are strongly polarized towards the interior of the polyanion by $p\pi$-$d\pi$ interactions. These oxygen atoms have little or no basic character and discourage extensive protonation and further polymerization of the polyanion [1a, 2, 3, 4a, 120].

---

[9] This z range corresponds to the range $1 \leqq Z^+ = p/q < 2$ for isopolyoxometalate(VI) ions formed by p H^+ and q MO_4^{2-} ions.

[10] The electronic possibilities of oxygen and of the metal obviously correspond perfectly to these geometrical requirements of the M–O frameworks (see the discussion on the symmetry properties of the orbitals on the M and O atoms in Sect. 3.6.2) and therefore exert no influence themselves. This, however, is apparently a consequence of the only minor changes associated with the extension of the tetrahedral to an octahedral coordination (see Sects. 3.1 and 4.6.3) and hence leads back to the mechanistic view.

This view is not in accord with the fact that the basicity of the first polymetalate(VI) ions occurring on acidification of metalate(VI) solutions is actually even larger than that of the starting MO_4^{2-} ions (for instance, the protonation constants of $Mo_7O_{24}^{6-}$ and $W_7O_{24}^{6-}$ amount to ca. $10^{4.5}$ [15t, 121, 122a], whereas those of MoO_4^{2-} and WO_4^{2-} are ca. $10^{3.8}$ [15q, 122a], compare Sect. 3.6.7). Another misleading interpretation is the conclusion that the main feature for the stability (and formation) of polyoxo anion structures is the existence of (terminal) O–M π-bonding arising from pπ-dπ overlap [1c, 2, 4a, 123]. This feature is irrelevant since it already applies to the starting monomeric metalate ions (compare Scheme (23); see also Sect. 4.6.3) so that, from this point of view, there would be no reason for the monomeric MO_4 species to undergo aggregation reactions. The same applies to the statements

- that terminal *cis*-MO_2 groups tend to increase the stability of polyoxo complexes [16, 82],
- that the tendency toward *cis* disposition of oxo groups is the result of the maximum utilization of dπ-acceptor orbitals on M by the π-donating oxo groups [1a, 4a, 124], and
- that bonding within a polyoxometalate cluster is controlled by the dominating influence of the strong bonds in the *cis*-$M(Ot)_2$ groups of the polymetalate ions [19]. [Surprisingly, these authors state in the same paper that M–Ob bonds (Ob: bridging oxygen atoms) ensure the stability of the polyanion structures.]

The bonding power present in the M–O–M bridges, which is enhanced by pπ-dπ bonding, is decisive for the stability of the polyoxometalate ions as polymeric entities, but ultimately even this factor is not the real reason for the stability of the systems in which the polyoxometalate ions form: this is the fulfillment of Le Chatelier's principle for the reaction between metalate and H^+ ions, see the discussion on Schemes (16) and (17) in Sect. 3.6.7.

The views in the literature cited above do not take into account, as already mentioned, that in the starting species (MO_4^{w-} and protonated forms) the *cis*-(*fac*-)arrangement of the terminal oxygen atoms and their multiple-bond character are already realized: *trans*-(*mer*-)related positions of the terminal oxygen atoms in the polyoxometalate ions would imply "hard" formation mechanisms (involving major changes of the M–O bonds, e.g., cleavage and new formation of M–O bonds of the monomeric species) on the way from the monomeric MO_4 species to the polyoxometalate ions. Correspondingly, the structure types would be quite different. According to our view, which is in harmony with mechanistic concepts on the formation of other complexes (compare Sects. 2.4 and 2.5), the *cis*-(*fac*-)related positions of the terminal oxygen atoms result from the original monomeric MO_4 species as a consequence of the only minor structural changes associated with the proposed formation mechanisms ("soft" mechanisms) [15u, 22, 35, 36a, 62] (cf. reaction scheme (23), Fig. 5 in Ref. 22a, and the statements on the geometrically and electronically exactly defined routes in Sect. 3.1) and therefore are not available to explain the stability or formation of the polyoxometalate ions. Hence in the present case the trans influence simply labels the bond length (bond valence) difference between covalent and coordinate M–O bonds. Coordinate bonds are generally much weaker than covalent bonds [48]. Due to the polymeric character of the polyoxometalate ions and, as a consequence, the occurrence of differently coordinated oxygen atoms, a mutual approach of their bond characters is possible only with restrictions whereas in the case of small, symmetrical molecules (e.g., BF_4^-, SiF_6^{2-}) a complete approach of the bond characters occurs. Coordinate (dative) bonds in polyoxometalate ions have also been assumed by Schröder [79]. In this connection it is surprising that nearly no objections have been raised against the often used formulation $Mo(OH)_6$ with six equal M–O bonds for the diprotonated monomeric molybdate ion (compare Ref. 43) although there is consensus with respect to the occurrence and explanation of short, medium-sized, and long M–O bonds in the MO_6 octahedra of the polyoxometalate ions.

Based on the ideas of Cotton and Wing [125] about the access and use of $d\pi$ orbitals of the M atoms for terminal M-O bonds, some authors [79, 86, 126] even assume bond orders (in our sense: bond valences; cf. footnote 1) > 2 (for instance, a bond order of 2.2 [79] is assigned to the Mo-O bond length of 1.67 Å).

4.6.3
Comparison of the Polyoxometalates of Groups VI (and V) with Polytellurates and -periodates

A main feature of the bond model described above is the retention of the major characteristic of the monomeric MO_4 species in the polyoxo anions, namely – disregarding the (two) additionally coordinated oxygen atoms – the bond patterns (for $E = M = M^{VI}$)

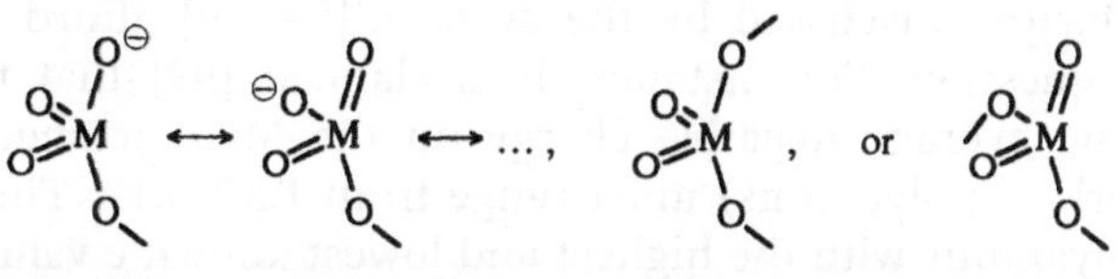

(24)

(cf. formulae (VI) and (VII)), which are precisely the same as that of the MO_4^{2-} ion and its protonated forms [43], in the latter again disregarding the (two) additionally coordinated oxygen atoms (water molecules), if present (e.g., in $MoO_2(OH)_2(OH_2)_2$). This feature supposes that all steps of the routes from the monomeric to the polymeric species occur with a minimum of disruption of M–O bonds, see above (Sects. 3.1 and 4.6.2).

Comparing the structures of the polytellurates and polyperiodates (or the respective acids) with those of the monotellurates and monoperiodates (or the respective acids, $Te(OH)_6$ and $IO(OH)_5$), we can state that the fundamental steps of the aggregation processes are obviously polycondensation reactions of the type (for $E = E^{VI}$)

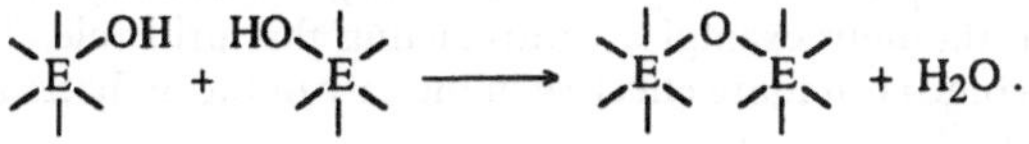

(25)

Independent of the mechanisms which actually occur, one observes again that the major bond characteristic of the monomeric telluric and periodic acids is retained in the polytelluric and -periodic acids (or mono- and polytellurate and -periodate ions). An explanation of the differences in structure and bonding between the polymolybdates, -tungstates, etc. on the one hand and polytellurates and -periodates on the other hand [1a] thus merely requires the explanation of the structural differences between the *mono*molybdate etc. and *mono*tellurate etc. ions; see also Sect. 4.6.1 in Ref. 61.

4.6.4
Closed Loops of -M–O–M–O- Bridges Around the (X-)M–O Frameworks and Magnitude of the Negative Charge of the Polyanion As Stabilizing Factors?

Nomiya and Miwa [97, 98] believe that interpenetrating closed loops -M–Ob–M–Ob- around the M–O frameworks, regarded as a kind of macrocyclic π-bonding system (the origin of the π-bond character in the -M–Ob–M–Ob- bridges has not been explained), mainly contribute to the stability of the M–O frameworks. Although the basic assumption that bridging oxygen atoms link the MO_6 octahedra and therefore contribute to the stability of the M–O frameworks is undoubtedly correct (see Scheme (16)), the conclusion that *the number of interpenetrating closed loops* -M–Ob–M–Ob- around the M–O frameworks is the governing factor is obviously incorrect for the isopolyoxo anions. At least for $M_7O_{24}^{6-}$ and $M_{12}O_{40}(OH)_2^{10-}$ (para-A and para-B polymetalate ions), two polymetalate types of similar stability [15w, 122b], the concept leads to quite different values for the "structural stability index" (η), thus indicating the invalidity of the concept for isopolymetalate ions [15v].

Additional rules and factors discussed by the authors [98, 99] afford no answers for the case in question. For instance, it is claimed [99] that the "electric condition" e (the average negative charge on the MO_6 octahedra constituting the framework of polyanions) must range from 1.33–0.17. These limits characterize the polyanions with the highest and lowest known e values; but a theoretical justification was not given. Polymetalate ions of M^{VI} and M^{V} have not been distinguished. Thus the above e range for polymetalate ions of M^{VI} represents a range of $Z^+ \equiv p/q = 8 - n - e$ from 0.67–1.83 and for polymetalate ions of M^{V} a range from 1.67–2.83. Polyoxometalate ions with $Z^+ < 1$ in general, however, do not exist [22a][11]. Their existence would require an aggregation of unprotonated MO_4^{2-} ions (no possibility for the extension of the coordination sphere, limited possibilities for the elimination of H_2O molecules) together with HMO_4^- ions.

[11] The polytungstate ions $HW_5O_{19}^{7-}$ and $W_4O_{16}^{8-}$ in the compounds $K_7[HW_5O_{19}] \cdot 10H_2O$[127] and "$7Li_2WO_4 \cdot 4H_2O$" [128, 129] [formally a monotungstate ($Z^+ = 0$!)], respectively, which form under extreme conditions, are the only examples contradicting this principle. This might be due to tendencies which become even more effective in the case of the niobates and tantalates; compare footnote 5 in Ref. 61.

4.6.5
Tendency of M to Achieve Neutrality in Charge?

The increase in coordination number has also been viewed as a consequence of the tendency of M^{VI} to achieve near neutrality in charge. Whereas in MO_4^{2-} four strongly σ- and π-donating oxo ligands suffice to neutralize the positive charge on M^{VI}, in the mono- or more highly protonated monometalate ions the OH groups are less effective as a donor, leaving the M atom with more residual positive charge. Thus, it must take up additional ligands to reapproach neutrality by formation of isopolymetalate ions or, if possible, by chelate complex formation [124]. In the former case (formation of isopolymetalate ions) it is supposed that the additional oxygen ligands, although simultaneously bound to up to five other M atoms in the isopolymetalate ions, have a greater σ- and π-donating power than an OH group. Note also that the "additional" oxygen ligands are not ligands additionally present. In fact, in polyoxometalate ions composed of MO_6 octahedra the ratio of O to M atoms is much smaller than in the protonated monometalate species.

4.6.6
Controversial Discussion About the Distribution of the Ionic Charge Over the Oxygen Atoms

It is generally assumed that the ionic charge is accepted by the oxygen atoms, see, for example, Refs. 77, 117, and 130a. However, its distribution over the different types of oxygen atoms is one of the most controversially discussed topics of polyoxometalate chemistry [15a, 130b].

Since according to X-ray photoelectron spectroscopic investigations [131] the energy of the O 1s level of a bridge oxygen atom is higher than that of a terminal one, oxygen atoms of high coordination number bear a lower effective negative charge. This means that the decrease in the transfer of electron density from the oxygen atom to each of its bonds with the metal is compensated by an increase in the number of its bonds: the overall negative charge on the oxygen atom decreases [77]. From the sum of the bond valences about the oxygen atoms in several polymolybdate and -tungstate structures [8–13] it has been concluded that the ionic charge (to be compensated by the cations) is mainly distributed over the terminal oxygen atoms. On average, terminal and bridging oxygen atoms take up the available charge, referred to equal numbers of terminal and bridging oxygen atoms, in a ratio of approximately 85:15 [14a, 15a]. More recent and more detailed results obtained with this method are described in Ref. 61. Assuming that protonation occurs at the most basic oxygen atoms [according to recent investigations [61] this assumption is not valid for polyoxometalate ions built by extended MO_k polyhedra (MO_6 octahedra)], ^{51}V-NMR [132], Raman [133], and IR spectroscopic [134] investigations also indicate that terminal oxygen atoms bear the main part of the ionic charge. Location of the charge electrons (mainly) on the terminal oxygen atoms of the polyoxometalate ions has also been considered in theoretical studies (besides that of the present article)

- to explain some features of heteropoly compounds (terminal oxygen atoms behave as if they are bigger than the other oxygen atoms and thus have a decisive influence on the heteropolymetalate structure that is formed) [21],
- to account for the more space available on terminal oxygen atoms (cf. the VSEPR model) [14a, 15a, 22a],
- to keep the charge electrons as far as possible from each other (Coulombic repulsion) [22, 35, 36a, 87–89], and
- to account for the necessary contacts between cations and anions in ion pairs [10, 11, 20] and salts [92].

Contrary to the above views, ^{51}V-NMR [135] investigations, ^{17}O-NMR investigations [1e, 136, 137], and protonation sites [136, 138] have been interpreted to indicate that the ionic charge is mainly distributed over the bridging oxygen atoms. This was also concluded from bond length-bond valence considerations by Pope and Klemperer [1f, 26, 136]. The interpretation errors made in the latter papers are discussed in Ref. 61. According to X-ray photoelectron spectroscopic investigations the negative charges on bridging and terminal oxygen atoms are quite similar [117, 139]. Fuchs [140] assumed also an extensive equidistribution of the ionic charge over the oxygen atoms of the polyoxometalate ions and, additionally, O–M bond valences of equal size about the individual oxygen atoms. This is not in accord with the application of Eqs. (20) and (21) to structural data of polyoxo anions and is even inconsistent in itself [113]. The statement that the bond length-bond valence (bond order) relationship represents a stability condition for polyoxometalate ions [140] is not incorrect but is insufficient.

4.6.7
Polyoxometalate Ions Composed of a Small Anionic Species and a Neutral Polymeric Part?

Long M–O distances are considered by many authors to characterize ionic bonds [23–28, 141] that can undergo facile cleavage [26–28]. Thus a number of polyoxometalate ions are described to be composed of an O^{2-} or other anion surrounded by or otherwise combined with an $(MO_3)_\zeta$ cage, ring, or similar structure; for example $(O^{2-})(M_6O_{18})$ $(= M_6O_{19}^{2-})$, $(O^{2-})_2(Mo_4O_{12})_2$ $(= (\beta\text{-})$ $Mo_8O_{26}^{4-})$, or $(MoO_4^{2-})_2(Mo_6O_{18})$ $(= \alpha\text{-}Mo_8O_{26}^{4-})$ [3, 26–28, 142]. Such structures are in direct contrast to the view presented in this article: the structure-stabilizing effect of the inner bond valence is reduced and the large distance between the charges of the cations and polyanions that would result for salts of most of such species do not allow the necessary charge compensation according to the relevant bond length-bond valence functions. This view has also been questioned [32, 33a, 34a] for several reasons in the case of $XM_{12}O_{40}^{m-}$ heteropolyoxo anions, where an XO_4^{m-} tetrahedron is assumed to be surrounded by an $M_{12}O_{36}$ cage [1g, 26, 29–31, 52]. Interpretation errors leading to the above formulations are discussed in Sect. 4.8.1 in Ref. 61. In an alternative description of the $M_6O_{19}^{2-}$ ion, delocalized multicenter bonds involving the six metal atoms and the central oxygen atom have been assumed [23]. This approaches the view of the bond model presented in this article.

4.6.8
Importance of the Coulombic Repulsion Between the Metal Cations?

Kepert [90] has explained the long, inner and short, peripheral M–O bonds (and hence the typical distortion patterns of the MO_6 octahedra in the polyoxometalate ions) by the electrostatic repulsion between the highly positive charges of the metal cations in the centers of the MO_6 octahedra, and many authors [67, 81, 82, 86, 143, 144] agreed with this view. This repulsion and the size of the M cations have also been used to explain the arrangements of the MO_6 octahedra in polyoxo anions, i.e., their structures or structure types. This view is now largely rejected [1a, 15j, 33b]; see also Sect. 3.6.4. Alternatively, the repulsive forces between the M cations are assumed to be the reason for the shortening of shared edges (O···O distances) between neighboring MO_k polyhedra [145].

4.6.9
Electrostatic Repulsion and Attraction Between the Atoms of the Polyanions

An electrostatic model that also describes bond distances has been proposed by Björnberg [146]. The distortions of the MO_k coordination polyhedra are seen as a consequence of electrostatic repulsion and attraction between the atoms of the molecule. A computer program calculates the magnitude and direction of the electrostatic force on each atom in the polyanion as exerted by the other atoms in the polyanion, each atom being assigned a charge equal to its formal charge. Starting with a molecule of arbitrary shape, atoms are then allowed to move along the direction of the force by an amount proportional to the magnitude of the force, stopping when an empirical constraint is reached, i.e., when the attractive force on M from terminal oxygen equals the total force on M. The resulting (calculated) bond lengths parallel the observed (experimental) values, but the model tends to yield too short terminal M–O bonds.

4.6.10
Other Electrostatic Arguments?

The distortion of the MO_6 octahedra has also be seen as an extreme local cation (M^{n+}) to anion (O^{2-}) oversaturation and undersaturation [147] and as the need to balance charges on the inner, multiply linked oxygen atoms [82].

4.6.11
Differences in the Polyoxomolybdate and Polyoxotungstate Structure Types As a Direct Consequence of Differences of the Two Metals?

As already mentioned, two types of MO_6 octahedra are observed depending on the type of the metal atom displacement from the center of the MO_6 octahedra. In type I the metal atom is displaced towards one, always terminal, oxygen atom, in type II towards two *cis*, usually but not always, terminal oxygen atoms

[1d, 3, 76]. The structural data for polyanions have been interpreted to indicate that Mo^{VI} (d^0) forms preferably type II distortions while W^{VI} (d^0) forms preferably type I distortions [1h, 33b, 76, 148–150]. This means that the structural differences are seen as a more or less *direct* consequence of differences of the properties of the two metals. In the bond model described above there is no room for views of this kind. In our opinion the differences between the polymolybdate and polytungstate systems have their origin in the higher ability of the protonated tungstate ion for complex formation (a view that is generally accepted and explained by the greater extension and hence better overlapping of the d orbitals of tungsten with the p orbitals of the donor atoms [85, 148–150]). Accordingly, as a consequence of the law of mass action, the resulting pH level at which the aggregation reactions in the polytungstate system occur is *higher* (in the tungstate case the aggregation reactions start at pH values 1.5–2 units higher than in the molybdate case) [14e, 15cc, 22a, 36c, 95, 151] (compare also the discussion on the effects of Le Chatelier's principle in Sect. 3.6.7). This *indirect* effect leads to the initiation of further, slower reactions in the tungstate case under formation of type I MO_6 octahedra (e.g., in the para-B and metatungstate ions); details have been discussed elsewhere [15x, 95, 151].

4.6.12
π-Bond Character of the Si–O and Ge–O Bonds in Keggin-Type Heteropolyoxometalate Ions?

In our bond model X–O bonds of central heteroatoms in highly symmetrical heteropolyoxometalate ions have multiple bond character if the oxidation number of X exceeds its coordination number (compare resonance scheme (8)). Hence the Si–O and Ge–O bonds in the 12-molybdosilicate and -germanate ions have only single-bond character (s = 1 v.u.). In contrast to this view, IR and Raman spectroscopic investigations have been interpreted to indicate that these bonds have a considerable π-bond character [85]. The reason for this difference is the erroneous assignment of vibrational frequencies of X–O bonds in the reference compounds, i.e., for the reference compounds assumptions violating the valence sum rule (Eq. (21)) have been made due to the non-observance of a considerable coordinate bond character of bonds; see also Sect. 14 in Ref. 48.

4.6.13
Aspects of the Structure Types of the Polyoxometalate Ions

Aspects of the structure types of the polyoxometalate ions occurring in relation to the size and the oxidation number of M and to the conditions in the solution in which the polyoxometalate ions form or from which the insoluble polyoxometalates separate (concentration of M, degree of acidification P, pH, reaction time, and others) have been discussed comprehensively in Refs. 14b and 15b. Since their theoretical basis is essentially that of the bond model presented in this article (compare Sects. 1 and 3.1), both investigations on the structure types of the polyoxometalates (their formation mechanisms, the respective driving forces, and directing factors) and on bonding and charge distribution in polyoxometalates are compatible. Introduction of elements of the present bond model into the concept of the structure types of the polyoxometalate ions should improve and simplify the latter. This possibility had not been recognized at that time.

4.7 Other Aspects of Structure and Bonding of Polyoxometalate Ions Considered in the Literature

The aspects of structure and bonding described in this section have no *direct* connection to the bond model presented in Sect. 3 and discussed in Sects. 4.1–4.5. They serve merely to complete the picture on bonding (and structure) in (poly)oxometalates.

4.7.1 *Monometalate Ions*

Bonding in the $4d^0$ tetrahedral molybdate ion has been treated by molecular orbital approaches. The oxo ligands are involved in both σ and π bonding with Mo. An energy level diagram in T_d symmetry has been established, and electronic, vibrational, and other spectra have been studied, see Refs. 4e, 15z and 52.

4.7.2 *Polymetalate Ions*

Molecular orbital schemes for type I MO_6 octahedra (the octahedra have one terminal oxygen atom; their symmetry is approximately C_{4v}) and for type II MO_6 octahedra (the octahedra have two *cis*-terminal or -pseudoterminal oxygen atoms; their symmetry is approximately C_{2v}) have been described [1i, 4c]. Type I MO_6 octahedra can be reduced with only minor structural changes (bond length alterations) since the electron added to M enters an orbital that is predominantly non-bonding (the LUMO in type I octahedra is a non-bonding, mainly metal-centered d_{xy} orbital). In type II MO_6 octahedra the added electron must enter an antibonding orbital and will result in large structural changes (the LUMO in type II octahedra is strongly antibonding) [1i, 3]. Electronic absorption, photoelectron, and other spectra have been studied [1j, 4d, 15y].

5 Summary

Adequate and *consequent* application of standard concepts of chemistry to polyoxometalate ions and the network of bonds in their structures, namely

- Lewis's octet rule, extended to the decet and dodecet rule for M^V and M^{VI}, respectively,
- the concept of the pπ-dπ M–O double bond,
- the concept of the coordinate bond (dative bond),
- the resonance concept,
- the concept of the resonance bond number (or the valence sum rule and the bond valence concept),
- the concept of the polydentate ligands and the chelate effect,
- the concept of the greater space requirements of unshared electron pairs (cf. the VSEPR model),
- the model of multicenter pπ-dπ multiple bonds for certain μ-oxo bridges between metal atoms,
- Brønsted's acid/base concept and acid/base equilibria,
- Pauling's rules about the acid/base strength of the (monomeric) oxoacids/oxoanions,
- Le Chatelier's principle (the law of mass action), and others,

leads to a set of Lewis-type resonance structures in which chainlike $M_{\xi}O_{3\xi+1}$ ($\xi = 1$ to q; q = degree of aggregation) and occasionally ring-shaped $M_{\xi}O_{3\xi}$ building groups, composed of strongly distorted MO_4 tetrahedra sharing corners, are arranged such that the MO_4 units simultaneously act as mono- and bidentate and even as multiply bridging ligands and make available *cis*-located coordination sites to other MO_4 units of the polyanion by appropriate three-dimensional bending of the chains and rings, thus meshing the MO_4 units and forming MO_6 octahedra (or, in rare cases, other MO_k polyhedra) (Figs. 1(a)–(h) and 2). In the case of the polyoxometalate ions with an O:M ratio $\leqq 3$ there are cyclic or oligocyclic $M_qO_{\leqq 3q}$ entities (sometimes also containing short chain-like parts) of strongly distorted, corner-shared MO_4 tetrahedra that form the polyoxometalate ion according to the same coordination mode (Fig. 1(i), (j)). (This description is not equivalent to a similar description sometimes given in the literature for type II polyoxometalate structures.)

The substantial feature of the fictitious meshing process (Fig. 1) is the transformation of terminal into bridging M–O bonds by which the cohesion of the MO_4 building groups and hence the stability of the polyoxometalate ions is strongly enhanced ("meshing effect"). The average stabilization by the meshing effect results in up to 2.85 v.u. [valence units] per M atom for the M^{VI} and up to 1.90 v.u. per M atom for the M^V isopolyoxo species hitherto known. Additionally there is the condensation of H_2O molecules by which the fictitious $M_{\xi}O_{3\xi+1}$, $M_{\xi}O_{3\xi}$, and $M_qO_{\leqq 3q}$ entities are formed. The average stabilization by this effect ("normal bonding") results in 0.40 to 2.00 v.u. per M atom for the M^{VI} and 1.67 to 3.00 v.u. per M atom for the M^V species hitherto known.

For the MO_6 octahedra (MO_k polyhedra) of the M–O frameworks three types of resonance can be distinguished:

- A first resonance type leads to a certain equalization of the inner M–O bonds and explains the cohesion of the MO_4 building units in the polyoxometalate ions (Scheme (5), Fig. 2(a)).

- A second resonance type leads to the distribution of the (negative) ionic charge over (nearly) all types of oxygen atoms and explains the equalization of the outer and of some inner M–O bonds and the cohesion of the MO_4 building units in the polyoxometalate ions (Scheme (6), Fig. 2(b)).
- The third resonance type is connected with a charge separation and explains the occurrence of positively charged oxygen atoms, the strengthening of some inner M–O bonds, the enhanced basicity of the polyoxometalate ions, and again the cohesion of the MO_4 building units (Scheme (7), Fig. 2(c)).

Central heteroatoms of heteropolyoxometalate ions lead to an additional type of resonance if the coordination number and the oxidation number of the heteroelement are different (Scheme (8)).

From the M–O bond valences as indicated by the resonance formulae for the differently coordinated O atoms (Fig. 3) one can directly derive that

- terminal M–O bonds have multiple-bond character (1.5–2 v.u. for M^{VI} species and 1–2 v.u. for M^{V} species, depending on the structure and the Z^+ value of the polyoxometalate species and on the weight of the different resonance structures);
- pseudoterminal M–O bonds have also multiple-bond character (approaching that of terminal M–O bonds); the other M–O bond is (usually) a coordinate (weak) bond;
- the bonds of two-coordinate oxygen atoms (except those in which pseudoterminal oxygen atoms are involved) have in both bonds (approximately) single-bond character;
- the bonds of three-coordinate oxygen atoms (M_3O bridges) have in general partly (approximately) single-bond character, partly a character of coordinate bonds;
- the bonds of five- and six-coordinate oxygen atoms have a character of coordinate bonds. The character of the bonds of four-coordinate oxygen atoms lies between that of the three- and five-coordinate oxygen atoms.

Other factors have only a small influence on the M–O bond character.

The other factors have, however, a significant influence on the distribution of the charge over the oxygen atoms since the charge results as the small difference between the valence of oxygen as defined by its oxidation number and the sum of the bond valences of the oxygen atom under consideration. The distribution of the charge is conducted

- by the favorable acceptance of the (negative) ionic charge by the terminal (and pseudoterminal) oxygen atoms due to
  - the high statistical weight of the resonance structures with charged terminal oxygen atoms compared to the small statistical weight of the resonance structures with charged bridging oxygen atoms – according to the coordination number of oxygen (compare Figs. 3 and 4);
  - the stabilization of the polymeric structure if there is only small negative charge on the bridging oxygen atoms (high bonding power in the M–O–M bridges according to Scheme (16)), in combination with Le Chatelier's principle;

  - the enhanced basicity of the polyoxometalate ions and hence in acidified solutions increased consumption of protons, in combination with Le Chatelier's principle;
  - the space requirements of unshared electron pairs in which the charge electrons are merged in (cf. the VSEPR model);
  - the large distances between the negative ionic charges in the polyanion and hence only small Coulombic repulsion between them;
  - the necessary contacts to the cations in the case of solid salts (compare formulae (XII) and (XIII)).
- by the favorable acceptance of the negative charge, generated through charge separation, by the terminal oxygen atoms due to
  - the additional stabilization of the polymeric structure through bond strengthening in multicenter $p\pi$-$d\pi$ multiple bonds and positively charged oxygen atoms in suitable (i.e., approximately linear) M–O–M bridges according to resonance scheme (17), in combination with Le Chatelier's principle;
  - the enhanced basicity of the polyoxometalate ions and hence in acidified solutions increased consumption of protons, in combination with Le Chatelier's principle;
  - the ionic interaction between positively and negatively charged oxygen atoms which leads to a further additional stabilization of the structures if the geometric conditions for an interaction are given.
- by the stoichiometric requirements for Lewis-type resonance structures as a fundamental principle, i.e.,
  - as a minimum requirement the resonance structures have to correspond to the stoichiometric minimum of the ionic charge on bridging O atoms.
- by the geometrical requirements of the (X–)M–O framework due to (great influence since inner M–O and/or X–O bonds are concerned)
  - the interdependence of the M–O bond lengths and hence bond valences, which is the greater the more compact and the more symmetrical (the more rigid) a structure is;
  - given, fixed bond lengths (e.g., of central X–O bonds).
- by the geometrical and charge-wise requirements of the packing of cations, oxometalate ions, and molecules of water of crystallization in the crystalline state due to (small influence only since a great part of the adaptation can be performed among the terminal M–O bonds)
  - their size and shape;
  - the necessity of the realization of the bond length-bond valence function and the valence sum rule (Eqs. (20) and (21)) (by which in addition to the size and shape the charge is considered) for the (ionic) bonds between cations and negatively charged oxygen atoms of the oxometalate ions or water molecules.
- by (the requirements of) the capacity of the terminal O atoms for the acceptance of unshared electron pairs in relation to that of the bridging O atoms which causes in the case of polyoxometalate structures with a high density of the ionic charge like some of M^V
  - the repression of the charge acceptance by terminal O atoms;

- the suppression of charge separation processes with acceptance of the generated negative charge by terminal O atoms.

The last four items (geometrical and charge-wise requirements) mean that the first two items (favorable acceptance of the negative charge by the terminal oxygen atoms) can only be fulfilled to a degree still allowing the realization of the (X–)M–O frameworks and, for solid salts, the realization of the packing of their constituents with respect to the necessary bond lengths and charge distribution (hierarchal principle).

The ability of the oxygen atoms to accept variable parts of negative and even positive charge is thus responsible on the one hand for the realization of the M–O (and X–O) bond valences, as required by the stoichiometry of the polyoxo species, by the interdependence of the bond lengths of the (X–)M–O frameworks, by the size of the heteroatoms X, and by the packing of cations, oxometalate ions, and molecules of water of crystallization, and on the other hand for a far-reaching stabilization of the polyoxometalate structures and of the oxometalate/H^+ systems (formation of the more stable and/or more basic polyoxo species to consume the H^+ ions according to Le Chatelier's principle). This property of oxygen allows also the partial substitution of atoms of the addenda element M and the substitution of the heteroelement X in the structures of polyoxometalate ions by other elements in spite of differences in oxidation number (ionic charge of the polyanion) and size of the addenda- and heteroelements. Moreover, the appearance of positively charged oxygen atoms favors the amphoteric behavior of the species.

Polyoxometalate chemistry presents an example of how the law of mass action can influence the bond lengths in a structure: The consumption of protons in acidified metalate solutions is favored according to Le Chatelier's principle by formation of more stable and more basic (poly)oxometalate species. These bear the ionic charge and the additional negative charge generated by charge separation preferably on terminal oxygen atoms and the equivalent positive charge produced by charge separation on certain bridging oxygen atoms, hence increasing the inner and decreasing the outer M–O bond valences and shortening the inner and lengthening the outer M–O bonds to a degree just allowing the building of the structures from the viewpoints of possible Lewis formulae, of the interdependence of the bond lengths in the structures, and of the charge-wise balance.

The protonation behavior is such as to keep necessary bond lengths and bond valence alterations as small as possible. Since protonation of the terminal O atoms would convert the short, strong terminal M–O bonds into medium-sized, middle-strong bonds, this has to be seen as a strong modification (change of the type of the MO_6 octahedron). Otherwise, protonation of bridging O atoms opens two (in the case of two-coordinate bridging) or three (in the case of three-coordinate bridging O atoms) paths for the necessary bond valence and bond lengths compensation. Hence protonation is expected to occur at bridging O atoms.

MO_6 octahedra with three unshared O atoms exist usually with one of these O atoms as an OH group or a coordinated H_2O molecule to keep the inner bond valence and hence the stability and kinetic inertness of the structures and of the MO_4^{w-}/H^+ system as great as possible.

Electronically driven asymmetries of the coordination environments of M as effect of the symmetry properties of the orbitals on the M and O atoms, if present, coincide with the numerous factors specified above to determine bonding in polyoxometalates and polyoxometalate ions.

Experimental results (M–O and X–O bond lengths of polyoxometalate structures and the M–O and X–O bond valences and charges on the O atoms derived from them) correspond with the above view about bonding in polyoxometalate ions [61], which is merely based on standard concepts of chemistry. Hence the formalism of the description of the bonds in the polyoxometalate structures used in this article [formulation of Lewis-type resonance structures with M–O (and X–O) bond orders two, one, and zero] and statements on the weight of the resonance formulae reproduce the bond lengths as found by X-ray crystal investigations. The weight of the resonance formulae can partly (in principle) be calculated (for instance, that due to the interdependence of the bond lengths), partly only be estimated (for instance, that due to the charge separation and to the influence of the [negative] charge density already present on the terminal oxygen atoms on the charge separation).

Frequently expressed views about partial questions on bonding in polyoxometalate ions in the literature disagree with this bond model:

- The "trans influence" mentioned in the literature regarding polyoxometalates [1a, 19, 77, 78, 81] is on the one hand simply the result of the necessity to realize the bond length-bond valence function and the valence sum rule (Eqs. (20) and (21)) for the connectivity of the atoms given through the M–O framework and on the other hand is a mark of the bond length (bond valence) difference between M–O bonds of predominantly covalent and predominantly coordinate bond character. Coordinate bonds are generally much weaker than covalent bonds [48]. Due to the polymeric character of the polyoxometalate ions with the consequence of differently coordinated oxygen atoms, a mutual approach of their bond characters by resonance is only possible with restrictions, in comparison with the case of small, symmetrical molecules (e.g., BF_4^-, SiF_6^{2-}) where a complete approach of the bond characters is attained. An influence of the orbital symmetries appears possible but cannot be recognized.
- The *cis*-terminal MO_2 groups of the M–O frameworks cannot control the bonding or be responsible for the stability or formation of the polyoxometalate ions, as is claimed in the literature [1a, 16, 19, 78, 81, 82], because such groups are already present in the starting monomeric MO_4 species. The bonding power of (the sum of the bond valences about) the *bridging* oxygen atoms is important for the stability of the polymeric structures and hence for the stability of the oxometalate/H^+ systems, although the *cis*-terminal MO_2 groups indeed contain sometimes most of the bond valence in the polyoxometalate ions.
- The statements that the ionic charge is mainly located on the bridging oxygen atoms [1e, 26, 135–137] contradict other experimental results,

especially a bond-valence treatment of structural data [61], and theoretical considerations, in particular stability arguments based on the bonding power of the bridging oxygen atoms (see Scheme (16)) and arguments of the charge compensation in the solid state between cations and polyoxometalate ions.

- The characterization of terminal oxygen atoms as having little or no basic character [1a, 2, 26, 120] rests obviously on the assumption that terminal M–O bonds have the bond order two and therefore bear no charge. In contrast to this view the just mentioned [61] experimental investigations and theoretical considerations (maximization of the inner M–O bond valence, necessary contacts to the cations in the solid salts, space requirements of unshared electron pairs, etc.) indicate that the ionic charge is mainly located on the terminal oxygen atoms which therefore are the most basic ones.
- The assumption of bond orders (bond valences) >2 for terminal M–O bonds [79, 86, 126] contradicts the valence sum rule.
- The formulations of many polyoxometalate ions as anions (O^{2-}, MO_4^{2-}, and others) surrounded by or otherwise combined with uncharged $(MO_3)_\zeta$ cages, rings, or something similar (e.g., $(O^{2-})(M_6O_{18}) = M_6O_{19}^{2-}$, $(O^{2-})_2(Mo_4O_{12})_2 = (\beta\text{-})Mo_8O_{26}^{4-}$, $(MoO_4^{2-})_2(Mo_6O_{18}) = \alpha\text{-}Mo_8O_{26}^{4-}$, $(PO_4^{3-})(M_{12}O_{36}) = PMo_{12}O_{40}^{3-}$) [1g, 26–31, 142] are in direct contrast to the bond model presented in this article; they rest on some wrong interpretations of bond lengths and bond valences [34a, 61] (see also Refs. 32 and 33a). An additional important point is the large distance between the charges of the cations and anions that would result for salts of most of such species and would not allow the necessary charge compensation according to the relevant bond length-bond valence functions.
- The statements that the preferential formation of type II polyoxomolybdate and type I polyoxotungstate ions is a *direct* effect of differences of the properties of the two metals [1h, 33b, 76, 148] are not in accord with the bond model proposed in this article. Other explanations based on the law of mass action have been offered [15x, 95, 151].
- The differences in structure and bonding between the polymolybdates, -tungstates, etc. on the one hand and polytellurates and -periodates on the other hand [1a] have their origin in the differences in structure and bonding between the *mono*molybdates, -tungstates etc. on the one hand and *mono*tellurates and -periodates on the other hand; that means that only the latter have to be explained.

6
List of Symbols

A cation of a salt

B parameter of the exponential bond length-bond valence function for a definite element pair

C_M total concentration of M in an acidified metalate solution

c charge (generally), given in charge units (c.u.) [examples: the charge on the terminal O atom is –1 c.u. or $1-(1\ominus)$; the formal ionic charge of $Mo_7O_{24}^{6-}$ is –6 c.u. or 6–]

c_j charge on the jth oxygen atom of the structure (species)
$\tilde{c}$ average charge on the O atoms with the coordination number k in the collection of resonance formulae specified in Fig. 3 assuming equal probability for all resonance structures of the M_kO groups (these quantities represent proportional numbers)
$\tilde{c}_j$ probability quantity for the charge on the jth O atom of the structure (species) (corresponds to the $\tilde{c}$ value of the respective O atom in Fig. 3)
d E–O, M–O, or X–O distance (bond length)
d_0 parameter of the (exponential or power) bond length-bond valence function for a definite element pair (equals the "single-bond" distance)
E electropositive element (M, X, etc.)
e "electric condition" (average negative charge on the MO_6 octahedra of a structure; in the original paper [99] designated as "p"); identical with z
h number of OH groups in a polyoxometalate structure (species)
h_u number of OH groups as unshared O atoms (of Oth atoms) in the structure
K_p protonation constant
k coordination number of E, M, X, or O
M group VI or group V transitional metal ("addenda" atom)
m (positive) number, in combination with the charge symbols (m−; m+) named the ionic charge (formal ionic charge) of the polyoxo species
N parameter of the bond length-bond valence power function for a definite element pair
n oxidation number, (atom) valence of the element
Ot terminal O atom ($M{-}O^{\ominus}$, M=O)
Ob bridging (inner) O atom without any distinction
O2a 2-coordinate O atom in an angular M–O–M bridge
O3a 3-coordinate O atom in angular M–O–M bridges
O4l 4-coordinate O atom in an (approximately) linear M–O–M bridge; the simultaneous presence of angular M–O–M bridges is irrelevant
O5l 5-coordinate O atom participating in two (approximately) linear M–O–M bridges; the simultaneous presence of angular M–O–M bridges is irrelevant
O6l 6-coordinate O atom participating in three linear M–O–M bridges; the simultaneous presence of angular M–O–M bridges is irrelevant
o number of coordinated H_2O molecules in a structure (species)
P degree of acidification of a metalate solution: molar ratio of H^+ ions *introduced* to MO_4^{w-} ions *introduced*
p stoichiometric coefficient of H^+ in the overall equation for formation of the polyoxo species
pH negative decadic logarithm of the H^+ *concentration*
q stoichiometric coefficient of MO_4^{w-} in the overall equation for formation of the polyoxo species; degree of aggregation of the polyoxo species
r stoichiometric coefficient of eliminated (condensed) H_2O in the overall equation for formation of the polyoxo species
S_n sum of the inner bond valences in a structure caused by "normal bonding"

S_m sum of the inner bond valences in a structure caused by the "meshing effect"

S_i sum of the inner bond valences in a structure if the formal ionic charge resides exclusively on terminal oxygen atoms

s bond valence (in the older literature also named bond strength or bond order) (generally)

s_i bond valence of the ith M–O or X–O bond about the M, X, or O atom considered

$\tilde{s}$ average bond valence of the M–O bonds of the O atoms with the coordination number k in the collection of resonance formulae specified in Fig. 3 assuming equal probability for all resonance structures of the M_kO groups (these quantities represent proportional numbers)

$\tilde{s}_i$ probability quantity for the bond valence of the ith M–O or X–O bond about the M or X atom, respectively, considered (corresponds to the $\tilde{s}$ value of the respective O atom in Fig. 3)

t number of terminal O atoms in a polyoxo structure

u number of O atoms in a polyoxo structure (species with OH groups and coordinated H_2O molecules require a modified approach)

w charge of the unprotonated monomeric metalate ion

X heteroatom, heteroelement

Z degree of protonation of a metalate solution: molar ratio of *reacted* H^+ ions to MO_4^{w-} ions *introduced*

Z^+ ratio of the stoichiometric coefficients of H^+ and MO_4^{w-} in the overall equation for formation of an isopolyoxometalate ion ($Z^+ \equiv p/q$)

z average negative charge per MO_6 octahedron (MO_k polyhedron) in a structure

ζ fictitious degree of aggregation of partial structures

η "structural stability index" (mean value of the number of closed -M-O-M-O- loops passing through each MO_6 octahedron) [97–99]

ξ fictitious, non-committed stoichiometric quantity for M (possible values, mainly depending on the Z^+ value and on the connectivity of the atoms in the structure: $\xi = 1$ to q)

Running indices:

i for the k O atoms about an M atom

j for the u O atoms of the polyoxometalate ion $M_q^nO_u^{m-}$

Acknowledgment. The author is indebted to the Fonds der Chemischen Industrie for financial support.

7 References

1. Pope MT (1983) Heteropoly and Isopoly Oxometalates, Springer, Berlin. (a) pp 128–132, 136–141, (b) p 23, (c) p 137, (d) p 19, (e) pp 11, 21, 37, (f) pp 20–21, (g) p 72, (h) pp 48, 140–141, (i) pp 101–102, (j) pp 109–117

2. Pope MT (1983) 29th IUPAC Congress, Cologne, Abstracts of Papers, p 22
3. Pope MT, Müller A (1991) Angew Chem 103: 56–70; Angew Chem Int Ed Engl 30: 34–48
4. Pope MT (1992) Progr Inorg Chem 39: 181–257. (a) pp 182–186, (b) p 211, (c) pp 201–202, (d) pp 186–189, (e) p 210
5. Brown ID (1978) Chem Soc Rev 7: 359–376
6. Donnay G, Donnay JDH (1973) Acta Crystallogr B 29: 1417–1425
7. Allmann R (1975) Monatsh Chem 106: 779–793
8. Böschen I, Buss B, Krebs B (1974) Acta Crystallogr B 30: 48–56
9. Vivier H, Bernard J, Djomaa H (1977) Rev Chim Minerale 14: 584–604
10. D'Amour H, Allmann R (1972) Z Kristallogr 136: 23–47
11. Allmann R (1971) Acta Crystallogr B 27: 1393–1404
12. D'Amour H, Allmann R (1973) Z Kristallogr 138: 5–18
13. Allmann R, D'Amour H (1975) Z Kristallogr 141: 161–173
14. Tytko KH (1977) Habilitationsschrift, University Göttingen. (a) pp 104–107 (see Ref. 15a), (b) pp 78–163, (c) pp 151–161, (d) pp 137–150, (e) p 152
15. Tytko KH (1987) Oxomolybdenum(VI) Species in Aqueous Solution, In: Gmelin Handbook of Inorganic Chemistry, 8th edn, Molybdenum Suppl Vol B 3a, pp 67–358. (a) pp 253–254, 319, (b) pp 344–352, (c) pp 291–295, (d) pp 306–307, (e) pp 295–297, (f) pp 298–299, (g) pp 289–290, (h) pp 316–318, (i) pp 306–308, (j) pp 283–284, 328–330, (k) pp 310–311, (l) pp 321–322, 333–334, 335–337, 348–349, (m) p 210, (n) pp 80, 88–90, (o) pp 321–322, (p) p 79, (q) pp 230–233, (r) pp 321–322, 348–349, (s) pp 273–274, (t) pp 257–258, (u) pp 289–299, (v) pp 357–358, (w) pp 201–208, (x) pp 288–289, 334–335, (y) pp 244–272, (z) pp 217–244, (aa) p 320, (bb) pp 174–181, 196–200, (cc) p 334
16. Evans HT Jr, Gatehouse BM, Leverett P (1975) J Chem Soc Dalton Trans 505–514
17. Sjöbom K, Hedman B (1973) Acta Chem Scand 27: 3673–3691
18. Swallow AG, Ahmed FR, Barnes WH (1966) Acta Crystallogr 21: 397–405
19. Wilson AJ, McKee V, Penfold BR, Wilkins CJ (1984) Acta Crystallogr C 40: 2027–2030
20. Druskovich DM, Kepert DL (1975) Aust J Chem 28: 2365–2372
21. Linnett JW (1961) J Chem Soc 3796–3803
22. (a) Tytko KH, Glemser O (1971) Z Naturforsch B 26: 659–678; (b) Tytko KH (1971) Angew Chem 83: 935–936; Angew Chem Int Ed Engl 10: 860
23. Rocchiccioli-Deltcheff C, Thouvenot R, Fouassier M (1982) Inorg Chem 21: 30–35
24. Mattes R, Bierbüsse H, Fuchs J (1971) Z Anorg Allg Chem 385: 230–242
25. Fuchs J (1973) Z Naturforsch B 28: 389–404
26. Day VW, Klemperer WG (1985) Science 228: 533–541
27. Filowitz M, Klemperer WG, Shum W (1978) J Am Chem Soc 100: 2580–2581
28. Day VW, Fredrich MF, Klemperer WG, Shum W (1977) J Am Chem Soc 99: 952–953
29. Clark CJ, Hall D (1976) Acta Crystallogr B 32: 1545–1547
30. Boeyens JCA, McDougal GJ, Smit J van R (1976) J Solid State Chem 18: 191–199
31. Fuchs J, Thiele A, Palm R (1982) Z Naturforsch B 37: 1418–1421
32. Brown GM, Noe-Spirlet MR, Busing WR, Levy HA (1977) Acta Crystallogr B 33: 1038–1046
33. Spitsyn VI, Kazanskii LP, Torchenkova EA (1981) Soviet Sci Rev B 3: 111–196. (a) p 120, (b) pp 113–114
34. Tytko KH (1989) Reactions of Oxomolybdenum(VI) Species in Aqueous Solution, In: Gmelin Handbook of Inorganic Chemistry, 8th edn, Molybdenum Suppl Vol B 3b, pp 1–207. (a) pp 118–119, (b) pp 36–40, 40–62, (c) pp 36–40, 62–67, 69–73, (d) pp 115–117
35. Tytko KH, Glemser O (1969) Chimia 23: 494–502
36. Tytko KH, Glemser O (1976) Adv Inorg Chem Radiochem 19: 239–315. (a) pp 294–305, (b) p 241, (c) p 309
37. Tytko KH (1972) 1st Meeting Intern Soc Study Solute-Solute-Solvent Interact, Marseille, 1972, Abstr No 15, pp 1–16
38. Cotton FA, Wilkinson G (1982) Anorganische Chemie, 4th edn, Verlag Chemie, Weinheim. (a) pp 214–215, (b) p 167, (c) pp 238–239
39. Kepert DL (1972) The Early Transition Elements, Academic Press, London; pp 288–289

40. Kepert DL (1973) Isopolyanions and Heteropolyanions, In: Bailar JC Jr et al (eds), Comprehensive Inorganic Chemistry, vol 4, Pergamon Press, Oxford, pp 607–672; pp 623–624, 636–637
41. Clark GM, Morley R (1976) Chem Soc Rev 5: 269–295; pp 286–295
42. Greenwood NN, Earnshaw A (1984) Chemistry of the Elements, Pergamon Press, Oxford. (a) pp 1262–1264, (b) pp 54–55
43. Tytko KH (1986) Polyhedron 5: 497–503; (1985) 5th Intern Conf Chem Uses Molybdenum, Newcastle upon Tyne, Proceedings, pp 107–108
44. Pauling L (1973) Die Natur der chemischen Bindung, 3rd edn, Verlag Chemie, Weinheim. (a) p 7, (b) pp 307–310
45. Basolo F, Johnson RC (1964) Coordination Chemistry, W A Benjamin, New York; p 23
46. Dickersen RE, Gray HB, Haight GP (1978) Prinzipien der Chemie, Walter de Gruyter, Berlin; p 412
47. Riedel E (1988) Anorganische Chemie, Walter de Gruyter, Berlin; pp 95–96
48. Haaland A (1989) Angew Chem 101: 1017–1032; Angew Chem Int Ed Engl 28: 992
49. Krasochka ON, Sokolova YuA, Atovmyan LO (1975) Zh Strukt Khim 16: 696–698; J Struct Chem 16: 648–650
50. Gillespie RJ (1963) J Chem Educ 40: 295–301; (1970) 47: 18–23
51. Coomber R, Griffith WP (1968) J Chem Soc A 1128–1131
52. Stiefel EI (1987) Molybdenum(VI), In: Wilkinson G et al (eds), Comprehensive Coordination Chemistry, vol 3, Pergamon Press, Oxford, pp 1375–1420; pp 1376–1377
53. Flynn CM Jr, Stucky GD (1969) Inorg Chem 8: 335–344
54. Golubev AM, Muradyan LA, Kazanskii LP, Torchenkova EA, Simonov VI, Spitsyn VI (1977) Koord Khim 3: 920–925; Soviet J Coord Chem 3: 715–720
55. Knowles PF, Diebler H (1968) Trans Faraday Soc 64: 977–985
56. Diebler H, Timms RE (1971) J Chem Soc A 273–277
57. Gilbert K, Kustin K (1976) J Am Chem Soc 98: 5502–5512
58. Freedman ML (1958) J Am Chem Soc 80: 2072–2077
59. Schwarzenbach G, Meier J (1958) J Inorg Nucl Chem 8: 302–312
60. Kepert DL (1962) Progr Inorg Chem 4: 199–274; pp 260–263
61. Tytko KH, Mehmke J, Fischer S (1999) Struct Bonding 93: 129–321
62. Tytko KH (1983) Chem Scr 22: 201–208
63. Tytko KH, Schönfeld B (1975) Z Naturforsch B 30: 471–484
64. Tytko KH (1975) Chemiedozententagung, Düsseldorf, Referateband, p A 40
65. Tytko KH (1974) 16th Intern Conf Coord Chem, Dublin, Proceedings, Ref R 8
66. Tytko KH, Mehmke J (1983) Z Anorg Allg Chem 503: 67–86
67. Goiffon A, Spinner B (1975) Rev Chim Minerale 12: 316–327
68. Goiffon A, Spinner B (1975) Bull Soc Chim France 2435–2441
69. Pope MT (1994) Polyoxoanions, In: King RB (ed), Encyclopedia of Inorganic Chemistry, vol 6, Wiley, Chichester, 3361–3371. (a) pp 3361–3362, (b) p 3361
70. Tytko KH (1976) Z Naturforsch B 31: 737–748
71. Kustin K, Liu ST (1973) J Am Chem Soc 95: 2487–2491
72. Honig DS, Kustin K (1973) J Am Chem Soc 95: 5525–5528
73. Brown ID (1992) Acta Crystallogr B 48: 553–572. (a) pp 561–562, (b) p 560
74. Boisson MB, Gibbs GV, Zhang ZG (1988) Phys Chem Mineral 15: 409–415
75. Brown ID (1994) Bond Length-Bond Valence Relationships in Inorganic Solids, In: Bürgi HB, Dunitz JD (eds), Structure Correlation, vol 2, VCH, Weinheim, pp 405–429. (a) pp 414–415, (b) pp 423–424
76. Pope MT (1972) Inorg Chem 11: 1973–1974
77. Porai-Koshits MA, Atovmyan LO (1975) Koord Khim 1: 1271–1281; Soviet J Coord Chem 1: 1065–1074
78. Shustorovich EM, Porai-Koshits MA, Buslaev YuA (1975) Coord Chem Rev 17: 1–98; pp 67–81
79. Schröder FA (1975) Acta Crystallogr B 31: 2294–2309
80. Porth D (1991) Diplomarbeit, University Göttingen; pp 30–33

81. Porai-Koshits MA, Atovmyan LO (1981) Zh Neorgan Khim 26: 3171–3180; Russ J Inorg Chem 26: 1697–1703
82. Evans HT Jr (1971) Perspect Struct Chem 4: 1–59; pp 53–56
83. Cruickshank DWJ (1961) J Chem Soc 5486–5504
84. Cruickshank DWJ (1985) J Mol Struct 130: 177–191
85. Lange G, Hahn H, Dehnicke K (1969) Z Naturforsch B 24: 1498–1507
86. Shao M, Wang L, Zhang Z, Tang Y (1984) Sci Sin Ser B (Engl Ed) 27: 137–148
87. Chojnacki J (1963) Bull Acad Polon Sci Ser Sci Chim 11: 365–368
88. Cruywagen JJ, Rohwer EFCH (1975) Inorg Chem 14: 3136–3137
89. Sillén LG (1954) Acta Chem Scand 8: 299–317; p 304
90. Kepert DL (1969) Inorg Chem 8: 1556–1558
91. Tytko KH (1973) Z Naturforsch B 28: 272–275
92. Wells AF (1984) Structural Inorganic Chemistry, 5th edn, Clarendon Press, Oxford; pp 27–328
93. Tytko KH, Schönfeld B, Buss B, Glemser O (1973) Angew Chem 81: 305–307; Angew Chem Int Ed Engl 12: 330
94. Tytko KH, Baethe G, Cruywagen JJ (1985) Inorg Chem 24: 3132–3136
95. Tytko KH, Baethe G, Hirschfeld ER, Mehmke K, Stellhorn D (1983) Z Anorg Allg Chem 503: 43–66
96. Cotton FA, Wilkinson G (1988) Advanced Inorganic Chemistry, 5th edn, Wiley, New York; pp 104–106
97. Nomiya K, Miwa M (1984) Polyhedron 3: 341–346
98. Nomiya K, Miwa M (1985) Polyhedron 4: 89–95
99. Nomiya K, Miwa M (1985) Polyhedron 4: 675–679, 1407–1412
100. Ma L, Liu S, Zubieta J (1989) Inorg Chem 28: 175–177
101. Krebs B, Paulat-Böschen I (1976) Acta Crystallogr B 32: 1697–1704
102. Böschen I, Krebs B (1974) Acta Crystallogr B 30: 1795–1800
103. Brown ID, Shannon RD (1973) Acta Crystallogr A 29: 266–282
104. Brown ID, Wu KK (1976) Acta Crystallogr B 32: 1957–1959
105. Donnay G, Allmann R (1970) Am Mineral 55: 1003–1015
106. Pyatenko YuA (1972) Kristallografiya 17: 773–779; Sov Phys - Crystallogr 17: 677–682
107. Trömel M (1983) Acta Crystallogr B 39: 664–669; (1984) Acta Crystallogr B 40: 338–342; (1986) Acta Crystallogr B 42: 138–141
108. Brown ID, Altermatt D (1985) Acta Crystallogr B 41: 244–247
109. Brown ID (1981) The Bond Valence Method: An Empirical Approach to Chemical Structure and Bonding, In: O'Keeffe M, Navrotsky A (eds), Structure and Bonding in Crystals, vol 2, Academic Press, New York, pp 1–30
110. O'Keeffe M (1989) Struct Bonding 71: 161–190
111. Perloff, A (1970) Inorg Chem 9: 2228–2239
112. Müller A, Penk M, Krickemeyer E, Bögge H, Walberg HJ (1988) Angew Chem 100: 1787–1789; Angew Chem Int Ed Engl 27: 1719
113. Tytko KH, Mehmke J, Kurad D (1999) Struct Bonding 93: 1–66
114. Sanderson RT (1983) Polar Covalence, Academic Press, New York. (a) pp 179–183, (b) pp 194
115. Downs AJ, Adams CJ (1973) Chlorine, Bromine, Iodine, Astatine, In: Bailar JC Jr et al (eds), Comprehensive Inorganic Chemistry, vol 2, Pergamon Press, Oxford, pp 1107–1594; pp 1353–1361
116. Brown DH, Perkins PG, Stewart JJP (1972) J Chem Soc Dalton Trans 2243–2246
117. Kazanskii LP, Spitsyn VI (1976) Dokl Akad Nauk SSSR 227: 140–143; Dokl Phys Chem Proc Acad Sci USSR 226/231: 225–227
118. Hawthorne FC (1992) Z Kristallogr 201: 183–206
119. Waugh JLT, Shoemaker DP, Pauling L (1954) Acta Crystallogr 7: 438–441
120. Baker LCW, Lebioda L, Grochowski J, Mukherjee AG (1980) J Am Chem Soc 102: 3274–3276

121. Tytko KH, Cordis V, Mehmke K, Hirschfeld ER (1985) U.S.-Japan Seminar on the Catalytic Activity of Polyoxoanions, Shimoda, Abstracts, pp 35–39
122. Mehmke K (1988) Dissertation, University Göttingen. (a) pp 110, (b) pp 110, 113–115, 151–154, 159
123. Howarth OW, Pettersson L, Andersson I (1989) J Chem Soc Dalton Trans 1915–1923
124. Stiefel EI (1977) Progr Inorg Chem 22: 1–223; pp 34–35
125. Cotton FA, Wing RM (1965) Inorg Chem 4: 867–873
126. Krebs B (1972) Acta Crystallogr B 28: 2222–2231
127. Fuchs J, Palm R, Hartl H (1996) Angew Chem 108: 2820–2822; Angew Chem Int Ed Engl 35: 2651–2653
128. Hüllen A (1964) Naturwissenschaften 51: 508; Angew Chem 76: 588
129. Jahr KF, Fuchs J (1966) Angew Chem 78: 725–735; Angew Chem Int Ed Engl 5: 689–699
130. Tytko KH (1985) Molybdate Hydrates with Alkali Metals and Ammonium and with Alkaline Earth Metals, In: Gmelin Handbook of Inorganic Chemistry, 8th edn, Molybdenum Suppl Vol B 4, pp 1–213. (a) p 37, (b) pp 30–32, 37
131. Tisley DG, Walton RA (1973) J Mol Struct 17: 401–409
132. Howarth OW, Jarrold MJ (1978) J Chem Soc Dalton Trans 503–506
133. Griffith WP, Lesniak PJB (1969) J Chem Soc A 1066–1071
134. Corigliano F, Di Pasquale S (1975) Inorg Chem Acta 12: 99–101
135. Kazanskii LP, Spitsyn VI (1975) Dokl Akad Nauk SSSR 223: 381–384; Dokl Phys Chem Proc Acad Sci USSR 220/225: 721–723
136. Klemperer WG, Shum WJ (1977) J Am Chem Soc 99: 3544–3545
137. Klemperer WG, Shum WJ (1978) J Am Chem Soc 100: 4891–4893
138. Day VW, Klemperer WG, Maltbie DJ (1987) J Am Chem Soc 109: 2991–3002
139. Kazanskii LP, Saprykin AS, Golubev AM, Spitsyn VI (1977) Dokl Akad Nauk SSSR 233: 405–408; Dokl Phys Chem Proc Acad Sci USSR 232/237: 282–284
140. Fuchs J, Knöpnadel I (1982) Z Kristallogr 158: 165–179
141. Freeman MA, Schultz FA, Reilley CN (1982) Inorg Chem 21: 567–576
142. Klemperer WG (1990) Inorg Synth 27: 71–135; pp 71–74
143. Böschen I, Buss B, Krebs B, Glemser O (1973) Angew Chem 85: 409; Angew Chem Int Ed Engl 12: 409
144. Enjalbert R, Galy J (1986) Acta Crystallogr C 42: 1467–1469
145. Baur WH (1972) Am Mineral 57: 709–731
146. Björnberg A (1980) Dissertation, University Umeå; pp 41–45 (see also Reference 1, p 19)
147. Moore PB (1974) Neues Jahrb Mineral Abhandl 120: 205–227; pp 220–221
148. Henry M, Jolivet JP, Livage J (1992) Struct Bonding 77: 153–206; pp 163–165
149. D'Amour H, Allmann R (1976) Z Kristallogr 143: 1–13
150. D'Amour H (1976) Acta Crystallogr B 32: 729–740
151. Cordis V, Tytko KH, Glemser O (1975) Z Naturforsch B 30: 834–841

Bonding and Charge Distribution in Isopolyoxometalate Ions and Relevant Oxides – A Bond Valence Approach

K. H. Tytko, J. Mehmke, S. Fischer

Institut für Anorganische Chemie, Universität Göttingen,
Tammannstraße 4, D-37077 Göttingen, Germany

Law of simple synthesis and laborious analysis:
It is easier to devise a mechanism which produces a specific behavior than to identify the mechanism underlying an observed behavior.

Valentin Braitenberg [341]

Refined bond length-bond valence relationships for M–O bonds have been applied to isopolyoxometalate and some oxide structures ($M = Mo^{VI}$, W^{VI}, V^{V}, Nb^{V}, Ta^{V}). The M–O bond valences and the distribution of the (negative) ionic charge over the different types of oxygen atoms have thus been obtained. Additional negative charge and an equivalent positive charge were found to exist on certain types of oxygen atoms. The reasons for the observed distribution of the bond lengths and bond valences in the M–O frameworks and of the charge over the oxygen atoms have been analyzed. The main factors determining the distribution are the stabilization of the polymeric structures by strengthening of the inner M–O bonds of the M–O frameworks through acceptance of the negative charge by the terminal oxygen atoms and of the positive charge by (approximately) linearly bridging oxygen atoms and/or (angularly) bridging OH groups and coordinated H_2O molecules. Both charge situations are restricted by the necessity to fulfill simultaneously the geometrical conditions of the M–O frameworks (interdependence of the M–O bond lengths), the bond length-bond valence relationships of the M–O bonds, and the valence sum rule for the M and O atoms and by avoidance, as far as possible, of very high negative charge on terminal oxygen atoms. Similar but much less detailed calculations and analyses were undertaken by some authors to locate the positions of OH groups and coordinated H_2O molecules as well as to elucidate the hydrogen bridge system of polyoxometalates. Other authors have attempted to define, among other features, the charge distributions and basicities of the oxygen atoms in polyoxometalate ions. Our results differ considerably from those of the second group of authors, and reasons for the differences are discussed. Additionally, a number of general questions are treated in detail: the conditions for the extension of the coordination sphere of the protonated tetrahedral monometalate ions and the reasons for the occurrence of distorted MO_6 octahedra (MO_k polyhedra); the question of the coordination number of the addenda elements M in special cases; the amphoteric character of the species; the protonation sites; the relationship between charge distribution and basicity (as defined by the protonation constants) of polyoxometalate ions; the strength of the integration of the different types of MO_6 octahedra (MO_k polyhedra) within the M–O frameworks (kinetic inertness of the structures); the stabilization of the polymeric character of the different M–O frameworks by the "meshing effect" and by "normal" bonding; the question of a "trans influence"; the question of ions (guests) encapsulated by (uncharged or charged) M–O frameworks (hosts); and others. In summary, the bond valence model, which is strictly fulfilled within the accuracy of the bond lengths for polyoxo species, proves to be a powerful tool for studying bonding questions and yields a fascinating picture of the polyoxometalate structures.

Keywords: polyoxometalates, structure and bonding, bond model, bond valence model (method), bond length-determing factors, charge-determining factors, stability of polyoxometalates

1 Introduction

The distribution of the ionic charge over the oxygen atoms in iso- and heteropolyoxometalate ions ($M = Mo^{VI}$, W^{VI}, V^{V}, Nb^{V}, Ta^{V}) is the subject of controversy in the literature. A number of research groups believe, on the basis of experimental data and theoretical considerations, that the charge is mainly distributed over the terminal oxygen atoms of the polyoxometalate ions [1–6]. In contrast, other groups, again on the basis of experimental data and theoretical considerations, contend that the charge is mainly distributed over the bridging oxygen atoms [7–12, 13a]. These problems have been reviewed [14a, 15a, 16].

Our interest in the distribution of the ionic charge over the oxygen atoms in polyoxometalate ions originates from our theoretical investigations on the geometrical course of aggregation reactions of protonated monometalate ions leading to polymetalate ions, on the driving forces for these reactions, and on the factors directing their course [14b]. The formal ionic charge (maximum number of protons that can formally be taken up) and the basicity (ease of protonation as expressed by the protonation constants) of mono- and polyoxometalate ions are the major "inner" parameters while the availability of protons, the metalate concentration, and the resulting pH value are the major "outer" parameters for the reaction processes in the polyoxometalate systems MO_4^{w-}/H^+. Protonation of the monometalate ion generally allows the primary extension of the tetrahedral to an octahedral coordination of the M atom and hence a primordial aggregation [4, 17–19] ("addition mechanism"; see also Ref. 16). Further aggregation processes require the elimination of oxygen by condensation of H_2O from two OH groups (at least formally) in each case ("condensation mechanism" [2, 4, 20] and other condensation processes [14c]; see also Ref. 16). The importance of protonation reactions of mono- and polyoxometalate ions for aggregation processes led us to propose relationships between the number of the terminal oxygen atoms of the $M^{VI}O_6$ octahedra or the average charge on the terminal oxygen atoms of polyoxometalate ions (those mainly bearing the negative ionic charge) and the protonation constant of polyoxometalate ions [2, 4] in analogy to the well-known [21, 22] relationship for monomeric oxoanions of the main group elements E (or the pK_a values of their conjugate oxoacids $EO_\mu(OH)_\nu$, respectively).

The present investigation was mainly undertaken to confirm and/or refine our previous proposals for the basicity of polyoxometalate ions and to clarify the discrepancies in the literature with regard to the distribution of ionic charge over the oxygen atoms of polyoxometalate ions. Furthermore, application of the method has allowed us to test our recently proposed bond models for polyoxometalate ions [16, 23, 24].

2
Applications of the Method: The Bond Valence Model

2.1
Determination of the Valence of the M–O Bonds

It has been shown by several authors [25–39] that the bond valence ("bond strength", "bond order" [25, 28, 29, 37]) s of the M–O bonds (or of the bonds between other bond partners) depends on the bond length d according to

$$\log s = (d_o - d)/B \quad \text{or} \quad s = (d_o/d)^N \qquad (1a, 1b)$$

($d_0 \approx 1.90$ Å for Mo^{VI}, W^{VI}, Nb^V, and Ta^V, 1.80 Å for V^V; $B \approx 0.8$ Å, $N \approx 5$) and that the sum of the bond valences about the M atoms (or about another bond partner) amounts to the absolute amount of their oxidation number n, i.e.,

6 v.u. for M^{VI} and 5 v.u. [valence units] for M^{V} (the sum of the "bonding power" of the atoms is constant):

$$\sum_{i=1}^{k_M} s_i = |n| \quad (\text{'valence sum rule'}) \tag{2}$$

(*i*: index for the M–O bonds about an atom, k: coordination number of the atom indicated). These relationships mean that one valence unit corresponds to the bonding power of two electrons [25, 29, 35, 36, 38, 40a], i.e., the bond valence states the (fractional) number of electron pairs that form the corresponding bond.

The equations have been used, among others, to check the sum of the bond valences about the M atoms in polyoxometalate ions, to identify oxygen atoms that constitute an OH group or H_2O molecule in polyoxometalate ions, to characterize the oxygen atoms having contacts to cations, or to elucidate the hydrogen bridge system of polyoxometalates [41–48]. The oxygen atoms in polyoxometalates having contacts to cations are those bearing the negative ionic charge in the free (e.g., dissolved) polyoxometalate ions.

In the present investigation we have used Eq. (1a) since it has been shown to give a somewhat better description of the bond length-bond valence relationship than Eq. (1b) for M–O bonds of Mo^{VI} and V^{V} [35]. This equation has also been used for the M–O bonds of W^{VI}, Nb^{V}, and Ta^{V}. The parameters B (and d_0) used were derived from the structural data of a large number of mono- and polyoxometalates and oxides; to obtain values as exact as possible, other types of compounds were not considered [35]. The (general [35]) parameters B and d_0 and the corresponding standard deviations for the sum of the bond valences about the M atoms, for the bond valence of a single M–O bond, and for the charge on the oxygen atoms are given in Tables 1 and 2 of Ref. 35.

The precision of the results can be enhanced, when not the general d_o, but, for each individual investigation of a structure, an individually fitted d_o, d_{0i}, based on the (general) B parameter, is used [29, 35]. In this way the standard deviations for the sum of the bond valences about the M atoms, for the bond valence of the M–O bonds, and for the charge on the oxygen atoms come out somewhat smaller (by ca. 38%) than those based on the general B and d_o parameters [35] (Table 1; see also Sect. 2.3.2.3 and Table 6 in Ref. 35). The average sum of the bond valences about the M atoms amounts exactly to the oxidation number of M and the sum of the charges on the oxygen atoms exactly to the formal ionic charge of the polyoxo species. According to our experience it is absolutely necessary to operate with d_{oi} instead of d_o if reliable results are to be obtained [35]. This corresponds with observations by Allmann [29]. In the literature there is rather the tendency to generalizations (use of 'universal' B or N values for a group of element–O bonds [26, 33, 49], use of uniform d_o and B or N values for different oxidation states of M [32, 50–52]). The d_{oi} values for the individual structure determinations are given in Tables 3–16, 18, and 20–36 together with the structural data. The difference between d_{oi} and d_o corresponds to (small) *systematic* bond lengths errors (standard deviation of this error: 0.005 Å) with structure determinations; the use of d_{oi} is hence fully justified (see also Sect. 2.3.2.2 in Ref. 35).

Table 1. General parameters B and d_0 and range of the individually fitted parameters d_{oi} of Eq. (1a) for M–O bonds in reliably investigated polyoxometalate ions and oxides and corresponding estimated standard deviations for the sum of the bond valences about the M atoms ($\sigma_{\Sigma i}$), for the valence of the M–O bonds (σ_{si}), for the charge on the oxygen atoms (σ_{ci}), and for the bond lengths (σ_{di}) (compiled from Ref. 35)

M	Mo^{VI}	W^{VI}	V^{V}	Nb^{V}	Ta^{V}
B in Å	0.9530	0.9482	0.8506	0.8592	0.85[a]
d_0 in Å	1.915	1.916	1.799	1.917	1.921
d_{0i} in Å[b]	1.904–1.926	1.910–1.919	1.787–1.811	1.910–1.920	1.921
$\sigma_{\Sigma i}$ in v.u.[c]	0.07 (1.2%)		0.07 (1.3%)		
σ_{si} in v.u.[c]	0.012		0.012		
σ_{ci} in c.u.[c]	0.021		0.018		
σ_{di} in Å[c,d]	0.010		0.010		

[a] Since all available oxotantalate structures are built by TaO_6 octahedra, this value was estimated on the basis of the B values for the M–O bonds of Mo^{VI}, W^{VI}, V^{V} and Nb^{V} (see Ref. 35).

[b] Individual d_{oi} values are given in the following tables describing the structures.

[c] The small number of reliably investigated structures does not allow the derivation of reliable statistical data for W^{VI}, Nb^{V}, and Ta^{V} polyoxometalates.

[d] These values correspond to σ_{si} and $\sigma_{\Sigma i}$ and have been obtained by back calculation using the bond length-bond valence function.

It proved necessary to test the structural data (bond lengths) of the individual compounds for their reliability. In particular, data on oxotungstates and oxoniobates are usually of poor quality [35] (data on oxotantalates hardly exist). This test was performed by determining the average (related to 1 M atom) error square for the sum of the bond valences about the M atoms, the result should not exceed 0.05 $(v.u.)^2$; this value corresponds to $2\sigma_\Sigma$ (the statistical 95% range) with $\sigma_\Sigma = 0.12$ v.u. taken from the results for the sets of reference structures investigated by using the general d_o [35] and statistically tested for outliners.

In crystal structures the ideal symmetry of a polyoxometalate ion is often not realized due to packing effects in the crystal. In solution, however, the free ion can be expected to have the ideal symmetry [53, 54]. In the present investigation we have averaged the bond lengths and bond valences of all symmetry-equivalent M–O bonds[1] assuming ideal symmetry and thus obtained also average charges for the symmetry-equivalent oxygen atoms.

Many polyoxometalate types have repeatedly been investigated, usually as salts with different cations, sometimes as salts with different numbers of

[1] Since the bond length-bond valence relationships are non-linear, average bond valences and bond valences appropriate to average bond lengths do not correspond. Therefore, initially the individual bond valences have to be calculated from the bond lengths and only then can they be averaged. This procedure must be followed if the bond lengths differences are authentic, which can be assumed if the errors (standard deviations) of the bond lengths are small. If the bond lengths differences are not authentic, which can be assumed if the errors are great (as is the case with many structures of the polytungstates), it is more reasonable to average the bond lengths and then to calculate the bond valences since in this way the bond lengths errors are at least partially compensated.

molecules of water of crystallization and the same cation. The standard deviations based on these multiply characterized M–O bonds include also the averaging for the symmetry-equivalent M–O bonds. They are the most important standard deviations since they indicate the reproducibility of the results obtained by different authors and/or the reproducibility of the construction of the M–O frameworks with different cations or different numbers of molecules of water of crystallization. They are given in parenthesis directly following the numerical value of the respective quantity.

It should be noted that structures containing only one type of M–O bond (this is the case for the tetrahedral monometalates if only one crystallographically independent $M^nO_4^{w-}$ ion is present) always appear to be reliable if individually fitted d_o values, d_{oi}, are used. In this case the average bond valences of the M–O bonds (and average charges on the oxygen atoms) have the expected values, namely s = n/4 v.u. (and c = −w/4 c.u.), and consequently the standard deviations with respect to the different investigations of the structures are zero. Thus the reliability of the structural data (bond lengths) of the tetrahedral monometalate ions can only be checked by use of the general d_o values.

Similarly, structures containing only one type of MO_k polyhedra (e.g., $M_6O_{19}^{m-}$, $W_{12}O_{38}(OH)_2^{6-}$) always appear as reliable if averaged bond lengths and the individually fitted parameters d_{oi} are used.

The standard deviations for the bond valences of the M–O bonds and for the charge on the oxygen atoms are solely caused by the standard deviation of the M–O bond lengths, corresponding to 0.010 Å for the set of reference structures. The mathematical form of the bond length-bond valence function does not contribute to the standard deviations (this has been proven for the oxomolybdates and oxovanadates) [35].

2.2 Determination of the Charge on the Oxygen Atoms

The difference between the sum of the M–O bond valences about an oxygen atom and the oxidation number two (absolute amount) of oxygen is, as a consequence of the valence sum rule, the (usually negative) charge c on that oxygen atom [55]:

$$c = \left(\sum_{i=1}^{k_O} s_i\right) - 2 \quad \text{(valence sum rule for oxygen, modified for the charge)} \qquad (3)$$

If this charge (given in charge units (c.u.)) is negative, it defines unbalanced bonding power of the O atom in question which is usually compensated by the cations of the salt, and often via molecules of water of crystallization.

Although the improvement of the standard deviations for the sum of the bond valences about the M atoms, for the valence of the M–O bonds, and for the charge on the oxygen atoms is rather moderate by the use of individually fitted d_{oi} values (see Sect. 2.1), the values for the charge distribution about the oxygen atoms are considerably improved: The sum of all charges amounts exactly to the formal ionic charge of the polyoxometalate ions [35].

The following scale is used to describe the magnitude of a charge:

- 0.00 to 0.02 c.u. = unproved;
- 0.03 to 0.05 c.u. = very small;
- 0.06 to 0.12 c.u. = small;
- 0.13 to 0.24 c.u. = medium;
- 0.25 to 0.49 c.u. = high;
- $\geqq$0.50 c.u. = very high.

2.3
Treatment of Oxometalate Ions Containing OH Groups or Coordinated H_2O Molecules

These oxometalate ions have first been treated in the completely deprotonated form and subsequently the O–H bonds have been added and assumed to have bond valences of 1 v.u., i.e., hydrogen bonding has been neglected. Although it is possible in many cases to assign more realistic bond valences to the O–H bonds on the donor side, this is more difficult for the acceptor side of the hydrogen bridges. For reasons of the necessary charge balance we assume throughout the calculations a value of 1 v.u. for the O–H bonds and make some extra considerations regarding charge compensation by hydrogen bonding.

2.4
Cation Coordination by Oxygen

The cations of the polyoxometalates are coordinated by oxygen atoms of the polyoxo anions and of the molecules of water of crystallization in a usually less regular manner, for which obviously also bond length-bond valence relationships have to be fulfilled [29, 42–44]; see also Refs. 31 and 56. So far the bond valence model seems to make no distinction between covalent and ionic bonding [39d] with the consequence that, among others, the characterization of the ionic bond as accomplished by *undirected* binding forces cannot be supported.

Cation-oxygen bonding is not quantitatively investigated in this article.

3
Results of the Calculation of Bond Valences and Charge Distributions and Special Discussion

3.1
Oxometalate Types Investigated

In the following sections the results obtained for the most important isopolyoxometalate types and relevant oxides are discussed in the sequence

- mono- and polymetalate ions exclusively with tetrahedral structures: MoO_4^{2-}, WO_4^{2-}, VO_4^{3-}; $Mo_2O_7^{2-}$, $V_2O_7^{4-}$; $V_3O_{10}^{5-}$; $[VO_3^-]_\infty$; $V_5O_{14}^{3-}$
- polymetalate ions with octahedral structures of the block type: $Mo_7O_{24}^{6-}$; (β-)$Mo_8O_{26}^{4-}$; $Mo_6O_{19}^{2-}$, $W_6O_{19}^{2-}$, $Nb_6O_{19}^{8-}$, $Ta_6O_{19}^{8-}$; $V_{10}O_{28}^{6-}$, $Nb_{10}O_{28}^{6-}$, $H_2V_{10}O_{28}^{4-}$, $H_3V_{10}O_{28}^{3-}$; $Mo_8O_{26}(OH)_2^{6-}$, $Mo_{10}O_{34}^{8-}$(I), $Mo_{10}O_{34}^{8-}$(II)[2], $[Mo_8O_{27}^{6-}]_\infty$; $[Mo_6O_{20}^{4-}]_\infty$(o)[2]
- polymetalate ions with octahedral structures of the cage type: $W_{12}O_{40}(OH)_2^{10-}$; $W_{12}O_{38}(OH)_2^{6-}$; $W_{10}O_{32}^{4-}$; α-$Mo_8O_{26}^{4-}$; $Mo_{36}O_{112}(H_2O)_{16}^{8-}$
- other types: $[Mo_4O_{14}^{4-}]_\infty$; $[Mo_6O_{20}^{4-}]_\infty$ (o,tp)[2]; $[V_2O_6^{2-}]_\infty$; "MoO_3" (= $[Mo_2O_6]_\infty$); "(α-)$MoO_3 \cdot H_2O$" (= $[Mo_2O_6(H_2O)_2]_\infty$); "$MoO_3 \cdot 2H_2O$" (= $\{[Mo_{16}O_{48}(H_2O)_{16}] \cdot 16H_2O\}_\infty$); "$V_2O_5$" (= $[V_4O_{10}]_\infty$).

Some structure types have not been considered because of excessive errors in the M–O bond lengths (see the 0.05 (v.u.)^2 reliability test [35] mentioned above), in other cases (most of the two- and three-dimensional polymers, usually prepared from melts) the structures are too difficult to analyse or we could not unequivocally assign the oxygen atoms. The last item refers to a surprisingly large number of inadequate structural descriptions. The structure of the so-called decamolybdates [47] has not been evaluated because of effects due to the statistical non-occupancy of Mo positions (see footnote g of Table 46). Aspects of this structure have, however, occasionally been taken into consideration to complete the discussion on certain topics in Sect. 4. The same holds for the recently prepared $HW_5O_{19}^{7-}$ penta- [57] and $H_3W_6O_{22}^{5-}$ hexatungstate [58] ions. The latter have borderline properties (see Sect. 4.6.12); the respective numerical data are given in the Tables within obtuse-angled (⟨ ⟩) brackets.

Unfortunately, the visual presentation of the polyoxo structures in the literature is often very questionable. Three examples are discussed briefly:

- $Mo_8O_{26}(OH)_2^{6-}$: In the original paper [59] the positions of the OH groups for this structure are not shown in the coordination polyhedral model (although in the paper a presentation of this model is indeed given). Instead, only the conventional atom-and-bond model is given under a very unfavorable viewing angle. This is obviously the reason for the incorrect adoption of the OH positions in the coordination polyhedral model of a frequently cited monograph [13h] about polyoxoanions.
- $[Mo_6O_{20}^{4-}]_\infty$(o,tp): In this case the M–O framework is incorrectly represented in one of the original papers [60] and in the mentioned monograph [13 h] (the errors are different in each case).
- $[Mo_2O_6]_\infty$ and other extended, in particular oxide structures. Due to the common practice of presenting structural sections with complete coordination polyhedra for M, it is often very difficult to identify the

[2] In rare cases a formula corresponds to different structures for which a terminological distinction has as yet not been made: (o) structure built up exclusively of MO_6 octahedra; (o,tp) structure built up of MO_6 octahedra and tetragonal MO_5 pyramids; (I), (II) structures with different connection sites of the MO_4 tetrahedra present.

coordination numbers and coordination modes of the peripheral O atoms of the sections which appear as terminal O atoms. Therefore in this article we show the sections in extended structures with cut polyhedra.

In general, the bond lengths given in the literature have been used. For some recalculations the sometimes somewhat deviating bond lengths of the Inorganic Crystal Structure Database [61] or of the Cambridge Structural Database [62] have been applied. This refers particularly to those structures for which the influence of very long M···O distances has been studied.

3.2
Occurring Structural Elements; Terminology

The M atoms of the oxometalate ions investigated in this study have tetrahedral (MoO_4, WO_4, VO_4), tetragonal-pyramidal (MoO_5, VO_5), trigonal-bipyramidal (VO_5), octahedral (MoO_6, WO_6, VO_6, NbO_6, TaO_6) (most frequent coordination mode), and pentagonal-bipyramidal (MoO_7) coordination by oxygen (see Fig. 1). This description includes M–O bonds down to 0.2 v.u. (see Tables 38 and 39).

The MO_k polyhedra share edges (in particular within the block-type structures or partial structures), corners (in particular in chains and rings of MO_4 tetrahedra) or both (in particular in the cage-type structures and in chains built by block-type subunits). Corner-sharing of MO_6 octahedra (MO_k polyhedra) as an inevitable consequence of edge-sharing is not mentioned in the description of the structures.

The MO_6 octahedra can be of type I (characterized by one unshared oxygen atom [63] or, more precisely, one short, terminal, four medium-sized bridging, and one long bridging M–O bond in *trans* position to the short one [64, 65]) or type II (characterized by two *cis* unshared oxygen atoms [63] or, more precisely, two short *cis*, two medium-sized *trans*, and two long *cis* M–O bonds in *trans* position to the short ones [64, 65]) (Fig. 1). The (usually long) bonds in *trans* position to short ones are marked by an asterisk (*) in the structural descriptions. The short *cis* M–O bonds of the type II octahedra can be formed by terminal or by bridging (two-coordinate) oxygen atoms. The latter are termed 'pseudoterminal' [64] or 'quasiterminal' [13f, 66] oxygen atoms. The distinction according to type I and type II can also be made for the other MO_k polyhedra with $k > 4$ (see Fig. 1 and Sects. 4.3.1 and 4.3.2). (For the type III octahedra see Sect. 4.6.12.).

The oxygen atoms of polyoxometalate ions differ fundamentally (i) in their coordination number against M (one to six), (ii) in the connection mode of the neighboring MO_k polyhedra to which they belong (for geometrical reasons both factors influence the M–O–M angles which range from ca. 85° to 180°), and (iii) in the character of the oxygen atoms in the M–O–M bridges (virtually terminal - pseudoterminal - oxygen atoms or not). The types of oxygen atoms occurring in (iso)polyoxometalate ions based on these criteria are characterized in Fig. 2.

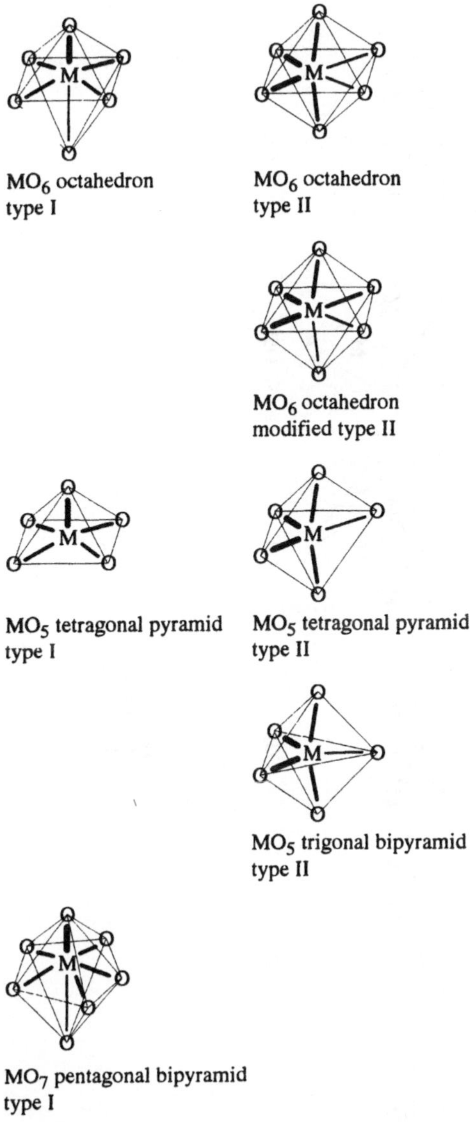

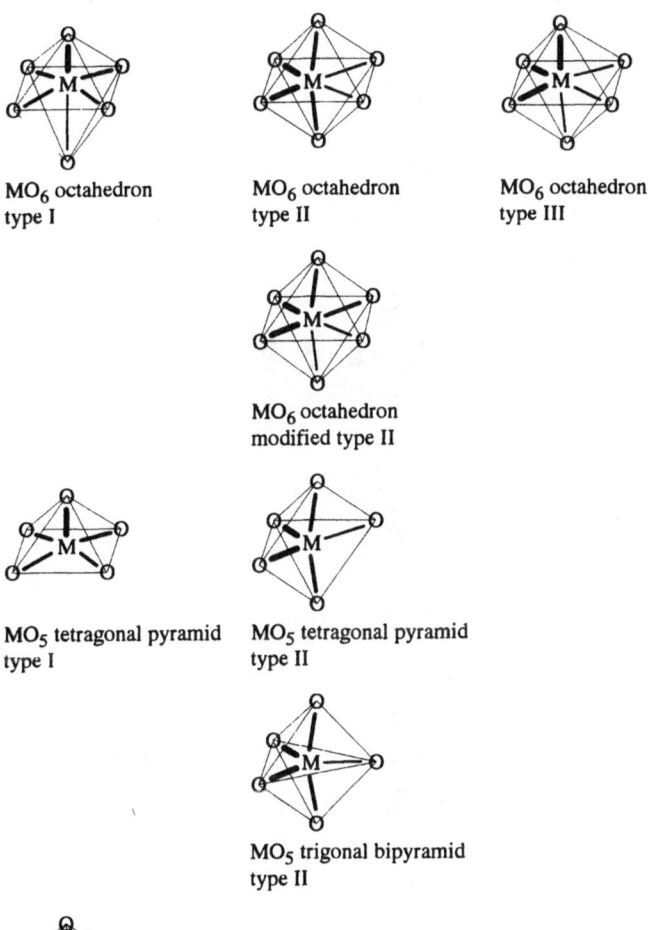

Fig. 1. Types of MO_k ($k > 4$) polyhedra occurring in polyoxometalate ions or oxides of Mo^{VI}, W^{VI}, V^{V}, Nb^{V}, and Ta^{V}. ▬ short (strong), — medium-sized (middle-strong), —— long (weak) M–O bonds

The first index of the oxygen atoms states their coordination number against M: 't' (for terminal) defines the one-coordinate oxygen atoms, 'p' (for pseudoterminal) and '2' the two-coordinate oxygen atoms, and '3', '4', '5', and '6' the three- to six-coordinate oxygen atoms (bridging oxygen atoms without any distinction are designated by 'b'). The second index of the non-terminal

The oxygen atoms are part of MO_6 octahedra (or of other MO_k polyhedra with $k > 4$):

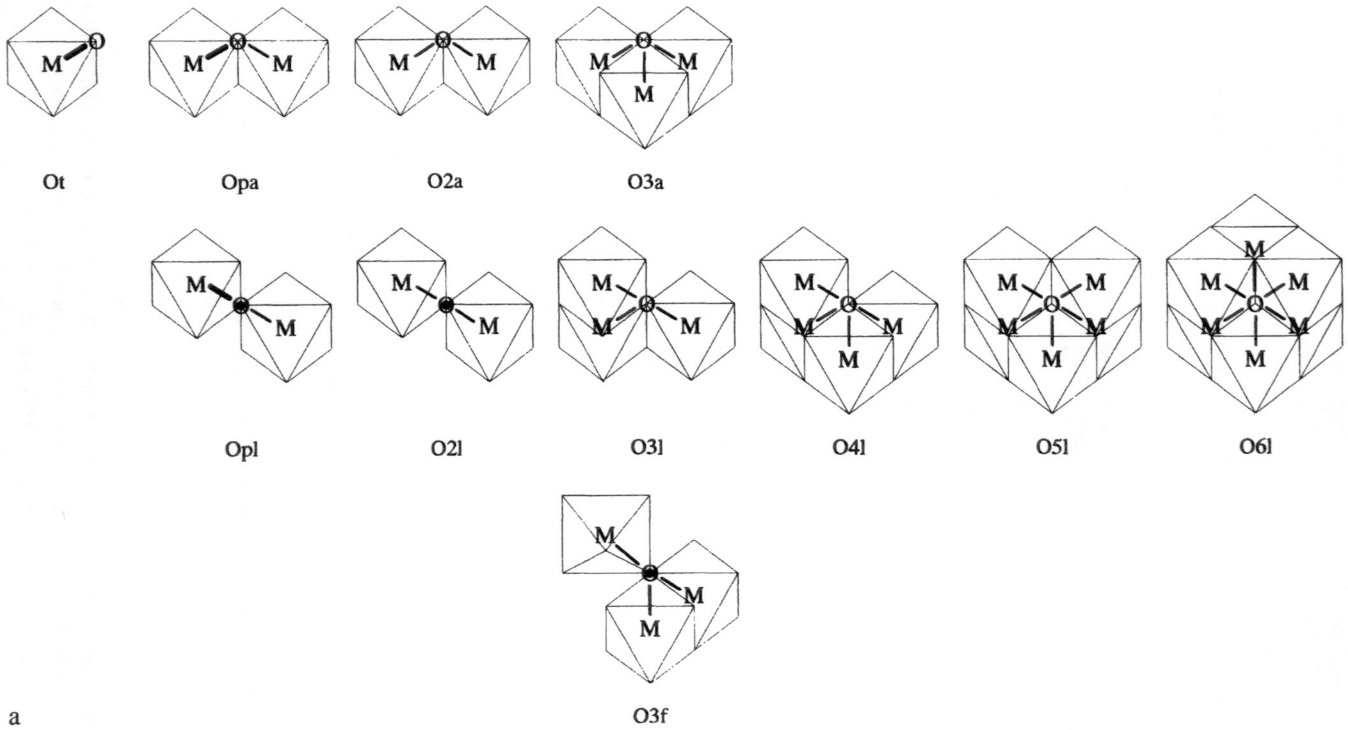

Fig. 2. Types of oxygen atoms occurring in (iso)polyoxometalate ions and oxides of Mo^{VI}, W^{VI}, V^{V}, Nb^{V}, and Ta^{V} as defined by their coordination number against M, by the connection mode of the neighboring MO_k polyhedra, and, if true, by a pseudoterminal character. The oxygen atoms are found at the corners, the metal atoms near to the centers of the distorted MO_k polyhedra. ▬ strong (short), — middle-strong (medium-sized) or weak (long) M–O bonds; ● the M–O–M angle to the corner-shared polyhedron can change over a wide range and the corner-shared polyhedron can rotate about the M–O axis. Configuration O4f exists in heteropolyoxometalate ions (X: heteroelement) only.

The oxygen atoms are also part of MO_4 tetrahedra:

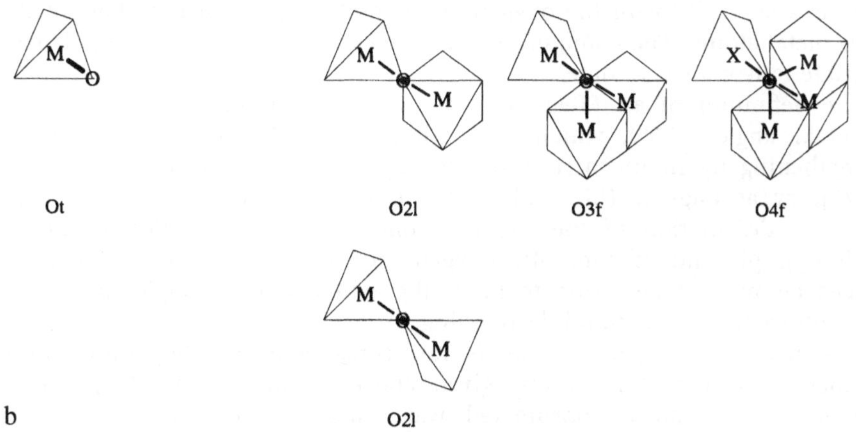

Fig. 2. (Continued)

oxygen atoms designates the connection mode of the neighboring MO_k polyhedra and hence, together with the coordination number, the magnitude of the M–O–M angles. Oxygen atoms which are part of a common edge of neighboring MO_k polyhedra ($k > 4$) are marked by ‘a’ if corner-sharing of MO_k polyhedra is absent. Oxygen atoms which form the connection site of corner-sharing neighboring MO_k polyhedra ($k \geqq 4$) are labelled by ‘l’, independent of the fact that the criterion ‘a’ may also be present. Oxygen atoms which form the connection site of neighboring MO_k polyhedra ($k \geqq 4$) where one polyhedron shares a corner with two or three polyhedra sharing edges are marked by ‘f’, independent of the fact that the criterion ‘a’ is also present. The next index, if present, is a number to distinguish between different oxygen atoms of the same type.

The ideal geometries shown in Fig. 2 are in reality more or less distorted, see the above definition of the type I and type II MO_6 octahedra. The M–O–M angles of the “a series” (only angular M–O–M bridges with two- or three-coordinate oxygen atoms are present) in the configurations pa, 2a, and 3a are smaller than ca. 125° (in the range ca. 85° to 125°). The M–O–M angles of the “l series” (linear or approximately linear M–O–M bridges with two- to six-coordinate oxygen atoms are present; in the case of the three- to six-coordinate oxygen atoms there are inevitably also angular M–O–M bridges) range from ca. 120° to 180°, the greatest variability occurring with the two-coordinate oxygen atoms. Thus the configurations pl and 2l can form angles from ca. 120° to 180°; configuration 3l forms a ‘T’, distorted to give a flat ‘Y’; configuration 4l is a strongly distorted tetrahedron (a “seesaw”-like structure as in SF_4); configuration 5l forms a tetragonal pyramid (with the central oxygen inside the pyramid); and configuration 6l forms a rather regular octahedron. In the case of the “f series” the oxygen atom is something like the site of a furcation; inevitably there are also angular M–O–M bridges. In the configuration 3f the large M–O–M angles range from ca. 120° to 140° and thus there is a sharp 'Y' with the oxygen atom somewhat out of the plane of the three metal atoms. The configuration 4f also exists but not in the *iso*polyoxometalate ions known to date.

The definition of an M–O–M bridge as linear or approximately linear (M–O–M angle $\gtrapprox$125°), which is a criterion for the possibility of bond strengthening by multicenter pπ-dπ multiple bonds in the bridge [23, 24, 67–71], or as angular (M–O–M angle $\lessapprox$125°) can hence directly be seen from the designation of the oxygen atoms with the exception of the 2l (including pl) and 3f (and 4f) oxygen atoms. Therefore only for these oxygen atoms is it necessary to know the actual M–O–M angles for rating the multicenter pπ-dπ multiple bonding.

Protonation of oxygen atoms is not considered for the coordination number of oxygen, for reasons given above. However, an OH group is marked by 'h' and a coordinated water molecule by 'w'. Notice that protonation of a terminal oxygen atom breaks its terminal character with respect to the M–O bond valence but not in respect of the coordination number against M. Thus the term 'terminal' used to characterize the short bonds in type I and type II (and type III) MO_k polyhedra applies only to Ot atoms (M–O). Oth atoms form medium-sized (M–OH) and Otw atoms long ($M–OH_2$*) bonds. Ot, Oth, and Otw atoms together are named 'unshared' O atoms (Ou).[3]

The structural descriptions in Tables 2–36 and Figs. 3 and 5–29 are given for the "free state" of the polyoxo species, i.e., the highest possible symmetry of the M–O frameworks was taken as a basis and corresponding averaged bond lengths, bond valences, and charges have been determined. This largely eliminates the possible deforming influence of the cations and molecules of water of crystallization in the solid-state structure. For tetrahedral structural parts free rotation about the relevant M–Ob axis (Ob: O2l) was assumed. Symmetry-equivalent M atoms are marked by superscript Roman numbers.

For the terminological distinction between O atoms of neighboring structural groups (long range bonding) see Sect. 3.3.

3.3
Consideration of Very Long M···O Distances As M–O Bonds?

In the literature the longest M···O distances in isopolyoxometalates defining an MO_k polyhedron correspond to M–O bonds of 0.2 v.u. ($Nb_{10}O_{28}^{6-}$: 0.19 v.u.; $H_3V_{10}O_{28}^{3-}$: 0.20 v.u.; $Mo_7O_{24}^{6-}$: 0.23 v.u.), if *averaged* values for the symmetry-equivalent bonds are considered (see also Sect. 4.6.2). Since the next longer M···O distances in the polyoxometalates appeared to comply with small but real bond valences and since the bond valence sums for some MO_k polyhedra with $k < 6$ showed a deficit against the oxidation number of M, a small

[3] Unfortunately, in the literature Oth and Otw atoms are sometimes also named "terminal" O atoms.

collection of the polyoxometalate structures was investigated in greater detail. The selection of the structures was realized according to the structure type, an average and an extreme character, the availability of the necessary structural data in the literature, and the reliability of the data.

For the following structural descriptions (Sects. 3.4–3.28) we have neglected in general the very long M–O bonds for several reasons:

- Apart from some of the structures comprising MO_k polyhedra with $k < 6$, the errors are only very small, $\leqq +0.03$ v.u. and $\leqq |\pm 0.03|$ c.u. (see Sect. 4.1, in particular Table 37).
- The structures would become very difficult to survey and to describe.
- The additional arithmetic expenditure is a multiple of the "simple" case.
- (For some structures the necessary atomic coordinates are not given in the literature.)

Thus the definition of an MO_k polyhedron considers only M–O bonds down to 0.2 v.u. However, for the structures comprising MO_k polyhedra with $k < 6$ we made generally some extra calculations (as far as possible).

In the following the additional oxygen atoms forming very long M···O bonds are marked by primes if their origin is a neighboring monomeric or polymeric entity: O′ belongs to a neighboring monomeric unit in any direction; O″ belongs to neighboring polymeric building units in the longitudinal direction of a polymeric unit, including those in a chain; O‴ belongs to lateral neighboring polymeric building units, chains, or layers. Only the shortest additional M···O bonds are identified in this way.

3.4 The Monometalate Ions MoO_4^{2-}, WO_4^{2-}, and VO_4^{3-}

The tetrahedral monomolybdate ions are usually more or less distorted. All kinds of distortions (four different bond lengths; two short, two long bonds; etc.) and nearly undistorted MoO_4 tetrahedra exist. The Mo–O bond valences in 15 reliable[4] structures [72–86] range from 1.19–1.86 v.u. (expected average: 1.50 v.u.) and the charge on the (terminal) oxygen atoms correspondingly from −0.81 to −0.14 c.u. (expected average: −0.50 c.u.), see Table 2a. The bond lengths of ten further monomolybdate structures [86–95] do not fulfill the 0.05 $(\text{v.u.})^2$ $[(2\sigma_\Sigma)^2]$ [35] reliability criterion. A more comprehensive study of six [72, 76, 79, 87, 91, 92] monomolybdate structures (see Table 37) revealed that up to 2.9% of the bond valence of the Mo atoms is directed to oxygen atoms of neighboring MoO_4^{2-} ions, to molecules of water of crystallization, and/or to oxygen atoms of ligands (18-crown-6) of the cations. The largest individual bond valences of these sorts reach 0.08 v.u. (Mo···Ot′; 1.3%) [87]; 0.02 v.u. (Mo···Ow; 0.4%) [92]; 0.01 v.u. (Mo···O(18-crown-6); 0.1% of the

[4] That means in the case of the monometalates consisting of only one crystallographically independent $M^nO_4^{w-}$ ion that the *general* d_o value has to be used. If individual d_o values, d_{oi}, are used, all structures appear as reliable since there is only one type of M–O bond (compare Sect. 2.1).

total bond valence) [92]. Neglect of these additional oxygen atoms leads to small errors for the M–O bond valences of up to +0.05 v.u. and for the charges of up to +0.05/–0.06 c.u. Contrary to our expectations, not one of the unreliable structures tested became reliable by this measure; see also Sect. 4.1. A large number of inorganic and organic cations have been reported.

The monotungstate ions are also usually more or less distorted. Only one [96] of the available structures (cation: Sn^{2+}) fulfills the reliability condition. The bond lengths of four other monotungstate structures [97–100] do not fulfill this condition. The ranges of the W–O bond valences (expected average: 1.50 v.u.) and of the charges on the oxygen atoms (expected average: –0.50 c.u.) (see Table 2b) are somewhat smaller than those for the monomolybdates, which is most probably due to the only small number of available structure determinations.

The monovanadate ions show more frequently a regular-tetrahedral coordination. The V–O bond valences of 25 reliable[4] structures [101–120] range from 1.00 to 1.53 v.u. (expected average: 1.25 v.u.) and the charge on the

Table 2. Average bond lengths d and bond valences s of the M–O bonds and average charges c on the oxygen atoms in monometalate ions

Table 2a. Data for 15 reliable structures of MoO_4^{2-}

M–O bond	O type frequency	d[a] in Å	s[a] in v.u.	c[a] in c.u.
M–Ot	4×	1.750(7)	1.500(0)	–0.500(0)
Sum of the negative charges (formal ionic charge):				–2.000 c.u.

(Average) bond valences about the M atoms:
M (1×): 1.500–1.500–1.500–1.500 v.u.[a] (regular or distorted MO_4 tetrahedra);
Σs_i = 6.000 v.u.

[a] The observed individual values range from 1.836 to 1.651 Å, 1.187 to 1.856 v.u., and –0.813 to –0.144 c.u. The bond length corresponding to s = 1.500 v.u. is d = 1.747 Å (effect of the hyperbola-like form of the bond length-bond valence function).

Table 2b. Data for WO_4^{2-} (only one reliable structure available[a])

M–O bond	O type frequency	d[a] in Å	s[a] in v.u.	c[a] in cu.
M–Ot	4×	1.751	1.500	–0.500
Sum of the negative charges (formal ionic charge):				–2.000 c.u.

(Average) bond valences about the M atoms:
M (1×): 1.500–1.500–1.500–1.500 v.u.[a] (regular or distorted MO_4 tetrahedra);
Σs_i = 6.000 v.u.

[a] The range of the observed individual values for the (reliable and unreliable) structures is, due to the only small number of structures investigated (available), somewhat smaller than that for the reliable monomolybdate structures.

Table 2c. Data for 25 reliable structures of VO_4^{3-}

M–O bond	O type frequency	d^a in Å	s^a in v.u.	c^a in c.u.
M–Ot	4×	1.719(8)	1.250(0)	−0.750(0)
Sum of the negative charges (formal ionic charge):				−3.000 c.u.

(Average) bond valences about the M atoms:
M (1×): 1.250–1.250–1.250–1.250 v.u.[a] (regular or distorted MO_4 tetrahedra);
Σs_i = 5.000 v.u.

[a] The observed individual values range from 1.809 to 1.636 Å, 0.996 to 1.532 v.u., and −1.004 to −0.468 c.u. The bond length corresponding to s = 1.250 v.u. is d = 1.717 Å (effect of the hyperbola-like form of the bond length-bond valence function).
Note that for the case of only one crystallographically independent MO_4 tetrahedron in the structure, individually fitted d_0 values (d_{oi}) lead necessarily for all $M^{VI}O_4$ tetrahedra to the average s = 1.500 v.u. and the average c = −0.500 c.u. and for all $M^{V}O_4$ tetrahedra to the average s = 1.250 v.u. and the average c = −0.750 c.u., independent of the reliability of the structural data (bond lengths); see Sect. 2.1. This is a consequence of the fact that only one type of M–O bond exists. The unreliability of the structural data can be seen from the deviating d_{oi} values ($|d_{oi} - d_o| > 0.012$ Å) or, in the case of $|d_{oi} - d_o| < 0.012$ Å and presence of at least two crystallographically independent MO_4 tetrahedra in the structure, from the $(2\sigma_\Sigma)^2 > 0.05$ (v.u.)^2 criterion [35].

oxygen atoms correspondingly from −0.97 to −0.50 c.u. (expected average: −0.75 c.u.), see Table 2c. The occurrence of more regular tetrahedra as compared to MoO_4^{2-} is a consequence of the smaller distance of the average of the bond valences (VO_4^{3-}: 1.25 v.u.; MoO_4^{2-}: 1.50 v.u.) to the limiting value of 1 v.u. The bond lengths of 15 further monovanadate structures [106, 120–133] do not fulfill the reliability criterion. A more comprehensive study of six monovanadate structures [101–103, 121, 122, 124] (see Table 37) revealed that up to 3.9% of the bond valence of the V atoms are directed to oxygen atoms of neighboring VO_4^{3-} ions and/or to molecules of water of crystallization. The largest individual bond valences of these sorts reach 0.07 v.u. (V ··· Ot′; 1.5%) [124]; 0.01 v.u. (V ··· Ow; 0.1% of the total bond valence) [102]. Neglect of these additional oxygen atoms leads to small errors for the M–O bond valences of up to +0.05 v.u. and for the charges of up to +0.04/−0.04 c.u. Contrary to our expectations, not one of the unreliable structures tested became reliable by this measure; see also Sect. 4.1. A large number of inorganic cations were reported.

The deviations from the ideal tetrahedral symmetry originate obviously from effects of the packing of cations, anions, and molecules of water of crystallization in the crystal.

3.5 The Dimetalate Ions $Mo_2O_7^{2-}$ and $V_2O_7^{4-}$

For the dimolybdate ion of the only known dimolybdate [134] (Fig. 3 and Table 3) the average Mo-Ot bond valence amounts to 1.63 v.u. (individual values from 1.61–1.66 v.u.), corresponding to a high charge on the terminal

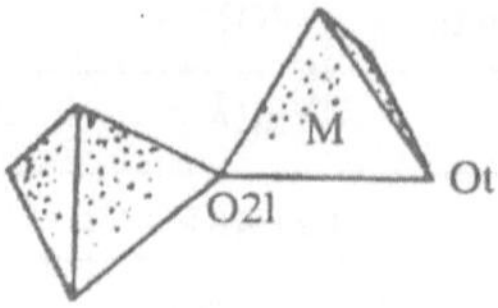

Fig. 3. Structure of the dimetalate ions $M_2^{VI}O_7^{2-}$ and $M_2^{V}O_7^{4-}$ and characterization of the crystallographically independent M and O atoms (the M-O-M angle may change and the MO_4 tetrahedra may rotate about the O2l-M axis).

oxygen atoms of −0.37 c.u. (individual values from −0.39 to −0.34 c.u.). The Mo-O2l bond valence of 1.11 v.u. corresponds to a charge on the approximately linearly bridging oxygen atom (Mo–O–Mo angle 154°) of +0.22 c.u. The cation is tetra-*n*-butylammonium. The influence of very long M-O bonds in this structure could not be studied because the atomic coordinates are not stated in the literature.

The divanadate ions in 18 reliable divanadate structures [135–148] (Fig. 3 and Table 4) show average V–Ot bond valences of 1.29–1.37 v.u., corresponding to a very high average charge on the terminal oxygen atoms of −0.71 to −0.63 c.u., and average V–O2l bond valences of 0.90–1.13 v.u., corresponding to a charge on the bridging oxygen atoms of −0.21 to +0.27 c.u. The bond lengths of three further divanadate structures [149–151] do not fulfill the reliability criterion. The cations are Na^+, Rb^+, Mg^{2+}, Ca^{2+}, Sr^{2+}, Ba^{2+}, Mn^{2+}, Co^{2+}, Ni^{2+}, Cu^{2+}, Zn^{2+}, and Cd^{2+}.

For the 14 divanadates of the *divalent* cations the negative or positive charge on the bridging oxygen atoms (−0.21 to +0.27 c.u.) and the valence of the

Table 3. Average bond lengths d and bond valences s of the M-O bonds and average charges c on the different types of oxygen atoms in $Mo_2O_7^{2-}$

M-O bond	O type frequency	M-O-M angle	d in Å	s[a] in v.u.	c in c.u.
M-Ot[b]	6×		1.716	1.631	−0.369
M-O2l M^i-O2l	1×	153.6°	1.876 1.876	1.108 1.108	+0.216
Sum of the negative charges:					−2.216 c.u.
Sum of the positive charges:					+0.216 c.u.
Total (formal ionic charge):					−2.000 c.u.

Bond valences about the M atoms:
M (2×): 1.631–1.631–1.631–1.108[c] v.u. (distorted MO_4 tetrahedra);
Σs_i = 6.000 v.u.

[a] d_{oi} = 1.9184 Å.
[b] Only averaged M-Ot bond lengths are given in the literature; individual values: 1.722 to 1.710 Å; 1.607 to 1.655 v.u.; −0.393 to −0.345 c.u.
[c] Bonds of bridging O atoms.

Table 4. Average bond lengths d and bond valences s of the M–O bonds, average charges c on the different types of oxygen atoms (in parentheses: O type frequency), M–O–M bond angles and conformation in 18 reliable structures of $V_2O_7^{4-}$

			M–Ot[b] (6×)			M–O2l (1×)		
Cation	V–O–V angle	Cf.[a]	d in Å	s[c] in v.u.	c in c.u.	d inÅ	s[c] in v.u.	c in c.u.
Rb^+	180.0°	s	1.675(22)	1.360(80)	−0.640(80)	1.819	0.920	−0.160
Na^+	149.4°	e	1.677(5)	1.353(20)	−0.647(20)	1.812	0.941	−0.118
Na^+	133.4°	s	1.680(12)	1.365(43)	−0.635(43)	1.831	0.906	−0.189
Na^+	125.6°	e	1.677(6)	1.367(24)	−0.633(24)	1.833	0.898	−0.204
Cd^{2+}	180.0°	s	1.702(21)	1.301(76)	−0.699(76)	1.764	1.098	+0.196[d]
Mn^{2+}	180.0°	s	1.688(11)	1.307(38)[d]	−0.693(38)	1.759	1.077	+0.155[d]
Zn^{2+}	149.3°	s	1.696(29)	1.314(103)	−0.686(103)	1.775	1.059	+0.117
Cu^{2+}	147.8°	s	1.697(36)	1.289(130)	−0.711(130)	1.743	1.133	+0.267[d]
Cu^{2+}	145.0°	s	1.714(62)	1.291(225)	−0.709(225)	1.758	1.128	+0.257
Mg^{2+}	140.6°	s	1.689(36)	1.339(132)[d]	−0.661(132)	1.801	0.984	−0.032[d]
Ca^{2+}	139.1°	e	1.685(22)	1.341(82)	−0.659(82)	1.802	0.976	−0.047
Ba^{2+}	125.6°	e	1.683(15)	1.356(54)	−0.644(54)	1.821	0.933	−0.134
	123.7°	e	1.685(14)	1.356(53)	−0.644(53)	1.824	0.931	−0.137
Sr^{2+}	123.0°	e	1.689(40)	1.348(144)	−0.652(144)	1.815	0.956	−0.088
	123.0°	e	1.706(30)	1.332(104)	−0.668(104)	1.810	1.003	+0.007
Pb^{2+}	122.1°	e	1.693(20)	1.347(72)	−0.653(72)	1.817	0.960	−0.080
Ni^{2+}	117.6°	e	1.692(27)	1.368(105)	−0.632(105)	1.847	0.897	−0.206
Co^{2+}	117.5°	e	1.695(25)	1.368(98)[d]	−0.632(98)	1.850	0.897	−0.207

[a] Conformation of the two VO_4 tetrahedra: e: eclipsed, s: staggered.
[b] The M–Ot bonds of the MO_4 tetrahedra have been averaged. In parentheses: standard deviation for these bonds. Smallest and highest individual values: 1.089 and 1.604 v.u., −0.911 and −0.396 c.u.
[c] $\bar{d}_{oi}$ = 1.7969 Å; $d_{oi\,min}$ = 1.7866 Å, $d_{oi\,max}$ = 1.8108 Å.
[d] Inclusion of the very long M···O bonds leads to the following values including changes > 0.03 v.u. or >0.03 c.u.: Cd^{2+}: O2l +0.168 c.u.; Mn^{2+}: M–Ot 1.271 v.u., O2l +0.094 c.u.; Co^{2+}: M–Ot 1.336 v.u.; Cu^{2+}: O2l +0.229 c.u.; Mg^{2+}: M–Ot 1.295 v.u., O2l −0.093 c.u.

bridging V–O bonds (0.90–1.13 v.u.) (as well as the length of the V–O–V bridges [70, 144] or bond valence sum about the bridging oxygen atoms [70]) depend on the V–O–V angle (117° to 180°) (Fig. 4(a)) which can be explained by the quality of the overlapping of the pπ orbitals of the oxygen atoms with the dπ orbitals of the metal atoms to give multicenter pπ-dπ multiple bonds. The conformation of the two VO_4 tetrahedra will certainly also play a rôle [70]; in general the staggered conformation occurs if the V–O–V angle is greater than ca. 140° whereas the eclipsed conformation occurs if the V–O–V angle is less than ca. 140° (see Table 4). The cations are mainly coordinated to the terminal oxygen atoms; if the V–O–V angle is less than ca. 140°, there are also interactions between cations and bridging oxygen atoms, as is indicated by the negatively charged bridging oxygen atoms. In the literature cation coordination to the bridging oxygen atoms is assumed for V–O–V angles <148° [144] and 117–118° [70].

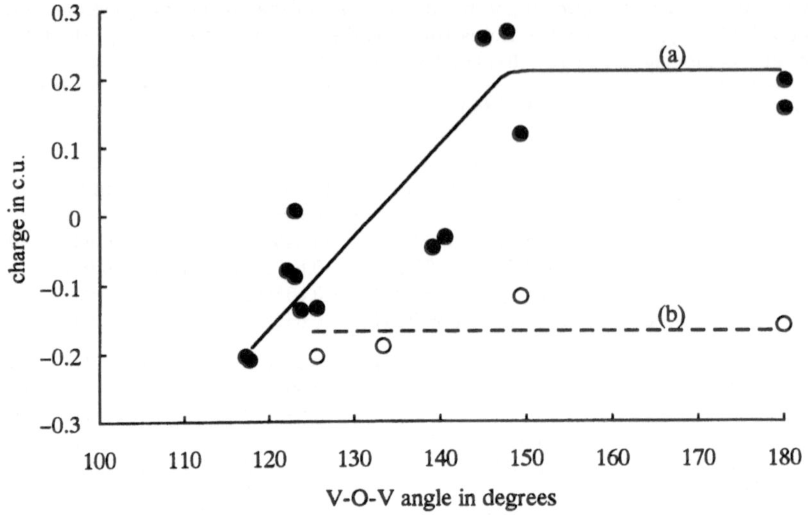

Fig. 4. Dependence of the charge on the bridging oxygen atoms of the $V_2O_7^{4-}$ ions on the V–O–V angle. (a) Divalent cations, (b) univalent cations.

A more comprehensive study of six of the divanadate structures in question [138, 142–145, 147] (see Table 37) revealed that up to 3.6% of the bond valence of the V atoms are directed to additional oxygen atoms of the same and/or of neighboring divanadate ions and/or of molecules of water of crystallization. The largest individual bond valences of these sorts – as far as it was possible to identify them – reaches 0.02 v.u. ($V^i \cdots Ot$; 0.4%); 0.17 v.u. ($V \cdots Ot''$; 3.4%); 0.01 v.u. ($V \cdots Ot'''$; 0.1%) [138]; 0.01 v.u. ($V \cdots Ow$; 0.2% of the bond valence of V) [139]. Neglect of these additional oxygen atoms leads to errors in the M–O bond valences of up to +0.05 v.u. and in the charges of up to +0.06/ −0.14 c.u. See also Sect. 4.1.

In the four divanadates of the *univalent* cations, which bear a medium negative charge of −0.20 to −0.12 c.u. on the bridging oxygen atoms and display comparatively long V–O distances in the bridge (0.90 to 0.94 v.u.), the range of the V–O–V angles is also large (180° to 126°). Nevertheless in this case there is no dependence between bond angle and charge on the oxygen bridge (Fig. 4(b)). This contrary behavior may be coupled with the fact that univalent cations are more suitable to coordinate with the bridging oxygen atoms of pyroanions [70].

In the dimolybdate and divanadate ions the ionic charge is mainly located on the terminal oxygen atoms, as expected. Occurrence of positive charge on the bridging oxygen atoms is, in accordance with our bond model for polyoxometalate ions forming chains or rings of corner-sharing MO_4 tetrahedra [23, 24] (see also footnote 8 in Ref. 16), a stabilizing factor for the polymeric structures through bond strengthening of inner (bridging) M–O bonds and is obviously produced by charge separation, see resonance structure (4b). Occurrence of negative charge on the bridging oxygen atoms appears to be less favorable, as resonance structure (4c) indicates: The polymeric character of the anion is destabilized through bond weakening of inner (bridging) M–O bonds.

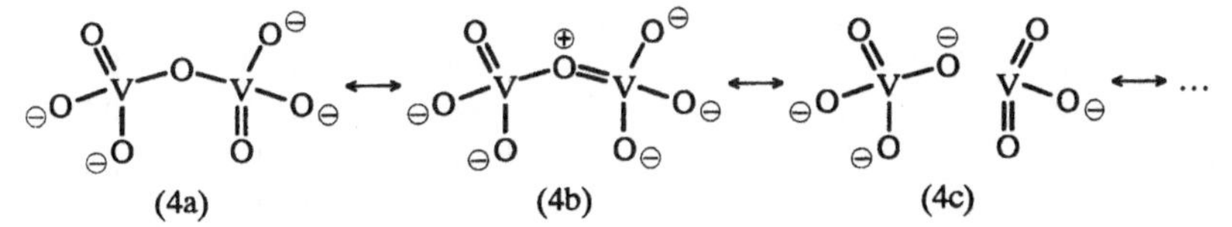
(4)

3.6 The Trivanadate Ion $V_3O_{10}^{5-}$

The chain-like trivanadate ion of the sodium salt [152] (Fig. 5, Table 5) resembles the divanadate ions of the salts of the univalent (alkali metal) cations. The outer VO_4 tetrahedra show average V1–Ot1 bond valences of 1.41 v.u. (individual values from 1.38 to 1.44 v.u.), corresponding to an average charge on the outer terminal oxygen atoms of −0.59 c.u. (individual values from −0.62 to −0.56 c.u.), and V1–O2l bond valences of 0.81 v.u. The inner VO_4 tetrahedron is characterized by V2–Ot2 bond valences of 1.45 v.u., corresponding to a charge on the inner terminal oxygen atoms of −0.55 c.u., and by V2–O2l bond valences of 1.00 v.u. The resulting charge on each of the bridging oxygen atoms is −0.19 c.u. As expected, the ionic charge is again mainly located on the terminal oxygen atoms. The bond lengths of another trivanadate [153] do not fulfill the reliability criterion.

A more comprehensive study (see Table 37) revealed that 0.9% of the bond valence of the V atoms are directed to additional oxygen atoms of the same trivanadate ion and of molecules of water of crystallization. The largest individual bond valence of these sorts reaches 0.01 v.u. (V2 ··· Ot1; 0.1%) and 0.01 v.u. (V1 ··· Ow; 0.2% of the bond valence of V). Neglect of these additional oxygen atoms leads to very small errors in the M–O bond valences of up to +0.01 v.u. and in the charges of up to +0.01 c.u. See also Sect. 4.1.

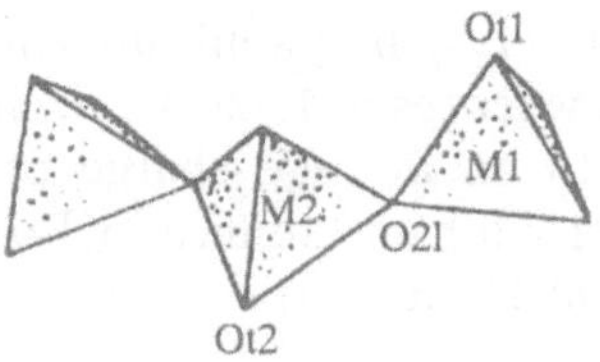

Fig. 5. Structure of the $V_3O_{10}^{5-}$ ion and characterization of the crystallographically independent M and O atoms (the M–O–M angles may change and the MO_4 tetrahedra may rotate about the O2l–M axes).

Table 5. Average bond lengths d and bond valences s of the M–O bonds and average charges c on the different types of oxygen atoms in $V_3O_{10}^{5-}$

M–O bond	O type frequency	M–O–M angle	d in Å	s[a] in v.u.	c in c.u.
M1–Ot1[b]	6×		1.669	1.413	−0.587
M2–Ot2[c]	2×		1.659	1.451	−0.549
M1–O2l M2–O2l	2×	122.8°	1.876 1.795	0.807 1.004	−0.189
Sum of the negative charges (formal ionic charge):					−5.000 c.u.

Bond valences about the different M atoms:
M1 (2×): 1.413–1.413–1.413–0.807[d] v.u. (distorted MO_4 tetrahedra);
$\Sigma s_i = 5.044$ v.u.
M2 (1×): 1.451–1.451–1.004[d]–1.004[d] v.u. (distorted MO_4 tetrahedron);
$\Sigma s_i = 4.911$ v.u.
Average Σs_i for the MO_4 tetrahedra: 5.000 v.u.

[a] d_{oi} = 1.7966 Å.
[b] The M1–Ot bonds of these MO_4 tetrahedra have been averaged; individual values: 1.678, 1.668, 1.662 Å; 1.379, 1.416, 1.440 v.u.; −0.621, −0.584, −0.560 c.u.
[c] The M2–Ot bond lengths of this MO_4 tetrahedron do not differ.
[d] Bonds of bridging O atoms.

3.7 The Metavanadate Ion $[VO_3^-]_\infty$

Metavanadate ions $[VO_3^-]_q$ (the metavanadate stage corresponds to $Z^+ = 2$[5]) occur in three structure types: rings (q = 3?, 4, 5, 6? [24, 154–156]) of corner-sharing VO_4 tetrahedra; chains ($q = \infty$) of corner-sharing VO_4 tetrahedra; double chains ($q = \infty$) of edge-sharing VO_5 trigonal bipyramids. The bond lengths of a tetrameric ring [154] do not fulfill the reliability criterion. The chains of corner-sharing VO_4 tetrahedra are treated in this section and the double chains of edge-sharing VO_5 trigonal bipyramids in Sect. 3.24.

Reliable structural data of twelve alkali metal or mixed alkali metal salts of the tetrahedral metavanadate chains [157–161] are known (Fig. 6, Table 6). The bond lengths of two further investigations [158, 162] do not fulfill the reliability criterion. Salts of divalent cations are not described. All VO_4 tetrahedra of the chains are crystallographically equivalent.

The average V–Ot bond valences of 1.52(1) v.u. (individual values from 1.44 to 1.59 v.u.) correspond to an average charge of −0.48(1) c.u. (individual values from −0.56 to −0.41 c.u.) on the terminal oxygen atoms. The average V–O2l bond valences of 0.98(1) v.u. correspond to a very small average charge

[5] Z^+ is (at least formally) the ratio of the stoichiometric coefficients of H^+ and $M^nO_4^{w-}$ in the gross equation of formation for an isopolyoxometalate ion.

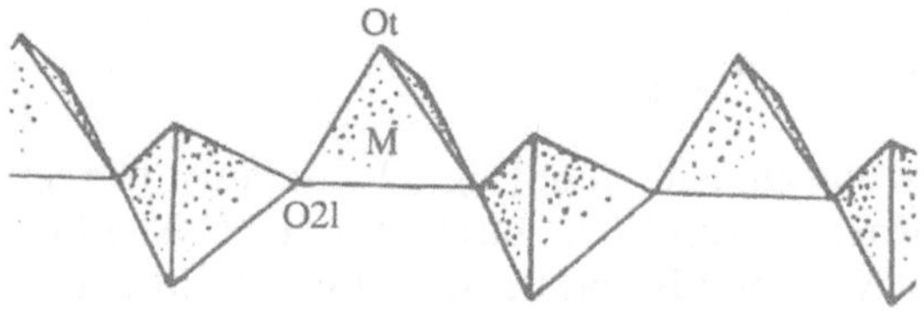

Fig. 6. Structure of the $[VO_3^-]_\infty$ ion and characterization of the crystallographically independent M and O atoms (the M–O–M angles may change and the MO_4 tetrahedra may rotate about the O2l–M axes).

of −0.04(1) c.u. on the bridging oxygen atoms. The V–O–V angles of the oxygen bridges range (as far as stated) from 141° to 147°.

A more comprehensive study of one of the investigations [159] (see Table 37) revealed that 0.9% of the bond valence of the V atoms are directed to additional oxygen atoms of the same and of neighboring metavanadate chains, the greatest individual bond valences of these reaching 0.01 v.u. (V···Ot″; 0.1%) and 0.01 v.u. (V···Ot‴; 0.2% of the bond valence of V). Neglect of the additional oxygen atoms leads to very small errors in the M–O bond valences of up to +0.01 v.u. and in the charges of up to +0.01/−0.01 c.u. See also Sect. 4.1.

The cation–O(bridge) distances point to a cation coordination including the bridging oxygen atoms [157–162]. However, according to our above calculations the corresponding bond valence amounts for *all* cation–O(bridge) bonds together to 0.04 v.u. only.

Table 6. Average bond lengths d and bond valences s of the M–O bonds and average charges c on the different types of oxygen atoms in $[VO_3^-]_\infty$ (average of twelve reliable structural investigations)

M–O bond	O type frequency	M–O–M angle	d in Å	s^a in v.u.	c in c.u.
$M\text{–}Ot^b$	2×		1.641(4)	1.521(6)	−0.479(6)
M–O2l	1×	140.6–147.2°[c]	1.804(2)	0.979(6)	−0.043(12)
M^i–O2l			1.804(2)	0.979(6)	
Sum of the negative charges (formal ionic charge):					−1.001 c.u.

Bond valences about the M atoms:
M (1×): $1.521–1.521–0.979^d–0.979^d$ v.u. (distorted MO_4 tetrahedra);
Σs_i = 5.000 v.u.

[a] $\bar{d}_{oi}$ = 1.7957 Å; $d_{oi\,min}$ = 1.7913 Å, $d_{oi\,max}$ = 1.7992 Å.
[b] The M–Ot bonds of the MO_4 tetrahedra have been averaged; extreme individual values: 1.662, 1.625 Å; 1.437, 1.589 v.u.; −0.563, −0.411 c.u.
[c] As far as given in the literature.
[d] Bonds of bridging O atoms.

3.8
The Pentavanadate Ion $V_5O_{14}^{3-}$

The cage-like pentavanadate ion (Fig. 7) contains two crystallographically independent, corner-sharing VO_4 tetrahedra [163]. The bond valences of the different types of V–O bonds and the distribution of the ionic charge over the different types of oxygen atoms are given in Table 7. The cation is tetra-*n*-butylammonium. Referred to the small d_{oi} (= 1.776 Å), the structure belongs actually to the unreliable ones; however, the bond lengths are rather uniformly shortened and hence the standard deviation for the sum of the bond valences about the M atoms does indeed exceed the $2\sigma_{\Sigma}$ but not the $2\sigma_{\Sigma i}$ limit (this is the only case of this kind observed in this investigation).

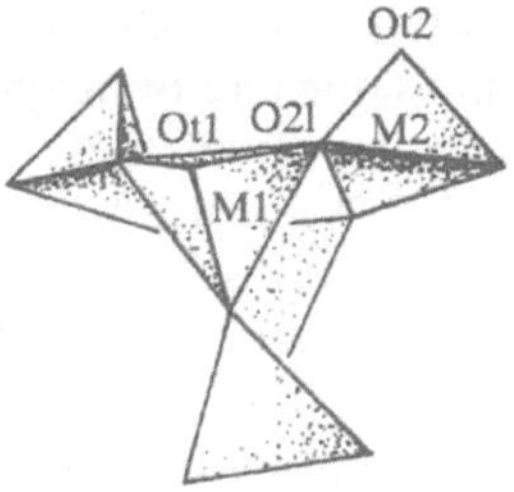

Fig. 7. Structure of the $V_5O_{14}^{3-}$ ion and characterization of the crystallographically independent M and O atoms.

Table 7. Average bond lengths d and bond valences s of the symmetry-equivalent M–O bonds and average charges c on the different types of oxygen atoms in $V_5O_{14}^{3-}$

M–O bond	O type frequency	M–O–M angle	d in Å	s^a in v.u.	c in c.u.
M1–Ot1	2×		1.580	1.679	−0.321
M2–Ot2	6×		1.601	1.586	−0.414
M1–O21 M2–O21	6×	143.3°	1.723 1.818	1.140 0.881	+0.021
Sum of the negative charges:					−3.126 c.u.
Sum of the positive charges:					+0.126 c.u.
Total (formal ionic charge):					−3.000 c.u.

Bond valences about the different M atoms:
M1 (2×): $1.679–1.140^b–1.140^b–1.140^b$ v.u. (distorted MO_4 tetrahedra);
Σs_i = 5.098 v.u.
M2 (3×): $1.586–1.586–0.881^b–0.881^b$ v.u. (distorted MO_4 tetrahedra);
Σs_i = 4.934 v.u.
Average Σs_i for the MO_4 tetrahedra: 5.000 v.u.

[a] d_{oi} = 1.7756 Å.
[b] Bonds of bridging O atoms.

The eight terminal oxygen atoms (two sorts) are highly negatively charged (−0.32 and −0.41 c.u., altogether −3.13 c.u., which is more than the formal ionic charge). The six approximately linearly bridging (143°) two-coordinate oxygen atoms (one sort) are insignificantly positively charged (+0.02 c.u.).

A more comprehensive study (see Table 37) revealed that 0.3% of the bond valence of the V atoms are directed to additional oxygen atoms of the same $V_5O_{14}^{3-}$ ion, the greatest individual bond valence of these reaching less than 0.005 v.u. ($V1^{i} \cdots O21$; 0.1% of the bond valence of V). Neglect of the additional oxygen atoms leads to errors in the M–O bond valences of less than +0.005 v.u. and in the charges of less than ±0.005 c.u. See also Sect. 4.1.

3.9 The Heptamolybdate (Paramolybdate) Ion $Mo_7O_{24}^{6-}$

The heptamolybdate ion (Fig. 8) contains three crystallographically independent, edge-sharing MoO_6 octahedra of type II. The bond valences of the different types of Mo–O bonds and the distribution of the ionic charge over the different types of oxygen atoms are given in Table 8 (average of eight [53, 164–168] reliable crystal structure determinations). The cations are Na^+, K^+, ammonium, *n*-propylammonium, isopropylammonium, 2-aminopyridinium, 4-aminopyridinium, and guanidinium.

The twelve terminal oxygen atoms (four sorts) bear most of the ionic charge (−0.35(4), −0.43(5), −0.38(4), and −0.42(5), altogether −4.76 c.u.), the two pseudoterminal, angularly bridging two-coordinate oxygen atoms (one sort) −0.24(4) c.u. The other six angularly bridging two-coordinate oxygen atoms (two sorts) bear −0.15(4) and −0.08(4) c.u. and the two angularly bridging three-coordinate oxygen atoms (one sort) −0.09(5) c.u. The two four-coordinate oxygen atoms (one sort), involved in (approximately) linear Mo–O–Mo bridges, bear a small but definite positive charge, +0.09(3) c.u.

A more comprehensive study of one of the heptamolybdate structures [166] showed that the errors due to the neglect of the very long $M \cdots O$ bonds are very small, see Sect. 4.1, in particular Table 37.

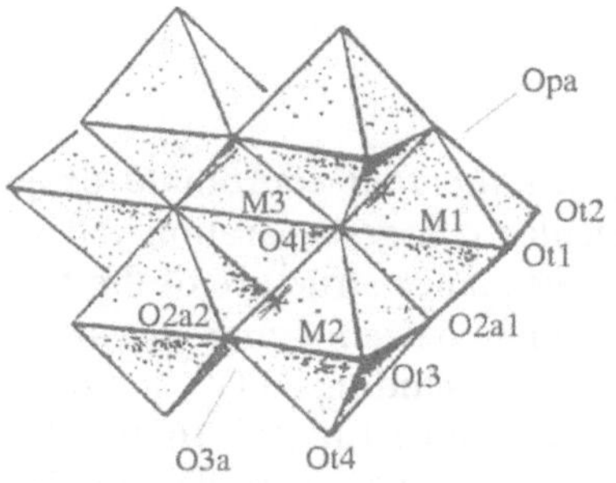

Fig. 8. Structure of the $M_7^{VI}O_{24}^{6-}$ ions and characterization of the crystallographically independent M and O atoms.

Table 8. Average bond lengths d and bond valences s of the symmetry-equivalent M–O bonds and average charges c on the different types of oxygen atoms in $Mo_7O_{24}^{6-}$ (average of eight reliable structural investigations)

M–O bond	O type frequency	d in Å	s^a in v.u.	c in c.u.
M1–Ot1	2×	1.708(10)	1.653(44)	−0.347(44)
M1–Ot2	2×	1.728(12)	1.569(49)	−0.431(49)
M2–Ot3	4×	1.717(13)	1.617(41)	−0.383(41)
M2–Ot4	4×	1.726(16)	1.580(51)	−0.420(51)
M3–Opa M1–Opa*	2×	1.739(8) 2.524(42)	1.533(29) 0.232(24)	−0.236(38)
M1–O2a1 M2–O2a1	4×	1.922(12) 1.976(19)	0.985(23) 0.867(47)	−0.148(40)
M2–O2a2 $M2^i$–O2a2	2×	1.934(6) 1.934(6)	0.958(19) 0.958(19)	−0.085(39)
M3–O3a M2–O3a* $M2^i$–O3a*	2×	1.896(14) 2.267(26) 2.267(26)	1.049(30) 0.430(30) 0.430(30)	−0.091(48)
M1–O4l* M2–O4l* $M2^{ii}$–O4l* M3–O4l*	2×	2.159(12) 2.162(12) 2.162(12) 2.262(13)	0.558(22) 0.551(13) 0.551(13) 0.433(15)	+0.092(31)
Sum of the negative charges:				−6.184 c.u.
Sum of the positive charges:				+0.184 c.u.
Total (formal ionic charge):				−6.000 c.u.

Bond valences about the different M atoms:
M1 (2×): 1.653–1.569–0.985–0.985–0.232^p–0.558 v.u. (type II MO_6 octahedra); Σs_i = 5.982 v.u.
M2 (4×): 1.617–1.580–0.867–0.958–0.430–0.551 v.u. (type II MO_6 octahedra); Σs_i = 6.003 v.u.
M3 (1×): 1.533^p–1.533^p–1.049–1.049–0.433–0.433 v.u. (type II MO_6 octahedron); Σs_i = 6.030 v.u.
Average Σs_i for the MO_6 octahedra: 6.001 v.u.

[a] $\bar{d}_{oi} = 1.9151$ Å; $d_{oi\,min} = 1.9087$ Å, $d_{oi\,max} = 1.9202$ Å.
[p] The strong bonds are pseudoterminal M–O bonds, the weak bonds are the complementary bonds of the pseudoterminal oxygen atoms.

The positive charge on the four-coordinate oxygen atoms is obviously produced by intramolecular charge separation (cf. resonance scheme (7) in Ref. 16). Remarkably, a resonance structure of the $M_7O_{24}^{6-}$ ion with a negative charge on the four-coordinate oxygen atoms cannot be formulated unless one accepts resonance structures including an O^{2-} ion or violation of coordination mode (V), see Sects. 3.1, 3.4, and 4.1 in Ref. 16. In this context the weakness of the M1–Opa bonds (0.23 v.u.) has been explained [16].

The bond lengths ascertained for the analogous $W_7O_{24}^{6-}$ ion [169, 170] do not fulfill the reliability criterion.

3.10
The (β-)Octamolybdate (Metamolybdate) Ion (β-)$Mo_8O_{26}^{4-}$

The (β-)octamolybdate ion (Fig. 9) contains three crystallographically independent, edge-sharing MoO_6 octahedra of type II. The bond valences of the different types of Mo–O bonds and the distribution of the ionic charge over the different types of oxygen atoms are given in Table 9 (average of 18 [42, 171–185] reliable crystal structure determinations). The cations are ammonium, alkylammonium, anilinium, substituted anilinium, substituted pyridinium, and $(C_6H_5)_3PCH_2COOC_2H_5^+$ besides diethylammonium.

The 14 terminal oxygen atoms (five sorts) bear most of the ionic charge (−0.30(4), −0.32(4), −0.32(2), −0.31(3), and −0.28(3), altogether −4.32 c.u., which is more than the formal ionic charge of -4 c.u.), the two pseudoterminal, angularly bridging two-coordinate oxygen atoms (one sort) −0.09(3) c.u. The other angularly bridging oxygen atoms bear a small but definite positive charge, the four two-coordinate ones (one sort) +0.05(3) and the four three-coordinate ones (one sort) +0.11(2) c.u. The two five-coordinate oxygen atoms (one sort), involved in (approximately) linear Mo-O-Mo bridges, bear a small but definite negative charge, −0.07(3) c.u.

A more comprehensive study of one of the (β-)octamolybdate structures [184] showed that the errors due to the neglect of the very long M···O bonds are very small, see Sect. 4.1, in particular Table 37.

A positive charge on oxygen atoms of angular M–O–M bridges (the two- and three-coordinate oxygen atoms) and a negative charge on oxygen atoms involved in linear M–O–M bridges (the five-coordinate oxygen atoms) are very uncommon (see the results for the other M^{VI} structures, Table 41, and Sect. 4.4.1). However, in the present case of the (β-)octamolybdate ion a resonance structure with a negative charge on the high(five)-coordinate oxygen atoms can indeed be formulated (which is not possible for, e.g., the four-coordinate oxygen atoms of $M_7O_{24}^{6-}$ and the six-coordinate oxygen atom of $M_6O_{19}^{2-}$ – unless one accepts resonance structures including an O^{2-} ion or violation of the above-mentioned coordination mode (V)), see Fig. 1(e) in Ref. 16. The positive charge on the angular Mo–O–Mo bridges is obviously produced by intramolecular charge separation and causes on the other hand

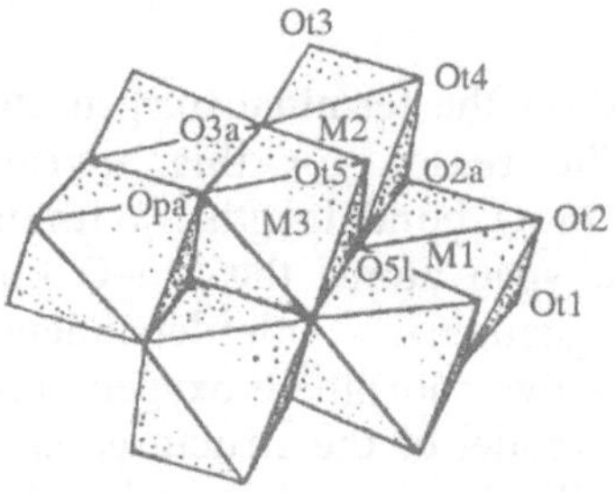

Fig. 9. Structure of the (β-)$Mo_8O_{26}^{4-}$ ion and characterization of the crystallographically independent M and O atoms.

Table 9. Average bond lengths d and bond valences s of the symmetry-equivalent M–O bonds and average charges c on the different types of oxygen atoms in $(\beta\text{-})Mo_8O_{26}^{4-}$ (average of 16 reliable structural investigations)

M–O bond	O type frequency	d in Å	s^a in v.u.	c in c.u.
M1–Ot1	2×	1.699(7)	1.703(37)	−0.297(37)
M1–Ot2	2×	1.704(12)	1.682(37)	−0.318(37)
M2–Ot3	4×	1.704(6)	1.682(22)	−0.318(22)
M2–Ot4	4×	1.703(10)	1.686(28)	−0.314(28)
M3–Ot5	2×	1.695(8)	1.719(25)	−0.281(25)
M3–Opa	2×	1.752(7)	1.497(22)	−0.089(25)
$M1^{i}$–Opa*		2.285(19)	0.414(20)	
M1–O2a	4×	1.925(5)	0.986(8)	+0.054(31)
M2–O2a		1.892(10)	1.068(31)	
M2–O3a		2.000(15)	0.823(14)	
M3–O3a	4×	1.950(6)	0.928(12)	+0.107(21)
$M2^{ii}$–O3a*		2.347(16)	0.356(12)	
M1–O5l*		2.452(42)	0.276(22)	
M2–O5l*		2.323(14)	0.377(12)	
$M2^{i}$–O5l*	2×	2.323(14)	0.377(12)	−0.074(26)
M3–O5l*		2.152(15)	0.570(25)	
$M3^{i}$–O5l*		2.384(26)	0.326(21)	
Sum of the negative charges:				−4.646 c.u.
Sum of the positive charges:				+0.644 c.u.
Total (formal ionic charge):				−4.002 c.u.

Bond valences about the different M atoms:
M1 (2×): 1.703–1.682–0.986–0.986–0.414^{p}–0.276 v.u. (type II MO_6 octahedra); $\Sigma s_i = 6.047$ v.u.
M2 (4×): 1.682–1.686–1.068–0.823–0.377–0.356 v.u. (type II MO_6 octahedra); $\Sigma s_i = 5.992$ v.u.
M3 (2×): 1.719–1.497^{p}–0.928–0.928–0.570–0.326 v.u. (type II MO_6 octahedra); $\Sigma s_i = 5.968$ v.u.
Average Σs_i for the MO_6 octahedra: 6.000 v.u.

[a] $\bar{d}_{oi} = 1.9186$ Å; $d_{oi\,min} = 1.9118$ Å, $d_{oi\,max} = 1.9259$ Å.
[p] The strong bonds are pseudoterminal M–O bonds, the weak bonds are the complementary bonds of the pseudoterminal oxygen atoms.

the high negative charge on the terminal oxygen atoms, which exceeds the formal ionic charge. The reason for this appearance is obviously the interdependence of the M–O bond lengths: Without a positive charge on the angularly bridging oxygen atoms the Mo–O bonds involved would be somewhat longer. The geometrical interdependence of the M–O bonds requires that those of the five-coordinate oxygen atoms must become longer too (see a ball-and-stick model of the structure, e.g., Fig. 66 in Ref. 186a or Fig. 1(e)–(f) in Ref. 16) with the consequence that the five-coordinate oxygen atoms would be still more negatively charged. This means that for the (β-)octamolybdate structure there is in principle no possibility for a charge on the five-coordinate oxygen atoms (which are involved in approximately linear M–O–M bridges and therefore favor a positive charge) that is more positive than the charge on the two- and three-coordinate oxygen atoms forming angular M–O–M bridges (which prefer a small negative charge).

3.11 The Hexametalate Ions $Mo_6O_{19}^{2-}$, $W_6O_{19}^{2-}$, $Nb_6O_{19}^{8-}$, and $Ta_6O_{19}^{8-}$

In the ideal case all six edge-sharing type I MO_6 octahedra of the hexametalate ions (Fig. 10) are crystallographically equivalent. The relevant structural data are given in Tables 10–13. They are based on seven [184, 187–191] reliable crystal structure determinations on hexamolybdates (cations: $HN_3P_3[N(CH_3)_2]_6^+$, triphenylbenzylphosphonium, tetrakis(diethyldithiocarbamato) molybdenum(V), K(18-crown-6)$^+$, $(C_6H_5)_3PCH_2COOC_2H_5^+$, tetra-*n*-butylammonium, tetraphenylarsonium), three [192, 193] on hexatungstates (cation: tetrabutylammonium), one [194] on a hexaniobate (cation: Na^+ combined with H_3O^+), and one [13b] on a hexatantalate (cation not stated). The bond lengths of two further hexatungstates (cations: $(t\text{-}C_4H_9NC)_7W^{2+}$ and not stated, respectively) [195, 196] do not fulfill the reliability criterion.

In the $M_6^{VI}O_{19}^{2-}$ ions the six terminal oxygen atoms (one sort) bear most of the ionic charge (hexamolybdate ion: −0.25(2), altogether −1.48 c.u.; hexatungstate ion: −0.27(2), altogether −1.63 c.u.). The twelve angularly bridging, two-coordinate oxygen atoms (one sort) bear a very small negative charge only (hexamolybdate ion: −0.06(1) c.u.; hexatungstate ion: −0.04(1) c.u.). The central, six-coordinate oxygen atoms, involved in linear M–O–M bridges, bear a rather high positive charge (hexamolybdate ion: +0.25(1) c.u.; hexatungstate ion: +0.18(1) c.u.).

A more comprehensive study of one of the hexamolybdate structures [184] showed that the errors due to the neglect of the very long M···O bonds are very small, see Sect. 4.1, in particular Table 37.

The occurrence of a positive charge on the central oxygen atom is in direct contrast to the frequently [9, 13b, 197, 198] expressed opinion that this oxygen atom can be viewed as an O^{2-} ion surrounded by a neutral (= uncharged)

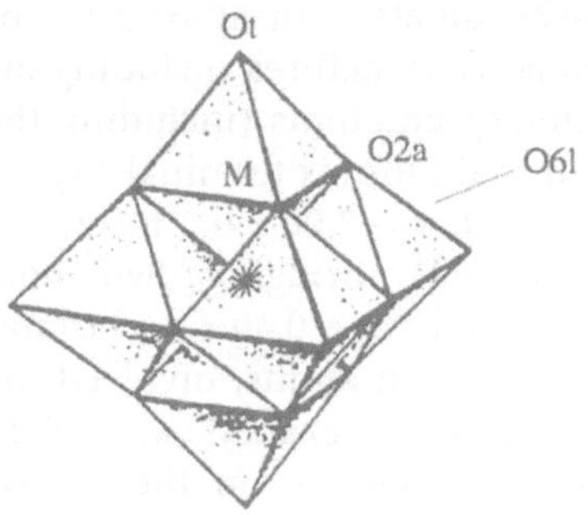

Fig. 10. Structure of the $M_6^{VI}O_{19}^{2-}$ and $M_6^{V}O_{19}^{8-}$ ions and characterization of the crystallographically independent M and O atoms.

Table 10. Average bond lengths d and bond valences s of the symmetry-equivalent M-O bonds and average charges c on the different types of oxygen atoms in $Mo_6O_{19}^{2-}$ (average of seven reliable structural investigations)

M-O bond[a]	O type frequency	d in Å	s^b in v.u.	c in c.u.
M-Ot	6×	1.679(4)	1.753(23)	−0.247(23)
M-O2a[c]	12×	1.924(5)	0.968(6)	−0.064(11)
M^i-O2a		1.924(5)	0.968(6)	
M-O6l*	1×	2.317(3)	0.375(2)	+0.250(12)
M^i-O6l*		2.317(3)	0.375(2)	
M^{ii}-O6l*		2.317(3)	0.375(2)	
M^{iii}-O6l*		2.317(3)	0.375(2)	
M^{iv}-O6l*		2.317(3)	0.375(2)	
M^v-O6l*		2.317(3)	0.375(2)	
Sum of the negative charges:				−2.250 c.u.
Sum of the positive charges:				+0.250 c.u.
Total (formal ionic charge):				−2.000 c.u.

Bond valences about the M atoms:
M (6×): 1.753–0.968–0.968–0.968–0.968–0.375 v.u. (type I MO_6 octahedra);
Σs_i = 6.000 v.u.

[a] Since for some of the investigations only averaged bond lengths are given, the bond valences for all structures have been calculated on the basis of averaged bond lengths. A model calculation shows that the bond valences and charges obtained in this way differ by less than 0.01 v.u. and 0.01 c.u., respectively, from those obtained by the usual procedure.
[b] $\bar{d}_{oi}$ = 1.9108 Å; $d_{oi\,min}$ = 1.9075 Å, $d_{oi\,max}$ = 1.9147 Å.
[c] In some of the structures M-O2a and M^i-O2a are rather different (up to 0.146 Å or 0.34 v.u.).

M_6O_{18} cage and therefore bears the entire ionic charge of the polyanion. The statements about the charge on the bridging and terminal oxygen atoms and about the basicity (ease of protonation) sequence of the various types of oxygen atoms in those papers [9, 10, 13b] also contradict the above results. Remarkably, a resonance structure of the $M_6O_{19}^{2-}$ ion fulfilling the condition (V) (or conditions (VI) and (VII)) described in Ref. 16 with a negative charge on the high(six)-coordinate, central oxygen atom cannot be formulated (unless one accepts resonance structures including an O^{2-} ion).

In the $M_6^VO_{19}^{8-}$ ions all oxygen atoms (including the O6l atoms) accept the negative ionic charge, but again the six terminal oxygen atoms bear the highest (hexaniobate: −0.50, altogether −3.00 c.u.; hexatantalate: −0.61, altogether −3.67 c.u.), the twelve angularly bridging, two-coordinate oxygen atoms a lower (but still high) (hexaniobate: −0.40 c.u.; hexatantalate: −0.34 c.u.), and the central, six-coordinate oxygen atoms, involved in linear M–O–M bridges, the least (hexaniobate: −0.25 c.u.; hexatantalate: −0.27 c.u.) negative charge.

The occurrence of negative charge on the central oxygen atoms of the hexaniobate and -tantalate ions is not seen in contradiction to the above statement that a resonance structure of the $M_6O_{19}^{n-}$ ion with a negative charge on the six-coordinate, central oxygen atom cannot be formulated. This

Table 11. Average bond lengths d and bond valences s of the symmetry-equivalent M–O bonds and average charges c on the different types of oxygen atoms in $W_6O_{19}^{2-}$ (average of three reliable structural investigations)

M–O bond	O type frequency	d in Å	s[a] in v.u.	c in c.u.
M–Ot	6×	1.689(6)	1.728(21)	−0.272(21)
M–O2a	12×	1.924(3)	0.978(5)	−0.045(10)
M^i–O2a		1.924(3)	0.978(5)	
M–O6l*	1×	2.331(4)	0.363(2)	+0.176(10)
M^i–O6l*		2.331(4)	0.363(2)	
M^{ii}–O6l*		2.331(4)	0.363(2)	
M^{iii}–O6l*		2.331(4)	0.363(2)	
M^{iv}–O6l*		2.331(4)	0.363(2)	
M^v–O6l*		2.331(4)	0.363(2)	
Sum of the negative charges:				−2.172 c.u.
Sum of the positive charges:				+0.176 c.u.
Total (formal ionic charge):				−1.996 c.u.

Bond valences about the M atoms:
M (6×): 1.728–0.978–0.978–0.978–0.978–0.363 v.u. (type I MO_6 octahedra);
Σs_i = 6.001 v.u.

[a] $\bar{d}_{oi}$ = 1.9138 Å; $d_{oi\,min}$ = 1.9100 Å, $d_{oi\,max}$ = 1.9164 Å.

Table 12. Average bond lengths d and bond valences s of the symmetry-equivalent M–O bonds and average charges c on the different types of oxygen atoms in $Nb_6O_{19}^{8-}$

M–O bond	O type frequency	d in Å	s[a] in v.u.	c in c.u.
M–Ot	6×	1.771	1.500	−0.500
M–O2a	12×	2.005	0.802	−0.396
M^i–O2a		2.005	0.802	
M–O6l*	1×	2.381	0.292	−0.248
M^i–O6l*		2.381	0.292	
M^{ii}–O6l*		2.381	0.292	
M^{iii}–O6l*		2.381	0.292	
M^{iv}–O6l*		2.381	0.292	
M^v–O6l*		2.381	0.292	
Sum of the negative charges (formal ionic charge):				−8.002 c.u.

Bond valences about the M atoms:
M (6×): 1.500–0.802–0.802–0.802–0.802–0.292 v.u. (type I MO_6 octahedra);
Σs_i = 5.000 v.u.

[a] d_{oi} = 1.9222 Å.

statement refers to metalate ions that need protonation for the extension of the tetrahedral to an octahedral coordination of the metal atom (M = Mo^{VI}, W^{VI}, V^V) [2, 4, 19]. For M = Nb^V and Ta^V it is supposed [13m, 18, 20, 199–201, 202a] that these elements occur always in the octahedral coordination state (or at

Table 13. Average bond lengths d and bond valences s of the symmetry-equivalent M–O bonds and average charges c on the different types of oxygen atoms in $Ta_6O_{19}^{8-}$

M–O bond	O type frequency	d^a in Å	s^b in v.u.	c in c.u.
M–Ot	6×	1.80	1.389	−0.611
M–O2a	12×	1.99	0.830	−0.339
M^i–O2a		1.99	0.830	
M–O6l*	1×	2.38	0.289	−0.268
M^i–O6l*		2.38	0.289	
M^{ii}–O6l*		2.38	0.289	
M^{iii}–O6l*		2.38	0.289	
M^{iv}–O6l*		2.38	0.289	
M^v–O6l*		2.38	0.289	
Sum of the negative charges (formal ionic charge):				−8.002 c.u.

Bond valences about the M atoms:
M (6×): 1.389–0.830–0.830–0.830–0.830–0.289 v.u. (type I MO_6 octahedra);
$\Sigma s_i = 5.000$ v.u.

[a] Only averaged bond lengths are given in the literature.
[b] $d_{oi} = 1.9214$ Å.

least with a coordination number >4) with oxygen, that means resonance formula (5c) in resonance scheme (5), which does not fulfill the condition of

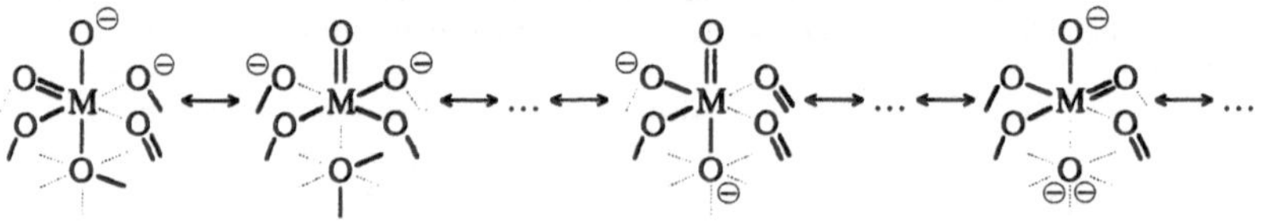

(5a) (5b) (5c) (5d) (5)

the elongation of the "beam" of the seesaw (cf. Scheme (V) in Ref. 16), can explain the negative charge on the central, six-coordinate oxygen atom[6]. Another possibility could be resonance formula (5d). The reason why only

[6]The statement that Nb^V and Ta^V occur always with octahedral oxygen coordination is based on the sequence of the octahedral radii of the M^{n+} transition metal cations (see, e.g., Ref. 13c: V^{5+} 0.68 Å, Mo^{6+} 0.73 Å, W^{6+} 0.74 Å, Nb^{5+} 0.78 Å, Ta^{5+} 0.78 Å; see also Ref. 14d). In the case of V^{5+}, Mo^{6+}, and W^{6+} the protonation of the tetrahedral oxometalate ions leads to an elongation of the corresponding M–O bond(s), and only then is the extension of the tetrahedral coordination sphere to an octahedral one possible [4, 17–19]. However, the octahedral radii of the M^{n+} transition metal cations are merely to a first, very rough approximation decisive for the occurrence of the octahedral coordination. The M–O bond *lengths* are the important factor for the extension of the tetrahedral coordination sphere, and these are also determined by the valence of the M–O bonds [e.g., for the unprotonated M–O bonds it is 1.67 v.u. in $(HO)M^{VI}O_3^-$, 1.33 v.u. in $(HO)VO_3^{2-}$, 1.5 v.u. in $(HO)_2VO_2^-$, and for the protonated M–O bonds 1 v.u. in each case]; see Sect. 4.6.1. There exist still other factors that have a diversifying influence on the extension of the coordination sphere of the M atoms, for instance the enhanced tendency of complex formation of W^{VI} as compared to Mo^{VI} [2, 4, 5] or of Ta^V and Nb^V as compared to V^V due to more suitable (more extended) d orbitals on the M atoms [203, 204] or the negative ionic charge of the monomeric building units [14n, 14q, 24, 205].

somewhat less than 50% of the ionic charge is located on the terminal oxygen atoms must be seen in the high charge (−8 c.u.) and small number of terminal (six), but large number of bridging (13) oxygen atoms; additionally, resonance structures with negative charge on bridging O atoms cannot be avoided. Moreover, the placement of the large number of cations (eight) requires a correspondingly large number of charged oxygen atoms to fulfill a detailed charge balance.

For the non-occurrence of a corresponding $V_6O_{19}^{8-}$ hexavanadate ion two reasons have been named in the literature: First, $V_6O_{19}^{8-}$ requires for its formation only 10 H^+ per 6 VO_4^{3-} and hence there is not the minimum of 2 H^+/VO_4^{3-} which is necessary for the expansion of the coordination sphere of the VO_4 tetrahedra [14q, 20, 24, 205] ($Nb_6O_{19}^{8-}$ and $Ta_6O_{19}^{8-}$ do not need 2 H^+/MO_4^{3-} for their formation, see the preceding paragraph; see also Sect. 4.6.1, this article, and Sect. 3.1 in Ref. 16). Secondly, under the pH conditions of a solution with 2 H^+/VO_4^{3-} a hypothetical $V_6O_{19}^{8-}$ or $H_2V_6O_{19}^{6-}$ would be so basic and hence become so strongly protonated that it would directly undergo further aggregation [24] ($Nb_6O_{19}^{8-}$ and $Ta_6O_{19}^{8-}$ form, due to the greater tendency of Nb^V and Ta^V for complex formation and for acceptance of an octahedral coordination - compare also footnote 6 - at much higher pH values where it is not necessary to be protonated).

3.12
The Decametalate Ions $V_{10}O_{28}^{6-}$ and $Nb_{10}O_{28}^{6-}$

These decametalate ions (Fig. 11) contain three crystallographically independent, edge-sharing MO_6 octahedra of type I and type II. The relevant structural data are given in Tables 14 and 15. They are based on seven reliable crystal structure determinations on decavanadates (cations: Na^+, K^+ and Zn^{2+}, Ca^{2+}, La^{3+}, Nd^{3+}, Er^{3+}, and Yb^{3+}) [206–212] and two on decaniobates (cations: tetramethylammonium and tetraethylammonium in combination with Na^+) [213]. The bond lengths of a further decavanadate structure (cation: Y^{3+}) [214] do not fulfill the reliability criterion.

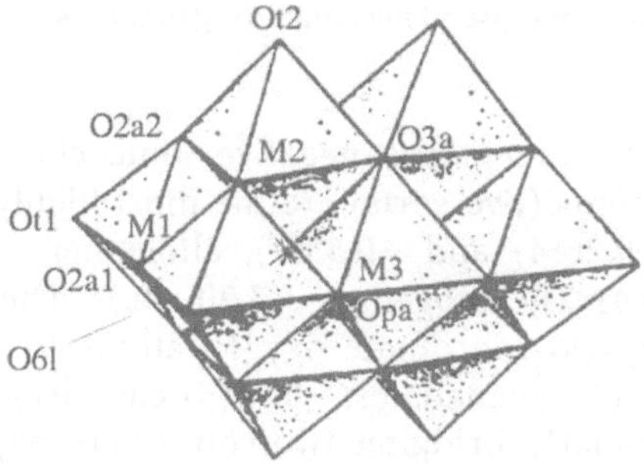

Fig. 11. Structure of the $M_{10}^VO_{28}^{6-}$ ions and characterization of the crystallographically independent M and O atoms.

Table 14. Average bond lengths d and bond valences s of the symmetry-equivalent M–O bonds and average charges c on the different types of oxygen atoms in $V_{10}O_{28}^{6-}$ (average of seven reliable structural investigations)

M–O bond	O type frequency	d in Å	s^a in v.u.	c in c.u.
M1–Ot1	4×	1.591(14)	1.742(39)	−0.258(39)
M2–Ot2	4×	1.611(23)	1.650(67)	−0.350(67)
M3–Opa $M1^i$–Opa	4×	1.685(6) 2.046(15)	1.349(19) 0.510(27)	−0.140(26)
M1–O2a1[b] $M1^i$–O2a1	2×	1.831(7) 1.831(7)	0.908(19) 0.908(19)	−0.185(37)
M1–O2a2 M2–O2a2	8×	1.871(10) 1.819(9)	0.816(17) 0.939(25)	−0.245(34)
M2–O3a $M2^i$–O3a M3–O3a	4×	2.004(10) 2.004(10) 1.917(12)	0.569(15) 0.569(15) 0.721(21)	−0.140(18)
M1–O6l* $M1^i$–O6l* M2–O6l* $M2^i$–O6l* M3–O6l* $M3^i$–O6l*	2×	2.322(11) 2.322(11) 2.224(16) 2.224(16) 2.121(10) 2.121(10)	0.241(14) 0.241(14) 0.316(19) 0.316(19) 0.416(17) 0.416(17)	−0.055(41)
Sum of the negative charges (formal ionic charge):				−6.000 c.u.

Bond valences about the different M atoms:
M1 (4×): 1.742–0.908–0.816–0.816–0.510[P]–0.241 v.u. (type I MO_6 octahedra);
$\Sigma s_i = 5.033$ v.u.
M2 (4×): 1.650–0.939–0.939–0.569–0.569–0.316 v.u. (type I MO_6 octahedra);
$\Sigma s_i = 4.982$ v.u.
M3 (2×): 1.349[P]–1.349[P]–0.721–0.721–0.416–0.416 v.u. (type II MO_6 octahedra);
$\Sigma s_i = 4.972$ v.u.
Average Σs_i for the MO_6 octahedra: 5.000 v.u.

[a] $\bar{d}_{oi} = 1.7960$ Å; $d_{oi\,min} = 1.7912$ Å, $d_{oi\,max} = 1.8011$ Å.
[b] In some of the structures M1–O2a1 and $M1^i$–O2a1 are rather different (up to 0.065 Å or 0.158 v.u.).
[P] The stronger bonds are pseudoterminal M–O bonds, the weaker bonds are the complementary bonds of the pseudoterminal oxygen atoms.

All oxygen atoms accept the negative ionic charge, but again the eight terminal oxygen atoms (two sorts) are the most highly charged oxygen atoms (decavanadates: −0.26(4) and −0.35(7), altogether −2.43 c.u.; decaniobates: −0.39(0) and −0.41(2), altogether −3.20 c.u.). The four pseudoterminal, angularly bridging two-coordinate oxygen atoms (one sort) bear a substantially lower charge (decavanadates: −0.14(3) c.u.; decaniobates: −0.10(1) c.u.). The other ten angularly bridging two-coordinate oxygen atoms (two sorts) bear a medium amount of the charge (decavanadates: −0.18(4) and −0.24(3) c.u.; decaniobates: −0.19(0) and −0.16(1) c.u.). The four angularly bridging three-coordinate oxygen atoms (one sort) bear a small amount of the

Table 15. Average bond lengths d and bond valences s of the symmetry-equivalent M–O bonds and average charges c on the different types of oxygen atoms in $Nb_{10}O_{28}^{6-}$ (average of two reliable structural investigations)

M–O bond	O type frequency	d in Å	s[a] in v.u.	c in c.u.
M1–Ot1	4×	1.739(6)	1.606(2)	−0.394(2)
M2–Ot2	4×	1.742(11)	1.593(23)	−0.407(23)
M3–Opa	4×	1.823(7)	1.280(7)	−0.098(5)
M1^i–Opa		2.093(2)	0.623(12)	
M1–O2a1	2×	1.952(6)	0.906(2)	−0.188(4)
M1^i–O2a1		1.952(6)	0.906(2)	
M1–O2a2	8×	1.983(5)	0.835(1)	−0.157(6)
M2–O2a2		1.912(3)	1.009(5)	
M2–O3a		2.123(1)	0.573(9)	
M2^i–O3a	4×	2.123(1)	0.573(9)	−0.082(7)
M3–O3a		2.012(11)	0.773(11)	
M1–O6l*		2.534(3)	0.191(2)	
M1^i–O6l*		2.534(3)	0.191(2)	
M2–O6l*	2×	2.416(16)	0.262(7)	−0.220(4)
M2^i–O6l*		2.416(16)	0.262(7)	
M3–O6l*		2.224(1)	0.437(8)	
M3^i–O6l*		2.224(1)	0.437(8)	
Sum of the negative charges (formal ionic charge):				−5.996 c.u.

Bond valences about the different M atoms:
M1 (4×): 1.606–0.906–0.835–0.835–0.623[p]–0.191 v.u. (type I MO_6 octahedra); Σs_i = 4.996 v.u.
M2 (4×): 1.593–1.009–1.009–0.573–0.573–0.262 v.u. (type I MO_6 octahedra); Σs_i = 5.019 v.u.
M3 (2×): 1.280[p]–1.280[p]–0.773–0.773–0.437–0.437 v.u. (type II MO_6 octahedra); Σs_i = 4.980 v.u.
Average Σs_i for the MO_6 octahedra: 5.002 v.u.

[a] $\bar{d}_{oi}$ = 1.9150 Å; $d_{oi\,min}$ = 1.9100 Å, $d_{oi\,max}$ = 1.9201 Å.
[p] The stronger bonds are pseudoterminal M–O bonds, the weaker bonds are the complementary bonds of the pseudoterminal oxygen atoms.

charge (decavanadates: −0.14(2) c.u.; decaniobates: −0.08(1) c.u.). The charge on the two central, six-coordinate oxygen atoms (one sort), involved in linear M–O–M bridges, is with the decavanadates −0.05(4) and with the decaniobates −0.22(0) c.u.

The statements about the charge on the various types of oxygen atoms and hence on their basicity sequence for the decavanadate ion in the literature [7, 13b, 215] contradict the above results. The reason why only ca. 40 to 50% of the ionic charge is located on the terminal oxygen atoms must be seen in the high charge (−6 c.u.) and small number of terminal (eight), but large number of bridging (20) oxygen atoms; additionally, resonance structures with negative charge on bridging O atoms cannot be avoided. The comparatively high negative charge on the central, six-coordinate oxygen atoms involved in linear M–O–M bridges of $Nb_{10}O_{28}^{6-}$ is obviously conditioned by geometrical constraints. The interdependence of the M–O bond lengths and hence bond valences in the M–O framework causes that the charge on the central oxygen atoms of this species is always more negative than that on the surrounding angularly bridging two- and three-coordinate oxygen atoms. Interestingly, because of the presence of bridging oxygen atoms in *trans* position to the central, six-coordinate oxygen atoms a resonance structure of the $M_{10}O_{28}^{6-}$ ion with a negative charge on the latter can be formulated, which is necessary for M = V but not for Nb (compare footnote 6). The only small charge on the angularly bridging pseudoterminal oxygen atoms is due to the fact that the second M–O bond of the pseudoterminal oxygen atom is not – as is the case for all structures of the M^{VI} species – a weak "*trans*" but a middle-strong M–O bond.

3.13
Doubly Protonated Decavanadate Ions, $H_2V_{10}O_{28}^{4-}$

The relevant structural data, based on one reliable crystal structure determination [216], are given in Table 16. They indicate that the protonation sites are two O2a2 atoms, centrosymmetrically situated to each other. The same protonation sites have been observed in two further structure investigations. However, in one of these investigations [217] bond lengths are not given and those of the other investigation [218, 219] do not fulfill the reliability criterion. The cations are 5,7-dimethyl-2,3-dihydro-1,4-diazepinium, adenosinium, and 4-methylpyridinium ions.

The eight terminal oxygen atoms (four sorts) accept most of the ionic charge (−0.29 to −0.24, altogether 2.17 c.u.). The four pseudoterminal, angularly bridging two-coordinate oxygen atoms (two sorts) bear a substantially lower charge (−0.15 and −0.10 c.u.). The other eight angularly bridging, unprotonated two-coordinate oxygen atoms (four sorts) bear a high to very small negative charge (−0.27 to −0.04 c.u.). The two protonated oxygen atoms (one sort) of otherwise the same type are positively charged (+0.21 c.u.), if an O–H bond valence of 1 v.u. is assumed (assuming the more realistic O–H bond valence of 0.8 v.u., the charge reduces to +0.01 c.u.; to the same degree, preferably on terminal and pseudoterminal oxygen atoms, the negative charges stated above are reduced too). The four angularly bridging three-coordinate oxygen atoms (two sorts) bear small negative charge (−0.12 and −0.04 c.u.). The charge on the two central, six-coordinate oxygen atoms (one sort), involved in linear M–O–M bridges, is −0.06 c.u.

The sum of the M–O bond valences remains, of course, unchanged upon protonation. However, some types of oxygen atoms are more strongly (+), other types more weakly (−) bound after protonation:

Ot1	Ot2	Opa	O2a1	O2a2	O2a2h	O3a	O6l
(4×)	(4×)	(4×)	(2×)	(6×)	(2×)	(4×)	(2×)
−0.08	+0.34	+0.07	+0.29	+0.24	−1.08	+0.24	−0.02 v.u. (total: 0.00 v.u.)

The terminal oxygen atoms are altogether 0.26 v.u. more strongly, the bridging oxygen atoms altogether 0.26 v.u. more weakly bound than in the unprotonated decavanadate ion; that means the protonation has a weakly destabilizing effect on the polymeric character of the structure in this respect (compare also the decrease of the quantity S_o/q in Table 46). The full effect on the inner bond valence by a diprotonation of bridging O atoms would be a decrease of 2 v.u. However, since the species becoming protonated ($V_{10}O_{28}^{6-}$) carries a high negative charge on bridging O atoms (see also Sect. 3.6.3 in Ref. 16) and since the geometrical interdependence of the inner bond lengths obviously permits the "activation" of this negative charge there remains only the above small change of 0.26 v.u. The weakening of the bridging of the M atoms takes place almost exclusively at the protonated oxygen atoms. However, the reduction of the ionic charge by protonation does not only take place at the protonation sites (O2a2h) but occurs more strongly delocalized:

Ot1	Ot2	Opa	O2a1	O2a2	O2a2h	O3a	O6l
(4×)	(4×)	(4×)	(2×)	(6×)	(2×)	(4×)	(2×)
−0.08	0.34	0.07	0.29	0.24	0.92	0.24	−0.02 c.u. (total: 2.00 c.u.)

(positive values mean a reduction of the ionic charge). This is a consequence of the valence sum rule (Eq. (2)): The weakening of the M–O bond(s) of a protonated oxygen atom requires the strengthening of other M–O bonds on the same MO_6 octahedron (see Table 17). This effect is for the most part propagated to the neighboring MO_6 octahedra by which the final result is a more or less strong alternate bond strengthening and weakening and a corresponding change of the charge. On an average 58% of the bond valence compensation on the protonated MO_6 octahedra take place via *trans* bond valence alternations (V1fr and V1bl: 69%; V2tr and V2ul: 48%). That means that the compensation in the (single) *trans* positions to the protonation sites occurs approximately six times more frequently than in the (four) *cis* positions. The reasons for the preference of *trans* bond valence alternations are apparently the smaller necessary structural changes in the MO_6 octahedra and their better propagation over the whole structure:

- Due to the approximate constancy of the bond valence sums and bond lengths sums for M–O bonds in *trans* position to each other (compare Sect. 4.3.2), the M atom in the MO_6 octahedron has essentially only to move a little along the corresponding O–M–O axis without greater alterations of the outer dimensions of the octahedron.
- The *cis* positions of the protonated bridging oxygen atoms are in part less suitable for the distribution of bond valence alterations, namely
  - the terminal M–O bonds because of the small capacity for changes due to the lack of propagation possibilities,
  - the central M–O bonds because of the small capacity for changes due to the smallness of the respective bond valences which means on the one hand again lack of propagation possibilities and on the other hand very large bond lengths changes owing to the small slope of the bond valence-bond length function in the region of small bond valences.

Table 16. Average bond lengths d and bond valences s of the symmetry-equivalent M–O bonds and average charges c on the different types of oxygen atoms of the $H_2V_{10}O_{28}^{4-}$ anion in 5,7-dimethyl-2,3-dihydro-1,4-diazepinium decavanadate

M–O bond[a]	O type frequency	d in Å	s[b] in v.u.	c in c.u.	Δs[c] in v.u.	Δc[c] in c.u.
M1bl–Ot1bl = M1fr–Ot1fr	2×	1.602	1.730	−0.270	−0.012	−0.012
M1fl–Ot1fl = M1br–Ot1br	2×	1.605	1.716	−0.284	−0.026	−0.026
M2tl–Ot2tl = M2ur–Ot2ur	2×	1.607	1.706	−0.294	0.056	0.056
M2tr–Ot2tr = M2ul–Ot2ul	2×	1.595	1.763	−0.237	0.113	0.113
M3b–Opabl = M3f–Opafr M1bl–Opabl = M1fr–Opafr	2×	1.679 2.061	1.404 0.499	−0.097	0.055 −0.011	0.043
M3b–Opabr = M3f–Opafl M1br–Opabr = M1fl–Opafl	2×	1.709 2.022	1.295 0.555	−0.151	−0.054 0.045	−0.011
M1bl–O2a1l = M1fr–O2a1r M1fl–O2a1l = M1br–O2a1r	2×	1.819 1.805	0.961 0.998	−0.040	0.053 0.090	0.145
M1bl–O2a2btl = M1fr–O2a2fur M2tl–O2a2btl = M2ur–O2a2fur	2×	1.810 1.870	0.985 0.837	−0.178	0.169 −0.102	0.067
M1fl–O2a2ftl = M1br–O2a2bur M2tl–O2a2ftl = M2ur–O2a2bur	2×	1.848 1.825	0.889 0.946	−0.166	0.073 0.007	0.079
M1bl–O2a2**h**bul = M1fr–O2a2**h**ftr M2ul–O2a2**h**bul = M2tr–O2a2**h**ftr	2×	2.014 1.966	0.567 0.646	+0.213[h]	−0.249 −0.293	0.458
M1fl–O2a2ful = M1br–O2a2btr M2ul–O2a2ful = M2tr–O2a2btr	2×	1.923 1.803	0.725 1.004	−0.271	−0.091 0.065	−0.026
M2tl–O3abt = M2ur–O3afu M2tr–O3abt = M2ul–O3afu M3b–O3abt = M3f–O3afu	2×	2.012 1.930 1.950	0.570 0.712 0.674	−0.044	0.001 0.143 −0.047	0.096
M2tl–O3aft = M2ur–O3abu M2tr–O3aft = M2ul–O3abu M3f–O3aft = M3b–O3abu	2×	1.958 2.016 1.958	0.660 0.564 0.660	−0.116	0.091 −0.005 −0.061	0.024
M1bl–O6ll* = M1fr–O6lr* M1fl–O6ll* = M1br–O6lr* M2tl–O6ll* = M2ur–O6lr* M2ul–O6ll* = M2tr–O6lr* M3b–O6ll* = M3f–O6lr* M3f–O6ll* = M3b–O6lr*	2×	2.272 2.331 2.306 2.246 2.180 2.067	0.282 0.240 0.257 0.303 0.362 0.491	−0.065	0.041 −0.001 −0.059 −0.013 −0.054 0.075	−0.010
Sum of the negative charges:				−4.426 c.u.		
Sum of the positive charges on OH groups:				+0.426 c.u.		
Total (formal ionic charge):				−4.000 c.u.		

Bond valences about the different M atoms:

M1bl = M1fr (2×): 1.730–0.499[p]–0.961–0.985–0.567[h]–0.282 v.u. (type I MO_6 octahedra); Σs_i = 5.024 v.u.

M1fl = M1br (2×): 1.716–0.555[p]–0.998–0.889–0.725–0.240 v.u. (type I MO_6 octahedra); Σs_i = 5.123 v.u.

M2tl = M2ur (2×): 1.706–0.837–0.946–0.570–0.660–0.257 v.u. (type I MO_6 octahedra); Σs_i = 4.976 v.u.

M2tr = M2ul (2×): 1.763–1.004–0.646[h]–0.712–0.564–0.303 v.u. (type I MO_6 octahedra); Σs_i = 4.990 v.u.

Table 16 (Continued)

M3b = M3f (2×): 1.404^{p}–1.295^{p}–0.674–0.660–0.362–0.491 v.u. (type II MO_6 octahedra);
$\Sigma s_i = 4.886$ v.u.
Average Σs_i for the MO_6 octahedra: 5.000 v.u.

[a] The additional indices to distinguish between the different M and O atoms of a type mean, related to the position of the $M_{10}O_{28}$ structure shown in Fig. 11: f: **f**ront; b: **b**ack; t: **t**op; u: bottom (**u**nder); l: **l**eft; r: **r**ight.
[b] $d_{oi} = 1.8044$ Å.
[c] Influence of the twofold protonation compared to the unprotonated decavanadate ion (Table 14). A negative Δs means a weaker, a positive Δs a stronger M–O bond after protonation. A negative Δc means a greater, a positive Δc a smaller negative charge on the oxygen atom after protonation.
[h] Protonated oxygen atoms; O–H bond valences of 1 v.u. have been assumed. Considering the commonly observed bond valence for O–H bonds of 0.8 v.u., there is no positive charge on this oxygen atom (+0.01 c.u.).
[p] The strong bonds are pseudoterminal M–O bonds, the weak bonds are the complementary bonds of the pseudoterminal oxygen atoms.

Table 17. Bond valence changes about the different MO_6 octahedra as a consequence of the diprotonation of the decavanadate ion at O2a2ftr and O2a2bul[a]

V1fr	-Ot1fr	−0.012 v.u. [b]	V1br	-Ot1br	−0.026 v.u.
	-O2a1r	+0.053 v.u.		-O2a1r	+0.090 v.u.
	-O2a2**h**ftr	−0.249 v.u.		-O2a2btr	−0.091 v.u.
	-O2a2fur	+0.169 v.u.		-O2a2bur	+0.073 v.u.
	-Opafr	−0.011 v.u.		-Opabr	+0.045 v.u.
	-O6lr	+0.041 v.u.		-O6lr	−0.001 v.u.
V2tr	-Ot2tr	+0.113 v.u.	V2ur	-Ot2ur	+0.056 v.u.
	-O2a2**h**ftr	−0.293 v.u.		-O2a2fur	−0.102 v.u.
	-O2a2btr	+0.065 v.u.		-O2a2bur	+0.007 v.u.
	-O3aft	−0.005 v.u.		-O3afu	+0.001 v.u.
	-O3abt	+0.143 v.u.		-O3abu	+0.091 v.u.
	-O6lr	−0.013 v.u		-O6lr	−0.059 v.u.
V3f	-Opafr	+0.055 v.u.			
	-Opafl	−0.054 v.u.			
	-O3aft	−0.061 v.u.			
	-O3afu	−0.047 v.u.			
	-O6lr	−0.054 v.u.			
	-O6ll	+0.075 v.u.			

[a] The remaining MO_6 octahedra are symmetry-equivalent to those listed.
[b] The brackets indicate M–O bonds in trans position to each other.

According to ^{17}O-NMR data the predominant protonation sites on $V_{10}O_{28}^{6-}$ *in aqueous solution* are O3a atoms, while O2a2 atoms are protonated to a lesser extent [7, 220].

In the literature [220], factors determining the protonation sites (in the solid state and in solution) are seen in the negative charge distribution on $V_{10}O_{28}^{6-}$, in hydrogen bonding effects, and particularly in nonlocal charge redistribution resulting from protonation. The latter shows a pattern of *trans* bond length alternations in (approximately planar) V_4O_4 rings containing a protonated oxygen atom by which electron density is not only withdrawn from the protonated oxygen atom but also from unprotonated oxygen atoms in the anion and thus may exert a restrictive effect on other possible protonation sites [220]. This view is, however, in part based on an assessment of the basicity sequence of the oxygen atoms [7, 220] that contradicts the results of the present study.

In our opinion the main reason for the preference of a particular protonation site is to keep the structural changes (necessary bond lengths alterations) of the M–O framework as a consequence of the protonation as small as possible and, together with the charge alterations, consistent with the packing conditions of polyanions, cations, and solvent molecules of crystallization. The mere reduction of the (negative) charge on the oxygen atom to be protonated to that in the protonated state (Δc in Table 45) does not cause any structural changes. Only the part of the protonic charge beyond that requires structural (bond lengths) changes and corresponding changes of the charge ($\Sigma(\Delta s)$ in Table 45) whose geometrical constraints can be considerably reduced by the better distribution of the necessary changes over the structure if the protonation site is a two- or even three-coordinate oxygen atom (Δs in Table 45). The occurrence of the bond lengths and bond valence changes in alternating sequences is a necessary consequence of the valence sum rule. Their occurrence in *trans* positions keeps the necessary structural changes small (since the outer dimensions of the MO_6 octahedra remain essentially unchanged and only M has to move a little on the corresponding O–M–O axis) and serves simultaneously to facilitate their transfer to more distant parts of the structures (see above). (See also Sects. 4.6.6 and 4.6.7.)

3.14
A Triply Protonated Decavanadate Ion, $H_3V_{10}O_{28}^{3-}$

The relevant structural data, based on one reliable crystal structure determination [220], are given in Table 18. They show that the protonation sites are two O2a2 atoms and one O3a atom in a colinear arrangement. The cation is tetraphenylphosphonium.

The eight terminal oxygen atoms accept most of the ionic charge (−0.39 to −0.22, altogether −2.12 c.u.). The four pseudoterminal, angularly bridging two-coordinate oxygen atoms bear a substantially lower charge (−0.20 to −0.06 c.u.). The other eight angularly bridging, unprotonated two-coordinate oxygen atoms bear a small amount of the charge (−0.15 to −0.08 c.u.). The two protonated oxygen atoms of otherwise the same type are positively charged (+0.23 and +0.27 c.u.), if an O–H bond valence of 1 v.u. is assumed (taking the more realistic O–H bond valence of 0.8 v.u., the charge reduces to +0.03 and +0.07 c.u., respectively; to the same degree, preferably on terminal and pseudoterminal oxygen atoms, the negative charges stated above are reduced too; see the next paragraph). The three angularly bridging, unprotonated three-coordinate oxygen atoms bear a very small positive charge (0.00 to

+0.05 c.u.). The protonated oxygen atom of the same type is positively charged (+0.26 c.u.), if an O–H bond valence of 1 v.u. is assumed (assuming the more realistic O–H bond valence of 0.8 v.u., the charge amounts to +0.06 c.u.; the negative charges stated above are reduced to the same degree further on). The charge on the two central, six-coordinate oxygen atoms, involved in linear M–O–M bridges, is −0.12 c.u.

The protonated and hence positively charged oxygen atoms are opposed to two bridging (Opa) and one terminal (Otl) oxygen atoms (in colinear arrangement) of the neighboring $V_{10}O_{28}$ unit (in the solid state structure). In this way the positive charge and an equivalent amount of negative charge are compensated by hydrogen bonds (see footnote h of Table 18).

The sum of the M–O bond valences remains unchanged upon protonation. Again some types of oxygen atoms are more strongly (+), other types more weakly (−) bound after protonation:

Otl	Ot2	Opa	O2al	O2a2	O2a2h	O3a	O3ah	O6l
(4×)	(4×)	(4×)	(2×)	(6×)	(2×)	(3×)	(1×)	(2×)
−0.12	+0.44	+0.05	+0.15	+0.71	−1.01	+0.51	−0.60	−0.13 v.u. (total: 0.00 v.u.)

The terminal oxygen atoms are altogether 0.32 v.u. more strongly, the bridging oxygen atoms altogether 0.32 v.u. more weakly bound than in the unprotonated decavanadate ion. This means the protonation has again a weakly destabilizing effect on the polymeric character of the structure. The full effect on the inner bond valance by a triprotonation of bridging O atoms would be a decrease of 3 v.u. The reasons discussed in Sect. 3.13 cause the only small change of 0.32 v.u. The weakening of the bridging of the M atoms takes place almost exclusively at the protonated oxygen atoms. The reduction of the ionic charge by protonation again does not only take place at the protonation sites (O2a2h and O3ah) but occurs more at strongly delocalized positions:

Otl	Ot2	Opa	O2al	O2a2	O2a2h	O3a	O3ah	O6l
(4×)	(4×)	(4×)	(2×)	(6×)	(2×)	(3×)	(1×)	(2×)
−0.12	0.44	0.05	0.15	0.71	0.99	0.51	0.40	−0.13 c.u. (total: 3.00 c.u.)

(positive values mean a reduction of the ionic charge). This is again a consequence of the valence sum rule (Eq. (2)): The weakening of the M–O bond(s) of a protonated oxygen atom requires the strengthening of other M–O bonds on the same MO_6 octahedron (see Table 19). This effect is for the most part propagated to the neighboring MO_6 octahedra by which the final result is a more or less strong alternate bond strengthening and weakening and a corresponding change of the charge. On an average 72% of the bond valence compensation on the protonated MO_6 octahedra take place via *trans* bond valence alternations (V1fl: 100%; V1fr: 76%; V2tl: 60%; V2tr: 60%; V3f: 84%). The compensation in the (single) *trans* positions to the protonation sites occurs on an average nine times more frequently than in the (V1fl, V1fr, and V3f: four; V2tl and V2tr: three) *cis* positions. The reasons for the preference of *trans* bond valence alternations have been discussed in the preceding section

Table 18. Bond lengths d and bond valences s of the M–O bonds and charges c on the oxygen atoms of the $H_3V_{10}O_{28}^{3-}$ anion in $[(C_6H_5)_4P]_3[H_3V_{10}O_{28}] \cdot 4CH_3CN$

M–O bond[a]	O type frequency	d in Å	s[b] in v.u.	c in c.u.	Δs[c] in v.u.	Δc[c] in c.u.
M1bl–Ot1bl	1×	1.596	1.730	−0.270	−0.012	−0.012
M1fl–Ot1fl	1×	1.622	1.612	−0.388	−0.130	−0.130
M1br–Ot1br	1×	1.598	1.721	−0.279	−0.021	−0.021
M1fr–Ot1fr	1×	1.585	1.782	−0.218	0.040	0.040
M2tl–Ot2tl	1×	1.591	1.754	−0.246	0.104	0.104
M2tr–Ot2tr	1×	1.589	1.763	−0.237	0.113	0.113
M2ul–Ot2ul	1×	1.585	1.782	−0.218	0.132	0.132
M2ur–Ot2ur	1×	1.594	1.739	−0.261	0.089	0.089
M3b–Opabl	1×	1.682	1.371	−0.106	0.022	0.034
M1bl–Opabl		2.038	0.523		0.013	
M3b–Opabr	1×	1.667	1.427	−0.058	0.078	0.082
M1br–Opabr		2.044	0.514		0.004	
M3f–Opafl	1×	1.707	1.281	−0.196	−0.068	−0.056
M1fl–Opafl		2.038	0.523		0.013	
M3f–Opafr	1×	1.687	1.352	−0.156	0.003	−0.016
M1fr–Opafr		2.061	0.491		−0.019	
M1bl–O2a1l	1×	1.804	0.985	−0.124	0.077	0.061
M1fl–O2a1l		1.841	0.891		−0.017	
M1br–O2a1r	1×	1.806	0.980	−0.099	0.072	0.086
M1fr–O2a1r		1.829	0.921		0.013	
M1bl–O2 a2btl	1×	1.963	0.641	−0.123	−0.175	0.122
M2tl–O2a2btl[d]		1.720	1.237		0.298	
M1fl–O2a2**h**ftl	1×	2.011	0.563	+0.231[h1]	−0.253	0.476
M2tl–O2a2**h**ftl		1.947	0.669		−0.270	
M1bl–O2a2bul	1×	1.840	0.894	−0.110	0.078	0.135
M2ul–O2a2bul		1.800	0.996		0.057	
M1fl–O2a2ful	1×	1.781	1.048	−0.150	0.232	0.095
M2ul–O2a2ful		1.880	0.802		−0.137	
M1br–O2a2btr	1×	1.954	0.656	−0.076	−0.160	0.169
M2tr–O2a2btr[d]		1.711	1.267		0.328	
M1fr–O2a2**h**ftr	1×	1.994	0.589	+0.273[h3]	−0.227	0.518
M2tr–O2a2**h**ftr		1.939	0.684		−0.255	
M1br–O2a2bur	1×	1.851	0.867	−0.153	0.051	0.092
M2ur–O2a2bur		1.806	0.980		0.041	
M1fr–O2a2fur	1×	1.795	1.009	−0.146	0.193	0.099
M2ur–O2a2fur		1.861	0.844		−0.095	
M2tl–O3abt		1.940	0.682		0.113	
M2tr–O3abt	1×	1.937	0.687	+0.045	0.118	0.185
M3b–O3abt		1.943	0.676		−0.045	
M2tl–O3a**h**ft*		2.132	0.405		−0.164	
M2tr–O3a**h**ft*	1×	2.118	0.421	+0.257[h2]	−0.147	0.397
M3f–O3a**h**ft		2.110	0.430		−0.291	

Table 18 (Continued)

M2ul-O3abu		1.937	0.687		0.118	
M2ur-O3abu	1×	1.967	0.634	+0.047	0.065	0.187
M3b-O3abu		1.917	0.726		0.005	
M2ul-O3afu		2.048	0.509		−0.060	
M2ur-O3afu	1×	2.039	0.521	−0.000	−0.048	0.140
M3f-O3afu		1.810	0.969		0.248	
M1bl-O6ll*		2.376	0.209		−0.032	
M1fl-O6ll*		2.286	0.267		0.026	
M2tl-O6ll*		2.228	0.313		−0.003	
M2ul-O6ll*	1×	2.342	0.230	−0.119	−0.086	−0.064
M3b-O6ll*		2.131	0.407		−0.009	
M3f-O6ll*		2.089	0.455		0.039	
M1br-O6lr*		2.392	0.201		−0.040	
M1fr-O6lr*		2.303	0.255		0.014	
M2tr-O6lr*		2.263	0.284		−0.032	
M2ur-O6lr*	1×	2.325	0.240	−0.119	−0.076	−0.064
M3b-O6lr*		2.135	0.402		−0.014	
M3f-O6lr*		2.056	0.498		0.082	

Sum of the negative charges:	−3.851 c.u.
Sum of the positive charges on OH groups:	+0.761 c.u.
Sum of the other positive charges:	+0.092 c.u.
Total (formal ionic charge):	−2.998 c.u.

Bond valences about the different M atoms:

M1bl(1×): 1.730−0.523^{p}−0.985−0.641^{p}−0.894−0.209 v.u. (type I MO_6 octahedron); $\Sigma s_i = 4.982$ v.u

M1fl(1×): 1.612−0.523^{p}−0.891−0.563^{h1}−1.048−0.267 v.u. (type I MO_6 octahedron); $\Sigma s_i = 4.905$ v.u.

M1br(1×): 1.721−0.514^{p}−0.980−0.656^{p}−0.867−0.201 v.u. (type I MO_6 octahedron); $\Sigma s_i = 4.939$ v.u.

M1fr(1×): 1.782−0.491^{p}−0.921−0.589^{h3}−1.009−0.255 v.u. (type I MO_6 octahedron); $\Sigma s_i = 5.048$ v.u.

M2tl^{d}(1×): 1.754−1.237^{p}−0.669^{h1}−0.682−0.405^{h2}−0.313 v.u. (type II MO_6 octahedron); $\Sigma s_i = 5.059$ v.u.

M2tr^{d}(1×): 1.763−1.267^{p}−0.684^{h3}−0.687−0.421^{h2}−0.284 v.u. (type II MO_6 octahedron); $\Sigma s_i = 5.107$ v.u.

M2ul(1×): 1.782−0.996−0.802−0.687−0.509−0.230 v.u. (type I MO_6 octahedron); $\Sigma s_i = 5.006$ v.u.

M2ur(1×): 1.739−0.980−0.844−0.634−0.521−0.240 v.u. (type I MO_6 octahedron); $\Sigma s_i = 4.959$ v.u.

M3b(1×): 1.371^{p}−1.427^{p}−0.676−0.726−0.407−0.402 v.u. (type II MO_6 octahedron); $\Sigma s_i = 5.009$ v.u.

M3f(1×): 1.281^{p}−1.352^{p}−0.430^{h2}−0.969−0.455−0.498 v.u. (type II MO_6 octahedron); $\Sigma s_i = 4.986$ v.u.

Average Σs_i for the MO_6 octahedra: 5.000 v.u.

[a] For the additional indices to distinguish between the different M and O atoms of a type see footnote a of Table 16.

[b] $d_{oi} = 1.7960$ Å.

[c] Influence of the threefold protonation compared to the unprotonated decavanadate ion (Table 14).

Table 18 (Continued)

[d] As a consequence of the protonation these bonds become pseudoterminal bonds and hence $M2tlO_6$ and $M2trO_6$ type II octahedra.

[h] Protonated oxygen atoms; O–H bond valences of 1 v.u. have been assumed. The positively charged protonated oxygen atoms are opposed to negatively charged oxygen atoms of a neighboring $H_3V_{10}O_{28}^{3-}$ unit. Thus the charges are at least partially compensated by hydrogen bonds:
h1: O2a2hftl +0.231 c.u., Opafr −0.156 c.u., Σc_i +0.075 c.u.
h2: O3ahft +0.257 c.u., Opafl −0.196 c.u., Σc_i +0.061 c.u.
h3: O2a2hftr +0.273 c.u.; Ot1fl −0.388 c.u., Σc_i −0.115 c.u.
Considering the commonly observed bond valence for O–H bonds of 0.8 v.u., there is a small positive charge on these O atoms: +0.03 c.u. on O2a2hftl, +0.07 c.u. on O2a2hftr, +0.06 c.u. on O3ahft.

[p] The strong bonds (≧ 1.237 v.u.) are pseudoterminal M–O bonds, the weak bonds (≦ 0.656 v.u.) the complementary bonds of the pseudoterminal oxygen atoms.

Table 19. Bond valence changes about the different MO_6 octahedra as a consequence of the triprotonation of the decavanadate ion at O2a2ftl, O2a2ftr, and O3aft

V1fl	-Ot1fl	−0.085 v.u. [b]	V1fr	-Ot1fr	+0.040 v.u.
	-O2all	−0.008 v.u.		-O2a1r	+0.076 v.u.
	-O2a2**h**ftl	−0.265 v.u.		-O2a2**h**ftr	−0.235 v.u.
	-O2a2ful	+0.228 v.u		-O2a2fur	+0.193 v.u.
	-Opafl	+0.022 v.u		-Opafr	−0.017 v.u.
	-O611	+0.046 v.u		-O6lr	+0.193 v.u.
V1bl	-Ot1bl	+0.002 v.u.	V1br	-Ot1br	−0.005 v.u.
	-O2al l	+0.081 v.u.		-O2a1r	+0.076 v.u.
	-O2a2btl	−0.177 v.u.		-O2a2btr	−0.160 v.u.
	-O2a2bul	+0.184 v.u.		-O2a2bur	+0.058 v.u.
	-Opabl	+0.022 v.u.		-Opabr	+0.012 v.u.
	-O6ll	−0.048 v.u		-O6lr	−0.063 v.u.
V2tl	-Ot2tl	+0.087 v.u.	V2tr	-Ot2tr	+0.094 v.u.
	-O2a2**h**ftl	−0.265 v.u.		-O2a2**h**ftr	−0.294 v.u.
	-O2a2btl	+0.273 v.u.		-O2a2btr	+0.299 v.u.
	-O3a**h**ft	−0.196 v.u.		-O3a**h**ft	−0.175 v.u.
	-O3abt	+0.132 v.u.		-O3abt	+0.138 v.u.
	-O6ll	+0.001 v.u		-O6lr	−0.041 v.u.
V2ul	-Ot2ul	+0.108 v.u.	V2ur	-Ot2ur	+0.027 v.u.
	-O2a2bul	+0.061 v.u.		-O2a2bur	+0.077 v.u.
	-O2a2ful	−0.126 v.u		-O2a2fur	+0.046 v.u.
	-O3abu	+0.138 v.u.		-O3abu	−0.084 v.u.
	-O3afu	−0.065 v.u		-O3afu	+0.079 v.u.
	-O6ll	−0.127 v.u.		-O6lr	−0.050 v.u.
V3f	-Opafl	−0.046 v.u.	V3b	-Opabl	−0.110 v.u.
	-Opafr	+0.012 v.u.		-Opabr	+0.073 v.u.
	-O3a**h**ft	−0.329 v.u.		-O3abt	−0.040 v.u.
	-O3afu	+0.253 v.u		-O3abu	+0.013 v.u.
	-O6ll	+0.057 v.u.		-O6ll	−0.007 v.u.
	-O6lr	+0.110 v.u.		-O6lr	−0.012 v.u.

[b] The brackets indicate M–O bonds in *trans* position to each other.

(smaller necessary structural changes and their better propagation over the whole structure). Surprisingly, the type I MO_6 octahedra of M2 change into type II ones on the protonated side of the polyanion.

^{17}O-NMR spectroscopic data indicate that $H_3V_{10}O_{28}^{3-}$ in 1:1 (v/v) CH_3CN/H_2O is protonated at one O3a and two O2a2 atoms too; however, the protonation sites are not necessarily those proven for the solid state [220].

In the literature, the reasons for the colinear arrangement of the protonation sites are seen, apart from hydrogen bonding effects, in particular in nonlocal charge redistribution resulting from protonation, as described in Sect. 3.13. Two (approximately planar) V_4O_4 rings in the structure containing protonated oxygen atoms must not intersect, i.e., must not have atoms in common in order that two protons do not compete for electron density from any of the same oxygen atoms as charge delocalization proceeds by *trans* bond alternation [220]. In our opinion the colinear arrangement of the O2a2 and O3a protonation sites is primarily an effect of the packing conditions of polyanions, cations, and molecules of acetonitrile of crystallization.

The reasons for the protonation of two- and three-coordinate, i.e., O2a- and O3a-oxygen atoms have been discussed in Sect. 3.13.

3.15
The Isopolyoxomolybdate Ions with Mo_8O_{28} Blocks: $Mo_8O_{26}(OH)_2^{6-}$, $Mo_{10}O_{34}^{8-}$ (I), $Mo_{10}O_{34}^{8-}$ (II), and $[Mo_8O_{27}^{6-}]_\infty$

The Mo_8O_{28} blocks of these polyoxomolybdate ions contain four crystallographically independent, edge-sharing MoO_6 octahedra of type II (Figs. 12–15). The relevant structural data are given in Tables 20–23. They are based on one reliable crystal structure determination in each case. Cations are isopropylammonium [59], ammonium [221], methylammonium [222], and ammonium [41], respectively.

The essentially identical parts of the Mo_8O_{28} blocks of the four polymolybdate types are characterized as follows. The in each case 14 terminal oxygen atoms (seven sorts) bear most of the negative ionic charge (dihydrogenoctamolybdate: −0.38, −0.40, −0.39, −0.39, −0.35, −0.35, −0.40, altogether −5.32 c.u.; decamolybdate I: −0.38, −0.49, −0.38, −0.40, −0.28, −0.30, −0.43, altogether −5.34 c.u.; decamolybdate II: −0.42, −0.41, −0.35, −0.39, −0.32, −0.45, −0.28, altogether −5.23 c.u.; polyoctamolybdate: −0.45, −0.32, −0.36, −0.43, −0.32, −0.39, −0.39, altogether −5.31 c.u.). The two pseudoterminal, two-coordinate angularly bridging oxygen atoms (one sort) bear a negative charge of medium size (dihydrogenoctamolybdate: −0.18 c.u.; decamolybdate I: −0.17 c.u.; decamolybdate II: −0.21 c.u.; polyoctamolybdate: −0.20 c.u.). The other four two-coordinate angularly bridging oxygen atoms (two sorts) bear a very small to medium-sized negative charge (dihydrogenoctamolybdate: −0.11 and −0.12 c.u.; decamolybdate I: −0.10 and −0.16 c.u.; decamolybdate II: −0.03 and −0.02 c.u.; polyoctamolybdate: −0.24 and −0.20 c.u.). The two three-coordinate angularly bridging oxygen atoms (one sort) bear a small to very small negative charge (dihydrogenoctamolybdate: −0.11 c.u.; decamolybdate I: −0.13 c.u.; decamolybdate II: −0.03 c.u.; polyoctamolybdate: −0.05 c.u.). The

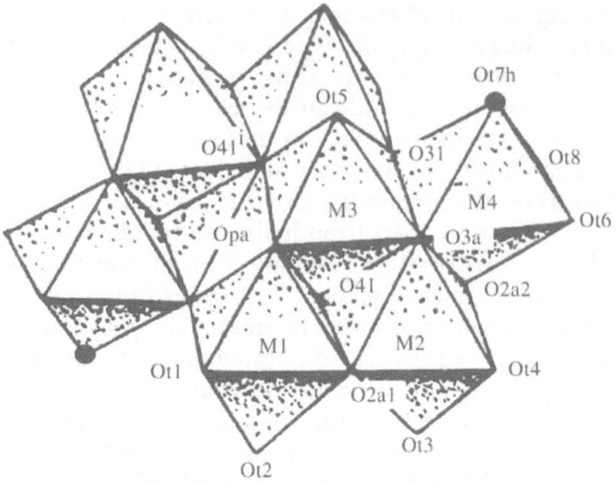

Fig. 12. Structure of the $Mo_8O_{26}(OH)_2^{6-}$ ion and characterization of the crystallographically independent M and O atoms (● OH groups). (In the structure presented in Ref. 13h the OH groups are incorrectly positioned.)

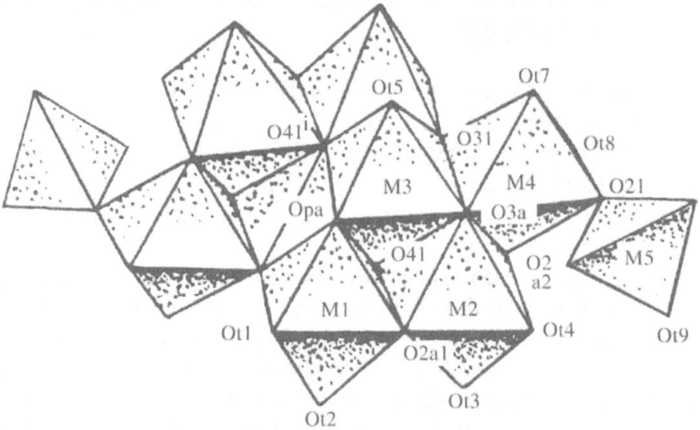

Fig. 13. Structure of the $Mo_{10}O_{34}^{8-}$ (I) ion and characterization of the crystallographically independent M and O atoms (in the free state the M5–O2l–M4 angle may change and the MO_4 tetrahedron may rotate about the O2l–M5 axis).

two three-coordinate oxygen atoms (one sort), involved in (approximately) linear Mo–O–Mo bridges, bear a positive charge (dihydrogenoctamolybdate: +0.16 c.u.; decamolybdate I: +0.22 c.u.; decamolybdate II: +0.17 c.u.; polyoctamolybdate: +0.10 c.u.). The two four-coordinate oxygen atoms (one sort), also involved in (approximately) linear Mo–O–Mo bridges, bear a positive charge too (dihydrogenoctamolybdate: +0.14 c.u.; decamolybdate I: +0.19 c.u.; decamolybdate II: +0.14 c.u.; polyoctamolybdate: +0.20 c.u.).

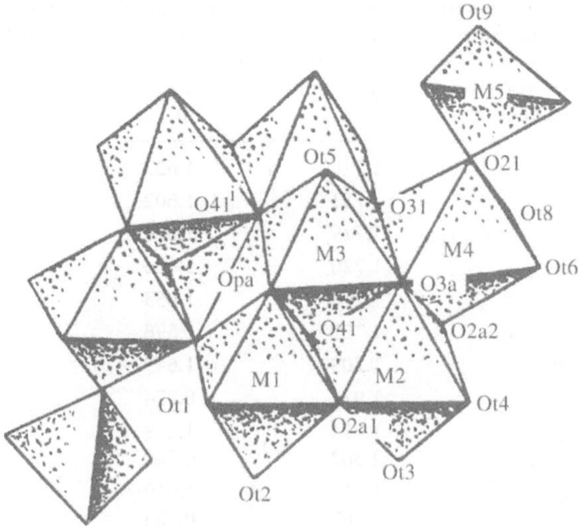

Fig. 14. Structure of the $Mo_{10}O_{34}^{8-}$ (II) ion and characterization of the crystallographically independent M and O atoms (in the free state the M5–O21–M4 angle may change and the MO_4 tetrahedron may rotate about the O21–M5 axis).

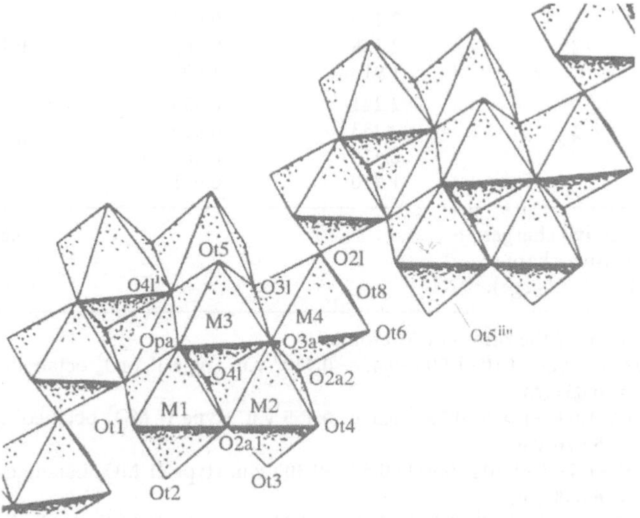

Fig. 15. Structure of the $[Mo_8O_{27}^{6-}]_\infty$ ion and characterization of the crystallographically independent M and O atoms.

The non-identical parts of the four polymetalate types are characterized as follows:

- *Dihydrogenoctamolybdate ion:* Assuming an O–H bond valence of 1 v.u., the protonated oxygen atoms bear –0.13 c.u. Since the hydrogen atoms of

Table 20. Average bond lengths d and bond valences s of the symmetry-equivalent M–O bonds and average charges c on the different types of oxygen atoms in $Mo_8O_{26}(OH)_2^{6-}$

M–O bond	O type frequency	d in Å	s[a] in v.u.	c in c.u.
M1–Ot1	2×	1.716	1.622	−0.378
M1–Ot2	2×	1.721	1.602	−0.398
M2–Ot3	2×	1.719	1.610	−0.390
M2–Ot4	2×	1.720	1.606	−0.394
M3–Ot5	2×	1.708	1.653	−0.347
M4–Ot6	2×	1.722	1.598	−0.402
M4–Ot8	2×	1.709	1.649	−0.351
M4–Ot7h	2×	1.972	0.874	−0.126[h]
M3–Opa M1–Opa*	2×	1.754 2.362	1.479 0.340	−0.180
M1–O2a1 M2–O2a1	2×	1.912 1.967	1.010 0.884	−0.106
M2–O2a2 M4–O2a2*	2×	1.910 1.975	1.015 0.867	−0.118
M2–O3a* M3–O3a M4–O3a	2×	2.300 1.875 2.308	0.396 1.104 0.388	−0.112
M3–O3l* M4–O3l* M1^i–O3l	2×	2.146 2.138 1.916	0.574 0.585 1.000	+0.159
M1–O4l* M2–O4l* M3–O4l* M3^i–O4l	2×	2.241 2.233 2.414 1.950	0.456 0.465 0.300 0.921	+0.143
Sum of the negative charges:				−6.602 c.u.
Sum of the positive charges:				+0.604 c.u.
Total (formal ionic charge):				−5.998 c.u.

Bond valences about the different M atoms:
M1 (2×): 1.622–1.602–1.010–1.000–0.456–0.340[p] v.u. (type II MO_6 octahedra); $\Sigma s_i = 6.031$ v.u.
M2 (2×): 1.610–1.606–1.015–0.884–0.396–0.465 v.u. (type II MO_6 octahedra); $\Sigma s_i = 5.976$ v.u.
M3 (2×): 1.653–1.479[p]–1.104–0.921–0.574–0.300 v.u. (type II MO_6 octahedra); $\Sigma s_i = 6.033$ v.u.
M4 (2×): 1.598–1.649–0.874[h]–0.867–0.388–0.585 v.u. (type II MO_6 octahedra); $\Sigma s_i = 5.962$ v.u.
Average Σs_i for the MO_6 octahedra: 6.000 v.u.

[a] $d_{oi} = 1.9161$ Å.
[h] Terminal OH groups; an O–H bond valence of 1 v.u. has been assumed.
[p] The strong bonds are pseudoterminal M–O bonds, the weak bonds are the complementary bonds of the pseudoterminal oxygen atoms.

Table 21. Average bond lengths d and bond valences s of the symmetry-equivalent M–O bonds and average charges c on the different types of oxygen atoms in $Mo_{10}O_{34}^{8-}$ (I) (cation: NH_4^+)

M–O bond	O type frequency	M–O–M angle[a]	d in Å	s^b in v.u.	c in c.u.
M1–Ot1	2×		1.717	1.619	−0.381
M1–Ot2	2×		1.747	1.506	−0.494
M2–Ot3	2×		1.716	1.623	−0.377[c]
M2–Ot4	2×		1.721	1.604	−0.396
M3–Ot5	2×		1.694	1.712	−0.288
M4–Ot7	2×		1.698	1.696	−0.304
M4–Ot8	2×		1.730	1.569	−0.431
M5–Ot9[d]	6×		1.729	1.580	−0.420
M3–Opa	2×		1.754	1.481	−0.172
M1–Opa*			2.355	0.347	
M1–O2a1	2×		1.842	1.197	−0.097
M2–O2a1			2.061	0.705	
M2–O2a2[g]	2×		1.798	1.332	−0.164
M4–O2a2*			2.200	0.504	
M4–O2l	2×	154.0°	1.933	0.961	+0.072
M5–O2l			1.874	1.111	
M2–O3a*			2.354	0.348	
M3–O3a	2×		1.879	1.095	−0.130
M4–O3a*			2.268	0.428	
M1[i]–O3l			1.958	0.905	
M3–O3l*	2×		2.181	0.528	+0.226
M4–O3l			2.012	0.794	
M1–O4l*			2.198	0.507	
M2–O4l*	2×		2.179	0.530	+0.197
M3–O4l*			2.421	0.296	
M3[i]–O4l			1.977	0.864	
Sum of the negative charges:					−8.990 c.u.
Sum of the positive charges:					+0.990 c.u.
Total (formal ionic charge):					−8.000 c.u.

Bond valences about the different M atoms:

M1 (2×): 1.619–1.506–1.197–0.905–0.507–0.347[p] v.u. (type II MO_6 octahedra); $\Sigma s_i = 6.081$ v.u.[e]

M2 (2×): 1.623–1.604–0.705–1.332–0.530–0.348 v.u. (type II MO_6 octahedra)[g]; $\Sigma s_i = 6.142$ v.u.[e]

M3 (2×): 1.712–1.481[p]–1.095–0.864–0.528–0.296 v.u. (type II MO_6 octahedra); $\Sigma s_i = 5.975$ v.u.[e]

M4 (2×): 1.696–1.569–0.961[f]–0.794–0.504–0.428 v.u. (type II MO_6 octahedra); $\Sigma s_i = 5.952$ v.u.[e]

M5 (2×): 1.580[d]–1.580[d]–1.580[d]–1.111[f] v.u. (distorted MO_4 tetrahedra); $\Sigma s_i = 5.850$ v.u.[e]

Average Σs_i for the MO_k polyhedra: 6.000 v.u.

Table 21 (Continued)

[a] Calculated by the present authors on the basis of literature data.
[b] d_{oi} = 1.9165 Å.
[c] Inclusion of the very long M···O bonds leads to the following values including changes >0.03 v.u. or >0.03 c.u.: Ot3 −0.304 c.u.
[d] The M–Ot bonds of the MO_4 tetrahedra have been averaged; individual values: 1.722, 1.767, 1.696 Å; 1.600, 1.435, 1.704 v.u.; −0.400, −0.565, −0.296 c.u.
[e] Consideration of the very long M···O distances leads to 6.050, 6.128, 5.999, 5.903, and 5.920 v.u., respectively.
[f] Bridging M–O bonds between MO_6 octahedra and MO_4 tetrahedra.
[g] O2a2 is quasi a second Opa atom, i.e., M2 forms quasi a type III octahedron!
[p] The strong bonds are pseudoterminal M–O bonds, the weak bonds are the complementary bonds of the pseudoterminal oxygen atoms.

Table 22. Average bond lengths d and bond valences s of the symmetry-equivalent M–O bonds and average charges c on the different types of oxygen atoms in $Mo_{10}O_{34}^{8-}$ (II) (cation: $(CH_3)NH_3^+$)

M–O bond	O type frequency	M–O–M angle	d in Å	s[a] in v.u.	c in c.u.
M1–Ot1	2×		1.728	1.585	−0.415
M1–Ot2	2×		1.726	1.593	−0.407
M2–Ot3	2×		1.711	1.652	−0.348
M2–Ot4	2×		1.722	1.608	−0.392
M3–Ot5	2×		1.704	1.680	−0.320
M4–Ot6	2×		1.737	1.551	−0.449
M4–Ot8	2×		1.695	1.717	−0.283
M5–Ot9[b]	6×		1.748	1.512	−0.488
M3–Opa M1–Opa*	2×		1.768 2.350	1.439 0.353	−0.208
M1–O2a1 M2–O2a1	2×		1.902 1.950	1.041 0.927	−0.032
M2–O2a2 M4–O2a2	2×		1.926 1.922	0.983 0.992	−0.025
M4–O2l M5–O2l	2×	145.8°	2.011 1.819	0.800 1.272	+0.072
M2–O3a* M3–O3a M4–O3a*	2×		2.286 1.869 2.271	0.412 1.128 0.427	−0.034
M1[i]–O3l M3–O3l* M4–O3l*	2×		1.910 2.142 2.158	1.021 0.583 0.561	+0.165
M1–O4l* M2–O4l* M3–O4l* M3[i]–O4l	2×		2.243 2.235 2.433 1.949	0.457 0.466 0.289 0.929	+0.141
Sum of the negative charges:					−8.754 c.u.
Sum of the positive charges:					+0.756 c.u.
Total (formal ionic charge):					−7.998 c.u.

Table 22 (Continued)

Bond valences about the different M atoms:
M1 (2×): 1.585–1.593–1.041–1.021–0.353[p]–0.457 v.u. (type II MO_6 octahedra); Σs_i = 6.050 v.u.
M2 (2×): 1.652–1.608–0.927–0.983–0.412–0.466 v.u. (type II MO_6 octahedra); Σs_i = 6.047 v.u.
M3 (2×): 1.680–1.439[p]–1.128–0.929–0.583–0.289 v.u. (type II MO_6 octahedra); Σs_i = 6.048 v.u.
M4 (2×): 1.717–1.551–0.992–0.800[c]–0.427–0.561 v.u. (type II MO_6 octahedra); Σs_i = 6.048 v.u.
M5 (2×): 1.510–1.496–1.529–1.272[c] v.u. (distorted MO_4 tetrahedra); Σs_i = 5.808 v.u.
Average Σs_i for the MO_k polyhedra: 6.000 v.u.

[a] d_{oi} = 1.9187 Å.
[b] The M–Ot bonds of the MO_4 tetrahedra have been averaged; individual values: 1.752, 1.748, 1.743 Å; 1.496, 1.510, 1.529 v.u.; −0.504, −0.490, −0.471 c.u.
[c] Bridging M–O bonds between MO_6 octahedra and MO_4 tetrahedra.
[p] The strong bonds are pseudoterminal M–O bonds, the weak bonds are the complementary bonds of the pseudoterminal oxygen atoms.

the O–H groups are most probably involved in hydrogen bonds to oxygen atoms of neighboring dihydrogenoctamolybdate ions – if necessary via NH_4^+ cations or molecules of water of crystallization –, the negative charge on the protonated oxygen atoms will still be somewhat higher and that on the (most probably terminal) acceptor oxygen atom(s) of the hydrogen bond correspondingly somewhat lower. (It should be noted that the positions of the H atoms were deduced [59] on the basis of the large length of two seemingly terminal Mo–O bonds (1.972 Å) and the, without protonation, high negative charge (−1.13 c.u.) on the corresponding oxygen atoms.) The M–OH bond valence (0.87 v.u.) does not correspond to the usual value [56a] of ≈1.2 v.u.

- *Decamolybdate I and II ions:* The approximately linearly bridging two-coordinate oxygen atoms between the octahedral Mo_8O_{28} blocks and the MoO_4 tetrahedra bear a small positive charge (+0.07 c.u. in both isomers). The three terminal oxygen atoms of the MoO_4 tetrahedra (three sorts) are highly negatively charged (I: −0.30 to −0.56, altogether −2.51 c.u.; II: −0.47 to −0.50, altogether −2.93 c.u.).
- *Polyoctamolybdate ion:* The linearly bridging two-coordinate oxygen atoms connecting the octameric groups by corner-sharing bear a positive charge (+0.11 c.u.).

A more comprehensive study (see Table 37) of the $Mo_{10}O_{34}^{8-}$ (I) ion revealed that 1.6% of the bond valence of the Mo atoms are directed to additional oxygen atoms of the same and – in particular in the region of the MoO_4 tetrahedra – of neighboring decamolybdate ions, the greatest individual bond valences of these sorts reaching 0.08 v.u. (Mo5 ··· Ot4; 1.4%); 0.03 v.u. (Mo3 ··· O2a1; 0.4%); 0.03 v.u. (Mo5 ··· Ot8″ or Mo5 ··· Ot8?; 0.5% of the bond valence of Mo). In this way the very long M ··· O distances lead to a

Table 23. Average bond lengths d and bond valences s of the symmetry-equivalent M–O bonds and average charges c on the different types of oxygen atoms in $[Mo_8O_{27}^{6-}]_\infty$

M–O bond	O type frequency	M–O–M angle	d in Å	s[a] in v.u.	c in c.u.
M1–Ot1	2×		1.732	1.546	−0.454
M1–Ot2	2×		1.697	1.682	−0.318
M2–Ot3	2×		1.707	1.642	−0.358
M2–Ot4	2×		1.725	1.572	−0.428
M3–Ot5	2×		1.697	1.682	−0.318
M4–Ot6	2×		1.714	1.614	−0.386
M4–Ot8	2×		1.716	1.606	−0.394
M3–Opa M1–Opa*	2×		1.753 2.370	1.469 0.331	−0.200
M1–O2a1 M2–O2a1	2×		1.935 1.999	0.946 0.811	−0.243
M2–O2a2 M4–O2a2	2×		1.883 2.046	1.073 0.724	−0.203
M4–O2l M4[i]–O2l	1×	b	1.889 1.889	1.058 1.058	+0.115
M2–O3a* M3–O3a M4–O3a*	2×		2.348 1.874 2.198	0.349 1.097 0.501	−0.053
M3–O3l* M4–O3l* M1[i]–O3l	2×		2.156 2.196 1.897	0.555 0.504 1.037	+0.096
M2–O4l* M3–O4l* M3[i]–O4l M1–O4l*	2×		2.196 2.401 1.952 2.212	0.504 0.307 0.908 0.485	+0.204

Sum of the negative charges: −6.710 c.u.
Sum of the positive charges: +0.714 c.u.
Total (formal ionic charge): −5.996 c.u.

Bond valences about the different M atoms:

M1 (2×): 1.546–1.682–0.946–1.037–0.331[p]–0.485 v.u. (type II MO_6 octahedra); $\Sigma s_i = 6.027$ v.u.

M2 (2×): 1.642–1.572–0.811–1.073–0.349–0.504 v.u. (type II MO_6 octahedra); $\Sigma s_i = 5.950$ v.u.

M3 (2×): 1.682–1.469[p]–1.097–0.908–0.555–0.307 v.u. (type II MO_6 octahedra); $\Sigma s_i = 6.018$ v.u.

M4 (2×): 1.614–1.606–0.724–1.058[c]–0.501–0.504 v.u. (type II MO_6 octahedra); $\Sigma s_i = 6.007$ v.u.

Average Σs_i for the MO_6 octahedra: 6.000 v.u.

[a] $d_{0i} = 1.9122$ Å.
[b] This angle is obviously close to 180°.
[c] Bridging M–O bonds between the octameric groups.
[p] The strong bonds are pseudoterminal M–O bonds, the weak bonds are the complementary bonds of the pseudoterminal oxygen atoms.

considerable improvement of the bond valence sums, in particular for the MO_4 tetrahedra (in this case from 5.85 to 5.92 v.u.). Neglect of these additional oxygen atoms leads to errors in the M–O bond valences of up to +0.02 v.u. and in the charges of up to +0.02/−0.07 c.u.

The influence of the very long bonds in $Mo_{10}O_{34}^{8-}$ (II) has not be studied because the structure is not listed in the databases.

A more comprehensive study of the $[Mo_8O_{27}^{6-}]_\infty$ ion showed that the errors due to the neglect of the very long M···O distances are very small, see Sect. 4.1, in particular Table 37.

3.16
The Polyhexamolybdate Ion $[Mo_6O_{20}^{4-}]_\infty$ (o)

In the chain structure of this polyanion [60, 223] (Fig. 16) two crystallographically independent, edge-sharing MoO_6 octahedra of type II are present. (The bond lengths of another block-type $[Mo_6O_{20}^{4-}]_\infty$ chain structure [224] are not reliably characterized. A third chainlike $[Mo_6O_{20}^{4-}]_\infty$ structure is built by MoO_6 octahedra and tetragonal MO_5 pyramids and is treated in Sect. 3.23.) The relevant structural data are given in Table 24. They are based on a crystal structure investigation of a rubidium salt [223]. In a further investigation on a mixed sodium-rubidium salt [60] the bond lengths do not fulfill the reliability criterion.

Most of the negative ionic charge is located on the twelve terminal oxygen atoms (two sorts) of the repeating unit (−0.28 and −0.34, altogether −3.82 c.u.). The four three-coordinate angularly bridging oxygen atoms bear a negative charge of medium size (−0.14 c.u.). The four three-coordinate oxygen atoms involved in (approximately) linear Mo–O–Mo bridges bear a small positive charge (+0.09 c.u.).

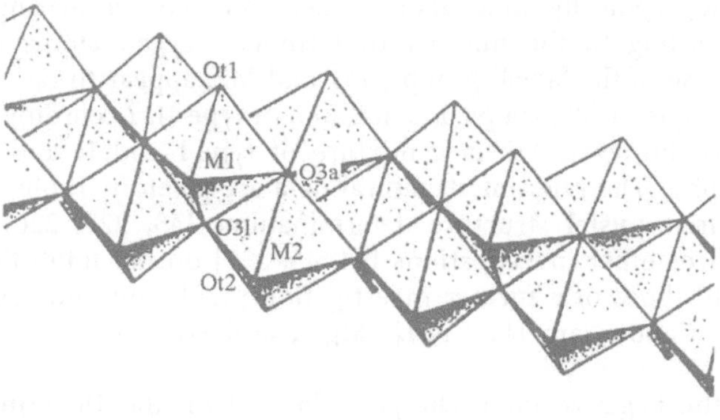

Fig. 16. Structure of the $[Mo_6O_{20}^{4-}]_\infty$ (o) ion and characterization of the crystallographically independent M and O atoms.

Table 24. Average bond lengths d and bond valences s of the symmetry-equivalent M–O bonds and average charges c on the different types of oxygen atoms in $[Mo_6O_{20}^{4-}]_\infty$ (o)

M–O bond	O type frequency	d[a] in Å	s[b] in v.u.	c in c.u.
M1–Ot1	4×	1.690	1.721	−0.278
M2–Ot2	8×	1.705	1.661	−0.339
M1–O3a		1.940	0.941	
M2–O3a*	4×	2.235	0.461	−0.137
$M2^i$–O3a*		2.235	0.461	
M1–O3l*[c]		2.515	0.301	
M2–O3l	4×	1.960	0.896	+0.094
$M2^{ii}$–O3l		1.960	0.896	
Sum of the negative charges:				−4.374 c.u.
Sum of the positive charges:				+0.375 c.u.
Total (formal ionic charge):				−3.999 c.u.

Bond valences about the different M atoms:

M1 (2×): 1.721–1.721–0.941–0.941–0.301–0.301 v.u. (type II MO_6 octahedra); $\Sigma s_i = 5.927$ v.u.

M2 (4×): 1.661–1.661–0.896–0.896–0.461–0.461 v.u. (type II MO_6 octahedra); $\Sigma s_i = 6.037$ v.u.

Average Σs_i for the MO_6 octahedra: 6.000 v.u.

[a] The individual values are given in the literature with two decimals.
[b] d_{0i} = 1.9147 Å.
[c] For the symmetry-equivalent M1–O3l bonds two strongly different bond lengths (2.82 and 2.21 Å) are given.

3.17 The Dodecatungstate (Paratungstate B, Paratungstate Z) Ion $W_{12}O_{40}(OH)_2^{10-}$

The structure of this cage-type polyanion is shown in Fig. 17. It contains four crystallographically independent edge- and corner-sharing WO_6 octahedra. According to the number of terminal oxygen atoms of the MO_6 octahedra those of the 'axial' groups, M1 and M2, appear to be of type I and those of the 'equatorial' groups, M3 and M4, of type II. According to the M–O bond lengths, however, M2 is a mixture of type I and II and only M1 is clearly of type I. The relevant structural data are given in Table 25. They are based on three crystal structure determinations [43, 225, 226]. The bond lengths of three other investigations [44, 45, 227] do not fulfill the reliability criterion, and those of a further investigation [228] could not unequivocally be assigned. Cations are Na^+, NH_4^+, Mg^{2+}, and H_3O^+ in combination with NH_4^+.

Most of the negative ionic charge is located on the 18 terminal oxygen atoms (five sorts, −0.37(7), −0.42(4), −0.48(3), −0.53(1), and −0.49(3), altogether −8.44 c.u.). The four corner-sharing (two-coordinate) oxygen

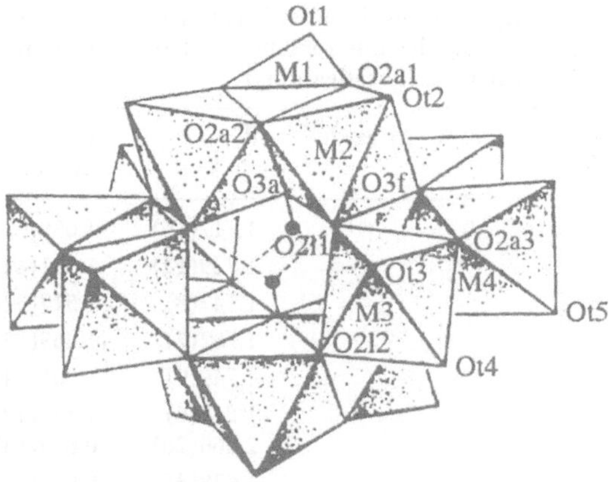

Fig. 17. Structure of the $W_{12}O_{40}(OH)_2^{10-}$ ion and characterization of the crystallographically independent M and O atoms (● H atoms as OH groups forming hydrogen bonds).

atoms (one sort), involved in approximately linear (136°) W–O–W bridges and having some pseudoterminal character, bear −0.17(2) c.u.; the other four corner-sharing (two-coordinate) oxygen atoms, also involved in approximately linear (146°) W–O–W bridges (one sort), bear −0.04(3) c.u. The ten angularly bridging two-coordinate oxygen atoms (three sorts) bear −0.19(1), −0.09(16!), and −0.05(1) c.u. The four three-coordinate oxygen atoms involved in a furcated W–O–W bridge (one sort; greatest W–O–W angle: 141°) bear −0.05(2) c.u. The two central, three-coordinate oxygen atoms (one sort) are protonated and bear +0.33(3) c.u., assuming an O–H bond valence of 1 v.u.

The high standard deviation for the charge on one of the edge-sharing two-coordinate oxygen atoms (O2a2) is due to an uncommon positive charge (+0.13 c.u.) obtained for one [226] of the investigations. We therefore also evaluated the data of the three crystal structure investigations that do not fulfill the reliability criterion. The six individual values for the charge on the oxygen atom in question (+0.13, −0.08, −0.15, −0.17, −0.19, −0.25 c.u.) indicate that the value +0.13 c.u. is obviously incorrect. Its elimination results in an average of −0.17(5) c.u. for the oxygen atom in question.

A more comprehensive study of one of the investigations [226] showed that the errors due to the neglect of the very long M···O distances are very small; see Sect. 4.1, in particular Table 37.

The hydrogen atoms of the OH groups are involved in intramolecular, furcated hydrogen bridges to neighboring oxygen atoms [44, 45, 229], those having some pseudoterminal character (O2l1). By this means the positive charge on the central oxygen atoms is compensated by the negative charge on the O2l1 oxygen atoms. (It should be noted that the positions of the H atoms were first deduced on the basis of the otherwise high negative charge (−0.67 c.u.) on the corresponding oxygen atoms and later confirmed by neutron diffraction studies [229].) The protonation of the two central oxygen atoms is a consequence of their exclusive participation in weak (long) M–O

Table 25. Average bond lengths d and bond valences s of the symmetry-equivalent M–O bonds and average charges c on the different types of oxygen atoms in $W_{12}O_{40}(OH)_2^{10-}$ (average of three reliable structural investigations)

M–O bond	O type frequency	M–O–M angle[a]	d in Å	s[b] in v.u.	c in c.u.
M1–Ot1	2×		1.717(16)	1.630(68)	−0.370(68)
M2–Ot2	4×		1.729(12)	1.581(42)	−0.419(42)
M3–Ot3	4×		1.747(8)	1.517(32)	−0.483(32)
M3–Ot4	4×		1.757(7)	1.469(13)	−0.531(13)
M4–Ot5	4×		1.749(9)	1.508(34)	−0.492(34)
M1–O2a1 M2–O2a1(*)	4×		1.867(6) 2.060(26)	1.132(19) 0.679(18)	−0.189(12)
M2–O2a2 M2^i–O2a2	2×		1.939(36) 1.939(36)	0.956(74) 0.956(74)	−0.087(161)[c]
M3–O2a3 M4–O2a3	4×		1.954(8) 1.907(6)	0.917(19) 1.028(14)	−0.055(10)
M2–O2(p)l1 M3–O2(p)l1*	4×	135.7°	1.801(2) 2.203(19)	1.330(5) 0.502(24)	−0.168(23)
M1^i–O2l2 M3–O2l2	4×	146.0°	1.961(12) 1.895(17)	0.900(28) 1.057(42)	−0.043(27)
M1–O3ah* M2–O3ah* M2^i–O3ah*	2×		2.286(26) 2.241(3) 2.241(3)	0.410(27) 0.458(0) 0.458(0)	+0.325(28)[h]
M3–O3f* M2–O3f M4–O3f*	4×	141.1° 123.8°	2.224(24) 1.891(11) 2.292(10)	0.476(26) 1.067(32) 0.404(11)	−0.053(16)

Sum of the negative charges: −10.646 c.u.
Sum of the positive charges on OH groups: +0.650 c.u.
Total (formal ionic charge): −9.996 c.u.

Bond valences about the different M atoms:
M1 (2×): 1.630−1.132−1.132−0.900−0.900−0.410[h] v.u. (type I MO_6 octahedra); $\Sigma s_i = 6.104$ v.u.
M2 (4×): 1.581−1.330[p]−0.956−1.067−0.679−0.458[h] v.u. (mixed type I/II MO_6 octahedra); $\Sigma s_i = 6.071$ v.u.
M3 (4×): 1.517−1.469−0.917−1.057−0.502[p]−0.476 v.u. (type II MO_6 octahedra); $\Sigma s_i = 5.938$ v.u.
M4 (2×): 1.508−1.508−1.028−1.028−0.404−0.404 v.u. (type II MO_6 octahedra); $\Sigma s_i = 5.880$ v.u.
Average Σs_i for the MO_6 octahedra: 6.000 v.u.

[a] Calculated by the present authors on the basis of literature [226] data.
[b] $\bar{d}_{0i} = 1.9175$ Å; $d_{0i\,min} = 1.9158$ Å, $d_{0i\,max} = 1.9185$ Å.
[c] A better value is −0.17(5) c.u., see text.
[h] These central oxygen atoms form OH groups; an O–H bond valence of 1 v.u. has been assumed. The positive charge is perfectly compensated by the negative charge on O2l1 via furcated hydrogen bridges; in this case an O–H bond valence of 0.67 v.u. is present.
[p] The strong bonds approximate to pseudoterminal M–O bonds, the weak bonds are the complementary bonds of the approximately pseudoterminal oxygen atoms.

bonds and small coordination number and of their inaccessibility: The small bond valence of "*trans*" M–O bonds and the small coordination number (three) of the respective oxygen atoms result in a very high negative charge for them. This charge must be neutralized by protonation because of their high basicity and because of the inaccessibility for a charge compensation by cations due to their central position.

The non-occurrence of positive charge on the corner-sharing two-coordinate O2l2 and on the three-coordinate O3f oxygen atoms, involved in approximately linear W–O–W bridges (W–O–W angles 146 and 141°, respectively), must be seen in connection with the high negative charge density of this dodecatungstate ion (see also Sect. 4.4.3.6 and the position of this species in Table 43).

3.18 The Dodecatungstate (Metatungstate) Ion $W_{12}O_{38}(OH)_2^{6-}$

The structure of this cage-type polyanion is shown in Fig. 18. All twelve edge- and corner-sharing WO_6 octahedra are crystallographically equivalent and of type I. The relevant structural data are given in Table 26 and are based on one crystal structure determination [230] only. The data of a further investigation [169] could not unequivocally be assigned since the two-coordinate edge- and corner-sharing oxygen atoms have not been distinguished. Cations are tetramethylammonium and tributylammonium.

Most of the negative ionic charge is located on the twelve terminal oxygen atoms (one sort) (−0.34, altogether −4.06 c.u.). The twelve angularly bridging (edge-sharing) two-coordinate oxygen atoms (one sort) bear −0.19 c.u. The twelve corner-sharing two-coordinate oxygen atoms (one sort) connecting the

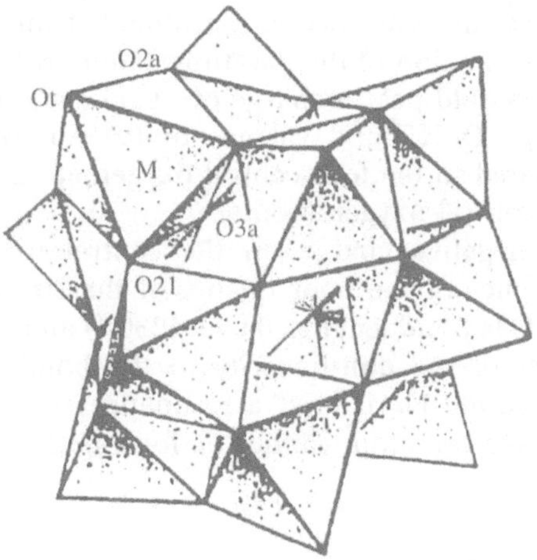

Fig. 18. Structure of the $W_{12}O_{38}(OH)_2^{6-}$ ion and characterization of the crystallographically independent M and O atoms (the two H atoms are bound to the four central O3a atoms forming a tetrahedron).

Table 26. Average bond lengths d and bond valences s of the symmetry-equivalent M–O bonds and average charges c on the different types of oxygen atoms in $W_{12}O_{38}(OH)_2^{6-}$

M–O bond	O type frequency	M–O–M angle	d^a in Å	s^b in v.u.	c in c.u.
M–Ot	12×		1.71	1.662	−0.338
M–O2a	12×		1.96	0.906	−0.189
M^i–O2a			1.96	0.906	
M–O2l	12×	145.3°	1.91	1.023	+0.045
M^{iii}–O2l			1.91	1.023	
M–O3ah*			2.22	0.482	
M^i–O3ah*	4×		2.22	0.482	-0.055^h
M^{ii}–O3ah*			2.22	0.482	
Sum of the negative charges:					−6.540 c.u.
Sum of the positive charges:					+0.542 c.u.
Total (formal ionic charge):					−5.998 c.u.

Bond valences about the M atoms:
M (12×): $1.662-0.906-0.906-1.023^c-1.023^c-0.482^h$ v.u. (type I MO_6 octahedra);
$\Sigma s_i = 6.000$ v.u.

[a] Only averaged M–O bond lengths are given in the literature.
[b] $d_{oi} = 1.9192$ Å.
[h] Twofold protonation of the four central oxygen atoms; an O–H bond valence of 1 v.u. has been assumed for each protonation, i.e., each O–H bond valence amounts on an average to 0.5 v.u.
[c] Bridging M–O bonds between the trimeric groups.

trimeric groups, involved in approximately linear W–O–W bridges (145°), bear a positive charge of +0.05 c.u. The four central (inaccessible), three-coordinate oxygen atoms (one sort) are protonated for the same reasons as discussed for the protonation of the paratungstate B ion (without protonation these oxygen atoms would bear a charge of −0.55 c.u.). They bear −0.05 c.u., assuming an average O–H bond valence of 0.5 v.u. (the two protons are statistically coordinated to the four central oxygen atoms and form hydrogen bonds to the other central oxygen atoms).

The very small negative charge on the central oxygen atoms is most probably a consequence of the poor quality of the structural data (standard deviation of the W–O3a bond lengths: 0.09 Å [230]!) and of the use of averaged bond lengths (effect of the non-linearity of the bond length-bond valence relationship, see footnote 1); at least a resonance structure with a negative charge on a central oxygen atom cannot be formulated.

3.19 The Decatungstate Ion $W_{10}O_{32}^{4-}$

The structure of this cage-type polyanion is shown in Fig. 19. It contains two crystallographically independent edge- and corner-sharing WO_6 octahedra of

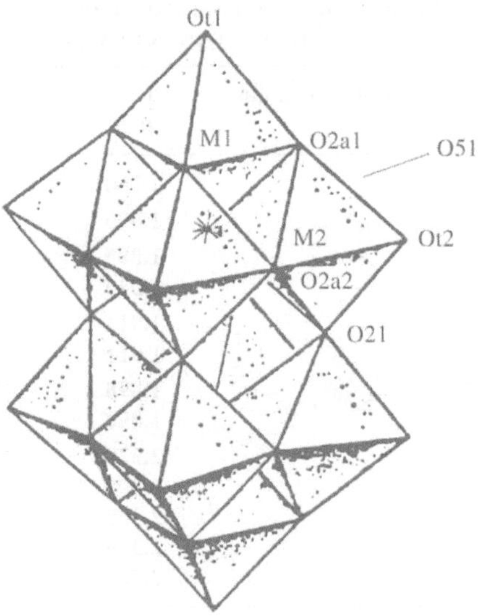

Fig. 19. Structure of the $W_{10}O_{32}^{4-}$ ion and characterization of the crystallographically independent M and O atoms.

type I. The relevant structural data are given in Table 27. They are based on one crystal structure determination [231] only. The cation is tri-*n*-butylammonium.

The ten terminal oxygen atoms (two sorts) bear most of the ionic charge (−0.53 and − 0.28, altogether −3.31 c.u.). The 16 two-coordinate angularly bridging oxygen atoms (two sorts) bear −0.16 and +0.04 c.u. The two central, five-coordinate oxygen atoms (one sort), involved in approximately linear W-O-W bridges, bear a small negative charge (−0.07 c.u.). The four approximately linearly (175°) bridging two-coordinate oxygen atoms (one sort) connecting the two pentameric building units bear a small positive charge (+0.11 c.u.).

According to our bond model [16], occurrence of negative charge (−0.07 c.u.) on the central, five-coordinate oxygen atoms is not allowed since coordination mode (V) (or (VI) and VII)) in Ref. 16 cannot be verified for the WO_6 octahedra of M1 and M2 if there is negative charge on the central oxygen atom(s); that is, a resonance formula with a negative charge on the central oxygen atoms cannot be formulated (unless with an O^{2-} ion or by violation of coordination mode (V)). However, since this charge is only small and the structure is that of a tungstate which reliably to investigate is difficult [35] (see Sect. 2.1), there is possibly no negative charge; an error of +0.02 Å (σ_d stated in the literature: 0.03 Å [231]) for the M-O5l bonds - all other bonds kept unchanged - can already explain the observed deviations since there are *four* identical M-O5l bonds which gives rise to the "arithmetic" error mentioned in Sect. 4.1.

Table 27. Average bond lengths d and bond valences s of the symmetry-equivalent M–O bonds and average charges c on the different types of oxygen atoms in $W_{10}O_{32}^{4-}$

M–O bond	O type frequency	M–O–M angle	d in Å	s[a] in v.u.	c in c.u.
M1–Ot1	2×		1.760	1.467	−0.533
M2–Ot2	8×		1.695	1.719	−0.281
M1–O2a1	8×		1.895	1.063	−0.159
M2–O2a1			2.023	0.778	
M2–O2a2	8×		1.911	1.018	+0.035
M2^{i}–O2a2			1.911	1.018	
M2–O21	4×	175.3°	1.898	1.055	+0.110
M2^{iv}–O21			1.898	1.055	
M1–O51*			2.320	0.377	
M2–O51*			2.308	0.389	
M2^{i}–O51*	2×		2.308	0.389	−0.067
M2^{ii}–O51*			2.308	0.389	
M2^{iii}–O51*			2.308	0.389	

Sum of the negative charges: −4.722 c.u.
Sum of the positive charges: +0.722 c.u.
Total (formal ionic charge): −4.000 c.u.

Bond valences about the different M atoms:
M1 (2×): 1.467–1.063–1.063–1.063–1.063–0.377 v.u. (type I MO_6 octahedra); $\Sigma s_i = 6.095$ v.u.
M2 (8×): 1.719–0.778–1.018–1.018–1.055[b]–0.389 v.u. (type I MO_6 octahedra); $\Sigma s_i = 5.976$ v.u.
Average Σs_i for the MO_6 octahedra: 6.000 v.u.

[a] d_{oi} = 1.9179 Å
[b] Bridging M–O bonds between the pentameric groups.

The reason for the very small positive charge on one sort of the two-coordinate angularly bridging oxygen atoms (O2a2) is obviously the same as that discussed for the (β-)octamolybdate ion (geometrical requirements due to the interdependence of the M–O bond lengths).

3.20
The α-Octamolybdate Ion α-$Mo_8O_{26}^{4-}$

The structure of this cage-type polyanion is shown in Fig. 20. There is one crystallographically independent MoO_6 octahedron and one such MoO_4 tetrahedron. The MoO_6 octahedra of type II share edges one with another, the MoO_4 tetrahedra share corners with the MoO_6 octahedra. The relevant structural data are given in Table 28. They are based on two crystal structure investigations, in one of which are given merely averaged Mo–O bond lengths [232]; the data of the other investigation [233] are only correctly described in

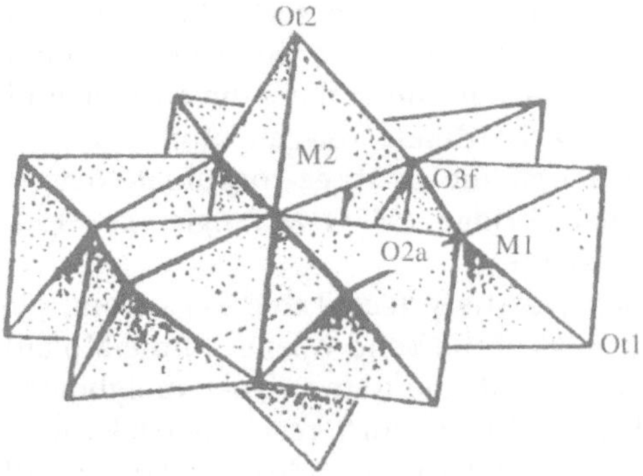

Fig. 20. Structure of the α-$Mo_8O_{26}^{4-}$ ion and characterization of the crystallographically independent M and O atoms.

Table 28. Average bond lengths d and bond valences s of the symmetry-equivalent M-O bonds and average charges c on the different types of oxygen atoms in α-$Mo_8O_{26}^{4-}$ (average of two reliable structural investigations)

M-O bond	O type frequency	M-O-M angle	d[a] in Å	s[b] in v.u.	c in c.u.
M1-Ot1	12×		1.694(2)	1.714(7)	−0.286(7)
M2-Ot2	2×		1.710(2)	1.650(8)	−0.350(8)
M1-O2a	6×		1.905(1)	1.030(2)	+0.060(5)
$M1^{i}$-O2a			1.905(1)	1.030(2)	
M1-O3f*			2.438(13)	0.285(7)	
$M1^{ii}$-O3f*	6×	130.5°	2.438(13)	0.285(7)	−0.038(7)
M2-O3f			1.780(3)	1.392(8)	
Sum of the negative charges:					−4.348 c.u.
Sum of the positive charges:					+0.348 c.u.
Total (formal ionic charge):					−4.000 c.u.

Bond valences about the different M atoms:

M1 (6×): 1.714−1.714−1.030−1.030−0.285[c]−0.285[c] v.u. (type II MO_6 octahedra); $\Sigma s_i = 6.058$ v.u.[d]

M2 (2×): 1.650−1.392[c]−1.392[c]−1.392[c] v.u. (distorted MO_4 tetrahedra); $\Sigma s_i = 5.824$ v.u.[d]

Average Σs_i for the MO_k polyhedra: 6.000 v.u.

[a] In one of the investigations [232] only averaged bond lengths are given.
[b] $\bar{d}_{oi} = 1.9167$ Å; $d_{oi\,min} = 1.9163$ Å, $d_{oi\,max} = 1.9172$ Å.
[c] Bonds of the bridging O atoms between MO_6 octahedra and MO_4 tetrahedra.
[d] Consideration of the very long M···O distances for the data of Ref. 233 leads to 6.024 and 5.930 v.u., respectively.

the Cambridge Structural DaTabase [62]. A further investigation confirms the structure, but no bond lengths are given [234]. The cations are *n*-propyltriphenylphosphonium and tetra-*n*-butylammonium.

The 14 terminal oxygen atoms (two sorts) bear most of the ionic charge, those of the MoO_6 octahedra −0.29 and those of the MoO_4 tetrahedra −0.35, altogether −4.13 c.u., which is more than the formal ionic charge. The six angularly bridging, two-coordinate oxygen atoms (one sort) bear a small (+0.06 c.u.) positive charge, the six three-coordinate oxygen atoms (one sort) involved in a furcated bridge (M–O–M angle: ca. 130.5°) a very small (−0.04 c.u.) negative charge.

A more comprehensive study (see Table 37) of one [233] of the investigations revealed that 1.3% of the bond valence of the Mo atoms are directed to additional oxygen atoms of the same α-octamolybdate ion, the greatest individual bond valences of these sorts reaching 0.03 v.u. (Mo2···O2a; 0.5%); 0.02 v.u. ($Mo2^i$···O3f; 0.3%); 0.01 v.u. ($Mo1^i$···Ot1; 0.2% of the bond valence of Mo). In this way the contribution of the very long M···O distances leads to a considerable improvement of the bond valence sums, in particular for the MO_4 tetrahedra (in this case from 5.82 to 5.93 v.u.). Neglect of these additional oxygen atoms leads to errors in the M–O bond valences of up to +0.02 v.u. and in the charges of up to +0.02/−0.01 c.u.

The reason for the uncommon positive charge (+0.06 c.u.) found for the angularly bridging, two-coordinate oxygen atoms is most probably the rigidity of the Mo–O framework, i.e., the interdependence of the M–O bond lengths as has been discussed for the (β-)octamolybdate and decatungstate ions.

3.21 The 36-Molybdate Ion $Mo_{36}O_{112}(H_2O)_{16}^{8-}$

The structure of this cage-type polyanion is shown in Fig. 21, the asymmetric unit (one half) in Fig. 22. There are nine crystallographically independent edge- and corner-sharing MoO_6 octahedra of type II and one such MoO_7 pentagonal bipyramid (of type I). A specialty is the occurrence of coordinated H_2O molecules which occupy the positions of unshared and of two-coordinate angularly bridging oxygen atoms. The relevant structural data are given in Table 29. They are based on two crystal structure determinations [55, 235]. The cations are Na^+ and K^+.

The 52 terminal oxygen atoms (13 sorts) bear most of the negative charge (−0.31(3), −0.33(1), −0.33(2), −0.41(5), −0.34(3), −0.36(7), −0.32(4), −0.30(2), −0.30(1), −0.35(4), 0.37(4), −0.35(0), −0.25(5)), altogether −17.21 c.u., which is by far more than the ionic charge of −8 c.u. Of the 16 pseudoterminal (two-coordinate) oxygen atoms twelve are involved in approximately linear bridges (three sorts) and four (those of the Mo_7 partial structures which form also in the $Mo_7O_{24}^{6-}$ structures the pseudoterminal oxygen atoms) in angular bridges (one sort). Those involved in the approximately linear bridges are uncharged or bear a small negative charge (−0.08(2), +0.01(1), and −0.09(0) c.u.) and those involved in the angular bridges a medium negative charge (−0.14(0) c.u.). The eight angularly bridging two-coordinate oxygen atoms (four sorts) bear a small or very small negative charge (−0.13(2), −0.02(1), −0.05(5), −0.03(4) c.u.). The four angularly bridging three-coordinate oxygen atoms (two sorts) bear a medium negative charge (−0.13(1) and −0.14(3) c.u.).

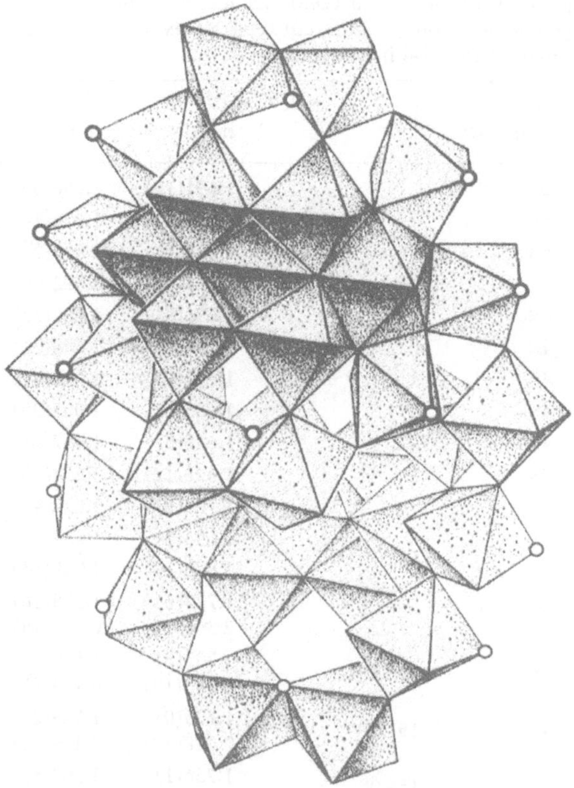

Fig. 21. Structure of the $Mo_{36}O_{112}(H_2O)_{16}^{8-}$ ion (O coordinated H_2O molecules).

The twelve two-coordinate oxygen atoms involved in approximately linear bridges (three sorts) bear a small but definite positive charge (+0.09(1), +0.07(1), and +0.10(0) c.u.). The 16 three-coordinate oxygen atoms (four sorts) involved in approximately linear bridges bear a medium positive charge (+0.17(1), +0.18(2), +0.20(0), and +0.24(1) c.u.). The four four-coordinate oxygen atoms (one sort) involved in approximately linear bridges bear a medium positive charge (+0.19(0) c.u.). The twelve water molecules located at a free corner (unshared O atoms) of an MoO_6 octahedron (three sorts) bear a high positive charge (+0.36(1), +0.33(2), and +0.33(1) c.u.), if O–H bond valences of 1 v.u. are assumed (taking the more realistic O–H bond valence of 0.8 v.u. the charge amounts to −0.04, −0.07, and −0.07 c.u., respectively; to the same degree, preferably on terminal and pseudoterminal oxygen atoms, the negative charges stated above are reduced too). The four water molecules forming an angular bridge between two molybdenum atoms (two sorts) bear a very high positive charge (+0.63(0) and +0.56(1) c.u.) if O–H bond valences of 1 v.u. are assumed (assuming the more realistic O–H bond valence of 0.8 v.u. the charge amounts to +0.23 and +0.16 c.u., respectively; again negative charges on terminal and pseudoterminal oxygen atoms are reduced to the same degree). (It should be noted that the positions of the H atoms were

Table 29. Average bond lengths d and bond valences s of the symmetry-equivalent M–O bonds and average charges c on the different types of oxygen atoms in $Mo_{36}O_{112}(H_2O)_{16}^{8-}$ (average of two reliable structural investigations)

M–O bond	O type frequency	M–O–M angle	d in Å	s[a] in v.u.	c in c.u.
M1–Ot1	4×		1.691(10)	1.686(35)	−0.314(35)
M1–Ot2	4×		1.694(4)	1.674(14)	−0.326(14)
M2–Ot3	4×		1.694(4)	1.669(22)	−0.331(22)
M2–Ot4	4×		1.714(13)	1.592(54)	−0.408(54)
M3–Ot5	4×		1.696(8)	1.662(35)	−0.338(35)
M4–Ot6	4×		1.701(17)	1.644(71)	−0.356(71)
M4–Ot7	4×		1.692(9)	1.678(41)	−0.322(41)
M5–Ot8	4×		1.685(6)	1.705(20)	−0.295(20)
M7–Ot9	4×		1.688(5)	1.697(14)	−0.303(14)
M8–Ot10	4×		1.698(12)	1.654(43)	−0.346(43)
M9–Ot11	4×		1.703(9)	1.633(41)	−0.367(41)
M10–Ot12	4×		1.698(2)	1.654(3)	−0.346(3)
M10–Ot13	4×		1.675(12)	1.749(46)	−0.251(46)
M6–Opa	4×		1.729(6)	1.535(16)	−0.135(2)
M5–Opa*			2.366(17)	0.330(15)	
M3–Opl1	4×	154.4°	1.733(3)	1.520(8)	−0.077(17)
M1–Opl1*			2.283(10)	0.403(9)	
M9–Opl2	4×	151.5°	1.725(0)	1.548(2)	+0.007(10)
M8–Opl2*			2.228(12)	0.459(12)	
M7–Opl3	4×	154.0°	1.736(1)	1.509(3)	−0.089(4)
M10–Opl3*			2.284(0)	0.402(2)	
M1–O2a1	2×		1.933(6)	0.937(11)	−0.126(22)
M1^i–O2a1			1.933(6)	0.937(11)	
M3–O2a2	2×		1.909(1)	0.992(5)	−0.016(10)
M3^i–O2a2			1.909(1)	0.992(5)	
M7–O2a3	2×		1.917(13)	0.976(27)	−0.048(54)
M7^i–O2a3			1.917(13)	0.976(27)	
M10–O2a4	2×		1.912(7)	0.987(18)	−0.026(36)
M10^i–O2a4			1.912(7)	0.987(18)	
M1–O2l1	4×	157.2°	1.923(2)	0.961(0)	+0.088(12)
M2–O2l1			1.857(6)	1.127(12)	
M4–O2l2	4×	158.4°	1.852(3)	1.138(11)	+0.066(5)
M8–O2l2			1.937(7)	0.928(16)	
M9–O2l3	4×	157.8°	1.930(5)	0.946(13)	+0.100(1)
M10–O2l3			1.847(5)	1.154(12)	
M3–O3a1*	2×		2.303(0)	0.384(1)	−0.129(7)
M3^i–O3a1*			2.303(0)	0.384(1)	
M6–O3a1			1.865(4)	1.104(8)	
M7–O3a2*	2×		2.278(11)	0.409(12)	−0.145(26)
M7^i–O3a2*			2.278(11)	0.409(12)	
M6–O3a2			1.890(0)	1.038(3)	
M2–O3l1	4×		2.046(24)	0.715(40)	+0.167(7)
M4–O3l1			2.071(6)	0.672(13)	
M5–O3l1			2.009(9)	0.780(26)	

Table 29 (Continued)

M2–O312*		2.158(16)	0.543(24)	
M3–O312	4×	1.982(8)	0.833(13)	+0.177(16)
M5–O312		1.998(15)	0.801(27)	
M9–O313		2.008(13)	0.783(21)	
M4–O313*	4×	2.150(15)	0.556(22)	+0.200(1)
M5–O313		1.968(2)	0.861(2)	
M9–O314*		2.104(8)	0.621(10)	
M7–O314	4×	1.984(2)	0.829(1)	+0.241(11)
M5–O314		2.004(2)	0.791(1)	
M3–O41*		2.110(21)	0.611(29)	
M7–O41*	4×	2.130(18)	0.583(27)	+0.192(4)
M5–O41		2.072(1)	0.671(3)	
M6–O41*		2.370(3)	0.327(4)	
M2–Otw1*	4×	2.333(9)	0.358(15)	+0.358(15)[b]
M4–Otw2*	4×	2.365(26)	0.332(20)	+0.332(20)[b]
M9–Otw3*	4×	2.365(6)	0.331(5)	+0.331(5)[b]
M1–O2aw1*	2×	2.388(2)	0.313(1)	+0.626(2)[b]
M1^{i}–O2aw1*		2.388(2)	0.313(1)	
M10–O2aw2*	2×	2.436(6)	0.279(3)	+0.557(5)[b]
M10^{i}–O2aw2*		2.436(6)	0.279(3)	

Sum of the negative charges:	−19.396 c.u.[b]
Sum of the positive charges except those on coordinated water molecules:	+4.952 c.u.
Sum of the positive charges on coordinated water molecules:	+6.450 c.u.[b]
Total (formal ionic charge):	−7.994 c.u.

Bond valences about the different M atoms:

M1 (4×): 1.686–1.674–0.961–0.937–0.403^{P}–0.313 v.u. (type II MO_6 octahedra); $\Sigma s_i = 5.974$ v.u.

M2 (4×): 1.669–1.592–1.127–0.715–0.543–0.358 v.u. (type II MO_6 octahedra); $\Sigma s_i = 6.004$ v.u.

M3 (4×): 1.662–1.520^{P}–0.992–0.833–0.611–0.384 v.u. (type II MO_6 octahedra); $\Sigma s_i = 6.002$ v.u.

M4 (4×): 1.678–1.644–1.138–0.672–0.556–0.332 v.u. (type II MO_6 octahedra); $\Sigma s_i = 6.020$ v.u.

M5 (4×): 1.705–0.780–0.801–0.671–0.791–0.861–0.330^{P} v.u. (type I pentagonal MO_7 bipyramids); $\Sigma s_i = 5.939$ v.u.

M6 (2×): 1.535^{P}–1.535^{P}–1.104–1.038–0.327–0.327 v.u. (type II MO_6 octahedra); $\Sigma s_i = 5.866$ v.u.

M7 (4×): 1.697–1.509^{P}–0.976–0.829–0.583–0.409 v.u. (type II MO_6 octahedra); $\Sigma s_i = 6.003$ v.u.

M8 (2×): 1.654–1.654–0.928–0.928–0.459^{P}–0.459^{P} v.u. (type II MO_6 octahedra); $\Sigma s_i = 6.082$ v.u.

M9 (4×): 1.633–1.548^{P}–0.946–0.783–0.621–0.331 v.u. (type II MO_6 octahedra): $\Sigma s_i = 5.862$ v.u.

M10 (4×): 1.749–1.654–1.154–0.987–0.402^{P}–0.279 v.u. (type II MO_6 octahedra): $\Sigma s_i = 6.225$ v.u.

Average Σs_i for the MO_k polyhedra: 6.000 v.u.

Table 29 (Continued)

[a] $\bar{d}_{oi} = 1.9056$ Å; $d_{0i\,min} = 1.9044$ Å, $d_{0i\,max} = 1.9068$ Å

[b] O–H bond valences of 1 v.u. have been assumed. Considering the commonly observed bond valence for O–H bonds of 0.8 v.u., the charge due to charge separation is +0.226 c.u. for O2aw1 and +0.157 c.u. for O2aw2; for Otw1 (−0.042 c.u.), Otw2 (−0.068 c.u.), and Otw3 (−0.069 c.u.) there is no charge separation. Correspondingly the sum of the negative charges is reduced by $16 \times 2 \times 0.2 = 6.4$ c.u., resulting in −12.996 c.u. The sum of the positive charges on coordinated water molecules is reduced to $[(0.226 + 0.157) \times 2 =]$ +0.766 c.u. and the new item "Sum of the negative charges on coordinated water molecules" amounts to $[-(0.042 + 0.068 + 0.069) \times 4 =] - 0.716$ c.u.

[p] The strong bonds are pseudoterminal M–O bonds, the weak bonds are the complementary bonds of the pseudoterminal oxygen atoms.

derived on the basis of the otherwise high negative charge on the corresponding oxygen atoms.)

The pattern of the Mo–O bonds of the MoO_7 pentagonal bipyramids corresponds to quasi type I MO_7 pentagonal bipyramids (compare Fig. 1). There are one strong (in axial position), five middle-strong (in the equitorial positions), and one weak bond (in *trans* position to the strong bond).

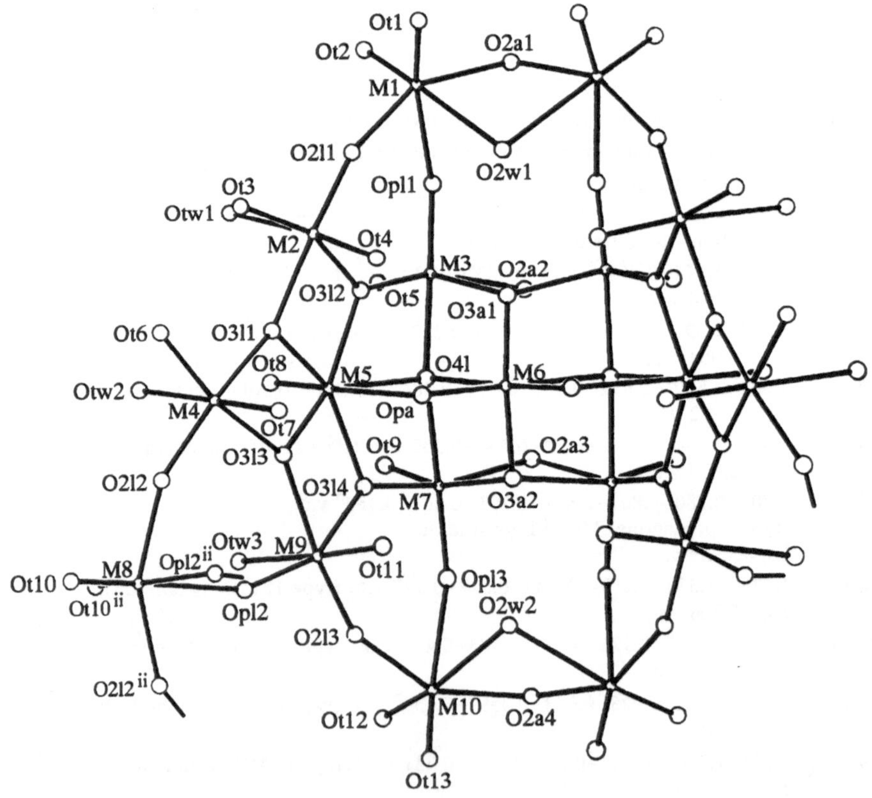

Fig. 22. Asymmetric unit (one half) of the $Mo_{36}O_{112}(H_2O)_{16}^{8-}$ ion and characterization of the crystallographically independent M and O atoms (this atom-and-bond model shows a view on the upper half of Fig. 21 from the plane of the paper).

The small negative charge on the (approximately) linearly bridging pseudoterminal oxygen atoms has to be seen as the result of a compensation of the usually medium-sized negative charge on pseudoterminal oxygen atoms by the usually small or medium-sized positive charge on linearly bridging oxygen atoms (see Table 41).

The hydrogen atoms of the coordinated water molecules are most probably involved in hydrogen bridges to neighboring oxygen atoms of the same or of neighboring Mo_{36} entities, as has already been intimated. The hydrogen bridges can also be arranged via molecules of water of crystallization. By this means the high or very high positive charge on the coordinated water-oxygen atoms is greatly reduced and simultaneously also the negative charge on the acceptor (usually terminal) oxygen atoms of the hydrogen bridges of the same or of neighboring Mo_{36} entities. The following charge balance for one of the structure investigations [55] has been described:

- Total negative charge: −19.72 c.u.; −18.75 c.u. on terminal and pseudoterminal, −0.97 c.u. on "normal" bridging oxygen atoms.
- −8.00 c.u. (the formal ionic charge) on the surface of the polyanion are balanced either by direct contact to the cations ($Na^{\oplus}\cdots{}^{\ominus}O\text{–}Mo$) or by indirect contact to the cations via hydrogen bonds of water of crystallization ($Na^{\oplus}\cdots\overset{H}{O}H\cdots{}^{\ominus}O\text{–}Mo$).
- −6.58 c.u. on suitable (terminal or bridging) oxygen atoms on the surface of the polyanion are balanced either by direct bonds to positively charged oxygen atoms of coordinated water molecules of the same or a neighboring Mo_{36} anion ($Mo\text{–}\overset{\oplus H}{O}H\cdots{}^{\ominus}O\text{–}Mo$) or via hydrogen bonds (via bridging hydrate water) ($Mo\text{–}\overset{\oplus H}{O}H\cdots\overset{H}{O}H\cdots{}^{\ominus}O\text{–}Mo$); this refers to the above-intimated hydrogen bridges.
- −5.16 c.u., mainly on terminal oxygen atoms of the polyanion, are intramolecularly balanced by the positive charge on bridging oxygen atoms generated by charge separation ($Mo\text{–}O^{\oplus}{=}Mo\text{–}O^{\ominus}$).

Note that the above formulas describing the charge balance represent in each case on the one hand sections of an extended framework accomplished by the interaction between cations, polyoxometalate ions, and molecules of water of crystallization and on the other hand only one of a very large number of canonical resonance formulas. This means that the cations, the oxygen atoms at the surface of the polyanions, and the molecules of water of crystallization all are multiply coordinated among one another (resulting in small bond valences) and thus form a network of weak bonds with the cations, polyanion-oxygens, and water-oxygens as nodes.

3.22
The Polytetramolybdate Ion $[Mo_4O_{14}^{4-}]_{\infty}$

The structure of the polymetalate chain is shown in Fig. 23. The tetrameric repeating unit contains two MoO_6 octahedra and two MoO_4 tetrahedra. All MoO_6 octahedra (type II) and all MoO_4 tetrahedra are crystallographically

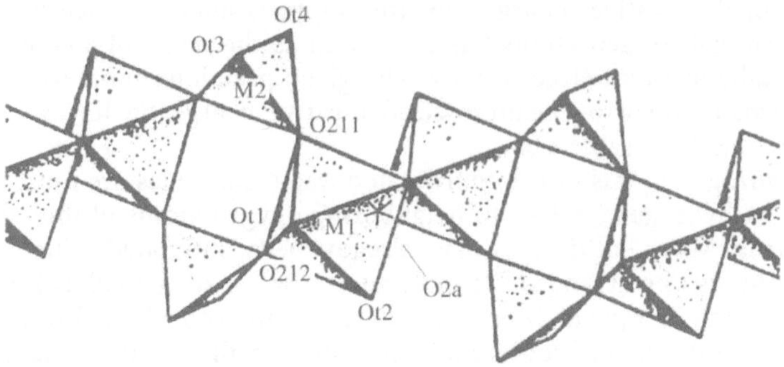

Fig. 23. Structure of the $[Mo_4O_{14}^{4-}]_\infty$ ion and characterization of the crystallographically independent M and O atoms.

equivalent. The MoO_6 octahedra share edges one with another, MoO_6 octahedra and MoO_4 tetrahedra share corners. The relevant structural data are given in Table 30. They are based on one reliable crystal structure determination [236]. Two other investigations [237, 238] are unsuited for an evaluation because the bond lengths do not fulfill the reliability criterion. The cations are ammonium and K^+ (the potassium salt was obtained from a melt).

The eight terminal oxygen atoms (two sorts on the MoO_6 octahedra and two sorts on the MoO_4 tetrahedra) bear most of the ionic charge, those four of the MoO_6 octahedra −0.44 and −0.39 c.u. and those four of the MoO_4 tetrahedra −0.48 and −0.36 c.u., altogether −3.33 c.u. The two angularly bridging two-coordinate oxygen atoms (one sort) between the MoO_6 octahedra have a medium-sized charge, −0.24 c.u. The four two-coordinate oxygen atoms bridging the MoO_6 octahedra and MoO_4 tetrahedra (two sorts) by Mo–O–Mo angles of 140 and 128° bear a small negative charge, −0.05 and −0.04 c.u.

A more comprehensive study (see Table 37) revealed that 1.9% of the bond valence of the Mo atoms are directed to additional oxygen atoms of the same and of neighboring chains, the greatest individual bond valences of these reaching 0.03 v.u. ($Mo2^i \cdots Ot1$; 0.4%); 0.07 v.u. ($Mo2 \cdots Ot4'''$; 1.2% of the bond valence of Mo). Neglect of these additional oxygen atoms leads to very small errors for the M–O bond valences of up to +0.03 v.u. and for the charges of up to +0.03/−0.05 c.u.

The absence of positive charge on the O211 oxygen atom forming a bridge of 140° between MoO_6 octahedron and MoO_4 tetrahedron has to be seen on the one hand in the relatively high negative charge density of the anion (small Z^+ value of only 1; see also Sect. 4.4.3.6 and the position of this species in Table 43) by which the capacity of acceptance of additional negative charge (formed by charge separation) by the terminal oxygen atoms is reduced. On the other hand this oxygen atom is involved in a weak M–O(*trans*) bond which makes it difficult for it to reach together with the second M–O bond (which is no pseudoterminal M–O bond!) bond valences of 2 v.u. (compare Sect. 4.3.4.4). The weak M–O(*trans*) bond and the approximate linearity of the M–O–M bridge in which O211 is involved are also the reason for the

Table 30. Average bond lengths d and bond valences s of the symmetry-equivalent M–O bonds and average charges c on the different types of oxygen atoms in $[Mo_4O_{14}^{4-}]_\infty$

M–O bond	O type frequency	M–O–M angle	d in Å	s[a] in v.u.	c in c.u.
M1–Ot1	2×		1.734	1.562	−0.438
M1–Ot2	2×		1.722	1.608	−0.392
M2–Ot3	2×		1.744	1.525	−0.475
M2–Ot4	2×		1.714	1.640	−0.360[b]
M1–O2a* M1^i–O2a	2×		2.166 1.840	0.550 1.209	−0.240
M1–O2l1*[c] M2–O2l1	2×	139.9°	2.311 1.735	0.388 1.559	−0.054
M1–O2l2 M2^i–O2l2	2×	127.8°	2.069 1.822	0.695 1.263	−0.041
Sum of the negative charges (formal ionic charge):					−3.999 c.u.

Bond valences about the different M atoms:
M1 (2×): 1.562–1.608–1.209–0.695[d]–0.550–0.388[d] v.u. (type II MO_6 octahedra); $\Sigma s_i = 6.013$ v.u.
M2 (2×): 1.525–1.640–1.559[d]–1.263[d] v.u. (distorted MO_4 tetrahedra); $\Sigma s_i = 5.988$ v.u.
Average Σs_i for the MO_k polyhedra: 6.000 v.u.

[a] $d_{oi} = 1.9187$ Å.
[b] Inclusion of the very long M···O bonds leads to the following values including changes >0.03 v.u. or >0.03 c.u.: Ot4 −0.314 c.u.
[c] O2l1 is quasi an Opl atom.
[d] Bridging M–O bonds between MO_6 octahedra and MO_4 tetrahedra.

unusually high bond valence of 1.56 v.u. for a bridging oxygen atom of an $M^{VI}O_4$ tetrahedron. The O2l2 oxygen atom forms only an angle of 128° and therefore does not favor a positive charge.

3.23
The Polyhexamolybdate Ion $[Mo_6O_{20}^{4-}]_\infty$ (o,tp)

The structure of the polymetalate chain is shown in Fig. 24. The identity period contains two crystallographically independent MoO_k polyhedra: MoO_6 octahedra of type II and tetragonal MoO_5 pyramids (of type II). The MoO_k polyhedra share edges. The relevant structural data are given in Table 31. They are based on two reliable crystal structure determinations [60, 239]. The data of an earlier investigation [240] have been interpreted with a different coordination of molybdenum. The cations are K^+ and Rb^+.

The four terminal oxygen atoms (one sort) of the two MoO_6 octahedra (−0.33(2) c.u.) and the eight ones (two sorts) of the four tetragonal MoO_5 pyramids (−0.49(1) and − 0.23(5) c.u.) in the hexameric identity period bear

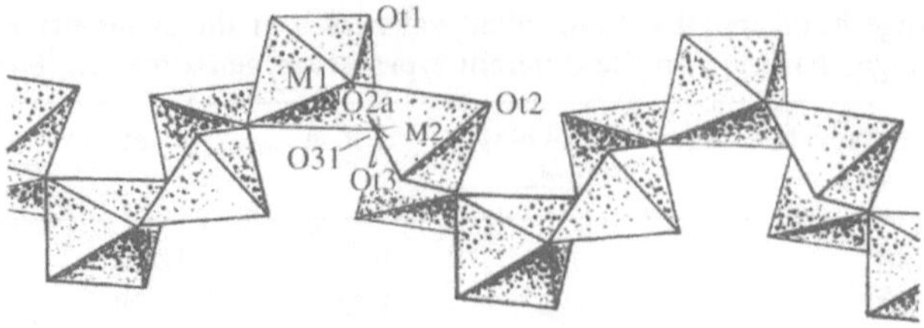

Fig. 24. Structure of the $[Mo_6O_{20}^{4-}]_\infty$ (o,tp) ion and characterization of the crystallographically independent M and O atoms. (The structures pictured in Refs. 60 and 13h are incorrect and that in Ref. 240 has been interpreted with a different coordination of Mo; only the structure presented in Ref. 239 is correct but does not consider the distortions of the MO_k polyhedra.)

Table 31. Average bond lengths d and bond valences s of the symmetry-equivalent M–O bonds and average charges c on the different types of oxygen atoms in $[Mo_6O_{20}^{4-}]_\infty$ (o,tp) (average of two reliable structural investigations)

M–O bond	O type frequency	d in Å	s[a] in v.u.	c in c.u.
M1–Ot1	4×	1.700(0)	1.665(15)	−0.335(15)
M2–Ot2	4×	1.740(0)	1.512(14)	−0.488(14)
M2–Ot3	4×	1.675(15)	1.770(49)	−0.230(49)
M1–O2a M2–O2a	4×	1.960(10) 1.920(10)	0.889(14) 0.979(32)	−0.132(19)
M1–O3l* M2–O3l* $M2^i$–O3l	4×	2.195(5) 2.025(25) 1.945(15)	0.504(2) 0.760(39) 0.922(42)	+0.186(2)
Sum of the negative charges:				−4.740 c.u.
Sum of the positive charges:				+0.744 c.u.
Total (formal ionic charge):				−3.996 c.u.

Bond valences about the different M atoms:
M1 (2×): 1.665–1.665–0.889–0.889–0.504–0.504 v.u. (type II MO_6 octahedra); $\Sigma s_i = 6.116$ v.u.
M2 (4×): 1.770–1.512[t]–0.979–0.922–0.760[tr] v.u. (quasi type II tetragonal MO_5 pyramids); $\Sigma s_i = 5.943$ v.u.
Average Σs_i for the MO_k polyhedra: 6.001 v.u.

[a] $\bar{d}_{oi} = 1.9110$ Å; $d_{0i\,min} = 1.9073$ Å, $d_{0i\,max} = 1.9147$ Å.
[t] Top of the MO_5 pyramid.
[tr] *Trans* position to the strongest M–O bond.

most of the ionic charge, altogether −4.21 c.u., which is more than the formal ionic charge. The four two-coordinate oxygen atoms (one sort) involved in angular Mo–O–Mo bridges bear a medium-sized negative charge (−0.13(2) c.u.). The four three-coordinate oxygen atoms involved in approximately linear Mo–O–Mo bridges bear a medium-sized positive charge of +0.19(0) c.u.

A more comprehensive study (see Table 37) of one [239] of the investigations revealed that 1.1% of the bond valence of the Mo atoms are directed to additional oxygen atoms of the same and of neighboring chains, the greatest individual bond valences of these reaching 0.01 v.u. (Mo2 ··· Ot2^{i}; 0.2%); 0.01 v.u. (Mo1 ··· Ot1$'''$; 0.1% of the bond valence of Mo). Neglect of these additional oxygen atoms leads to very small errors in the M–O bond valences of up to +0.01 v.u. and in the charges of up to +0.01/−0.01 c.u.

The pattern of the Mo–O bonds of the tetragonal MoO_5 pyramids corresponds to quasi type II tetragonal MO_5 pyramids. There are two strong (to the top and to a basis oxygen atom), two middle-strong (to basis oxygen atoms in *trans* position to each other), and a weak bond (in *trans* position to the strong bond of the basis oxygen atom); the second weak bond that should occur in *trans* position to the top oxygen atom is missing (see Fig. 1). Since there is only one weak bond, this is somewhat stronger than the (two) weak bonds of a type II MO_6 octahedron.

3.24 The Metavanadate Ion $[V_2O_6^{2-}]_\infty$

The structure of this polymetalate chain is shown in Fig. 25. The repeating unit contains two VO_5 trigonal bipyramids (of type II). They are crystallographically equivalent and share edges. The relevant structural data are given in Table 32. They are based on three crystal structure determinations [241–243]. The cations are Na^+ and Sr^{2+}.

The four terminal oxygen atoms (one sort) of the repeating unit bear most of the ionic charge (−0.53(1), altogether −2.11 c.u., which is more than the

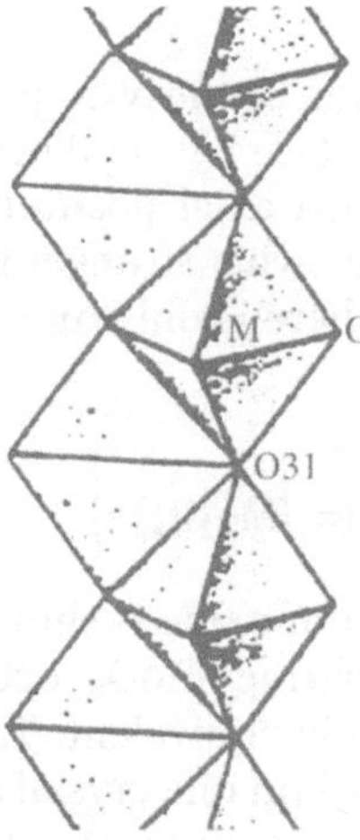

Fig. 25. Structure of the $[V_2O_6^{2-}]_\infty$ ion and characterization of the crystallographically independent M and O atoms.

Table 32. Average bond lengths d and bond valences s of the symmetry-equivalent M–O bonds and average charges c on the different types of oxygen atoms in $[V_2O_6^{2-}]_\infty$ (average of three reliable structural investigations)

M–O bond	O type frequency	d in Å	s[a] in v.u.	c in c.u.
M–Ot	4×	1.656(7)	1.472(7)	−0.528(7)
M–O3l		1.918(10)	0.724(15)	
M^i–O3l	2×	1.918(10)	0.724(15)	+0.056(14)
M^{ii}–O3l*		1.983(9)	0.607(21)	
Sum of the negative charges:				−2.112 c.u.
Sum of the positive charges:				+0.112 c.u.
Total (formal ionic charge):				−2.000 c.u.

Bond valences about the M atoms:
M (2×): $1.472^e–1.472^e–0.724–0.724–0.607^{e,tr}$ v.u.
(distorted type II trigonal MO_5 bipyramids);
$\Sigma s_i = 5.000$ v.u.

[a] $\bar{d}_{oi} = 1.7988$ Å; $d_{oi\,min} = 1.7928$ Å, $d_{oi\,max} = 1.8033$ Å.
[e] Equitorial positions of the trigonal MO_5 bipyramids.
[tr] *Trans* position to the strong M–O bond.

formal ionic charge). The two three-coordinate oxygen atoms (one sort), involved in approximately linear V–O–V bridges, are weakly positively charged (+0.06(1) c.u.).

A more comprehensive study (see Table 37) of one [241] of the $[V_2O_6^{2-}]_\infty$ structures revealed that 0.6% of the bond valence of the V atoms are directed to additional oxygen atoms (molecules of water of crystallization), the greatest individual bond valences reaching 0.01 v.u. (V···Ow; 0.1% of the bond valence of V). Neglect of these additional oxygen atoms leads to very small errors in the M–O bond valences of up to +0.01 v.u. and in the charges of up to +0.01 c.u.

The pattern of the V–O bonds of the VO_5 polyhedra corresponds to quasi type II trigonal VO_5 bipyramids (see Fig. 1). There are two strong (in equitorial positions), two middle-strong (in axial positions), and a weak (in equitorial position) bond. The latter is somewhat stronger than the (two) weak bonds of a type II MO_6 octahedron since there is only one bond of this kind.

3.25 Molybdenum Trioxide, "MoO_3" (= $[Mo_2O_6]_\infty$)

The structure of this molybdenum oxide is shown in Fig. 26. There are double layers of edge- and corner-sharing MoO_6 octahedra of type II. All MoO_6 octahedra are crystallographically equivalent. The relevant structural data are given in Table 33. They are based on one crystal structure determination [244].

The two terminal oxygen atoms (one sort) of the dimeric repeating unit bear a medium-sized negative charge of −0.20 c.u. The two pseudoterminal, approximately linearly bridging oxygen atoms (one sort) are virtually

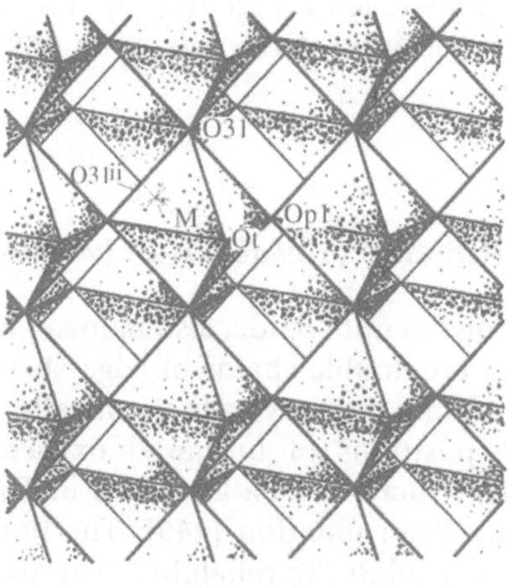

Fig. 26. Structure of molybdenum trioxide, $[Mo_2O_6]_\infty$("MoO_3"), and characterization of the crystallographically independent M and O atoms.

Table 33. Average bond lengths d and bond valences s of the symmetry-equivalent M-O bonds and average charges c on the different types of oxygen atoms in $[Mo_2O_6]_\infty$ (= "MoO_3")

M-O bond	O type frequency	M-O-M angle	d in Å	s^a in v.u.	c in c.u.
M-Ot	2×		1.671	1.801	−0.199
M-Opl	2×	b	1.734	1.547	−0.010
M^i-Opl*			2.251	0.443	
M-O3l			1.948	0.922	
M^i-O3l	2×		1.948	0.922	+0.209
M^{ii}-O3l*			2.332	0.365	
Sum of the negative charges:					−0.209·2 c.u.
Sum of the positive charges:					+0.209·2 c.u.
Total (formal ionic charge):					0.000·2 c.u.

Bond valences about the M atoms:
M (4×): $1.801-1.547^p-0.922-0.922-0.365-0.443^p$ v.u. (type II MO_6 octahedra);
$\Sigma s_i = 6.000$ v.u.

[a] $d_{oi} = 1.9145$ Å.
[b] This angle is obviously close to 180°.
[p] The strong bonds are pseudoterminal M-O bonds, the weak bonds are the complementary bonds of the pseudoterminal oxygen atoms.

uncharged (−0.01 c.u.). The two three-coordinate oxygen atoms (one sort), involved in approximately linear Mo-O-Mo bridges, bear a medium-sized positive charge of +0.21 c.u.

The layers are obviously hold together by the electrostatic attraction between the negatively and positively charged oxygen atoms which are opposed to each other.

3.26 Molybdenum Trioxide Hydrate, "(α-)$MoO_3 \cdot H_2O$" (= $[Mo_2O_6(H_2O)_2]_\infty$)

The structure of this molybdenum oxide, also named "white molybdic acid", is shown in Fig. 27. There are double chains of edge-sharing MoO_6 octahedra of type II. All MoO_6 octahedra are crystallographically equivalent. The water molecules occupy the positions of unshared oxygen atoms on the MoO_6 octahedra. The relevant structural data are given in Table 34. They are based on one crystal structure determination [245]. The bond lengths of a further investigation [246] do not fulfill the reliability criterion.

The four terminal oxygen atoms (two sorts) of the dimeric repeating unit bear a negative charge of −0.29 and −0.28 c.u. The two three-coordinate oxygen atoms (one sort), involved in approximately linear Mo–O–Mo bridges, bear a medium-sized positive charge of +0.22 c.u. The two water molecules (one sort) at free corners of the MoO_6 octahedra (unshared O atoms) bear a high positive charge (+0.35 c.u.), if an O–H bond valence of 1 v.u. is assumed.

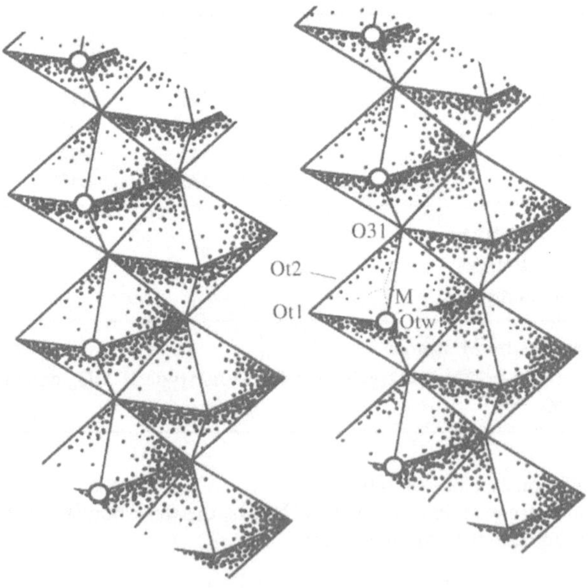

Fig. 27. Structure of the molybdenum oxide $[Mo_2O_6(H_2O)_2]_\infty$ ("(α-)$MoO_3 \cdot H_2O$") and characterization of the crystallographically independent M and O atoms (O coordinated H_2O molecules).

Table 34. Average bond lengths d and bond valences s of the symmetry-equivalent M-O bonds and average charges c on the different types of oxygen atoms in $[Mo_2O_6(H_2O)_2]_\infty$ (= "(α-)$MoO_3 \cdot 2H_2O$")

M–O bond	O type frequency	d in Å	s[a] in v.u.	c in c.u.
M–Ot1	2×	1.688	1.715	−0.285
M–Ot2	2×	1.687	1.719	−0.281
M–O31*		2.277	0.413	
M^i–O31	2×	1.959	0.890	+0.217
M^{ii}–O31		1.949	0.914	
M–Otw*	2×	2.346	0.350	+0.350[b]

Sum of the negative charges: −0.567·2 c.u
Sum of the positive charges except those on coordinated water molecules: +0.217·2 c.u.
Sum of the positive charges on coordinated water molecules: +0.350·2 c.u.
Total (formal ionic charge): 0.000·2 c.u.

Bond valences about the M atoms:
M (2×): 1.715–1.719–0.890–0.914–0.413–0.350 v.u. (type II MoO_6 octahedra);
$\Sigma s_i = 6.000$ v.u.

[a] $d_{oi} = 1.9112$ Å.
[b] O–H bond valences of 1 v.u. have been assumed. Considering the commonly observed bond valence for O–H bonds of 0.8 v.u., there is a very small negative charge on Otw (−0.050 c.u.), a small negative charge on Ot1 and Ot2 (−0.085 and −0.081 c.u., respectively), and a medium-sized positive charge on O31 (+0.217 c.u.).

The hydrogen atoms of the water molecules are involved in hydrogen bridges to the terminal oxygen atoms of the MoO_6 octahedra of neighboring double chains [245]. By this means the high positive charge on the water-oxygen atoms is strongly reduced (assuming the more realistic O–H bond valence of 0.8 v.u., the charge amounts to −0.05 c.u.) and simultaneously also the negative charge on the participating terminal oxygen atoms of the neighboring double chains (which then have charges of −0.09 and −0.08 c.u.). The hydrogen bridges connect the double chains with each other. The positive charge thus remaining (positive charge not compensated by the formation of hydrogen bonds) amounts to +0.22 c.u. Since this positive and the negative charge of neighboring double chains are opposed to each other, the holding together of the double chains can additionally be described by the electrostatic attraction of opposite charges.

3.27 Molybdenum Trioxide Dihydrate, "$MoO_3 \cdot 2H_2O$" (= $\{[Mo_{16}O_{48}(H_2O)_{16}] \cdot 16H_2O\}_\infty$)

The structure of this molybdenum oxide, also named "yellow molybdic acid", is shown in Fig. 28. There are layers of corner-sharing MoO_6 octahedra of type II

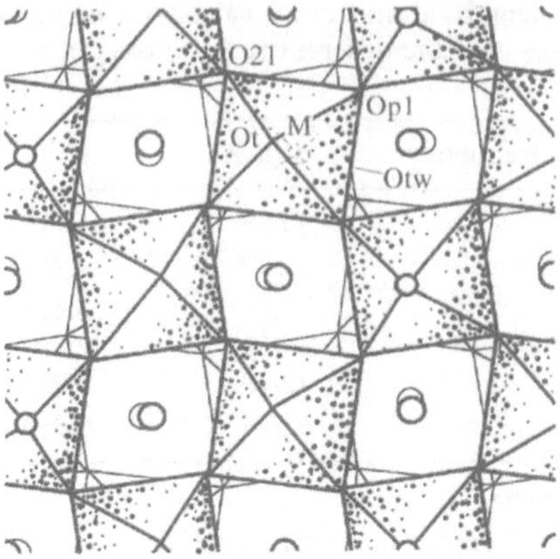

Fig. 28. Structure of the molybdenum oxide $\{[Mo_{16}O_{48}(H_2O)_{16}] \cdot 16H_2O\}_\infty$ ("$MoO_3 \cdot 2H_2O$") and characterization of the crystallographically independent M and O atoms (O ◌ coordinated H_2O molecules, ○ ○ water of crystallization between the layers).

(according to Ref. 9 they are of type I). The repeating unit contains four crystallographically independent MoO_6 octahedra which are, however, very similar to each other with regard to the bond lengths and are therefore taken as equivalent. One half of the H_2O molecules is coordinated to molybdenum and hence are constituents of the layers. The other half acts as hydrate water between the layers. The relevant structural data are given in Table 35. They are based on two reliable crystal structure determinations [247, 248].

The 16 terminal oxygen atoms (one sort) bear a negative charge of −0.34(1) c.u. The 16 pseudoterminal, approximately linearly bridging oxygen atoms (one sort) bear a small negative charge (−0.07(1) c.u.). The other 16 approximately linearly bridging two-coordinate oxygen atoms (one sort) are uncharged (+0.01(2) c.u.). The 16 water molecules (one sort) located at a free corner of the MoO_6 octahedra bear a positive charge of +0.40(1) c.u., if an O-H bond valence of 1 v.u. is assumed.

Both types of the water molecules and the terminal and pseudoterminal oxygen atoms of the MoO_6 octahedra are involved in hydrogen bridges [247, 248]. Assuming the more realistic O-H bond valence of 0.8 v.u., the positive charge on the water-oxygen atoms disappears, and the negative charge on the terminal and pseudoterminal oxygen atoms reduces to −0.01 c.u.

Surprisingly, type I MoO_6 octahedra do not appear although from a geometrical point of view the conditions for this type are absolutely fulfilled. Instead of this type II MoO_6 octahedra occur forming a network with alternatingly strong and weak (short and long) Mo–O bonds.

The assumption of a "formal bond order" greater than two [247] for the Mo–Ot bonds does not correspond with our calculations (s = 1.66 v.u.).

Table 35. Average bond lengths d and bond valences s of the symmetry-equivalent M–O bonds and average charges c on the different types of oxygen atoms in $\{[Mo_{16}O_{48}(H_2O)_{16}] \cdot 16H_2O\}_\infty$ (= "$MoO_3 \cdot 2H_2O$") (average of two reliable structural investigations)

M–O bond	O type frequency	M–O–M angle[a]	d in Å	s^b in v.u.	c in c.u.
M–Ot	16×		1.696(2)	1.655(12)	−0.345(12)
M–Opl M^i–Opl*	16×	149.1°	1.765(1) 2.166(10)	1.400(1) 0.533(13)	−0.067(14)
M–O2l M^{ii}–O2l	16×	150.4°	1.794(7) 2.052(3)	1.308(18) 0.701(3)	+0.009(20)
M–Otw*	16×		2.281(7)	0.403(6)	+0.403(6)[c]

Sum of the negative charges:	−0.412·16 c.u.
Sum of the positive charges except those on coordinated water molecules:	+0.009·16 c.u.
Sum of the positive charges on coordinated water molecules:	+0.403·16 c.u.
Total (formal ionic charge):	0.000·16 c.u.

Bond valences about the M atoms:

M (16×): 1.655–1.400[P]–1.308–0.701–0.533[P]–0.403 v.u. (type II MO_6 octahedra); $\Sigma s_i = 6.000$ v.u.

[a] Average values.

[b] $\bar{d}_{oi} = 1.9046$ Å; $d_{oi\,min} = 1.9037$ Å, $d_{oi\,max} = 1.9055$ Å.

[c] O–H bond valences of 1 v.u. have been assumed. Considering the commonly observed bond valence for O–H bonds of 0.8 v.u., there is no positive charge on Otw and no negative charge on Ot and Opl.

[P] The strong bonds are pseudoterminal M–O bonds, the weak bonds are the complementary bonds of the pseudoterminal oxygen atoms.

3.28 Divanadium Pentoxide, "V_2O_5" (= $[V_4O_{10}]_\infty$)

The structure of this oxide is shown in Fig. 29. There are layers of edge- and corner-sharing tetragonal VO_5 pyramids (of type I) which all are crystallographically equivalent. The relevant structural data are given in Table 36. They are based on one crystal structure determination [249]. In earlier investigations [250, 251] trigonal VO_5 bipyramids have been assumed.

The four terminal oxygen atoms (one sort) of the repeating unit bear a medium negative charge (−0.19 c.u.). The two approximately linearly bridging two-coordinate oxygen atoms bear a small positive charge (+0.09 c.u.), and the four three-coordinate oxygen atoms, involved in approximately linear M–O–M bridges, bear a medium positive charge (+0.15 c.u.).

A more comprehensive study (see Table 37) revealed that 2.3% of the bond valence of the V atoms are directed to additional oxygen atoms of the same and of the neighboring layers, the greatest individual bond valence of these reaching 0.07 v.u. (V · · · Ot‴; 1.3% of the bond valence of V). The next longer V–O distances correspond to 0.01 v.u. Neglect of the additional oxygen atoms

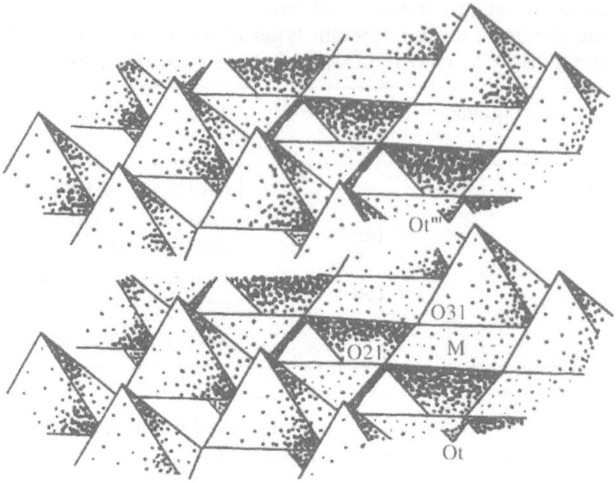

Fig. 29. Structure of divanadium pentoxide, $[V_4O_{10}]_\infty$ ("V_2O_5"), and characterization of the crystallographically independent M and O atoms.

leads to small errors in the M–O bond valences of up to +0.04 v.u. and in the charges of up to +0.05/−0.07 c.u.

The pattern of the V–O bonds of the tetragonal VO_5 pyramids corresponds to quasi type I tetragonal VO_5 pyramids (see Fig. 1). The strong bond points to

Table 36. Average bond lengths d and bond valences s of the symmetry-equivalent M–O bonds and average charges c on the different types of oxygen atoms in $[V_4O_{10}]_\infty$ (= "V_2O_5")

M–O bond	O type frequency	M–O–M angle	d in Å	s^a in v.u.	c in c.u.
M–Ot	4×		1.577	1.806^b	-0.194^b
M–O21	2×	[c]	1.779	1.045	$+0.091^b$
M^i–O21			1.779	1.045	
M–O31			1.878	0.800	
M^i–O31	4×		1.878	0.800	$+0.148^b$
M^{ii}–O31			2.017	0.549	
Sum of the negative charges:					−0.194·4 c.u.
Sum of the positive charges:					+0.194·4 c.u.
Total (formal ionic charge):					0.000·4 c.u.

Bond valences about the M atoms:

M (4×): 1.806^t–1.045–0.800–0.800–0.549 v.u. (quasi type I tetragonal MO_5 pyramids); $\Sigma s_i = 5.000$ v.u.

[a] $d_{oi} = 1.7954$ Å.

[b] Inclusion of the very long M···Ot''' bonds leads to the following values including changes >0.03 v.u. or >0.03 c.u.: M–Ot 1.767 v.u.; Ot −0.129 c.u.; O21 +0.056 c.u.; O31 +0.101 c.u.

[c] This angle is obviously close to 180°.

[t] Top of the pyramid.

the top, the four middle-strong bonds to the corners of the basis of the pyramid. The "*trans*" bond is virtually absent: The V···O distances between the layers (2.791 Å) do not correspond to V–O bonds [249]. According to our calculations this distance corresponds to 0.07 v.u. (1.3% of the bond valence of the V atoms), see above. The layers are hold together by the electrostatic attraction of the negatively and positively charged oxygen atoms which are opposed to each other.

4 General Discussion

4.1 Accuracy of the Results

The bond length-bond valence relationships used [35] (Eq. (1a) and the pertinent parameters of Table 1) and the relevant measuring data (the M–O bond lengths) for a large number of the (iso)polyoxometalate structures are sufficiently accurate to give reliable data for the M–O bond valences [estimated standard deviation when using the general B and the individually fitted d_{0i} parameters: 0.012 v.u., see Table 1; this value corresponds to a reasonable standard deviation (random error) for the bond lengths of 0.010 Å] and for the distribution of the ionic charge over the oxygen atoms of the polyanions (estimated standard deviation under the same conditions: 0.020 c.u.). Data that did not fulfill certain reliability criteria (those of a few polymolybdates and vanadates, but of the majority of the polytungstates; cf. Sect. 2.1 and Ref. 35) have only occasionally (in the case of important structure types) been evaluated. Notice also that in certain cases (for monometalate and polymetalate ions, if only one type of MO_k polyhedra is present and averaged bond lengths and individually fitted d_{oi} parameters are used; see Sect. 2.1) the structures appear always reliable.

In the past such data and relationships have obviously been considered too uncertain [13b] and often been misinterpreted [9, 13b, 232, 234, 252] (see Sect. 4.8.1). They have mainly been used, as mentioned, to identify the oxygen atoms constituting an OH group or a coordinated H_2O molecule [43–45, 47, 48, 55, 59, 203, 235] or to evaluate the hydrogen bridge system [41, 42, 44, 45]. For these purposes the requirements on the accuracy of the structural data are considerably smaller.

In addition to the above errors there are the small ($\leq$0.07 v.u./c.u.) or even very small ($\leq$0.03 v.u./c.u.) errors due to the intentional neglect of the very long M···O bonds. (Only for two divanadates are there greater errors for the charge of 0.14 and 0.08 c.u.) The following cases can be distinguished (Table 37):

- Within the block-type structures and partial structures the typical next smaller bond valences belong to M and O atoms of neighboring MO_6 octahedra sharing edges (e.g., M3···O2a1 or M1···Ot3 in $M_7O_{24}^{6-}$; typical

Table 37. Very long M···O bonds in some selected mono- and polyoxometalate ions and influence of their neglect

Oxometalate ion		Fraction of the very long M···O bonds[a]	Shortest very long M···O bond of a type		Maximum influence of the neglect of the very long M···O bonds	
			Bond	v.u.	v.u.	c.u.
a. Data for some selected block-type structures:						
$Mo_7O_{24}^{6-}$	[166]	1.2%	Mo3···O2a1	0.02	+0.02	+0.02/−0.01
			Mo2···Ow	0.01		
(β-)$Mo_8O_{26}^{4-}$	[184]	1.3%	Mo3···O2a	0.03	+0.02	+0.02/−0.01
$Mo_6O_{19}^{2-}$	[184]	1.4%	Mo^{ii}···O2a	0.02	+0.02	+0.03/−0.01
b. Data for some selected structures comprising neighboring block-type partial structures connected by common corners:						
$[Mo_8O_{27}^{6-}]_\infty$	[41]	1.5%	Mo3···O2a1	0.03	+0.02	+0.02/−0.02
			Mo4···Ot5^{ii}″	0.02		
			Mo2···Ow	0.01		
$W_{12}O_{40}(OH)_2^{10-}$	[226]	2.0%	W2···O2a3	0.05	+0.03	+0.03/−0.03
			W1···Ow	0.01		
c. Data for some selected monometalate ions and for most of the polyoxometalate structures comprising MO_k polyhedra with k < 6 (as far as the atomic coordinates are given in the literature):						
MoO_4^{2-}[b]	[72]	2.5%	Mo···Ot′	0.05	+0.03	+0.02/−0.06[c]
MoO_4^{2-}	[76]	2.5%	Mo···Ot′	0.05	+0.03	+0.03/−0.02[c]
			Mo···Ow	0.02		
MoO_4^{2-}	[79]	0.7%	Mo···Ot′	0.01	+0.01	+0.01/−0.00[c]
MoO_4^{2-}	[87]	2.9%	Mo···Ot′	0.08	+0.05	+0.05/−0.04[c]
			Mo···Ow	0.02		
MoO_4^{2-}	[91]	1.8%	Mo···Ot′	0.02	+0.02	+0.02/−0.01[c]
			Mo···Ow	0.02		
MoO_4^{2-}	[92]	0.9%	Mo···Ow	0.02	+0.01	+0.01/[c]
			Mo···O(18-crown-6)	0.01		

VO^{3-b}	[101]	1.4%	V···Ot′	0.02	+0.02	+0.01/−0.02[c]
VO_4^{3-}	[102]	0.6%	V···Ow	0.01	+0.01	+0.01/[c]
VO_4^{3}	[103]	2.8%	V···Ot′	0.02	+0.04	+0.01/−0.01[c]
VO_4^{3-}	[121]	0.0%			+0.00	+0.00/−0.00[c]
VO_4^{3-}	[122]	3.3%	V···Ot′	0.01	+0.03	+0.02/−0.04[c]
VO_4^{3-}	[124]	3.9%	V···Ot′	0.07	+0.05	+0.04/−0.04[c]
$V_2O_7^{4-}$	[135–137, 139–141, 146]	≦1.5%	V···Ot[?d,e]	≦0.02	<+0.04	+0.04/−0.04[c]
			V···Ow	0.01		
$V_2O_7^{4-}$	[138]	3.6%	V···Ot″	0.17	+0.05	+0.06/−0.14[c]
			V[i]···Ot	0.02		
			V···Ot‴	0.01		
$V_2O_7^{4-}$	[142]	3.1%	V···Ot[?d,e]	0.11	+0.04	+0.06/−0.08[c]
$V_2O_7^{4-}$	[143]	2.5%	V···Ot[?d,e]	0.02	+0.04	+0.02/−0.01[c]
$V_2O_7^{4-}$	[144, 145]	1.9%	V···Ot[?d,e]	0.03	+0.02	+0.04/−0.02[c]
$V_2O_7^{4-}$	[147]	1.5%	V···Ot[?d,e]	0.05	+0.02	+0.03/−0.03[c]
$V_3O_{10}^{5-}$	[152]	0.9%	V2···Ot1	0.01	+0.01	+0.01/[c]
			V_1···Ow	0.01		
$[VO_3^-]_\infty$	[159]	0.9%	V···Ot″	0.01	+0.01	+0.01/−0.01
			V···Ot‴	0.01		
$V_5O_{14}^{3-}$	[163]	0.3%	V1[i]···O2l	0.00	+0.00	+0.00/−0.00
$Mo_{10}O_{34}^{8-}$ (I)	[221]	1.6%	Mo5···Ot4	0.08	+0.02	+0.02/−0.07
			Mo3···O2a1	0.03		
			Mo5···Ot8″ or Mo5···Ot8[d]	0.03		
α-$Mo_8O_{26}^{4-}$	[233]	1.3%	Mo1[i]···Ot1	0.01	+0.02	+0.02/−0.01
			Mo2[i]···O3f	0.02		
			Mo2···O2a	0.03		
$[Mo_4O_{14}^{4-}]_\infty$	[236]	1.9%	Mo2[i]···Ot1	0.03	+0.03	+0.03/−0.05
			Mo2···Ot4‴	0.07		

Table 37 (Continued)

Oxometalate ion		Fraction of the very long M···O bonds[a]	Shortest very long M···O bond of a type		Maximum influence of the neglect of the very long M···O bonds	
			Bond	v.u.	v.u.	c.u.
$[Mo_6O_{20}^{4-}]_\infty$(o,tp)	[239]	1.1%	Mo2···Ot2^{i}	0.01	+0.01	+0.01/−0.01
			Mo1···Ot1‴	0.01		
$[V_2O_6^{2-}]_\infty$	[241]	0.6%	V···Ow	0.01	+0.01	+0.01/
"V_2O_5"	[249]	2.3%	V···Ot‴	0.07	+0.04	+0.05/−0.07
			V···Ot$^{?d,f}$	0.01		

[a] Bond valence fraction of the very long M···O bonds beyond the coordination of M defined by k (calculated for M···O distances up to 4.5 Å).
[b] Three reliable and three unreliable structures of MoO_4^{2-} and VO_4^{3-} have been investigated.
[c] These values do not refer to *averaged* M–Ot bonds of the MO_4 tetrahedra as has been accepted for the data in Tables 2, 4, and 5.
[d] An unequivocal identification of the oxygen atoms was impossible because of an insufficient structural description in the literature and/or because of a missing possibility to distinguish between the origin of the oxygen atoms in the print of the Inorganic Crystal Structure Database [61].
[e] Possibilities: V^{i}···Ot, V···Ot″, and/or V···Ot‴ (for the terminology distinction by primes see Sect. 3.3).
[f] Next smaller bond valence.

$M^{VI}\cdots O$ distances in the range 3.4 to 4.0 Å), which amount to 0.03 to 0.01 v.u. Altogether 1.2 to 1.5% of the bond valence of the M^{VI} atoms are involved in bonds beyond the octahedral coordination. These bonds are the inevitable consequence of the accumulation of MO_k polyhedra as a building principle of the polyoxometalate ions, i.e., of the connection modes of the MO_k polyhedra in the case of $k>4$. Neglect of these supernumerary, very weak bonds leads to very small errors: The bond valences come out greater by the same percentage size, i.e., by up to +0.02 v.u.; the charge is individually changed by up to +0.02/−0.02 c.u., in rare cases +0.03/−0.02 c.u.

- A similar configuration as described for the block-type structures occurs between neighboring block-type partial structures connected by common corners (as in $[Mo_8O_{27}^{6-}]_\infty$, $W_{12}O_{40}(OH)_2^{10-}$, etc.). In this case single bond valences up to 0.05 v.u. (corresponding to 3.2 Å) have been observed as a consequence of an additional possibility for distortions (e.g., $M2\cdots O2a3$ in $W_{12}O_{40}(OH)_2^{10-}$). Altogether up to 2.0% of the bond valence of the M^{VI} atoms are involved in bonds beyond the octahedral coordination, and again these bonds are the inevitable consequence of the accumulation (and hence of the connection modes) of the MO_k polyhedra. Neglect of these supernumerary, very weak bonds leads again to very small errors of the same percentage size, up to +0.03 v.u.; the charge is individually changed by up to +0.03/−0.03 c.u.
- If MO_k polyhedra with $k < 6$ are present, the bond valences of the next longer $M\cdots O$ distances in mono- and polyoxometalates are *sometimes* somewhat greater; MoO_4^{2-}: up to 0.08 v.u.; VO_4^{3-}: up to 0.07 v.u.; $V_2O_7^{4-}$: up to 0.17 v.u.; $Mo_{10}O_{34}^{8-}$ (I): 0.08 v.u.; $[Mo_4O_{14}^{4-}]_\infty$: 0.07 v.u.; "V_2O_5": 0.07 v.u. Altogether up to 2.9% (MoO_4^{2-}), 3.9% (VO_4^{3-}), 3.6% ($V_2O_7^{4-}$), 1.6% ($Mo_{10}O_{34}^{8-}$(I)), 1.9% ($[Mo_4O_{14}^{4-}]_\infty$), 2.3% ("V_2O_5") of the bond valence of the M atoms are involved in bonds beyond the polyhedral coordination defined by k; see also Sect. 4.6.2. In the *poly*oxo species of them part of these bonds is certainly again the inevitable consequence of the accumulation (and hence of the connection modes) of the MO_k polyhedra. However, part of these bonds has also to be considered as means to complete (extend) yet the coordination sphere of the MO_k polyhedra with $k<6$ which hence improves the sum of the bond valences about M (e.g., improvement of Σs_i for the MO_4 tetrahedra in $Mo_{10}O_{34}^{8-}$ (I) from 5.85 to 5.92 v.u.; in α-$Mo_8O_{26}^{4-}$ from 5.82 to 5.93 v.u.). Neglect of the supernumerary, very weak bonds in this category leads to small or very small errors, up to +0.05 v.u. and +0.06/−0.14 c.u.

  In some structures of this category, however, the bond valence of M involved in bonds beyond the polyhedral coordination defined by k is surprisingly small (MoO_4^{2-}, VO_4^{3-}, $V_2O_7^{4-}$, $V_3O_{10}^{5-}$, $[VO_3^-]_\infty$, $V_5O_{14}^{3-}$, $[Mo_6O_{20}^{4-}]_\infty$ (o,tp), $[V_2O_6^{2-}]_\infty$: <1.5%), and the bond valences of the next longer $M\cdots O$ distances do rarely exceed 0.01 v.u.; hence in these cases the neglect of the supernumerary, very weak bonds leads to only very small errors, up to +0.02 v.u. and +0.02/−0.02 c.u.

The M–O bond valences come always out too great at the expense of those of the neglected very long bonds. The charges on the oxygen atoms show positive and negative deviations which sum up to zero; the sometimes strongly unsymmetrical pattern of the positive and negative deviations is a consequence of a great difference of the frequencies of the corresponding types of O atoms.

The additional bond valence actually available for the M atoms by the additional bonds does not reduce the surprisingly large number of unreliable monometalate structures (for the bad performance of the monometalate structures see also Sect. 3.1 in Ref. 35).

In structures containing oxygen atoms forming many M–O bonds of the same sort there can be an additional, "arithmetic" error. The most conspicuous example is the M_6O_{19} structure. Although the valences of the relevant M–O bonds (M–O6l) differ by no more than 0.01 v.u. for $Mo_6O_{19}^{2-}$ and $W_6O_{19}^{2-}$, the charges on the O6l atoms differ by 0.07 c.u. (compare Tables 10 and 11). See also Sects. 3.19 and 4.6.11.2.

4.2
Packing Effects, Influence of the Cations

The M–O bonds of the polyoxometalate structures built by MO_k polyhedra with $k > 4$ (i.e., built by *extended* MO_4 tetrahedra, mainly MO_6 octahedra; in these structures the inner M–O bond lengths and hence bond valences are dependent of each other for geometrical reasons whereas in the structures and structural parts composed of MO_4 tetrahedra all M–O bond lengths can be adjusted independently of each other) show in regard of different cations, different numbers of molecules of water of crystallization with the same cation, or different authors an average standard deviation of 0.015 v.u. for the inner (bridging) and of 0.032 v.u. for the outer M–O bonds (bonds of the unshared O atoms, nearly exclusively terminal M–O bonds). The value for the bridging M–O bonds is only slightly greater than the standard deviation for the M–O bonds estimated for the use of the general B and individual d_{oi} values (0.012 v.u., see Table 1). This small difference of the standard deviations indicates that the packing of the constituents of the polyoxometalate salts (cations, polymetalate ions, molecules of water of crystallization) has only very small effects on the inner M–O bonds, that means the size, shape, and charge of the cations and the number of molecules of water of crystallization in the crystal structure do not cause significant bond length alterations of the inner M–O bonds of the polyoxometalate structures. The main reason for this is obviously the interdependence of the inner M–O bond lengths and hence bond valences of the polyoxometalate frameworks built by MO_k polyhedra with $k > 4$, see Sect. 4.3.4.2. Thus we can also conclude that in the case of the soluble polyoxometalate ions the structures in the free, dissolved state exist with the same dimensions of the inner M–O bonds and most probably with the ideal symmetry.

On the other hand it is well known that there are in many aqueous polyoxometalate systems only few polymetalate structure types that form with a large number of common inorganic cations and organic cations (for example, in the polyoxomolybdate system only $Mo_7O_{24}^{6-}$, $[Mo_6O_{20}^{4-}]_\infty$ (o), $Mo_{36}O_{112}(H_2O)_{16}^{8-}$, and $[Mo_5O_{15}(OH)(H_2O) \cdot H_2O^-]_\infty$) [14i, 186b, 253a] and that different cations usually induce the formation of insoluble(!) polymetalates with quite different structures under otherwise the same preparation conditions [186b, 253a, 254]. For instance, from aqueous molybdate solutions with $P \approx Z = 1.5$[7] the following polyoxometalates form:

- the yellowish polymolybdate $Na_2O \cdot 4MoO_3 \cdot 6H_2O$ (for a proposal of the structure see Ref. 255);
- the voluminous, fibrous "trimolybdate" $[K_4Mo_6O_{20}{\cdot}6H_2O]_\infty$ (see Fig. 16);
- the white, fine-crystalline $(NH_4)_2O \cdot 4MoO_3$ (the structure of this compound [256] corresponds to that of "$Tl_2Mo_4O_{13}$" [257, 258]) and $(NH_4)_4[(\beta\text{-})Mo_8O_{26}] \cdot 4(5)H_2O$, forming efflorescent, large, colorless crystals (see Fig. 9);
- the powdery polymolybdate $[(n\text{-}C_4H_9)_4N]_4[\alpha\text{-}Mo_8O_{26}]$ [259] (see Fig. 20).

The reasons are obviously the better packing conditions for cations, polymetalate ions, and molecules of water of crystallization if the polymetalate type is changed with the cation. [In (aqueous [14j] and non-aqueous [186b]) solution no cation-dependent equilibria occur, contrary to other statements [202, 259, 260, 261d]!] Otherwise there are also examples in which other preparation conditions (e.g., metalate concentration, reaction time, temperature) have a strong influence on the polyoxometalate type that forms as a solid keeping the cation and P value constant [253a].

Thus we must conclude that the M–O frameworks of the polyoxometalate ions built by MO_k polyhedra with $k > 4$ are rather rigid constructions (this is not the case for the tetrahedral polyoxometalate ions). The flexibility of the M–O bonds, existing in principle and shown by the variability of the bond lengths and the ability of the oxygen atoms to accept variable charge [16], is largely exhausted when the polymetalate structure has formed. This is most probably due to the interdependence of the inner M–O bond lengths in the polyoxometalate structures in question, as has already been mentioned above (first paragraph of this section); compare also Sects. 4.3.4.2 and 4.4.3.1.

The terminal M–O bonds retain a greater flexibility, as is indicated by the greater average standard deviation for their bond valences (0.032 v.u.) and by the great differences of the terminal M–O bond lengths and bond valences (compare Tables 8–16, 18, 20–32) or charges on the terminal oxygen atoms (compare particularly Tables 41 and 42) in the polyoxo anions. Notice that a change of the terminal M–O bond lengths influences the spatial *and* the charge-wise conditions around the polyoxometalate structure and hence allows an optimum packing of the constituents of the salts (compare Sect. 4.4.3.2).

---

[7] The degree of acidification P of a metalate solution is the molar ratio of H^+ ions *introduced* to MO_4^{w-} ions *introduced*; the degree of protonation Z of a metalate solution is the molar ratio of *reacted* H^+ ions to MO_4^{w-} ions *introduced* [4].

The strong influence of the cations on the M–O bond lengths of the monometalates refers also to terminal M–O bonds and confirms the view that the flexibility of the terminal M–O bond lengths facilitates the packing of the constituents of the salts. In this case there is essentially no possibility to switch over to the formation of other structure types. Also, the small size of the monometalate ions requires a greater flexibility of the (terminal) M–O bond lengths to reach a comparably close and optimal packing as for the polyoxometalate ions with their inherently great areas of closely packed oxygen atoms.

In the literature [53] packing effects due to cations are quoted to be not large. This statement does not conflict with our above declarations since it refers obviously to the case that the structure type does not change with the cation. However, in certain cases packing effects can cause asymmetries in M–O–M bridges (for instance, in the $Mo_6O_{19}^{2-}$ structure of a large organic cation alternating M–O2a bond lengths of 1.86 and 2.01 Å are present [13e]; for other examples of this kind see footnotes b of Table 14 and c of Table 24 and Sect. 4.3.2).

4.3 M–O Bond Lengths and Bond Valences

4.3.1 *M–O Bond Lengths and Bond Valence Ranges Occurring in the Different Types of MO_k Polyhedra and M–O Bond Types Occurring with the Different Types of O Atoms*

In the $M^{VI}O_6$ (M = Mo, W) octahedra the short, strong (terminal or pseudoterminal) M–O bonds (1.67 to 1.77 Å) range from 1.80 to 1.40 v.u., the medium-sized, middle-strong (mostly low-coordinately bridging) bonds (1.79 to 2.07 Å) from 1.33 to 0.67 v.u., and the long, weak (mostly high-coordinately bridging) M–O bonds in *trans* position to the short ones (2.10 to 2.52 Å) from 0.62 to 0.23 v.u. (Table 38). There is no overlapping of the ranges for the three bond types, but also no wide interval between them. Type II (the most frequent type) and type I MO_6 octahedra show no significant differences of the bond lengths and bond valences; the (on average) slightly shorter and stronger M–Ot bonds of the type I octahedra are an effect of the (on average) higher Z^+ values of the species (compare Eq. (10a) in Sect. 4.3.3) and are lost in the effects of the other operating factors. The pseudoterminal M–O bonds (1.55 to 1.40 v.u.) are somewhat weaker than the terminal bonds (1.80 to 1.47 v.u.), and the long bonds forming the complementary bonds of the pseudoterminal oxygen atoms (0.53 to 0.23 v.u.) are also somewhat weaker than the other long bonds (0.62 to 0.28 v.u.). (The bond lengths of the untypical borderline compounds discussed in Sect. 4.6.12 have been omitted here.) Fig. 30 shows as example for $Mo_7O_{24}^{6-}$, (β-)$Mo_8O_{26}^{4-}$, $M_6O_{19}^{2-}$, and $[Mo_6O_{20}^{4-}]_\infty$(o) the correspondence between the estimated average M–O bond valences according to Eq. (11) of Ref. 16 and the actual (based on the bond lengths) M–O bond valences derived in the present investigation. If we neglect

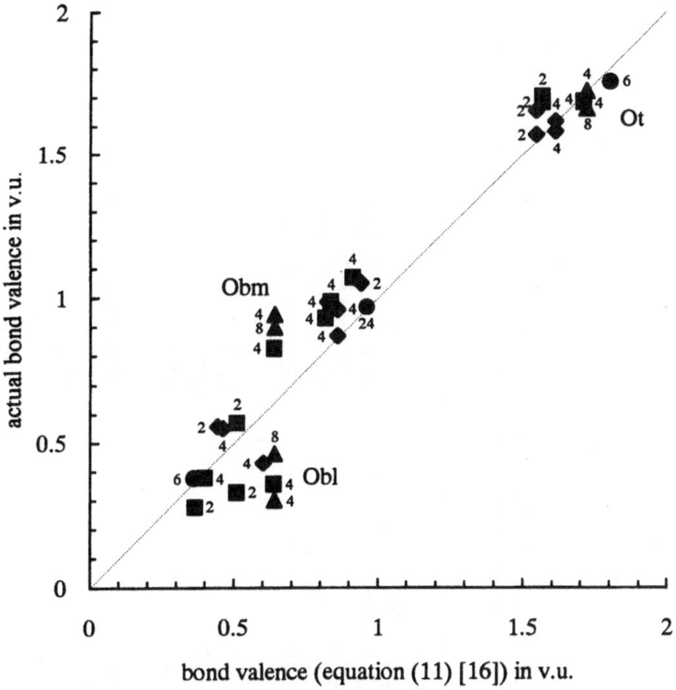

Fig. 30. Estimated average M–O bond valences according to Eq. (11) of Ref. 16 (which takes into consideration the high statistical weight of the resonance structures with charged terminal O atoms and the small statistical weight of the resonance structures with charged bridging O atoms) and actual bond valences (based on the bond lengths according to Eq. (1a)) for the individual M–O bonds (Ot: terminal, Obm: medium-sized bridging, Obl: long bridging M–O bonds) of the polyoxometalate ions $Mo_7O_{24}^{6-}$ (◆), (β-)$Mo_8O_{26}^{4-}$ (■), $Mo_6O_{19}^{2-}$ (●), and $[Mo_6O_{20}^{4-}]_\infty$ (o) (▲) as example (the numbers indicate the frequency of the respective bonds). Outliners in connection with the occurrence of pseudoterminal O atoms have been omitted.

the outliners in connection with the occurrence of the pseudoterminal O atoms, we see mainly that the bond valences of the middle-strong M–O bonds (in the range Obm) are on average greater and those of the weak M–O bonds (in the range Obl) on average smaller than expected according to Eq. (11) in Ref. 16. The main reason for this is the interdependence of the inner M–O bond lengths in the polyoxometalate frameworks as discussed in Sect. 4.3.4.2 in connection with the usually large number of bonds of the central O atoms. [The same is true if we compare the bond valences estimated according to Eq. (10a) or (10b), assuming an equidistribution of the available bond valence over the M–O bonds of the structures, with the actual M–O bond valences (see Sect. 4.3.3).]

The M–O bonds of the other, rarely occurring $M^{VI}O_k$ polyhedra with extended coordination spheres (pentagonal MO_7 bipyramids, tetragonal MO_5 pyramids) also fall under the above categories, except the long bonds of the type II tetragonal MO_5 pyramids (see Fig. 1) where one M–O bond (0.76 v.u.)

Table 38. Ranges of bond lengths and bond valences for the different types of $M^{VI}O_k$ polyhedra and bond types

Bond type[a]	d^b in Å	s in v.u.
MO_6 octahedra (41 different type II, five different type I and one mixed type I/II octahedron):		
M–Ot (short bonds)	1.671 to 1.760 ⟨1.780⟩	1.801 to 1.467 ⟨1.371⟩[d]
M–Op (short bonds)[c]	1.725 to 1.768 ⟨1.780⟩	1.548 to 1.400 ⟨1.371⟩
M–Ob (medium-sized bonds)	1.794 to 2.071	1.332 to 0.672[d]
M–O* (long bonds)	⟨2.070⟩ 2.104 to 2.452	⟨0.689⟩ 0.621 to 0.276[d]
M–O*c (long bonds)[c]	2.166 to 2.524	0.533 to 0.232
Pentagonal MO_7 bipyramids (one case of type I):		
M–Ot (short bonds)	1.685	1.705
M–Ob (medium-sized bonds)	1.968 to 2.072	0.861 to 0.671
M–O*c (long bonds)	2.366	0.330
Tetragonal MO_5 pyramids (one case of type II):		
M–Ot (short bonds)	1.675; 1.740	1.770; 1.512
M–Ob (medium-sized bonds)	1.920; 1.945	0.979; 0.922
M–O* (long bonds)	2.025	0.760
MO_4 tetrahedra with 1 Ot (one MO_4 tetrahedron):		
M–Ot	1.708	1.658
M–Ob	1.783	1.383
MO_4 tetrahedra with 2 Ot (one MO_4 tetrahedron):		
M–Ot	1.714; 1.744	1.640; 1.525
M–Ob	1.735; 1.822	1.559; 1.263

MO_4 tetrahedra with 3 Ot (three different MO_4 tetrahedra):		
M–Ot	1.696 to 1.766	1.704 to 1.439
M–Ob	1.819 to 1.876	1.272 to 1.108
MO_4 tetrahedra with 4 Ot (ca. 20 different MO_4 tetrahedra):		
M–Ot	1.651 to 1.836	1.856 to 1.187

[a] M–Ot: terminal M–O bonds; M–Op: pseudoterminal (bridging) M–O bonds; M–Ob: bridging M–O bonds except those of Op and O*; M–O*: bridging M–O bonds in *trans* position to terminal or pseudoterminal bonds; M–Oc: complementary M–O bonds of pseudoterminal O atoms.
[b] Since d_o (and B) for Mo–O and W–O bonds is virtually the same, Mo–O and W–O bonds can be lumped together.
[c] The pseudoterminal M–O bond of the mixed type I/II octahedron and the corresponding complementary M–O bond have not been included here for understandable reasons.
[d] Greatest and smallest values for the bond valences (check for the absence of unusually high or low values; ⟨ ⟩: untypical borderline cases discussed in Sect. 4.6.12):
M–Ot: 1.801 – 1.753 – 1.749 – ... – 1.506 – 1.469 – 1.467 ⟨– ... – 1.404 – 1.371⟩ v.u.
M–Ob: 1.332 – 1.308 – 1.209 – ... – 0.695 – 0.679 –0.672 v.u.
M–O*: ⟨0.689 – 0.673 – ... –⟩ 0.621 – 0.611 – 0.583 – ... – 0.289 – 0.279 – 0.276 v.u.

substitutes two long bonds of type II MO_6 octahedra of the range 0.62 to 0.28 v.u.

In the $M^{VI}O_4$ tetrahedra of polyanions the terminal M–O bonds (1.70 to 1.77 Å) range from 1.70 to 1.44 v.u. The bridging M–O bonds (1.74 to 1.88 Å) range from 1.56 to 1.11 v.u. and hence overlap with the terminal M–O bonds (for the reason of the overlapping – unusual occurrence of 1.56 v.u. in a bond of an M–O–M bridge – see Sect. 3.22). Isolated MO_4 tetrahedra (1.65 to 1.84 Å) show a wide bond valence range from 1.86 to 1.19 v.u.

In the M^VO_6 octahedra the short, strong (terminal or pseudoterminal) M–O bonds (V–O: 1.58 to 1.72 Å; Nb–O, Ta–O: 1.74 to 1.80 Å) range from 1.78 to 1.24 v.u., the medium-sized, middle-strong (mostly low-coordinately bridging) bonds (V–O: 1.78 to 2.11 Å; Nb–O, Ta–O: 1.91 to 2.12 Å) from 1.05 to 0.43 v.u., and the long, weak (mostly high-coordinately bridging) M–O bonds in *trans* position to the short ones (V–O: 2.06 to 2.39 Å; Nb–O, Ta–O: 2.22 to 2.53 Å) from 0.50 to 0.19 v.u. (Table 39). There is a weak overlapping of the ranges for the middle-strong and weak bond types. Type I (the more frequent type) and type II MO_6 octahedra show again no fundamental differences, although all short M–O bonds in the type II octahedra are of pseudoterminal character.

The M–O bonds of the other, rarely occurring M^VO_k polyhedra with extended coordination spheres (tetragonal MO_5 pyramids and trigonal MO_5 bipyramids) also fall under the above categories, except the long bonds of the type II VO_5 trigonal bipyramids where one M–O bond (0.61 v.u.) substitutes two long bonds of type II MO_6 octahedra of the range 0.50 to 0.19 v.u.

In the M^VO_4 tetrahedra (only VO_4 tetrahedra do exist) of polyanions the terminal M–O bonds (1.58 to 1.71 Å) range from 1.68 to 1.29 v.u. The bridging M–O bonds (1.72 to 1.88 Å) range from 1.14 to 0.81 v.u. and do not overlap with the terminal M–O bonds. Isolated VO_4 tetrahedra (1.64 to 1.81 Å) show a wide bond valence range from 1.53 to 1.00 v.u.

The *Opa* and *Opl* atoms in the M^{VI} structures form 1 strong and 1 weak M–O bond, in the M^V structures ($M_{10}O_{28}^{6-}$) 1 strong and 1 middle-strong M–O bond. The *O2a* and *O2l* atoms (O2aw atoms excluded for understandable reasons) form nearly exclusively 2 middle-strong M–O bonds; only in the case of O2a2 in $Mo_{10}O_{34}^{8-}$ (I) and O2l1 in $[Mo_4O_{14}^{4-}]_\infty$ are there quasi Opa and Opl atoms, respectively, forming 1 strong and 1 weak M–O bond. The *O3a* atoms exist with 0 to 3 middle-strong and 3 to 0 weak bonds and the *O3l* atoms with 1 to 3 middle-strong and 2 to 0 weak bonds. The *O4l* atoms form 0 or 1 middle-strong and 4 or 3 weak M–O bonds. *O5l* and *O6l* atoms form exclusively weak M–O bonds. These bond patterns have been expected (see Sect. 3.4 in Ref. 16).

The bond valences in the M^V structures are generally smaller than those in the M^{VI} structures. This is simply a consequence of the different oxidation numbers of the addenda elements M (see Eq. (10a) or (10b) in Sect. 4.3.3). The preferential occurrence of type II $M^{VI}O_6$ octahedra and of type I M^VO_6 octahedra is ultimately also an effect of the oxidation numbers of M, see Sect. 4.3.2.

The dimensions of the MO_6 octahedra of Mo^{VI} and W^{VI} on the one hand and of V^V on the other hand are nearly the same, as far as the inner (bridging)

M–O bonds are considered. The generally smaller bond valences (and correspondingly longer M–O bonds) in the V^V structures help to reach this goal (see Sect. 4.6.1). The differences of the size of V^V compared to Mo^{VI} and W^{VI} find expression in the shorter lengths of the terminal M–O bonds of V^V.

4.3.2
Type II and Type I MO_6 Octahedra and Relationships Between Them

The occurrence of strongly bound terminal (outer) oxygen atoms, of middle-strongly bound bridging (inner) oxygen atoms in general of low coordination number at the surface of the polyoxometalate structures, and of weakly bound bridging (inner) oxygen atoms in general of high coordination number in the interior of the M–O frameworks, i.e., the occurrence of type II and type I MO_6 octahedra or other MO_k polyhedra (compare Fig. 1), is an effect of the appearance of the polyoxometalate ions as structures (defined by the connectivity of the atoms and their oxidation number) with a definite surface. There is no other way to divide the available bond valence among the different M–O bonds of the structures, simultaneously fulfilling the bond length-bond valence relationship and the valence sum rule. MO_6 octahedra with three unshared oxygen atoms and an OH group or a coordinated H_2O molecule in an unshared position represent special cases, however, there are again type II octahedra. See also Sect. 4.3.3, in particular Eqs. (10a) and (10b).

The occurrence of pseudoterminal M–O bonds (pseudoterminal oxygen atoms) is apparently the consequence of an analogous integration (compare the differentiated changes of bond lengths, bond valence, and charge discussed in Sect. 4.6.7 which are necessary for the compensation of the effects of a protonation reaction to maintain these structural changes as small as possible) of the corresponding MO_6 octahedra within the M–O framework like those with terminal M–O bonds, required by the interdependence of the inner M–O bond lengths, the bond length-bond valence relationship, and the valence sum rule as discussed in Sect. 4.3.4.2. For the view of the differentiation of the M–O bond lengths to provide space for the M atoms in the MO_6 octahedra see Sect. 4.6.1, in particular Figs. 32a and 32b.

The sums of the bond lengths or bond valences of the M–O bonds in trans position to each other are very similar in the type I and type II MO_6 octahedra (this is not the case for W4 in the polytungstate $Na_5[H_3W_6O_{22}] \cdot 18H_2O$ [58], which is a modified type II octahedron, see Fig. 1). There are two kinds of *trans* positioned M–O pairs: short – long (or strong – weak) and medium-sized – medium-sized (or middle-strong – middle-strong).

The bond lengths and hence bond valences vary in a wide range for the types of the strong (short), middle-strong (medium-sized), and weak (long) M–O bonds of the individual MO_6 octahedra (compare Tables 38 and 39). This causes the occurrence of intermediate stages between type II and type I MO_6 octahedra. There are even bond patterns intermediate between type II and type I (e.g., for M2 in $W_{12}O_{40}(OH)_2^{10-}$). On admixing type II character to a type I octahedron, usually one of the middle-strong M–O bonds obtains some

Table 39. Ranges of bond lengths and bond valences for the different types of $M^{V}O_k$ polyhedra and bond types

Bond type[a]		d(V) in Å/d(Nb, Ta)[b] in Å	s in v.u.
MO_6 octahedra (seven different type II and 16 different type I octahedra):			
M–Ot	(short bonds)	1.585 to 1.622/1.739 to 1.80	1.782 to 1.389[c]
M–Op	(short bonds)	1.667 to 1.720/1.823	1.427 to 1.237
M–Ob	(medium-sized bonds)	1.781 to 2.110/1.912 to 2.123	1.048 to 0.430[c]
M–Obc	(medium-sized bonds)	2.022 to 2.061/2.093	0.656 to 0.491
M–O*	(long bonds)	2.056 to 2.392/2.224 to 2.534	0.498 to 0.191[c]
Tetragonal MO_5 pyramids (one case of type I):			
M–Ot	(short bonds)	1.577/–	1.767
M–Ob	(medium-sized bonds)	1.779 to 2.017/–	1.045 to 0.549
M–O*	(long bonds)	–/–	–
Trigonal MO_5 bipyramids (one case of type II):			
M–Ot	(short bonds)	1.656/–	1.472
M–Ob	(medium-sized bonds)	1.918/–	0.724
M–O*	(long bonds)	1.983/–	0.607
MO_4 tetrahedra with 1 Ot (one MO_4 tetrahedron):			
M–Ot		1.580/–	1.679
M–Ob		1.723/–	1.140
MO_4 tetrahedra with 2 Ot (14 different MO_4 tetrahedra):			
M–Ot		1.601 to 1.662/–	1.589 to 1.437
M–Ob		1.795 to 1.818/–	1.004 to 0.881

MO_4 tetrahedra with 3 Ot (17 different MO_4 tetrahedra):		
M–Ot	1.669 to 1.714/–	1.413 to 1.291
M–Ob	1.743 to 1.876/–	1.133 to 0.807
MO_4 tetrahedra with 4 Ot (ca. 30 different MO_4 tetrahedra):		
M–Ot	1.636 to 1.809/–	1.532 to 0.996

[a] See footnote a of Table 38.
[b] Since d_0 for Nb–O and Ta–O bonds is nearly the same, Nb–O and Ta–O bonds can be lumped together.
[c] Greatest and smallest values for the bond valences (check for the absence of unusually high or low values):
M–Ot: 1.782 – 1.782 – 1.763 – ... – 1.593 – 1.500 – 1.389 v.u.
M–Ob: 1.048 – 1.009 – 1.009 – ... – 0.563 – 0.509 – 0.430[h] v.u.
M–O*: 0.498 – 0.491 – 0.455 – ... – 0.209 – 0.201 – 0.191 v.u.
[h] Protonated oxygen atom.

pseudoterminal character and another one (in *trans* position to it) obtains the character of a weak bond (with some "*trans*" bond character). This is one of the reasons for the wide range of the middle-strong M–O bonds and for the only small empty intervals between strong, middle-strong, and weak M–O bonds.

The preferential occurrence of type II $M^{VI}O_6$ octahedra and of type I $M^{V}O_6$ octahedra is an effect of the oxidation numbers of M. Smaller oxidation numbers lead to more highly charged and, accordingly, to more basic species with the consequence of a relieved (because of the greater protonation constants) and enhanced (because of the higher charge) protonation of the (mono- and growing poly)metalate ions and hence more favorable possibilities for the formation of type I MO_6 octahedra which need, as a rule, a threeefold protonated monomeric species and a suitably threefold protonated growing polymeric species if formed according to the condensation mechanism [2, 14e, 14l, 20, 186c].

The preferential occurrence of type I WO_6 octahedra and of type II MoO_6 octahedra has been explained by the greater tendency of W^{VI} to expand its coordination sphere which thus leads to smaller concentrations of the protonated monometalate species at which they undergo the aggregation reactions; this is the reason why the aggregation reactions in the tungstate case start at 1.5 to 2 units higher pH values [2, 4, 5, 14kk, 262h, 263]. That means that in the tungstate case the small concentrations of the threefold protonated monometalate ion (and of the growing polymeric species) reach to form type I MO_6 octahedra, but they do not reach in the molybdate case. The smaller compressibility of the W^{6+} core compared with that of Mo^{6+} and the greater extension of the 5d compared with the 4d orbitals [14t, 203] have also been used to explain the differences, however, without particularization. The latter feature would just explain the greater tendency of W^{VI} to expand its coordination sphere as mentioned above.

The occurrence of alternatingly shorter and longer M–O bonds in type I octahedra of some poly*molybdate* ions (e.g., in $Mo_6O_{19}^{2-}$ [187] and, in particular, in heteropolymolybdates: $GeMo_{12}O_{40}^{4-}$ [13f, 264], $P_2Mo_{18}O_{62}^{6-}$ [203, 265], $As_2Mo_{18}O_{62}^{6-}$ [266]) can also be seen as a consequence of an admixture of type II to type I. In the $X_2Mo_{18}O_{62}^{6-}$ structures there is even a complete change to type II in the case of the twelve "equitorial" MO_6 octahedra. The reason in the special [187] case of $Mo_6O_{19}^{2-}$ and in some Keggin structures is obviously a packing effect in the solid state since this effect is not observed in structures with other cations. From a comparison of (iso- and heteropolyoxo) structures with different cations we conclude from the existence of regular and of distorted structures of the same type that packing effects can produce (alternating) bond lengths differences up to ca. 0.15 Å with M–O2a bonds. In the case of the cited $X_2Mo_{18}O_{62}^{6-}$ heteropolymolybdates, which show (alternating) bond lengths differences up to 0.39 v.u. but in the corresponding heteropolytungstates no distortions, it is possibly due to a small difference of the bond length-bond valence parameters for Mo–O and W–O bonds in combination with the rigidity of the M–O frameworks resulting from their high symmetry (compare Sect. 4.3.4.2). Hence the rigidity is lowered by formation of unsymmetrical M–O–M bridges or (which is equivalent) of unsymmetrical O–M–O coordinations and thus allows the adjustment of bond lengths or bond valence to the geometrical requirements of the M–O frameworks (see Sect. 4.6.3, in particular Table 44). A difference in d_o of 0.01 Å for Mo–O and W–O bonds (taking equal B parameters as a basis) can explain (alternating) bond lengths differences of 0.25 Å by a corresponding shift of the M atoms in the centers of the MO_6 octahedra along the relevant, approximately linear O–M–O axes, i.e., without appreciable alterations of the positions of the O atoms. (For the reason for the occurrence of *alternating* bond lengths alterations see Sect. 4.6.7.)

Surprisingly, there is one structure ("$MoO_3 \cdot 2H_2O$") in which, according to the connection mode of the MO_6 octahedra, type I octahedra are to be expected (see, e.g., Fig. 2 in Ref. 247). Nevertheless in reality type II octahedra occur. This fact and the aforementioned occurrence of distortions of the type I in the direction of type II octahedra indicate that type II octahedra are apparently the more adequate coordination polyhedra for M^{VI} and that type I octahedra are only forced by the symmetry properties of some structure types. In the literature this aspect is discussed in the reverse sense, i.e., the tendency to form type II or type I MO_6 octahedra is seen as a property of the addenda element M (Mo^{VI}: type II; W^{VI}: type I) [13g, 203, 267a].

In this connection it should also be noticed that an ideal type I MO_6 octahedron lies geometrically precisely in the middle between an ideal type II and a regular MO_6 octahedron (average M–O bond lengths in an average type II and an average type I $M^{VI}O_6$ octahedron: 2.01 and 1.96 Å, respectively; M–O bond lengths in a regular $M^{VI}O_6$ octahedron: 1.91 Å). Regular MO_6 octahedra have not yet been observed in polyoxo species. Possible M–O frameworks can only exist in extended three-dimensional form (case of exclusively regular MO_6 octahedra) or would require a central regular MO_6 octahedron (case of only one regular MO_6 octahedron). They obviously do not exist for spatial reasons, see Sect. 4.6.1, particularly Figs. 32a and 32b.

In one case of an isopolyoxo species (W4 in the recently prepared polytungstate $Na_5[H_3W_6O_{22}] \cdot 18H_2O$ [58]) the bond pattern resembles only faintly that of a type II octahedron. There are indeed two short, two medium-sized, and two long M–O bonds, however, the medium-sized bonds are not in *trans* position to each other but are *cis* located and hence one long M–O bond is not in a *trans* position to a short one.

The occurrence and preference of distorted, i.e., of type II and type I MO_6 octahedra is usually – often tacitly – explained by the favorable electronic configuration of the metal atoms for these cases [64, 268]. This is also seen as the reason why central, regular MO_6 octahedra (e.g., in a planar $M_7O_{24}^{6-}$ structure) cannot exist [64, 268]. The existence of mixed type I/II MO_6 octahedra, of the bond pattern of W4 in the polytungstate $Na_5[H_3W_6O_{22}] \cdot 18H_2O$ [58] (modified type II) and of W3 and W5 in the polytungstate $K_7[HW_5O_{19}] \cdot 10H_2O$ [57] (type III), as well as of the different types of MO_5 and MO_7 polyhedra, and of compounds with regular-octahedrally coordinated M^{VI} atoms (e.g., the compound MoF_6 [269]; for a commentary on the frequently postulated "molybdic acid" $Mo(OH)_6$ see Ref. 23) do not support this view. In our opinion the single-bond M–O distance is somewhat too small to allow a regular octahedral coordination of molybdenum by oxygen, see Sect. 4.6.1, in particular Figs. 32a and 32b.

Surprisingly not only the MO_6 octahedra but also the other MO_k polyhedra with an extended (k > 4) coordination sphere of M show the existence of type II and type I analogues. Tetragonal MoO_5 pyramids of type II (in $[Mo_6O_{20}^{4-}]_\infty$ (o,tp)), tetragonal VO_5 pyramids of type I (in "V_2O_5"), trigonal VO_5 bipyramids of type II (in $[V_2O_6^{2-}]_\infty$), and pentagonal MoO_7 bipyramids of type I (in $Mo_{36}O_{112}(H_2O)_{16}^{8-}$) occur in the structures under discussion (see Fig. 1).

4.3.3
Deviations of the M–O Bond Valences from "Average" M–O Bond Valences

From the expressions for the sum of all bond valences in a polyoxometalate ion $M_q^nO_u^{m-}$, formed by q $MO_4^{(8-n)-}$ and p H^+ (Eq. (6)), for the ionic charge of the polymetalate ion (Eq. (7)), for the quantity Z^+ (Eq. (8)), and for u (Eq. (9)):

$$nq = 2u - m \tag{6}$$

$$m = (8 - n)q - p \tag{7}$$

$$Z^+ = p/q \tag{8}$$

$$u = 4q - p/2, \tag{9}$$

the *average* bond valence of an M–O bond with a certain coordination number of the O atom results as

$$\bar{s}(k_O) = nq/uk_O = 2n/(8 - Z^+)k_O \tag{10a}$$

(n: oxidation number of M, k_O: coordination number of oxygen against M), if an equidistribution of the ionic charge over the oxygen atoms and of the bond valence over the M–O bonds (i.e., each M–O bond accepts an equal percentage cut from the maximally possible bond valence) is assumed. This means that there is no systematic influence on the distribution of the bond valence over the O–M bonds and/or on the distribution of the ionic charge over the oxygen atoms of the polymetalate ion and/or no charge separation.

The value of k_O (an integer) varies from 1 to 6, n (an integer) can be 6 (Mo^{VI}, W^{VI}) or 5 (V^V, Nb^V, Ta^V), and Z^+ (a decimal) varies from 1.00 to 1.80 for M^{VI} polyoxometalate ions (2.00 for the oxides) and from 1.00 to 2.40 for (unprotonated) M^V polyoxometalate ions (3.00 for "V_2O_5"). Accordingly, Eq. (10a) shows that the M–O bond valences depend very strongly on the coordination number of the oxygen atoms (corresponding to a factor of 6), but considerably less on the oxidation number of M (corresponding to a factor of 1.2) and on the Z^+ value of the special polyoxometalate ion including the oxides (corresponding to a factor of 1.17 for the M^{VI} and 1.40 for the M^V species, respectively).

Polyoxometalate ions that contain OH groups or coordinated H_2O molecules can be treated in the same way after a rearrangement of their formula according to $M^n_qO_{u-h}(OH)_h(H_2O)^{m-}_o$.[8] Now we have to use

$$\bar{s}(k_O) = nq/(u + h/2)k_O \tag{10b}$$

and, moreover, $\bar{s}(M \cdots OH_2) = 0$ v.u. and $\bar{s}(M-OH) = \bar{s}(k_O)/2$; examples: $Mo_8O_{26}(OH)_2^{6-}$: $\bar{s}(M-OH) = \bar{s}(k_O = 1)/2 = 0.889$ v.u.; $W_{12}O_{40}(OH)_2^{10-}$: $\bar{s}((M-)_3OH) = \bar{s}(k_O = 3)/2 = 0.293$ v.u.; $H_2V_{10}O_{28}^{4-}$: $\bar{s}(k_O = 2)/2 = 0.463$ v.u.

Table 40 shows the bond valences to be expected according to Eq. (10a) or (10b) for the isopolyoxometalate ions and oxides. By comparison with the bond valences, calculated from the bond lengths according to Eq. (1a), the following rules can be derived (valid for M^{VI} and M^V isopolyoxometalate ions and oxides in every coordination mode of M; also valid, of course, with application of Eq. (1b)):

- Terminal M–O bonds are always weaker (by down to 0.41 v.u.) than expected for an average M–Ot bond according to Eqs. (10). (Accordingly the sum of the bond valences of the bridging oxygen atoms is always greater than that expected for the occurrence of average M–O bond valences.)
- Pseudoterminal M–O bonds are on an average somewhat weaker than terminal ones, but are always considerably stronger (by 0.29 to 0.66 v.u.) than expected for the M–O bonds of an average two-coordinate O atom. The complementary M–O bonds of the pseudoterminal oxygen atoms are always considerably weaker (by 0.27 to 0.64 v.u.) than expected for an average two-coordinate M–O bond.
- Bridging M–O bonds of MO_4 tetrahedra are always stronger (by up to 0.77 v.u.) than expected for an average M–O bond of this type.
- The other bridging M–O bonds are usually stronger (by as much as 0.54 v.u.), in rare cases somewhat weaker (by down to 0.30 v.u.) than expected for an average M–O bond of a correspondingly coordinated oxygen atom, if the oxygen atom is not in a *trans* position to a terminal or pseudoterminal oxygen (middle-strong M–O bonds). They are usually weaker (by down to 0.47 v.u.), in rare cases somewhat stronger (by up to 0.20 v.u.) than expected for an average M–O bond of a correspondingly coordinated oxygen atom, if the oxygen atom is in a trans position to a terminal or pseudoterminal oxygen (weak M–O bonds).

[8] The underlying general overall equation for formation of the polyoxo species is
$p\,H^+ + q\,M^nO_4^{w-} + o\,H_2O \rightarrow M^n_qO_{4q-(p-h)/2-h}(OH)_h(H_2O)_o^{(wq-p)-} + (p-h)/2\,H_2O$
where $4q-(p-h)/2 = u$, $wq-p = m$, and $(p-h)/2 = r$.

Table 40. Average bond valences $\bar{s}$ of the M–O bonds according to Eqs. (10a) or (10b) for the polyoxometalate ions and oxides of M^{VI} and M^{V} on the basis of an equidistribution of the bond valence over the M–O bonds and of the charge over the oxygen atoms

$M_q^nO_{u-h}(OH)_h(H_2O)_o^{m-}$	Z^+	$\bar{s}$ in v.u. for k_O =					
		1	2	3	4	5	6
M^{VI} species:							
MO_4^{2-}	0	1.500					
$HW_5O_{19}^{7-}$	0.600	1.667	0.833[a]	0.556	0.417		
$Mo_2O_7^{2-}$	1.000	1.714	0.857				
$[Mo_4O_{14}^{4-}]_\infty$	1.000	1.714	0.857				
$Mo_7O_{24}^{6-}$	1.143	1.750	0.875	0.583	0.437		
$H_3W_6O_{22}^{5-}$	1.167	1.756[b]	0.878	0.585	0.439		
$W_{12}O_{40}(OH)_2^{10-}$	1.167	1.756	0.878	0.585[c]			
$Mo_{10}O_{34}^{8-}$ (I)	1.200	1.765	0.882	0.588	0.441		
$Mo_{10}O_{34}^{8-}$ (II)	1.200	1.765	0.882	0.588	0.441		
$Mo_8O_{26}(OH)_2^{6-}$	1.250	1.778[d]	0.889	0.593	0.444		
$[Mo_8O_{27}^{6-}]_\infty$	1.250	1.778	0.889	0.593	0.444		
$[Mo_6O_{20}^{4-}]_\infty$ (o)	1.333	1.800		0.600			
$[Mo_6O_{20}^{4-}]_\infty$ (o,tp)	1.333	1.800	0.900	0.600			
(β-)$Mo_8O_{26}^{4-}$	1.500	1.846	0.923	0.615		0.369	
α-$Mo_8O_{26}^{4-}$	1.500	1.846	0.923	0.615			
$W_{12}O_{38}(OH)_2^{6-}$	1.500	1.846	0.923	0.615[e]			
$W_{10}O_{32}^{4-}$	1.600	1.875	0.938			0.375	
$M_6O_{19}^{2-}$	1.667	1.895	0.947				0.316
$Mo_{36}O_{112}(H_2O)_{16}^{8-}$	1.778	1.929[f]	0.964[g]	0.643	0.482		
"MoO_3"	2.000	2.000	1.000	0.667			
"$MoO_3 \cdot H_2O$"	2.000	2.000[f]		0.667			
"$MoO_3 \cdot 2H_2O$"	2.000	2.000[f]	1.000				

M^V species:					
VO_4^{3-}	0	1.250			
$V_2O_7^{4-}$	1.000	1.429	0.714		
$V_3O_{10}^{5-}$	1.333	1.500	0.750		
$M_6O_{19}^{8-}$	1.667	1.579	0.789		0.263
$[VO_3^-]_\infty$	2.000	1.667	0.833		
$[V_2O_6^{2-}]_\infty$	2.000	1.667		0.556	
$V_5O_{14}^{3-}$	2.400	1.786	0.893		
$M_{10}O_{28}^{6-}$	2.400	1.786	0.893	0.595	0.298
$H_2V_{10}O_{28}^{4-}$	2.600	1.852	0.926[h]	0.617	0.309
$H_3V_{10}O_{28}^{3-}$	2.700	1.887	0.943[i]	0.629[j]	0.314
"V_2O_5"	3.000	2.000	1.000	0.667	

[a] $\bar{s}((M-)_2OH) = 0.417$ v.u.
[b] $\bar{s}(M-OH) = 0.878$ v.u.
[c] $\bar{s}((M-)_3OH) = 0.293$ v.u.
[d] $\bar{s}(M-OH) = 0.889$ v.u.
[e] $\bar{s}((M-)_3OH) = 0.308$ v.u.
[f] $\bar{s}(M \cdots OH_2) = 0$ v.u.
[g] $\bar{s}((M \cdots)_2OH_2) = 0$ v.u.
[h] $\bar{s}((M-)_2OH) = 0.463$ v.u.
[i] $\bar{s}((M-)_2OH) = 0.472$ v.u.
[j] $\bar{s}((M-)_3OH) = 0.314$ v.u.

4.3.4 Reasons for the Observed Deviations and Consequences

M–O bond lengths, M–O bond valences, and the charge on the oxygen atoms of the polyoxometalate ions are mutually dependent according to Eqs. (1) and (3) (Eq. (2) is included in Eq. (3)) and according to the connectivities in the M–O frameworks. This entails that the reasons for and consequences of bond length, bond valence, and charge alterations interact and that it is sometimes hard to recognize the primary reason for an observed effect – that means to recognize: What is the reason, what is the consequence?

4.3.4.1 Enhanced Sum of the Bond Valences of the M–O Bonds about the Bridging Oxygen Atoms

The generally enhanced sum of the bond valences of the M–O bonds about the bridging oxygen atoms favors the holding together of the MO_k building units and hence leads to a stabilization of the polymeric character of the structures with its favorable consequences, in particular an enhanced consumption of the protons introduced into the system (Le Chatelier's principle, see Eq. (16), Sect. 3.6.7 in Ref. 16).

This factor leads to negative charge on the bridging oxygen atoms that is as small as possible and, reversely, to negative charge on the terminal oxygen atoms as great as possible.

4.3.4.2 Interdependence of the Inner M–O Bond Lengths in Polyoxometalate Structures

The parts of the M–O frameworks built by MO_k polyhedra with $k>4$ display for geometrical reasons a more or less strong interdependence of the *inner* M–O bond lengths. This requires that the M–O bonds can accept the corresponding lengths and leads, because of the bond length-bond valence relationship (Eq. (1)) and the valence sum rule (Eq. (2) or (3)), to a restricted range for possible M–O bond valences and hence charges on the bridging oxygen atoms. The restrictions are the more pronounced the more compact and/or the more symmetrical (*the more rigid*) the M–O framework is [55, 270, 271]. That many inorganic solid-state structures are the result of a compromise between the conflicting requirements of chemical bonding and three-dimensional geometry (i.e., of bond length-bond valence relationship, valence sum rule, and the connectivity of the atoms) has recently also been postulated in the literature [36].

The consequence *can* be a charge on oxygen atoms that is different from the usual charge on that type of oxygen. Conspicuous examples are found in the polyoxo anions (β-)$Mo_8O_{26}^{4-}$ (O2a, O3a, O5l), $W_{10}O_{32}^{4-}$ (O2a2, O5l), α-$Mo_8O_{26}^{4-}$ (O2a), and $H_3V_{10}O_{28}^{3-}$ (O3abt, O3abu). For further discussion see Sect. 4.4.3.1.

4.3.4.3
Coincidence of Several Middle-Strong M–O Bonds about a Bridging Oxygen Atom

If several middle-strong M–O bonds coincide about an oxygen atom, the sum of the bond valences exceeds the oxidation number of oxygen (its absolute amount), and a positive charge (or, in the case of the M^V species, a less negative charge; see Sects. 4.4.3.5 and 4.4.3.6) results for that oxygen atom. To hold the charge in the customary range, it is necessary to reduce the bond valences (to increase the bond lengths) of the middle-strong M–O bonds which leaves weak middle-strong M–O bonds. (Another reason for the occurrence of weak middle-strong – along with strong middle-strong – M–O bonds is the appearance of mixed type I/II MO_6 octahedra, see Sect. 4.3.2.)

Three middle-strong M–O bonds about an oxygen atom are present in $Mo_{36}O_{112}(H_2O)_{16}^{8-}$ (O3l1), in $M_{10}O_{28}^{6-}$ (O3a), and in "V_2O_5" (O3l). Two middle-strong M–O bonds plus one weak M–O bond (O3a in (β-)$Mo_8O_{26}^{4-}$; O3l in $[Mo_6O_{20}^{4-}]_\infty$ (o); O3l in $[V_2O_6^{2-}]_\infty$; O3l2, O3l3, and O3l4 in $Mo_{36}O_{112}(H_2O)_{16}^{8-}$; O3l in $Mo_{10}O_{34}^{8-}$ (I); O3l in "(α-)$MoO_3 \cdot H_2O$"; O3l in "MoO_3") and even one middle-strong plus three or two weak M–O bonds can also result in positively charged oxygen atoms which requires the weakening of middle-strong M–O bonds. However, in most of the cited cases the oxygen atoms are also part of approximately linear M–O–M bridges which favors positively charged oxygen atoms too, see Sects. 4.4.1 and 4.4.2.2. The most prominent examples are the O3a atoms in the heteropolyanion $H_4As_4Mo_{12}O_{50}^{4-}$ [272], see Table 8 in Ref. 35.

This factor has not been and could not be recognized in our bond model for polyoxometalate ions [16] since it requires for each individual structure a separate investigation.

4.3.4.4
Coincidence of a Few and Weak M–O Bonds about a Bridging Oxygen Atom

In this case, which is the counterpart to the situation treated in the foregoing section, the sum of the bond valences of an oxygen atom cannot reach the oxidation number of oxygen (its absolute amount) with the result of a negative charge on that oxygen atom. The typical examples are pseudoterminal oxygen atoms participating in angular M–O–M bridges where the second bond is that of a "*trans*" oxygen atom (Opa atoms in $Mo_7O_{24}^{6-}$, (β-)$Mo_8O_{26}^{4-}$, $Mo_8O_{26}(OH)_2^{6-}$, $Mo_{10}O_{34}^{8-}$ (I), $Mo_{10}O_{34}^{8-}$ (II), $[Mo_8O_{27}^{6-}]_\infty$, and $Mo_{36}O_{112}(H_2O)_{16}^{8-}$) and the O3a atoms of the ring-shaped trimeric groups in $W_{12}O_{40}(OH)_2^{10-}$ and $W_{12}O_{38}(OH)_2^{6-}$ which are for all three bonds "*trans*" oxygen atoms. However, there are also examples of two-coordinate oxygen atoms forming a very strong medium-sized and a long "*trans*" M–O bond (O2a2 in $Mo_{10}O_{34}^{8-}$ (I)) and forming a (strong) "tetrahedral" and a long "*trans*" M–O bond (O2l1 in $[Mo_4O_{14}^{4-}]_\infty$), which are quasi pseudoterminal oxygen atoms (see Sect. 4.3.1).

In principle, even high-coordinate O atoms can belong to this case if the M–O bonds are weak. A typical example is the O5l atom in (β-)$Mo_8O_{26}^{4-}$.

Table 41. Distribution of the (negative) ionic charge and of the negative and positive charge generated by charge separation over the different types of oxygen atoms of the M^{VI} isopolyoxometalate ions and oxides

Species	Z^+	c or c_{max}/c_{min} in c.u.[a] for				
		Ot[e]	Opa	O2a	O3a	Opl
MO_4^{2-}	0	−0.81/−0.14				
$Mo_2O_7^{2-}$	1.00	−0.39/−0.34				
$[Mo_4O_{14}^{4-}]_\infty$	1.00	−0.48/−0.36		−0.24		
$Mo_7O_{24}^{6-}$	1.14	−0.43/−0.35	−0.24	−0.15/−0.08	−0.09	
$W_{12}O_{40}(OH)_2^{10-}$	1.17	−0.53/−0.37		−0.19/−0.05		−0.17[c] *136°*
$Mo_{10}O_{34}^{8-}$ (I)	1.20	−0.56/−0.29	−0.17	−0.16/−0.10	−0.13	
$Mo_{10}O_{34}^{8-}$ (II)	1.20	−0.50/−0.28	−0.21	−0.03/−0.02	−0.03	
$Mo_8O_{26}(OH)_2^{6-}$	1.25	−0.40/−0.35	−0.18	−0.12/−0.11	−0.11	
$[Mo_8O_{27}^{6-}]_\infty$	1.25	−0.45/−0.32	−0.20	−0.24/−0.20	−0.05	
$[Mo_6O_{20}^{4-}]_\infty$ (o)	1.33	−0.34/−0.28			−0.14	
$[Mo_6O_{20}^{4-}]_\infty$ (o,tp)	1.33	−0.49/−0.23		−0.13		
$(\beta\text{-})Mo_8O_{26}^{4-}$	1.50	−0.32/−0.28	−0.09	+0.05	+0.11	
$\alpha\text{-}Mo_8O_{26}^{4-}$	1.50	−0.35/−0.29		+0.06		
$W_{12}O_{38}(OH)_2^{6-}$	1.50	−0.34		−0.19		
$W_{10}O_{32}^{4-}$	1.60	−0.53/−0.28		−0.16/+0.04		
$Mo_6O_{19}^{2-}$	1.67	−0.25		−0.06		
$W_6O_{19}^{2-}$	1.67	−0.27		−0.04		
$Mo_{36}O_{112}(H_2O)_{16}^{8-}$	1.78	−0.41/−0.25	−0.14	−0.13/−0.02	−0.14/−0.13	−0.09/+0.01 *151/154°*
"MoO_3"	2.00	−0.20				−0.01 *ca.180°*
"$(\alpha\text{-})MoO_3 \cdot H_2O$"	2.00	−0.29/−0.28				
"$MoO_3 \cdot 2H_2O$"	2.00	−0.34				−0.07 *149°*

4.4 Charge on the Oxygen Atoms

4.4.1 *Distribution of the Charge Over the Various Types of Oxygen Atoms and Operating Rules*

Tables 41 and 42 give a survey on the distribution of the charge over the various types of oxygen atoms of polyoxometalate ions. In most cases the relationship between charge on and type of the oxygen atom is very clearly recognizable. Some cases demand a more detailed view of the charge distribution.

The terminal oxygen atoms bera bv far most of the negative charge. The

Table 41 (Continued)

O2l	O3l	O4l	O5l	O6l	O3f	Oh, Ow[b]
+0.22						
154°						
−0.05/−0.04						
140/128°						
		+0.09				
−0.04					−0.05	+0.33 O3ah[c]
146°					*141°*	
+0.07	+0.22	+0.19				
154°						
+0.07	+0.17	+0.14				
146°						
	+0.16	+0.14				−0.13 Oth
+0.11	+0.10	+0.20				
ca.180°						
	+0.09					
	+0.19					
			−0.07			
					−0.04	
					130°	
+0.05						[+0.45 O3ah[d]
145°						−0.55
+0.11			−0.07			
175°						
				+0.25		
				+0.18		
+0.07/+0.10	+0.17/+0.24	+0.19				+0.33/+0.36 Otw
151/158°						+0.56/+0.63 O2aw
	+0.21					
	+0.22					+0.35 Otw
+0.01						+0.40 Otw
150°						

[a] All values are averaged values with respect to symmetry-equivalent oxygen atoms and, if so, with respect to different structural investigations. For the different MO_k polyhedra no significant deviations occur.
[b] h: protonated oxygen atom; w: diprotonated oxygen atom (coordinated water molecule). O–H bond valences of 1 v.u. are taken as a basis. Considering the commonly observed bond valences for O–H bonds of 0.8 v.u., 0.2 c.u. have to be subtracted for OH groups and 0.4 c.u. for coordinated H_2O molecules.
[c] O2l atoms with some pseudoterminal character; their charge (2 × (−0.17) c.u.) is very precisely compensated in the central cave by bifurcated hydrogen bridges proceeding from O3ah (+0.33 c.u.).
[d] Two O3a atoms in the centre are assumed protonated (+0.45 c.u.), two others are unprotonated (−0.55 c.u.). Thus positive and negative charges in the central cavity are largely compensated (rest charge on O3a: −0.05 c.u.) by hydrogen bridges.
[e] ⟨The borderline structures $HW_5O_{19}^{7-}$ and $H_3W_6O_{22}^{5-}$ reach c_{max} = −0.63 and −0.57 c.u.,

$Z^+ = 2.00$). The individual values differ from the respective average by up to 0.20 c.u. ($W_{10}O_{32}^{4-}$), which comes close to the observations for the monometalate ions (up to 0.36 c.u.). For the M^V species the average charge ranges from −0.66 (divanadate ions, $Z^+ = 1.00$) to −0.19 c.u. (vanadium(V) oxide, $Z^+ = 3.00$); the individual values differ from the respective average by up to 0.26 c.u. ($V_2O_7^{4-}$) which again corresponds to the observations for the monometalate (vanadate) ions (up to 0.28 c.u.).

Note that the species containing many coordinated H_2O molecules (36-molybdate ion, molybdenum trioxide hydrates) show an excessive negative charge on the terminal oxygen atoms if an OH bond valence of 1 v.u. is assumed; in reality these terminal oxygen atoms are involved in intra- and intermolecular hydrogen bridges by which the charge is strongly reduced.

The pseudoterminal (two-coordinately bridging) oxygen atoms as a sort of terminal oxygen atoms bear medium-sized negative charge on principle. They can be divided into two groups according to the M–O–M angle.

Bridges with an M–O–M angle smaller than ca. 125° show a "normal" behavior, that is, their oxygen atoms bear a negative charge that is somewhat smaller (ca. 0.2 c.u.) than that on the terminal oxygen atoms because of their involvement in an additional, weak bond (M^{VI} species). The negative charge depends approximately on the Z^+ value of the polymetalate ion, as expected. For the M^{VI} ions it ranges from −0.24 ($Mo_7O_{24}^{6-}$, $Z^+ = 1.14$) to −0.09 c.u. ((β-)$Mo_8O_{26}^{4-}$, $Z^+ = 1.50$).

Bridges with an M–O–M angle from 180° to ca. 125° (linear and approximately linear bridges) bear a somewhat smaller negative charge or even no negative charge since such bridges favor a positive charge on the bridging oxygen atom, see below. The charge is an intermediate between that for "normal" pseudoterminal and linearly bridging oxygen atoms. For the M^{VI} species it ranges from −0.09 ($Mo_{36}O_{112}(H_2O)_{16}^{8-}$) to +0.01 c.u. ($Mo_{36}O_{112}(H_2O)_{16}^{8-}$).

In the M^V ions ($V_{10}O_{28}^{6-}$ and protonated forms, $Nb_{10}O_{28}^{6-}$) the charge on the angularly bridging pseudoterminal oxygen atoms ranges from −0.20 to −0.06 c.u. These somewhat smaller charges (in spite of the generally higher charge densities of the M^V species) are a consequence of the fact that the second bonds of the pseudoterminal oxygen atoms are not weak "*trans*" M–O bonds (as is the case for all M^{VI} species) but are middle-strong bonds. Thus this type of oxygen atoms really belongs to the group of the angularly bridging two-coordinate oxygen atoms.

The angularly bridging two- and three-coordinate oxygen atoms bear even smaller negative charges and in rare cases even small positive charges. In the M^{VI} ions they range from −0.24 ($[Mo_4O_{14}^{4-}]_\infty$, $[Mo_8O_{27}^{6-}]_\infty$) to +0.06 c.u. (α-$Mo_8O_{26}^{4-}$) for the two-coordinate and from −0.14 ($[Mo_6O_{20}^{4-}]_\infty$(o), $Mo_{36}O_{112}(H_2O)_{16}^{8-}$) to +0.11 c.u. (($\beta$-)$Mo_8O_{26}^{4-}$) for the three-coordinate oxygen atoms. In the M^V ions the charge ranges from −0.40 ($Nb_6O_{19}^{8-}$) to −0.04 c.u. ($H_2V_{10}O_{28}^{4-}$) for the two-coordinate and from −0.14 ($V_{10}O_{28}^{6-}$) to +0.05 c.u. ($H_3V_{10}O_{28}^{6-}$) for the three-coordinate oxygen atoms.

That the dependence on the Z^+ value of the polyoxometalate ions is largely lost is due to the fact that these charges result from the simultaneous balance

Table 42. Distribution of the (negative) ionic charge and of the negative and positive charge generated by charge separation over the different types of oxygen atoms of the M^V isopolyoxometalate ions and oxides

Species	Z^+	c or c_{max}/c_{min} in c.u.[a] for										
		Ot	Opa	O2a	O3a	Opl	O2l	O3l	O4l	O5l	O6l	Oh[b]
VO_4^{3-}	0	−1.00/−0.47										
$V_2O_7^{4-}$	1.00	−0.71/−0.63					−0.21/+0.27 *118/180°*					
$V_3O_{10}^{5-}$	1.33	−0.59/−0.55					−0.19 *123°*					
$Nb_6O_{19}^{8-}$	1.67	−0.50		−0.40							−0.25	
$Ta_6O_{19}^{8-}$	1.67	−0.61		−0.34							−0.27	
$[VO_3^-]_\infty$	2.00	−0.56/−0.41					−0.04 *141/147°*					
$[V_2O_6^{2-}]_\infty$	2.00	−0.53						+0.06				
$V_5O_{14}^{3-}$	2.40	−0.41/−0.32					+0.02 *143°*					
$V_{10}O_{28}^{6-}$	2.40	−0.35/−0.26	−0.14	−0.25/−0.18	−0.14						−0.06	
$Nb_{10}O_{28}^{6-}$	2.40	−0.41/−0.39	−0.10	−0.19/−0.16	−0.08						−0.22	
$H_2V_{10}O_{28}^{4-}$	2.60	−0.29/−0.24	−0.15/−0.10	−0.27/−0.04	−0.12/−0.04						−0.07	+0.21 O2ah
$H_3V_{10}O_{28}^{3-}$	2.70	−0.39/−0.22	−0.20/−0.06	−0.15/−0.08	±0.00/+0.05						−0.12	+0.23/+0.27 O2ah +0.26 O3ah
"V_2O_5"	3.00	−0.19					+0.09 *ca.180°*	+0.15				

[a] All values are averaged values with respect to symmetry-equivalent oxygen atoms and, if so, with respect to different structural investigations. For the different MO_k polyhedra no significant deviations occur.

[b] h: protonated oxygen atom. O–H bond valences of 1 v.u. are taken as a basis. Considering the commonly observed bond valences for O–H bonds of 0.8 v.u., 0.2 c.u. have to be subtracted for the OH groups.

of the inner M–O bond lengths and bond valences according to the geometrical conditions (connectivity of the atoms in the M–O frameworks) and the bond length-bond valence relationship and valence sum rule (see Sects. 4.3.4.2 and 4.4.3.1) and from the balance of the ionic charges according to the geometrical conditions of the packing of cations, oxometalate ions, and molecules of water of crystallization (in the solid state) or according to the arrangement of cations, oxometalate ions, and water molecules (in solution) (see Sect. 4.4.3.2) and thus are "rest charges". (The above factors affect also the dependence of the charge on the terminal oxygen atoms on the Z^+ value, but to a much lesser extent.)

Oxygen atoms that are part of a linear or approximately linear M–O–M bridge (at least one M–O–M angle is greater than ca. 125°) usually bear positive charge; the observed charges range up to +0.25 c.u. ($Mo_6O_{19}^{2-}$) and +0.27 c.u. ($V_2O_7^{4-}$). This is one of the most remarkable results of our calculations on the charge distribution over the oxygen atoms of the polyoxometalate ions. To balance charges, occurrence of positive charge requires the simultaneous appearence of an equal amount of negative charge on terminal oxygen atoms apart from that representing the formal ionic charge ("charge separation", see Sect. 4.4.2.2, below, and Sects. 3.2.3 and 3.6.7 in Ref. 16). Geometric strains as mentioned above (see Sects. 3.10, 3.19, 4.3.4.2) and treated below (Sect. 4.4.3.1) in greater detail and the weakness of the bonds (see Sect. 4.3.4.4) are obviously responsible for the non-occurrence of positive charge but rather of a small negative charge on the five-coordinate oxygen atoms in (β-)$Mo_8O_{26}^{4-}$ and $W_{10}O_{32}^{4-}$. (For the small negative charge on the five-coordinate oxygen atoms in $W_{10}O_{32}^{4-}$ (−0.07 c.u.) see also Sect. 4.7).

If the fictitious negative charge density for the terminal oxygen atoms (the formal ionic charge is assumed to be exclusively located on the terminal O atoms; compare Table 43 in Sect. 4.4.3.6) is very high (greater than ca. 0.5 to 0.7 c.u.) there is no longer positive charge on the oxygen atoms participating in (approximately) linear M–O–M bridges (suppression of the charge separation processes). This refers to the two-coordinate oxygen atoms O2l of $[Mo_4O_{14}^{4-}]_\infty$ (−0.05 and −0.04 c.u.), $W_{12}O_{40}(OH)_2^{10-}$ (−0.04 c.u.), $V_2O_7^{4-}$ (up to −0.21 c.u.), $V_3O_{10}^{5-}$ (−0.19 c.u.), $[VO_3^-]_\infty$ (−0.04 c.u.), and in particular to the six-coordinate oxygen atoms O6l of $M_6^VO_{19}^{8-}$ (−0.25, −0.27 c.u.) and $M_{10}^VO_{28}^{6-}$ (−0.06, −0.22 c.u.). In the case of $[Mo_4O_{14}^{4-}]_\infty$ there is additionally the weakness of a "*trans*" M–O bond that prevents the occurrence of positive charge on O2l1 (see Sects. 3.22 and 4.3.4.4).

Protonated oxygen atoms (OH groups, coordinated H_2O molecules) in exposed structural parts (i.e., with the possibility for unhindered charge compensation by formation of hydrogen bridges and hence adjustment of the standard O-H bond valence of 0.8 v.u. [36, 39e, 40b, 56a]) bear a positive charge if they bridge MO_6 octahedra (OH groups: +0.01 to +0.07 c.u. in $H_2V_{10}O_{28}^{4-}$ and $H_3V_{10}O_{28}^{3-}$; coordinated H_2O: +0.16 and +0.23 c.u. in $Mo_{36}O_{112}(H_2O)_{16}^{8-}$). They bear negative charge in the position of unshared O atoms (OH groups: −0.33 c.u. in $Mo_8O_{26}(OH)_2^{6-}$; coordinated H_2O: −0.04 to

−0.07 c.u. in $Mo_{36}O_{112}(H_2O)_{16}^{8-}$, −0.05 c.u. in "(α-)$MoO_3 \cdot H_2O$", 0.00 c.u. in "$MoO_3 \cdot 2H_2O$"), again assuming adjustment of the standard O–H bond valence of 0.8 v.u. (For transparency, in Tables 41 and 42 and in the basis structural Tables O–H bond valences of 1 v.u. – which means absence of hydrogen bonding – have been assumed so that bridging OH groups and coordinated H_2O molecules appear to bear higher positive charge. This procedure is necessary since otherwise a complete and unobjectionable evaluation of the hydrogen bridge system is required.)

The OH groups enclosed within structures ($W_{12}O_{40}(OH)_2^{10-}$, $W_{12}O_{38}(OH)_2^{6-}$) show a special, well-balanced system of hydrogen bridges (O–H bond valences of 0.17 (2×) and 0.67 v.u. in the para-B tungstate ion and 0.5 v.u. (2×) in the metatungstate ion per H atom) which leave no charge on the O atoms involved.

The above distribution of the charge over the different types of oxygen atoms of the polyoxo species allows the formulation of the following *rules*:

(1) Negative charge is preferentially accepted by terminal oxygen atoms.
(2) Pseudoterminal (two-coordinate) oxygen atoms behave approximately as terminal ones and bear somewhat less negative charge.
(3) Angularly bridging oxygen atoms bear only small negative and occasionally positive charge.
(4) Oxygen atoms participating in linear or approximately linear M–O–M bridges bear usually positive charge.
(5) Oxygen atoms forming an OH or OH_2 bridge are positively charged.
(6) Oxygen atoms on which two of the above structural features superimpose bear an intermediate charge.
(7) Bridging oxygen atoms of polyoxo species with a high negative charge density bear increased negative or reduced positive charge.

4.4.2
Reasons for the Operation of the Rules about the Charge Distribution

The distribution of the charge is, of course, a consequence of thermodynamically favorable and unfavorable factors. Favorable are those factors that stabilize the polyoxometalate structures and/or enhance their basicity, i.e., that increase the consumption of protons in the equation of formation of the polyoxometalate ions according to Le Chatelier's principle, see Sect. 3.6.7 in Ref. 16. A number of reasons for the mode of the distribution of the charge over the oxygen atoms of polyoxometalate ions and for its modification have been discussed in that theoretical study about bonding in polyoxometalate ions [16]. In Sects 4.3.4.1 to 4.3.4.4 there are also discussed some factors exercising an influence on the charge distribution. The following statements in this section and in Sect. 4.4.3 are mainly based on the argumentation given there [16].

4.4.2.1
The Preferential Acceptance of the (Negative) Ionic Charge by Terminal (and Pseudoterminal) Oxygen Atoms

The preferential acceptance of the (negative) ionic charge by the terminal (and pseudoterminal) oxygen atoms can be explained by [16]

- the high statistical weight of the Lewis-type resonance structures with charged terminal (and pseudoterminal) oxygen atoms compared to the small statistical weight of the resonance structures with charged bridging oxygen atoms – according to the coordination number of oxygen (see Fig. 3 in Ref. 16);
- the stabilization of the polymeric structures if there is no (or only a small) negative charge on the bridging oxygen atoms (full utilization of the bonding power of the bridging oxygen atoms for the holding together of the MO_k building groups of the species) according to Scheme (16) in Ref. 16 (see also Sect. 4.3.4.1);
- the enhanced basicity of the corresponding resonance structures (see Sect. 4.6.6) and hence in solution increased consumption of the protons according to Le Chatelier's principle (see Sect. 3.6.7 in Ref. 16).
- the space requirements of unshared electron pairs (cf. the VSEPR model) in which the charge electrons are merged in, which can be better fulfilled by terminal oxygen atoms;
- the large distances between the negative ionic charges within the polyanionic structures and hence small Coulombic repulsion between them if the charge resides on the terminal oxygen atoms;
- the necessary contacts to the cations in the crystal (see also Sect. 3.21, above, and formulas XII and XIII in Ref. 16).

For quantitative views see Fig. 31 and Sect. 4.6.8.

4.4.2.2
The Occurrence of Positive Charge on Linearly or Approximately Linearly Bridging Oxygen Atoms

The occurrence of positive charge on linearly or approximately linearly bridging oxygen atoms is made possible by "charge separation" processes. It is conducted [16] by

- the additional stabilization of the polymeric structure through bond strengthening by multicenter $p\pi$-$d\pi$ multiple bonds in the respective M–O–M bridges according to Scheme (17) in Ref. 16, for solutions in combination with Le Chatelier's principle, if the simultaneously generated negative charge is accepted by terminal oxygen atoms (otherwise the stabilizing effect of bond strengthening in the M–O–M bridges with positively charged oxygen atoms would be compensated by a destabilizing effect of bond weakening in M–O–M bridges with negatively charged oxygen atoms, see again Scheme (17) in Ref. 16);

- the intra- and/or intermolecular ionic interaction between positively and negatively charged oxygen atoms, which leads again to an additional stabilization of the polymeric structure (intramolecular interaction) and/or to a strengthening of the cohesion of the polymeric building units (intermolecular interaction in the solid state), if the positively and the simultaneously generated negatively charged oxygen atoms lie at the surface of the structure or can otherwise interact with each other (see formulas XV and XVI in Ref. 16);
- the enhanced basicity of the polyoxometalate ion as a consequence of the higher negative charge on it (see also Sect. 4.6.6) and hence in solution increased consumption of the protons according to Le Chatelier's principle, if the negative charge generated by charge separation is accepted by terminal oxygen atoms (see Sect. 3.6.7 in Ref. 16).

For quantitative views see Fig. 31 and Sect. 4.6.8.

4.4.2.3
The Occurrence of Positive Charge on Bridging OH Groups and on Bridging Coordinated H_2O Molecules

The positive charge on the bridging oxygen atoms forming an OH group or H_2O molecule as well as an equivalent amount of negative charge on the other (preferably terminal) oxygen atoms is, in part, a formal quantity only if we assign a bond valence of 1 v.u. to the O-H bonds. In reality the hydrogen atoms are involved in hydrogen bridge bonds to negatively charged oxygen atoms of the same or of a neighboring polyanion. This means that the OH groups or H_2O molecules bear a smaller positive charge and the usually terminal but in some cases also bridging acceptor-oxygen atoms, which participate in the hydrogen bridges, a correspondingly lesser negative charge than that stated with application of O–H bond valences of 1 v.u. The part due to ("true") charge separation can be estimated if we assume for the O–H bonds the commonly observed bond valence of 0.8 v.u. [36, 39e, 40b, 56a] and absence of factors influencing this value. Thus the remaining positive charge due to charge separation is +0.23 and +0.16 c.u. on the O2aw atoms of $Mo_{36}O_{112}(H_2O)_{16}^{8-}$ and +0.01 to +0.07 c.u. on the O2ah and O3ah atoms of $H_2V_{10}O_{28}^{4-}$ and $H_3V_{10}O_{28}^{3-}$. The central O3ah atoms of $W_{12}O_{40}(OH)_2^{10-}$ and $W_{12}O_{38}(OH)_2^{6-}$ are uncharged because of an exact charge compensation by intramolecular hydrogen bonding in the central, inaccessible cavities of these structures. Non-bridging coordinated water molecules do not cause (true) charge separation and are negatively charged or uncharged ($Mo_{36}O_{112}(H_2O)_{16}^{8-}$, "(α-)$MoO_3 \cdot H_2O$", "$MoO_3 \cdot 2H_2O$"). The same is true for non-bridging OH groups ($Mo_8O_{26}(OH)_2^{6-}$).

The occurrence of positive charge on *bridging* OH groups and H_2O molecules is facilitated by

- the additional stabilization of the polymeric structure through bond strengthening in the respective M–O–M bridges according to resonance scheme (11):

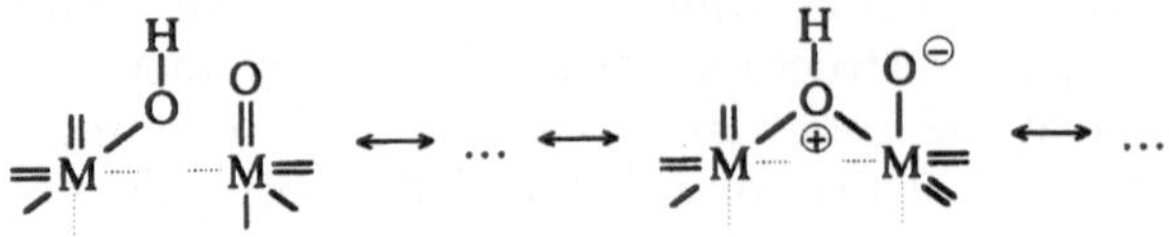

(11a)

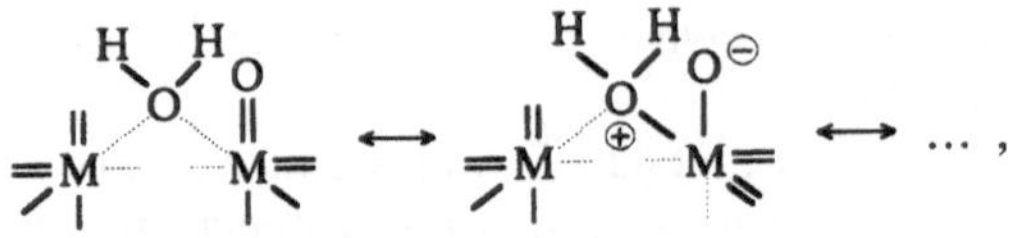

(11b)

for solutions in combination with Le Chatelier's principle, if the simultaneously generated negative charge is accepted by terminal oxygen atoms (otherwise the stabilizing effect of bond strengthening in the M–O–M bridges with positively charged protonated oxygen atoms would be compensated by a destabilizing effect of bond weakening in M–O–M bridges with negatively charged oxygen atoms);

- the intra- and/or intermolecular ionic interaction between positively and negatively charged oxygen atoms, which leads again to an additional stabilization of the polymeric structure (intramolecular interaction) and/or to a strengthening of the holding together of the polymeric building units (intermolecular interaction in the solid state) if the positive and the simultaneously generated negative charge on the oxygen atoms lie at the surface of the structure or can otherwise interact with each other, see formulas XV and XVI in Ref. 16;
- the enhanced basicity of the polyoxometalate ion as a consequence of the higher negative charge on it and hence in solution increased consumption of the protons according to Le Chatelier's principle if the negative charge generated by charge separation is accepted by terminal oxygen atoms (compare Sect. 3.6.7 in Ref. 16).

The above three items correspond to those in Sect. 4.4.2.2. For quantitative views see Fig. 31 and Sect. 4.6.8.

This factor has not been treated in our bond model for *poly*oxometalate ions [16]; however, positive charge on OH groups and coordinated H_2O molecules has been considered in our theoretical studies [23] on bonding in the monomeric protonated oxometalate ions.

The sum of the M–O bond valences of the bridging OH groups amounts to ca. 1.26 v.u. (1.21 to 1.32 v.u. in $H_2V_{10}O_{28}^{4-}$, $H_3V_{10}O_{28}^{3-}$, and $W_{12}O_{40}(OH)_2^{10-}$); 1.45 v.u. for the statistical half-protonation of the four central oxygens in $W_{12}O_{38}(OH)_2^{6-}$. This is a little bit more than the bond valence "on the anionic side of $(OH)^-$" of ca. 1.2 v.u. [56a]. The M–O bond valences of the bridging H_2O molecules sum up to ca. 0.59 v.u. (0.56 to 0.63 v.u. in $Mo_{36}O_{112}(H_2O)_{16}^{8-}$), which is somewhat more than the ca. 0.4 v.u. stated for "the anionic side of $(H_2O)^o$" [56a]. (For unshared [non-bridging] H_2O molecules, however, the value of ca. 0.4 v.u. agrees very well: 0.33 to 0.40 v.u. for $Mo_{36}O_{112}(H_2O)_{16}^{8-}$, "$MoO_3 \cdot H_2O$", and "$MoO_3 \cdot 2H_2O$".) The differences are caused solely by the stabilizing positive charge on the O atoms due to (true) charge separation as stated in the first paragraph of this section.

4.4.2.4
Coincidence of the Factors Coming into Force

The stability of polyoxometalate species (or of the polyoxometalate system in acidified aqueous solution) becomes optimal if the (negative) ionic charge and the negative charge generated by charge separation are exclusively spread over the terminal oxygen atoms and if a maximum of positive charge is generated and accepted by (approximately) linearly bridging oxygen atoms or by bridging OH groups or bridging coordinated H_2O molecules. These conditions, however, can only be fulfilled to a certain degree.

The factors named in Sects. 4.4.2.1–4.4.2.3 are already, in part, contradictory. The result is a superimposition of the effects. For example, the pseudoterminal oxygen atom of an (approximately) linear M–O–M bridge favors a medium-sized negative charge because of its pseudoterminal character, but also a medium-sized positive charge owing to its linearly bridging character (e.g., Opl1 to Opl3 in $Mo_{36}O_{112}(H_2O)_{16}^{8-}$). The pseudoterminal angularly bridging oxygen atoms (Opa) in the $M^{V}_{10}O_{28}^{6-}$ structures favor a high negative charge because of their pseudoterminal character and because of the generally high charge of the M^{V} species (see Sect. 4.4.3.5), but can only accept a small negative charge since the complementary (second) bond is not a weak bond in a *trans* position to a terminal or pseudoterminal M–O bond but is a middle-strong bond.

There are other factors that *can* hinder the optimal charge distribution over the oxygen atoms of the polyoxometalate ions. They are treated in Sect. 4.4.3.

4.4.3
Reasons for the Deviations from the Rules about the Charge Distribution

The favorable factors named in Sect. 4.4.2 are restricted by other factors that require a "harmonization" of the bond lengths, bond valences, and charges in the polyoxo species according to the bond length-bond valence relationship(s), the valence sum rule, the space requirements of the charge cloud, and others.

4.4.3.1
The Geometrical Requirements of the M–O Frameworks (Interdependence of the Inner M–O Bond Lengths)

This factor has already been treated in Sect. 4.3.4.2 under the main aspect of the bond lengths and bond valences.

In the example of the (β-)$Mo_8O_{26}^{4-}$ ion the small negative charge on the five-coordinate oxygen atoms instead of a small or even medium-sized positive charge to be expected cannot be overcome in principle: A more positive charge on this type of oxygen atoms requires also a more positive charge on those angularly bridging two- and three-coordinate oxygen atoms that form approximately a square around the five-coordinate oxygen atoms. This is apparently unfavorable and difficult to attain and the resulting charge distribution about the oxygen atoms in question is a compromise. [The reason for the negative charge on the five-coordinate oxygen atoms of (β-)$Mo_8O_{26}^{4-}$ is the accumulation of five weak (long) "*trans*" M–O bonds on the O51 atoms; compare also Sect. 4.3.4.4.]

Conversely, the uncommon occurrence of positive charge on angularly bridging oxygen atoms in (β-)$Mo_8O_{26}^{4-}$, α-$Mo_8O_{26}^{4-}$, $W_{10}O_{32}^{4-}$, and $H_3V_{10}O_{28}^{3-}$ is also a consequence of the interdependence of the M–O bond lengths in compact and/or highly symmetrical structures and usually coupled with an unusual feature on another oxygen atom; in the example of the preceding paragraph it is the negative charge on the five-coordinate oxygen atoms of (β-)$Mo_8O_{26}^{4-}$, involved in linear M–O–M bridges.

Thus we can state that the ability of oxygen to accept variable amounts of (negative and even positive) charge opens the possibility to fulfill simultaneously the bond length-bond valence relationship, the valence sum rule, and the connectivities in the *inner* M–O framework of the polyoxo species (see also Sect. 4.1 in Ref. 16).

4.4.3.2
The Geometrical and Charge-Related Requirements of the Packing of Cations, Oxometalate Ions, and Molecules of Water of Crystallization

This effect is analogous to that of the interdependence of the inner (bridging) M–O bond lengths (Sect. 4.4.3.1). The geometrical and charge requirements of the packing of cations, oxometalate ions, and molecules of water of crystallization have for the solid state a special influence on the distribution of the charge over the oxygen atoms of the oxometalate species. This special influence becomes apparent in the great variability of the *outer*, terminal M–O bond lengths and hence great variability of the charge on the terminal oxygen atoms of poly- and monooxometalate ions. The influence on the inner M–O bond lengths and hence on the charge of the bridging oxygen atoms is only small (see Sect. 4.2).

The following factors play a rôle and have simultaneously to be fulfilled:

- a packing of cations, (poly)oxometalate ions, and – if necessary – molecules of water of crystallization as close as possible under the conditions of the size, shape, and charge of the cations and the size, shape, charge and charge distribution about the oxygen atoms of the (poly)oxoanions;
- the realization of cation-oxygen(oxometalate ion) distances that fulfill Eq. (1) (or a suitable other equation) for the cation-O bonds and Eq. (2) (the valence sum rule) for the cations. (In this respect ionic bonds do not differ from covalent bonds, see Sect. 2.4, above, Sect. 4.4 in Ref. 16, and Refs. 29, 31, 36, 39d, 42–44.)

The fulfillment of the last-named item is a relevant reason for negative charge on bridging oxygen atoms at the surface of the polyoxo anions from this point of view; so far the optimum acceptance of the negative ionic charge by terminal oxygen cannot be expected. A problem is a higher negative charge on inner, unapproachable (e.g., O6l) oxygen atoms which is not intramolecularly compensated by positive charge on other oxygen atoms (e.g., in $Nb_6O_{19}^{8-}$, $Ta_6O_{19}^{8-}$, and $Nb_{10}O_{28}^{6-}$). This charge is hardly to balance by the cations according to Eqs. (1) and (2).

The rôle of the molecules of water of crystallization is not restricted to that of a filler of gaps in the crystal structure. Water molecules are suited to improve the fulfillment of Eqs. (1) and (2) by shifting the "effective points of gravity of the charges" (see also Sect. 3.21, Sect. 4.4 in Ref. 16, and Ref. 55). This corresponds to the view that H_2O, coordinated to a cation, carries the bond valence from the cation to a distant unsatisfied anion via a hydrogen bond, including the rôle of H_2O as a "bond valence transformer" [56b]. [An additional rôle of the water molecules, when acting as bond valence transformer, is seen in the moderation of the Lewis acidity of the cations in minerals such that it matches the Lewis basicity of the (usually polymeric) oxoanions to satisfy the "valence matching principle" [56c].]

4.4.3.3
Coincidence of Several Middle-Strong M–O Bonds about a Bridging Oxygen Atom

This case has already been treated in Sect. 4.3.4.3. Operation of this factor leads to positive charge on oxygen atoms that otherwise do not prefer positive charge so much (for examples see Sect. 4.3.4.3; the most impressive examples are +0.24 c.u. on the O3a atoms in the heteropolyanion $H_4As_4Mo_{12}O_{50}^{4-}$ [272], see Table 8 in Ref. 35) or, in the case of the M^V species with a high negative charge density, to a less negative charge (e.g., O3a in $V_{10}O_{28}^{6-}$, $H_2V_{10}O_{28}^{4-}$, and $Nb_{10}O_{28}^{6-}$).

4.4.3.4
Coincidence of a Few and Weak M–O Bonds about a Bridging Oxygen Atom

This case has already been treated in Sect. 4.3.4.4. Its operation leads to negative charge on certain oxygen atoms (for examples see Sect. 4.3.4.4).

4.4.3.5
Reduction of the Acceptance of the Ionic Charge by Terminal Oxygen Atoms because of a Very High Negative Charge Density on Them

A high formal ionic charge and a small number of terminal oxygen atoms, as applies to the octahedral isopolyoxometalate ions of M^V ($M_6O_{19}^{8-}$, $M_{10}O_{28}^{6-}$), lead to very highly charged terminal oxygen atoms, if the charge – considering the favorable conditions named in Sect. 4.4.2.1 – is exclusively accepted by the terminal oxygen atoms (see Table 43). This gives rise to a charge-related stress. Acceptance of a partial amount of the charge by bridging oxygen atoms reduces this stress (see Sect. 3.6.8 in Ref. 16).

Considering the many favorable effects of the acceptance of the formal ionic charge by terminal oxygen atoms (and of the operation of a charge separation process, see Sect. 4.4.2.2), it appears most likely that the reason for the charge-related stress is the spatial requirement of the charge cloud.

In Sections 3.6.3 and 3.6.8 in Ref. 16 it has been discussed that in some cases acceptance of ionic charge by bridging O atoms is required because corresponding resonance formulae for the polyoxo species cannot be avoided (polyoxometalate structures with $M_qO_{<3q}$ entities ($Z^+ > 2$), e.g., $M_{10}O_{28}^{6-}$, and with $m > t$, e.g., $M_6O_{19}^{8-}$). This factor has, of course, precedence over the charge distribution according to the type of oxygen atom.

4.4.3.6
Suppression of Charge Separation Processes because of a Very High Negative Charge Density on the Terminal Oxygen Atoms

Occurrence of a generally high negative charge density on the terminal oxygen atoms of the polyoxometalate ions acts suppressive on charge separation processes since the negative charge produced on this occasion has additionally to be placed on the terminal oxygen atoms (see Sect. 3.6.8 in Ref. 16).

This effect applies to the octahedral isopolyoxometalate ions of M^{VI} with the highest average negative charge on the terminal oxygen atoms, namely $W_{12}O_{40}(OH)_2^{10-}$ (−0.47 c.u.) and $[Mo_4O_{14}^{4-}]_\infty$ (−0.42 c.u.) (see also Table 43).

The effect is most pronounced for the isopolyoxometalate ions of M^V composed of MO_6 octahedra ($M_6O_{19}^{8-}$, $M_{10}O_{28}^{6-}$). They have a high ionic charge (because of the oxidation number 5 of M) and a small number of terminal oxygen atoms (because of their compactness). The high negative charge density leaves no room for still more negative charge on terminal oxygen atoms to be generated along with positive charge (compare Sect. 4.4.3.5). However, a charge separation process might indeed be active as can be demonstrated by a corresponding canonical resonance formula, but the importance of this formula in the set of canonical formulas becomes apparent only by a certain reduction and an equal rise of the negative charge on the respective oxygen atoms.

The M^V polyoxometalate types composed of corner-sharing MO_4 tetrahedra forming short and long chains ($V_2O_7^{4-}$, $V_3O_{10}^{5-}$, $[VO_3^-]_\infty$) or composed of edge-sharing trigonal VO_5 bipyramids ($[V_2O_6^{2-}]_\infty$) have a somewhat smaller negative charge density and therefore bear a positive charge on the approximately linearly bridging oxygen atoms in some cases. $V_5O_{14}^{3-}$ and the protonated decavanadate ions, having yet smaller negative charge densities, undergo a charge separation process. "V_2O_5" has no (negative) ionic charge and hence bears a medium positive charge on the approximately linearly bridging oxygen atoms.

Table 43 shows that the quotient −m/t – the basic density of the negative charge, if the entire ionic charge m⊖ would be located on the t terminal oxygen atoms – ranges for the M^{VI} polyanions from −0.15 ($Mo_{36}O_{112}(H_2O)_{16}^{8-}$) to

Table 43. Basic negative charge density −m/t (the formal ionic charge −m is assumed to be exclusively located on the t terminal oxygen atoms of the structure) and occurrence of charge separation for the isopolyoxo species investigated

Species	−m/t in c.u.	Positive charge on ··· oxygen atoms[a] linearly bridging	angularly bridging	protonated (OH, H_2O)[c]
"MoO_3"	0	+		
"$MoO_3 \cdot H_2O$"	0	+		−
"$MoO_3 \cdot 2H_2O$"	0	−[e]		−
"V_2O_5"	0	+		
$Mo_{36}O_{112}(H_2O)_{16}^{8-}$	−0.154	+		+
(β-)$Mo_8O_{26}^{4-}$	−0.286	−[b]	+[b]	
α-$Mo_8O_{26}^{4-}$	−0.286		+[b]	
$Mo_6O_{19}^{2-}$	−0.333	+		
$W_6O_{19}^{2-}$	−0.333	+		
$[Mo_6O_{20}^{4-}]_\infty$ (o)	−0.333	+		
$[Mo_6O_{20}^{4-}]_\infty$ (o,tp)	−0.333	+		
$Mo_2O_7^{2-}$	−0.333	+		
$V_5O_{14}^{3-}$	−0.375	+		
$H_3V_{10}O_{28}^{3-}$	−0.375	−[b]	+,−[b]	+
$W_{10}O_{32}^{4-}$	−0.400	+,−[b]	+,−[b]	
$Mo_{10}O_{34}^{8-}$ (I)	−0.400	+		
$Mo_{10}O_{34}^{8-}$ (II)	−0.400	+		
$Mo_8O_{26}(OH)_2^{6-}$	−0.429	+		−
$[Mo_8O_{27}^{6-}]_\infty$	−0.429	+		
$W_{12}O_{38}(OH)_2^{6-}$	−0.500	+[d]		−
$Mo_7O_{24}^{6-}$	−0.500	+[d]		
$H_3W_6O_{22}^{5-}$	−0.500	+[d]		−
$[Mo_4O_{14}^{4-}]_\infty$	−0.500	−[d,f]		
$H_2V_{10}O_{28}^{4-}$	−0.500	−[b,d]		+
$[VO_3^-]_\infty$	−0.500	−[d]		
$[V_2O_6^{2-}]_\infty$	−0.500	+[d]		
$W_{12}O_{40}(OH)_2^{10-}$	−0.556	−[d]		−
$V_3O_{10}^{5-}$	−0.625	−[d]		
$HW_5O_{19}^{7-}$	−0.636	+[d]		−
$V_2O_7^{4-}$	−0.667	+,−[d]		
$V_{10}O_{28}^{6-}$	−0.750	−		
$Nb_{10}O_{28}^{6-}$	−0.750	−		
$Nb_6O_{19}^{8-}$	−1.333	−		
$Ta_6O_{19}^{8-}$	−1.333	−		

[a] +: presence, −: absence of positively charged oxygen atoms of the respective type. A sign in the first and third column indicates simultaneously that the structural element is present in the structure.
[b] Geometrical requirements compel a negative charge on the linearly bridging oxygen atoms and/or a positive charge on the angularly bridging oxygen atoms.
[c] O–H bond valences of 0.80 v.u. have been assumed!
[d] Species of the transitional range $0.50 \leq m/t \leq 0.67$ c.u.
[e] A positive charge on the approximately linearly bridging oxygen atoms (Opl or O2l) could not interact with the negatively charged Ot atoms of the neighboring layer.
[f] This species is on the one hand at the border of the transitional range; on the other hand the oxygen atoms being in consideration are involved in a weak M–O(*trans*) bond which makes it difficult to reach together with the second M–O bond bond valences exceeding 2 v.u.

−0.56 c.u. ($W_{12}O_{40}(OH)_2^{10-}$), and for the M^V polyanions from −0.37 ($V_5O_{14}^{3-}$, $H_3V_{10}O_{28}^{3-}$) to −1.33 c.u. ($M_6O_{19}^{8-}$). For the range −m/t down to −0.50 c.u. charge separation is a general occurrence. In the transitional range −0.5 to −0.7 c.u. there occur in some cases positively charged approximately linearly bridging oxygen atoms. In the range $-m/t < -0.7$ c.u. charge separation does no longer take place. [Notice that $-m/t < -1$ c.u. means the significant occurrence of terminal O^{2-} ions in one of the resonance formulas (compare Schemes (5d) in Sect. 3.11, above, and (1) in Ref. 16) if there would be no negative charge on bridging oxygen atoms.]

4.4.4
"Origin" of the Negative Charge on the Terminal Oxygen Atoms

The "origin" of the negative charge on the terminal oxygen atoms of the polymetalate ions is shown in Fig. 31. There is a fictive basic charge on the oxygen atoms, accomplished by assuming a uniform distribution of the ionic charge over all oxygen atoms of the reacting species $(H^+)_p(MO_4^{w-})_q$ ($= -(qw-p)/4q$) (*A*). An additional charge results from the elimination of oxygen by condensation as r H_2O in the formation reaction of the polymetalate ion, assuming further on a uniform distribution of the charge over all oxygen atoms of the polymetalate ion $H_{p-2r}M_qO_{4q-r}^{(qw-p)-}$ ($= -\{[(qw-p)/(4q-r)] - (qw-p)/4q\}$) (*B*); coordinated H_2O molecules are equated to eliminated H_2O molecules, and two OH groups are equated to a coordinated H_2O molecule at this stage (thus assuming O–H bond valences of 1 v.u.). (The sum of both charges, namely $-(qw-p)/(4q-r)$, correlates to the average bond valences of the M–O bonds as described in Eq. (10), so that $(qw-p)/(4q-r) + k_O \cdot nq/uk_O = 2$.) For the terminal oxygen atoms an additional (average) charge results from the preferential acceptance of the ionic charge by terminal oxygen atoms (= x) (*C*) (connected with a corresponding decrease of charge on bridging oxygen atoms). Finally there is an additional (average negative) charge produced along with the positive charge by charge separation (= −Y/t; Y: positive charge in the structure produced by charge separation, known from the application of Eqs. (1) and (3) to the structural data; t: number of the terminal oxygen atoms in the polymetalate ion) (*D*), assuming a distribution of this charge merely on terminal oxygen atoms (see Sects. 4.4.2.2 and 4.4.2.3). Since the final (average negative) charge on the terminal oxygen atoms is known from the application of Eqs. (1) and (3) to the structural data of the polyoxo species (= −T/t; T: total negative charge on the terminal oxygen atoms of the structure), x(*C*) can be calculated according to $x = -T/t + [(qw-p)/(4q-r)] + Y/t$.

The average negative charge on the terminal oxygen atoms, if the formal ionic charge would exclusively be localized on terminal oxygen atoms, is also given ($= -(qw-p)/t$) (*E*).

Note that the additional charge due to the elimination of oxygen by condensation of H_2O and that due to the preferential acceptance of negative charge by the terminal oxygen atoms are relative quantities relating to the

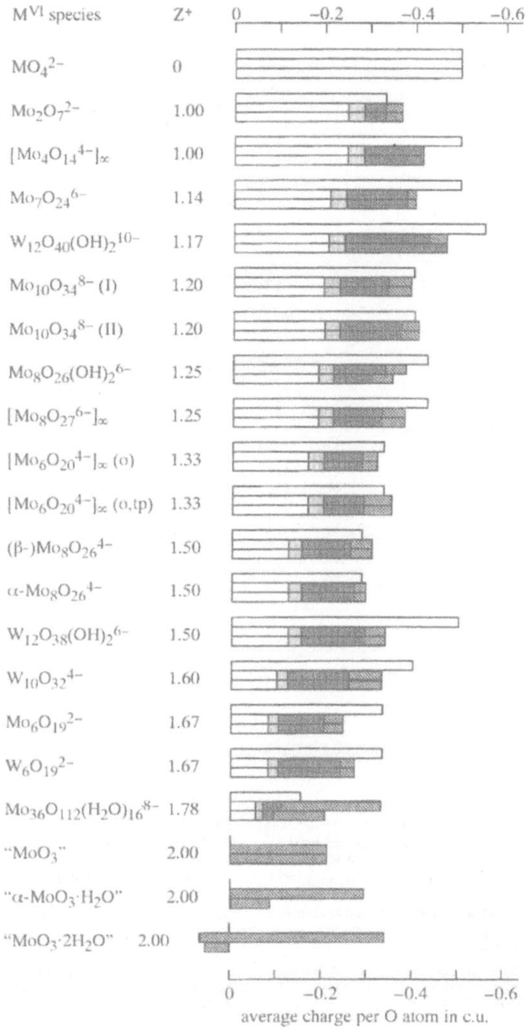

Fig. 31a. "Origin" of the negative charge on the terminal oxygen atoms of polyoxo species of the transition metal groups VI and V.

basic charge (thus approaching zero at $Z^+ = 2$ for $M = M^{VI}$ or $Z^+ = 3$ for $M = M^V$), whereas the additional charge due to charge separation is not related to any of the other varieties of the charge.

In Fig. 31 for $V_2O_7^{4-}$ two extreme examples are given.

In the lower beam of $W_{12}O_{40}(OH)_2^{10-}$ it is assumed that the positive charge on the two protonated O3ah atoms ($2 \times 0.325 = +0.65$ c.u.) compensates intramolecularly (within the cavity) the charge on the four (bridging) O2l1 atoms ($4 \times (-0.168) = -0.67$ c.u.) by formation of furcated hydrogen

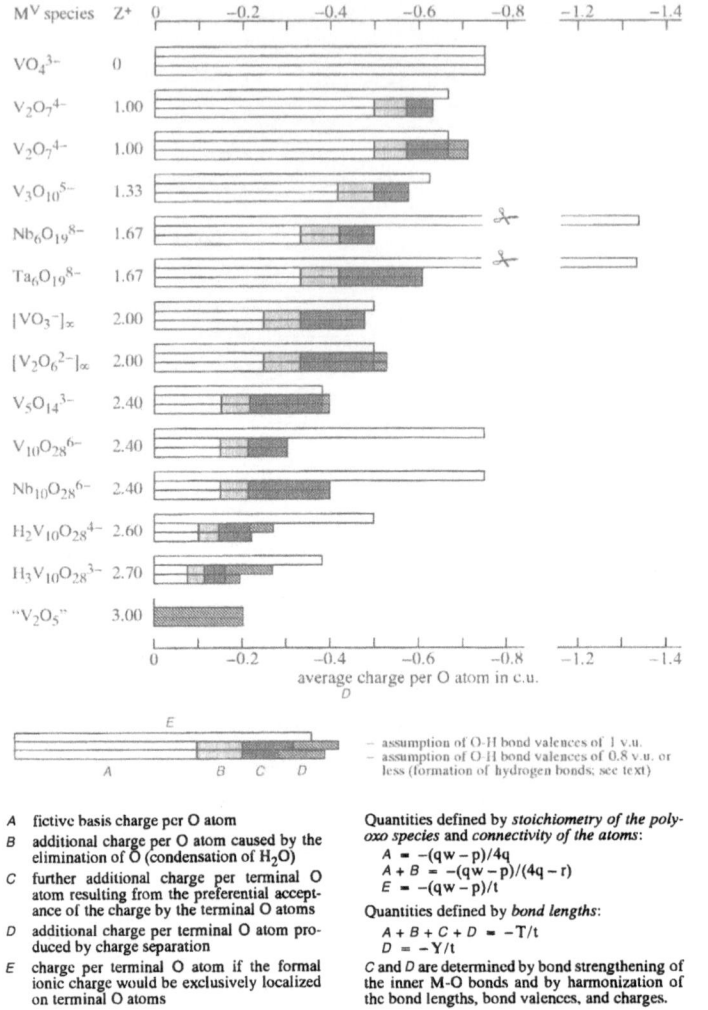

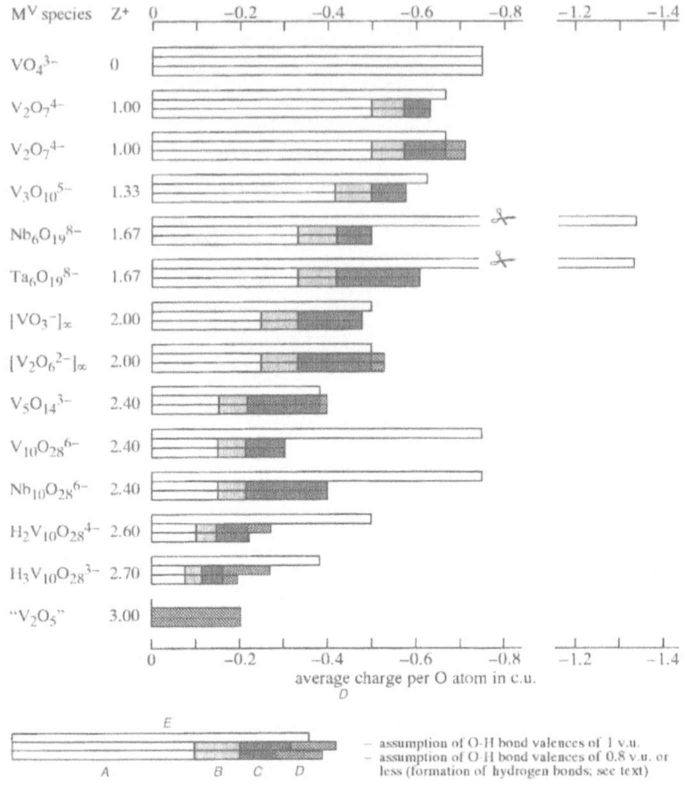

A fictive basis charge per O atom

B additional charge per O atom caused by the elimination of O (condensation of H_2O)

C further additional charge per terminal O atom resulting from the preferential acceptance of the charge by the terminal O atoms

D additional charge per terminal O atom produced by charge separation

E charge per terminal O atom if the formal ionic charge would be exclusively localized on terminal O atoms

Quantities defined by ***stoichiometry of the polyoxo species*** and ***connectivity of the atoms***:

$A = -(qw - p)/4q$
$A + B = -(qw - p)/(4q - r)$
$E = -(qw - p)/t$

Quantities defined by ***bond lengths***:

$A + B + C + D = -T/t$
$D = -Y/t$

C and *D* are determined by bond strengthening of the inner M-O bonds and by harmonization of the bond lengths, bond valences, and charges.

Fig. 31b.

bridges. The same is already assumed in Table 26 for the cavity in $W_{12}O_{38}(OH)_2^{6-}$ (hydrogen bridges to O3a atoms, all O–H bond valences 0.5 v.u.).

In the lower beams of $Mo_8O_{26}(OH)_2^{6-}$, $Mo_{36}O_{112}(H_2O)_{16}^{8-}$, "$MoO_3 \cdot H_2O$", "$MoO_3 \cdot 2H_2O$", $H_2V_{10}O_{28}^{4-}$, and $H_3V_{10}O_{28}^{3-}$ O–H bond valences of 0.8 v.u. [36, 39e, 40b, 56a] and hydrogen bridges to, ultimately, terminal oxygen atoms have been assumed.

The range beyond the basis line for "$MoO_3 \cdot 2H_2O$" of the upper beam is due to the fact that the negative charge generated along with the positive charge by charge separation is to that extent accepted by bridging oxygen atoms (compare Table 35). That of the lower beam corresponds to a (small) positive charge on the terminal oxygen atoms (Ot) due to hydrogen bonds between coordinated water molecules and terminal oxygen atoms if O–H bond valences of 0.8 v.u. are assumed; it would disappear if O–H bond valences of 0.83 v.u. and a rest charge on Otw of +0.06 c.u. are accepted.

4.5 Hierarchy of the Factors Determining the Bond Valence and Charge Distribution

The regularities discussed in Sects. 4.3 and 4.4 imply the following hierarchy of the factors determining the bond valence and charge distribution about a polyoxometalate structure:

- Starting point are the requirements of Lewis-type resonance formulae for the polyoxo structures; the restricting factor is the necessary minimum of (negative) ionic charge on bridging O atoms in the case of certain M^V structures (Sect. 4.1 in Ref. 16).
- The *lengths* of the *inner* (bridging) M–O bonds of the M–O framework have clearly primarily to be adjusted according to the connectivity of the atoms in the M–O framework, the bond length-bond valence relationship, and the valence sum rule (Sects. 4.3.4.2 and 4.4.3.1), proceeding, for example, with the average bond valences defined by Eq. (10) (Sect. 4.3.3). The lengths of the pseudoterminal M–O bonds (including the situation of coincidence of few and weak M–O bonds) (Sects. 4.3.4.4 and 4.4.3.4) and those of the bonds of the oxygen atoms on which several middle-strong M–O bonds coincide (Sects. 4.3.4.3 and 4.4.3.3) have also to be discussed here.
- In the case of solid-state structures the requirements of the packing of cations, polyoxometalate ions, and molecules of water of crystallization demand the next priority (Sect. 4.4.3.2). This refers primarily to the *lengths* of the *outer* (essentially terminal) M–O bonds of the M–O framework; however, there may also be necessary some (small) readjustments of inner M–O bond lengths to realize a suitable charge distribution at the surface of the polyanion for the interaction with the cations in accordance with the relevant bond length-bond valence relationship and valence sum rule. When these conditions are fulfilled, there is still a more or less large amount of free play to change the bond lengths (and hence bond valences) for most of the M–O bonds according to favorable parameters.[10]
- Within this degree of freedom the *bond valences* of the *inner* M–O framework are maximized, thereby maximizing the negative (ionic) charge on the terminal oxygen atoms (Sects. 4.3.4.1 and 4.4.2.1).
- At this stage very high negative charge on the terminal oxygen atoms is, according to the capacity of the terminal O atoms for the acceptance of

[10] Remember that it might have been necessary to change the primordial polyoxometalate ion to achieve a suitable packing of cations, polyoxometalate ions, and molecules of water of crystallization (compare Sect. 4.2). The main requirement of the new polyoxometalate structure is – apart from a suitable geometry – a Z^+ value close to that existing before.

unshared electron pairs in relation to that of the bridging O atoms, repressed (Sect. 4.4.3.5).

- Within the remaining degree of freedom the *bond valences* of the *inner* M–O framework are maximized further by charge separation if linear or approximately linear M–O–M bridges, bridging coordinated water molecules, or bridging OH groups are present (Sects. 4.4.2.2 and 4.4.2.3). Since very high negative charge on the terminal oxygen atoms is avoided, in such a case charge separation is suppressed (Sect. 4.4.3.6).
- On all stages a readjustment of the previous adjustments may be necessary within the degree of freedom of the bond lengths.

For further hierarchal and some quantitative considerations see Sects. 4.6.8.2 and 4.4.4. Notice in particular that the effects of the stabilization of the polyoxometalate structures by inner M–O bond valence based on the factors discussed in Sects. 4.3 and 4.4 and in this section are only small compared to the stabilization by inner M–O bond valence based on other factors, that nevertheless the above factors play a decisive rôle since the other factors are invariably determined by the stoichiometric coefficients and by the connectivity of the atoms (see Sect. 4.6.8.2).

4.6 Some Special Points

4.6.1 *The Geometric and Bond-Related Requirements for the Extension of the Coordination Sphere of the MO_4 Tetrahedra to that of the MO_6 Octahedra (MO_k Polyhedra); Different Behavior of the Te^{VI} and I^{VII} Oxospecies*

The occurrence of MO_k polyhedra with $k > 4$ requires the extension of the MO_4 tetrahedra of the original monometalate ions which is initiated by the protonation of an oxygen atom and thereby produces an *elongation* of the corresponding M–O bond of the MO_4 tetrahedron. In this way space is gained for the coordination of the usually two additional oxygen atoms [4, 17–19]. Using the bond length-bond valence relationships, this point can now be considered in detail.

Due to the valence sum rule, the decrease of the bond valence of the protonated M–O bond is coupled with an increase of the bond valence of the remaining M–O bonds and hence with their *shortening*. As a consequence of the hyperbola-like form of the bond length-bond valence function (for other consequences of this form of the function see footnote 1 in Sect. 2.1 and Sect. 4.6.3), however, there is indeed a small overall increase of the M–O bond lengths and hence a gain in space: The six valence units of Mo in MoO_4^{2-} result from four Mo–O bonds of 1.5 v.u. and 1.747 Å (summed up: 6.988 Å); those in $(HO)MoO_3^-$ result from three Mo–O bonds of 1.67 v.u. and 1.704 Å and one Mo–O bond of 1 v.u. and 1.915 Å (summed up: 7.027 Å). Diprotonation enhances the difference still more; this is obviously a reason why aggregation of the smaller vanadate ion requires at least diprotonation to attain octahedral coordination [14p, 20, 24, 205]. Thus the five valence units of V in VO_4^{3-} result from four V–O bonds of 1.25 v.u. and 1.717 Å (summed up: 6.868 Å); those in $(HO)_2VO_2^-$ result from two V–O bonds of 1.5 v.u. and 1.649 Å and two V–O bonds of 1 v.u. and 1.799 Å (summed up: 6.896 Å).

For a more realistic assessment of the situation one has to consider that the two additional oxygen atoms approach the MO_4 tetrahedron capable of extension its coordination sphere by an edge (see, e.g., Figs. 11, 14, and 15 in Ref. 4 or Scheme 23 in Ref. 16); the respective oxygen atoms of the tetrahedron form the medium-sized M–O bonds in *trans* position to each other of a type II MO_6 octahedron. Hence we have to compare on the one hand two Mo–O bonds of 1.747 Å (summed up: 3.494 Å) for MoO_4^{2-} and on the other hand Mo–O bonds of 1.915 and 1.704 Å (summed up: 3.619 Å) for $(HO)MoO_3^-$. Analogously we have to compare two V–O bonds of 1.717 Å (summed up: 3.433 Å) for VO_4^{3-} with two bonds of 1.799 Å (summed up: 3.598 Å) for $(HO)_2VO_2^-$; if only monoprotonated monovanadate ions are considered we have to compare the two V–O bonds of 1.717 Å (summed up: 3.433 Å) for VO_4^{3-} with one V–O bond of 1.693 Å (1.33 v.u.) and one of 1.799 Å (1 v.u.) (summed up: 3.492 Å) for $(HO)VO_3^{2-}$. Thus monoprotonation of the monovanadate ion produces merely conditions that are met for the unprotonated monomolybdate ion and hence should not allow extension of the tetrahedral coordination sphere of vanadium(V). Only diprotonation of the monovanadate ion creates conditions comparable to those of the monoprotonated monomolybdate ion.

Note that an explanation of the differences of the aggregation behavior of Mo or W compared to V concerning the expansion of their coordination spheres with the "smaller size" of V (meaning, e.g., the ionic radii of M [13j, 14p, 20, 24] or the d_o values of the M–O bonds) is only partially correct. Most of the size difference is already compensated by an effect of the different oxidation numbers of the M atoms, that means, by the different M–O bond valences: in MoO_4^{2-} the average bond valence amounts to 1.5 v.u., corresponding to 1.747 Å, and in VO_4^{3-} only to 1.25 v.u., corresponding to 1.717 Å. Thus there is merely a difference of 0.030 Å for the M–O bond lengths of the unprotonated MO_4^{w-} ions of Mo^{VI} and V^{V} that has to be overcome by additional protonation and not 0.05 or 0.06 Å as would appear according to the ionic radii (compare footnote 6) (or even 0.12 Å according to the d_o values, compare Table 1). (This discussion shows impressively the problems which are connected with the use of the idea of the 'ionic radii' in this field.)

The above description applies to the initial phase of the expansion process. As soon as the new, long M–O bonds form, the bonds present before are weakened and lengthened. For an average idealized type II MoO_6 octahedron in a polyoxomolybdate ion the typical bond valences and bond lengths are 1.600 (2×) – 1.000 (2×) – 0.400 (2×) v.u. and 1.720 (2×) – 1.915 (2×) – 2.294 (2×) Å, respectively (see Fig. 32a). The M–O bonds belonging to the relevant edge of the MoO_4 tetrahedron have lengths of 1.915 Å (2×) (summed up: 3.830 Å), which is indeed considerably more than the 3.494 Å for MoO_4^{2-}. Analogously for an idealized average type I VO_6 octahedron in a

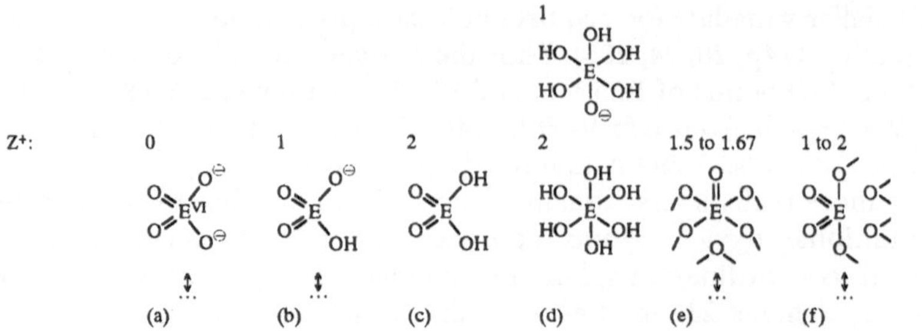

	(a)	(b)	(c)	(d)	(e)	(f)
	4×1.500 v.u.	3×1.667 v.u. 1×1.000 v.u.	2×2.000 v.u. 2×1.000 v.u.	6×1.000 v.u.	1×1.700 v.u. 4×0.980 v.u. 1×0.380 v.u.[a]	2×1.600 v.u. 2×1.000 v.u. 2×0.400 v.u.[a]
Cr^{VI} [b]:	4×1.644 Å	3×1.605 Å 1×1.794 Å	2×1.538 Å 2×1.794 Å	6×1.794 Å	1×1.598 Å 4×1.802 Å 1×2.152 Å	2×1.620 Å 2×1.794 Å 2×2.133 Å
	ø 1.644 Å	ø 1.652 Å	ø 1.666 Å	ø 1.794 Å	ø 1.826 Å	ø 1.849 Å
Mo^{VI}, W^{VI} [c]:	4×1.747 Å	3×1.704 Å 1×1.915 Å	2×1.628 Å 2×1.915 Å	6×1.915 Å	1×1.695 Å 4×1.923 Å 1×2.316 Å	2×1.720 Å 2×1.915 Å 2×2.294 Å
	ø 1.747 Å	ø 1.756 Å	ø 1.772 Å	ø 1.915 Å	ø **1.951 Å**	ø **1.977 Å**
Te^{VI}:				ø **1.921 Å**[d]		

Fig. 32a. Expected size of different $M^{VI}O_k$ ($E^{VI}O_k$) polyhedra, characterized by the average M–O bond length (ø), in dependence on their protonation and coordination state. Bold types: Octahedral coordination of this type is proven. [a] Average bond valences for the structure type in case of Mo^{VI} and W^{VI}; [b] d_0 = 1.794 Å, B = 0.852 Å [33, 49]; [c] for d_0 and B see Table 1; [d] average Te^{VI}–O single bond distance in $(NH_4)_6[TeMo_6O_{24}] \cdot Te(OH)_6 \cdot 7H_2O$ [273]; the bond length-bond valence parameters d_0 = 1.917 Å, B = 0.852 Å [33, 49] lead to 1.917 Å.

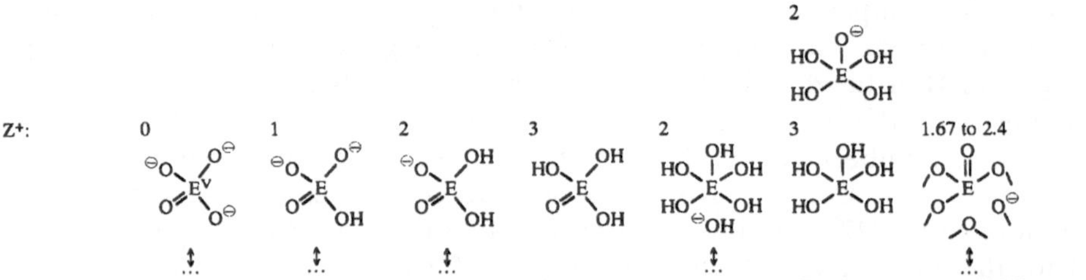

	(a)	(b)	(c)	(d)	(e)	(f)	(g) V / Nb
	4×1.250 v.u.	3×1.333 v.u. 1×1.000 v.u.	2×1.500 v.u. 2×1.000 v.u.	1×2.000 v.u. 3×1.000 v.u.	6×0.833 v.u.	5×1.000 v.u.	1×1.720 / 1.570 v.u. 2×0.900 / 0.900 v.u. 2×0.610 / 0.690 v.u. 1×0.260 / 0.250 v.u.[a]
V^V [b]:	4×1.717 Å	3×1.693 Å 1×1.799 Å	2×1.649 Å 2×1.799 Å	1×1.543 Å 3×1.799 Å	6×1.866 Å	5×1.799 Å	1×1.599 Å 2×1.838 Å 2×1.982 Å 1×2.297 Å
	ø 1.717 Å	ø 1.719 Å	ø 1.724 Å	ø 1.735 Å	ø 1.866 Å	ø 1.799 Å	ø **1.922 Å**
Nb^V (Ta^V) [b]:	4×1.834 Å	3×1.810 Å 1×1.917 Å	2×1.766 Å 2×1.917 Å	1×1.658 Å 3×1.917 Å	6×1.985 Å	5×1.917 Å	1×1.749 Å 2×1.956 Å 2×2.055 Å 1×2.434 Å
	ø 1.834 Å	ø 1.837 Å	ø 1.841 Å	ø 1.852 Å	ø **1.985 Å**	ø 1.917 Å	ø **2.034 Å**

Fig. 32b. Expected size of different M^VO_k (E^VO_k) polyhedra, characterized by the average M–O bond length (ø), in dependence on their protonation and coordination state. Bold types: Octahedral coordination of this type is proven. [a] Average bond valences for the structure type; [b] for d_0 and B see Table 1.

polyoxovanadate ion the typical bond valences and bond lengths are 1.720 (1×) – 0.900 (2×) – 0.610 (2×) – 0.260 (1×) v.u. and 1.599 (1×) – 1.838 (2×) – 1.982 (2×) – 2.297 (1×) Å, respectively (compare Fig. 32b). The M–O bonds belonging to the relevant edge of the VO_4 tetrahedron have lengths of 1.838 and 1.982 Å (summed up: 3.820 Å), which is again considerably more than the 3.433 Å for VO_4^{3-}.

The above statements concern primarily the expansion process from the MO_4 tetrahedra to the MO_6 octahedra. In Figs. 32a and b we consider in dependence on the protonation and coordination state the final state of the various MO_k polyhedra with regard to the individual bond lengths and to their average. We see that apparently the lower limit for the formation of an MO_6 octahedron amounts to 1.92 Å for the average M–O bond length and that *the O atoms* form rather ideal octahedra in that regard that the sums of the *trans*-located M–O bonds are quite similar, see Sect. 4.3.2. The dimensions of these octahedra correspond to a minimum of 2.72 Å for the non-bonded O···O distances within the O_6 octahedra. VO_6 octahedra and the VO_k polyhedra that form "VO_5 pyramids" (with a sixth very long V–O bond, compare Sect. 4.6.11.2) have a less regular shape.

We see also that for Nb^V and Ta^V there exists obviously a second type of MO_6 octahedron with six equal M–O bonds, $M(OH)_6^-$. Such species have indeed been reported in the literature [274c].

EO_6 octahedra of this second type have also been observed for E = Te^{VI} and I^{VII} ($Te(OH)_6$ and deprotonated forms [274d, 275c], $IO(OH)_5$ and deprotonated/protonated forms [275d]); compare also Sect. 4.6.3 in Ref. 16. Their occurrence is explained by the inability of the post-transition cations to form strong $p\pi$–$d\pi$ bonds with terminal oxygen atoms [13j]. The existence of the tetrahedral TeO_4^{2-} [275c] and IO_4^- [275d] ions and bond valence calculations on octahedral and square-pyramidal periodate ions [275d] by our group show that Te–O bond valences up to 1.5 and I–O bond valences up to 1.75 v.u. (the M–O bond valences in the polyoxo species reach 1.80 v.u.) and rather wide bond valence ranges (e.g., for $I_2O_9^{4-}$ from 0.80 to 1.53 v.u.) occur (B = 0.852 Å; d_o = 1.917 Å for Te^{VI}–O [33, 49] and 1.93 Å for I^{VII}–O [49] bonds) so that this line of argumentation appears to be rather unsatisfactory. An explanation has surely to consider on the one hand why TeO_4^{2-} and IO_4^- or their protonated forms do not form analogues of $MO_2(OH)_2(OH_2)_2$ with strong (O), middle-strong (OH), and weak (H_2O) M–O bonds (from which the structural motif of the polyoxometalate ions is derived, see Eq. (23) in Ref. 16). An adequate answer is that TeO_4^{2-} and IO_4^- are rather weak bases that cannot easily be protonated (compare Pauling's acidity/basicity rules [21]), and in particular cannot be diprotonated. However, a regular octahedral coordination of Te^{VI} and I^{VII} can be achieved via a mechanism for which no propositions have been made. On the other hand we have to explain why MO_4^{2-} or its protonated forms do not form the analogues of orthotelluric acid, $MO_2(OH)_4^{2-}$, $MO(OH)_5^-$, and $M(OH)_6$. Here we can state that $MO_2(OH)_2(OH_2)_2$ is obviously the better solution since it provides more space for M [see Fig. 32a, (f); the ortho-analogues (Fig. 32a, (d)) provide only space according to the d_o bond length-bond valence parameter for the M–O bonds].

To summarize briefly the results of this section: Octahedral coordination of M (or E) by oxygen requires a minimum average M–O (E–O) bond length of 1.92 Å. Factors having an influence upon that are (in order of hierarchy):

- a sufficient size of M in itself, e.g., as defined by the single bond distance d_o (the ionic radii of M and O are unsuitable quantities since they are not satisfactorily defined);
- a *suitable oxidation number of M (for geometrical reasons)*: smaller oxidation numbers cause smaller bond valences and hence longer M–O bonds;
- a protonation of the monometalate ion (and in the later aggregation phase the substitution of H by M), which leads to a diversification of the M–O bond types and hence bond lengths and, furthermore, as a consequence of the non-linearity of the bond length-bond valence function, to an increase of the *average* M–O bond length. The protonation in turn is relieved by a *suitable oxidation number of M (for reasons of the basicity and of the charge number)*: smaller oxidation numbers cause higher charge numbers of the mono- and growing polymetalate ions and thus greater protonation constants and higher degrees of protonation for them.

4.6.2
The Coordination Number of the Addenda Elements M

The assignment of a realistic and practicable coordination number and coordination polyhedron to M produces in general no difficulties. In a few cases, however, there are contradictory descriptions in the literature for MO_k polyhedra due to different opinions about long M–O distances to be considered as M–O bonds.

The range of the M–O distances in the polyoxometalate ions starts at 1.67 Å for Mo^{VI} and W^{VI} and 1.58 Å for V^V compounds and is open at the upper end.

For the great majority of the species there is a common unoccupied interval from 2.5 to 3.4 Å (0.23 to 0.03 v.u.) for Mo^{VI} and W^{VI} and from 2.4 to 3.3 Å (0.20 to 0.02 v.u.) for V^V if averaged symmetry-equivalent M–O bonds are considered (see Tables 37–39). This refers in the first place to the octahedral block-type polyoxometalate structures but also to the majority of the other structures. The very small [$\leqq$0.03 v.u., in rare, special cases 0.05 v.u. ($W_{12}O_{40}(OH)_2^{10-}$)] bond valences at the upper (with regard to the M···O distances) interval border are the inevitable consequence of the edge-sharing of the MO_6 octahedra in the block-type structures and hence are structurally of no relevance (with the exception of their consideration in Eq. (2)). The bond valences at the lower (with regard to the M···O distances) border of the unoccupied interval, however, are – in spite of their smallness (0.2 v.u.) – absolutely necessary for the existence of the structures. These bonds cannot be much stronger than observed due to the coordination mode of the relevant oxygen atoms: M–O6l in $M_6O_{19}^{w-}$ amounts to 0.33 v.u. (6 × 0.33 v.u. = 2 v.u.), M–O4f in $X^VM_{12}O_{40}^{3-}$ (X: heteroelement) to 0.25 v.u.

(3 × 0.25 v.u. + 1 × 1.25 v.u. = 2 v.u.) [253b], if the oxygen atoms under discussion bear no charge. In another approach we can start with Eq. (10), thus assuming an equidistribution of the available bond valence over the M–O bonds: M–O6l in $M_6O_{19}^{8-}$ yields 0.26 v.u., M–O6l in $M_{10}O_{28}^{6-}$ 0.30 v.u. (see Table 40). Hence the assignment of a realistic and practicable coordination number and coordination polyhedron to M (MO_6 octahedra) offers in general no difficulties for the block-type structures and structures composed of block-type structural parts.

The structures consisting of MO_k polyhedra or containing single MO_k polyhedra smaller than MO_6 octahedra show sometimes M–O bonds in the commonly unoccupied interval which can surely in some cases be regarded as occasion to obtain yet an increased coordination number for M, however, may in other cases only be a consequence of the packing conditions for cations, (poly-)oxoanions, and molecules of water of crystallization. For example, the greatest bond valences in this range (outside the range commonly regarded as coordination polyhedron) belong to two divanadate structures (0.17 and 0.11 v.u.), whereas those for the other divanadates investigated are in the range 0.01 to 0.05 v.u.; hence the large values are obviously the consequence of packing effects (compare Table 37). The next largest bond valences amount to 0.08 v.u. ($Mo_{10}O_{34}^{8-}$ (I), MoO_4^{2-}), 0.07 v.u. ("V_2O_5", $[Mo_4O_{14}^{4-}]_\infty$, VO_4^{3-}) and 0.06 v.u. (MoO_4^{2-}) (see Table 37). For the polyoxo species of them it can be assumed that there is a tendency to increase yet the coordination number of M. For $Mo_{10}O_{34}^{8-}$ (I) and α-$Mo_8O_{26}^{4-}$ there is a considerable improvement of the sum of the bond valences about the tetrahedrally coordinated Mo atoms, compare Sect. 4.1. Otherwise there is a somewhat smaller standard deviation $\sigma_{\Sigma i}$ (1.1% compared to 1.2%) in the investigated cases if the very long bonds are taken into account. Both items speak for the reality of the bond valence contribution due to the very long M···O bonds. However, the contribution of these bonds is, on the one hand, as a rule not necessary for the existence of the species and, on the other hand, comparatively so small that they should not be included in the definition of the coordination polyhedra for reasons discussed above (Sect. 3.3).

In an investigation concerning the coordination number and the range of bonding about hydrogen with reference to oxygen Trömel [276] concludes similarly that the range of bonding and consequently the coordination number to be considered in bond lengths-bond valence and valence sum calculations has to be chosen very widely. Only in this case is there a good correspondence between bond valence sum and oxidation number for hydrogen. (See also Sect. 2.2.2.4 in Ref. 35.)

There is yet another type of polyoxometalate ions, namely the heteropoly-oxovanadate ions with "unusual" central heteroelement "ions" or molecules and fractional and variable oxidation numbers for V from five to four. In this case the small bond valences of the very long M···OX and M···X bonds are a consequence of the small ionic charge of the original building group now forming the central "hetero part" and the large number of addenda atoms in the surrounding M–O cage and hence are a substantial structure-determining factor (see Sect. 4.6.11.2). The fractional and variable oxidation numbers of M facilitate the fine bondwise adjustment between central "hetero part" and surrounding M–O cage.

4.6.3
The Variability of the Bond Lengths and Bond Valences and of the Charge in M–O–M Bridges

The inner (bridging) M–O bond lengths and hence bond valences of the M–O frameworks of the polyoxometalate ions are, for geometrical reasons, more or less dependent of each other. This can lead to constrained structures, all the more the more compact and/or the more symmetrical (*the more rigid*) the M–O frameworks are (see Sects. 4.3.4.2 and 4.4.3.1).

One possibility to adjust bond lengths to the geometrical requirements and hence to reduce the constraints is the acceptance of variable amounts of the negative ionic charge by bridging oxygen atoms (see Sect. 4.4.3.1). The negative charge must not become large because of its destabilizing effect on the polymeric structure. The effect is, in relation to the optimal conditions for this situation (no negative charge on the bridging oxygen atoms), directed towards an *elongation of M–O bonds* of the bridges. It is without doubt present in many negatively charged M–O–M bridges and is one of the reasons why bridging oxygen atoms show no distinct dependence of the negative charge upon the Z^+ value of the polyoxometalate ion. (A second reason for negative charge on bridging oxygen atoms can be the cation-anion charge balance, see Sect. 4.4.3.2.)

Another possibility to adjust bond lengths to the geometrical requirements is the operation of the charge separation process. Since the negative charge generated in this way is accepted by terminal oxygen atoms there is so far no influence of the negative charge on the inner M–O frameworks. However, the positive charge generated is accepted by bridging oxygen atoms and works towards a *shortening of the M–O bonds*. This effect is operative at least in those less flexible M–O frameworks with a positive charge on angularly bridging oxygen atoms ((β-)$Mo_8O_{26}^{4-}$, $W_{10}O_{32}^{4-}$, α-$Mo_8O_{26}^{4-}$, $H_3V_{10}O_{28}^{3-}$; compare Sect. 4.3.4.2). [A second reason for positive charge on (approximately linearly) bridging oxygen atoms is the stabilizing effect of positive charge on the polymeric structure, compare Sect. 4.4.2.2.]

Both possibilities to adjust bond lengths in M–O–M bridges just described are coupled with a change of bond valences and hence with a change of the charge on the bridging oxygen atoms. The following possibility permits the adjustment of M–O bond lengths in M–O–M bridges without a change of the sum of the bond valences in the bridge and hence without a change of the charge on the bridging oxygen atom: formation of a more or less symmetrical M–O–M bridge. Due to the non-linear relationship between bond length and bond valence every deviation from the symmetry in the M–O–M bridge leads to an increase of the M $\cdots$ M distance, i.e., to an *elongation of the average M–O bonds at constant sum of the bond valences in the bridge and constant charge on the bridging oxygen atom.* Conversely, keeping the length of the M–O–M bridge (the M $\cdots$ M distance) constant, an unsymmetrical bridge comprises a

Table 44. Variability of bond lengths and bond valences in M–O–M bridges due to the hyperbolic form of the bond length-bond valence curves

1. Total bond valence in the M–O–M bridge and charge on the bridging oxygen atom are constant:
An asymmetric bridge is longer.
Example:

	Mo–O–Mo		Mo–O—Mo	
d in Å	1.915 1.915	($\Sigma d = 3.830$)	1.839 2.008	($\Sigma d = 3.847$)
s in v.u.	1 1	($\Sigma s = 2$)	1.2 0.8	($\Sigma s = 2$)

2. The length of the M–O–M bridge (the M ··· M distance) is constant:
An asymmetric bridge
- involves greater bond valence
- involves smaller negative or higher positive charge on the bridging oxygen atom.

Example:

			0,065 ⊕	
	Mo–O–Mo		Mo–O—Mo	
d in Å	1.915 1.915	($\Sigma d = 3.830$)	1.815 2.015	($\Sigma d = 3.830$)
s in v.u.	1 1	($\Sigma s = 2$)	1.271 0.794	($\Sigma s = 2.065$)

greater total bond valence for the bridging oxygen atom (see Table 44). (Hence a symmetrical M–O–M bridge of given length embraces always a minimal total bond valence.) Note that in the last case (constant M···M or, which is approximately equivalent, O···O distance) the M atom has only to move a little along the corresponding O–M–O axis in the MO_6 octahedron (see also Sects. 4.3.2 and 4.6.7). This mechanism is a third possibility to adjust bond lengths (or bond valence) to the geometrical requirements of the M–O frameworks [270, 271]. The effect is, however, only small (see Table 44). A similar result ("distortion theorem"), neglecting charge considerations, has been obtained by Brown [36].

For a fourth possibility to adjust bond lengths (or bond valence) to the geometric requirements of the M–O frameworks (occurrence of fractional oxidation numbers) see Sects. 4.6.2 and 4.6.11.2.

4.6.4
Charge Separation with the Oxides; Relationship to Pseudoterminal Oxygen Atoms

The occurrence of short, strong (multiple-bonded) terminal and of longer, weaker inner (bridging) M–O bonds (occurrence of type II and type I MO_6 octahedra or other MO_k polyhedra) in the polyoxometalate ions is an effect of the appearance of their structures (as defined by the connectivity of the atoms and their oxidation number) as discrete species with a definite surface (see Sect. 4.3.2; see also Sect. 4.8.4).

The MO_6 octahedra or MO_k polyhedra in the oxides and oxide hydrates ("MoO_3", "(α-)$MoO_3 \cdot H_2O$", "$MoO_3 \cdot 2H_2O$", "V_2O_5") show also type II or type I characteristics. In this case the distortion of the MO_6 octahedra (MO_k

polyhedra) results from the construction of the M–O frameworks as layers or chains of closely packed oxygen atoms by which again structures with a definite ("inner") surface are formed. The reason for the formation of the "inner surfaces" through the bond length (bond valence) differentiation is the necessity to increase the (average) M–O bond lengths in the oxides and oxide hydrates to have enough space for the central M atoms of the MO_6 octahedra (see Sect. 4.6.1, in particular Figs. 32a and 32b). Charge separation processes enable the layers and chains with short, terminal M–O bonds (negatively charged O atoms) and long M–O bonds to bridging or water-oxygen atoms (positively charged O atoms) to be formed. The charge separation leads additionally to electrostatic binding forces between the layers since the positively charged oxygen atoms are at (or near to) the surface of the structure (layer, chain) (see Sects. 4.4.2.2 and 4.4.2.3). These electrostatic binding forces compensate in part for the loss of bond valence of inner M–O bonds in the oxides (compare the quantity S_o/q in Table 46) which would, without bond length differentiation, amount for the M^{VI} compounds to 6 and for the M^{V} compounds to 5 v.u. per M atom if a regular octahedral cooordination of M by oxygen could be realized. In a theoretical approach (Sect. 3.6.6 in Ref. 16) it has been argued that the bond lengths differentiation is a prerequisite for a close packing of the O atoms in the crystal structure.

The above description of the oxide structures (formation of inner surfaces and increase of the average bond length about the M atoms by differentiation of bond lengths through charge separation) is entirely analogous to the description of the pseudoterminal oxygen atoms (analogous intergration of the MO_6 octahedra with pseudoterminal O atoms within the M–O frameworks and increase of the average bond length about the M atoms by differentiation of bond lengths through a suitable distribution of the available bond valence over the M–O bonds and charge over the O atoms).

Another reason for the alternation of bond lengths in M–O–M sequences (resulting in the formation of the pseudoterminal M–O bonds of 1.73 Å and a second bond of 2.25 Å in the case of "MoO_3") is seen in the alleviation of Mo···Mo repulsions caused by the edge-sharing of MoO_6 octahedra, which places adjacent Mo atoms relatively closely together [261c] (note that this argument can only explain part of the bond length-bond valence differentiation in the structures).

4.6.5
Oxonium Salts and Amphoteric Character as Consequences of Charge Separation Processes

The positively charged oxygen atoms correspond to oxonium ($M{=}\overset{\oplus}{O}{-}M$) or hydroxonium ($M{-}\underset{H}{\overset{\oplus}{O}}{-}M$, $M{-}\overset{\oplus}{O}H_2$, etc.) ions, and hence the corresponding polyoxometalate ions and oxides or oxide hydrates can be regarded as *inner* oxonium or hydroxonium polyoxometalates. In this sense there is a certain justification for the designation of the compounds "MoO_3", "$MoO_3 \cdot H_2O$", and "$MoO_3 \cdot 2H_2O$" as "molybdic acids" (see Ref. 277). The previously described 36-molybdic acid $(H_3O)_8[Mo_{36}O_{112}(H_2O)_{16}] \cdot 25\text{–}29H_2O$ [278] is hence simultaneously a normal hydroxonium polymolybdate and an inner hydroxonium and oxonium polymolybdate; the same applies to the "decamolybdic" acid $[(H_3O)Mo_5O_{15}(OH)(H_2O) \cdot H_2O]_\infty$ [278, 279].

The simultaneous occurrence of positively and negatively charged oxygen atoms gives the oxides and oxide hydrates an amphoteric [275a, 280, 281a, 282, 283] character.

4.6.6
Charge Distribution and Basicity of the Polyoxometalate Ions

Starting point of the present investigation was an approach to estimate the basicity (as defined by the protonation constant K_p; for other uses of the term "basicity" in this field see Refs. 14h and 253c) of polyoxometalate ions used in our group [2, 4]. The basicity is decisive for the protonation behavior of the (mono- and poly)oxometalate ions, and protonation of the oxometalate ions is a prerequisite for the extension of the coordination sphere of the metal atoms [17–19] and for the elimination of oxygen atoms by condensation of water molecules [2, 4, 20], as has repeatedly been mentioned.

The protonation constants K_p of the monomeric oxoanions (or the acid constants K_a of their conjugate oxoacids $EO_\mu(OH)_\nu$) of the main group elements E show a clear dependence upon μ, the number of the terminal oxygen atoms of the oxo species [21, 22] (see also Refs. 14o and 23). The protonation constants of the monomeric transition metal oxoanions are two to four orders of magnitude greater than those of the corresponding main group elements [23]. This has been explained by (an enhanced) charge separation which is stabilized by resonance according to

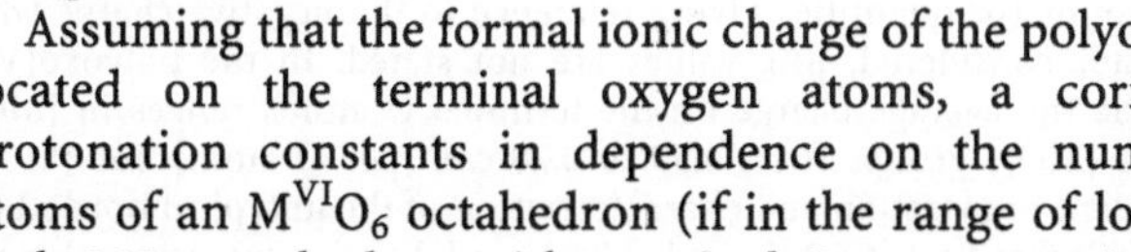
(12)

The charge separation is favored by the greater size of the transition metal atoms since in this case there is a less strong repulsion between the like charges [23].

Assuming that the formal ionic charge of the polyoxometalate ions is mainly located on the terminal oxygen atoms, a corresponding sequence of protonation constants in dependence on the number of terminal oxygen atoms of an $M^{VI}O_6$ octahedron (if in the range of low degrees of acidification each MO_6 octahedron with terminal O atoms carries a charge of exactly one negative charge unit) [2, 4, 14a] or in dependence on the average number of terminal oxygen atoms per negative charge unit (if the number of MO_k polyhedra having terminal oxygen atoms exceeds the charge number) [5, 14a,

14h] has been proposed.[11] The present study shows that the following factors not yet considered could play an important rôle for the protonation constants of the polyoxometalate ions:

- the acceptance of considerable and variable amounts of the formal ionic charge by bridging oxygen atoms, as discussed in Sect. 4.4.3.5, in particular in the octahedral M^V polyanions (see also Ref. 14f);
- the increase of the negative charge on terminal oxygen atoms through charge separation;
- the fact that bridging oxygen atoms have to be considered as protonation sites (thus leaving more terminal oxygen atoms in the protonated polyoxometalate ions), compare Sect. 4.6.7;
- stabilizing or destabilizing effects of the protonation on the M–O frameworks (e.g., the necessity for favorable or unfavorable bond length changes or the weakening of bridging protonated bonds, compare Sects. 3.13, 3.14, and 4.6.7); and others.

Because of this long list it appears difficult to establish a simple relationship between structural parameters of (poly)oxometalate ions and their protonation constants to make estimations for the latter. Figure 33 shows that the actual average negative charge on the terminal oxygen atoms (for the solid state) is a fundamental structural parameter (we have chosen the *average* negative charge on the terminal oxygen atoms since in solution most probably a wide equalization of these charges[12] is to be expected) determining the protonation behavior. However, since the species existing in solution are for the most part not characterized in regard of their structures (oxometalate species existing in solution and in crystals are to a large extent not identical) and/or in regard of their protonation constants, the number of examples is only small. A fully satisfying explanation of the relationships of Fig. 33 (scattering of the data about the straight lines; splitting of the data according to the building principle of the oxometalate species) cannot be given at present, although they

[11] A recent [261b, 284] characterization of the basicity sequence according to the presence of [terminal] $\{MO_3\}$ (strongly basic), $\{\{MO\}_2\}$, and $\{MO\}$ (weakly basic) groups is obviously based on the protonation of one of the terminal O atoms of an $\{MO_3\}$ group in the trimeric $Mo_3O_7(hmmp)_2^{2-}$ (H_3hmmp: 2-hydroxomethyl-2-methylpropane-1,3-diol) chelate complex ion [285] containing $\{MO_3\}$ and $\{MO_2\}$ groups (protonation with chloroacetic acid [285], reaction medium not stated). In this case the protonation may be conducted by the formation of a type II octahedron which becomes more strongly integrated within the M–O framework, see Sect. 4.6.9. The negative charge on these groups (or quantities giving reference to the negative charge with the necessary particularity) is not considered, pK_b values are not stated. In the polyoxo(VI) species investigated in this article the average charge on the terminal O atoms ranges in $\{MO\}$ groups from 0.40 to 0.20 and in $\{MO_2\}$ groups from 0.50 to 0.28 c.u. [per O atom]; for $\{MO_3\}$ groups there is no example [a recent preparation and characterization of the untypical borderline compound $K_7HW_5O_{19} \cdot 10H_2O$ [57] (see Sect. 4.6.12) yielded 0.53 c.u.]. For the untrustworthiness of the relationship between protonation site and basicity in octahedral polyoxo species see Sect. 4.6.7.

[12] Terminal O atoms at the same M atom can perfectly equalize their bond valence and hence their charge without any change of inner M–O bonds. Terminal O atoms at neighboring M atoms can equalize their bond valence to a certain degree with slight changes in the μ-oxo bridge(s) of the M atoms.

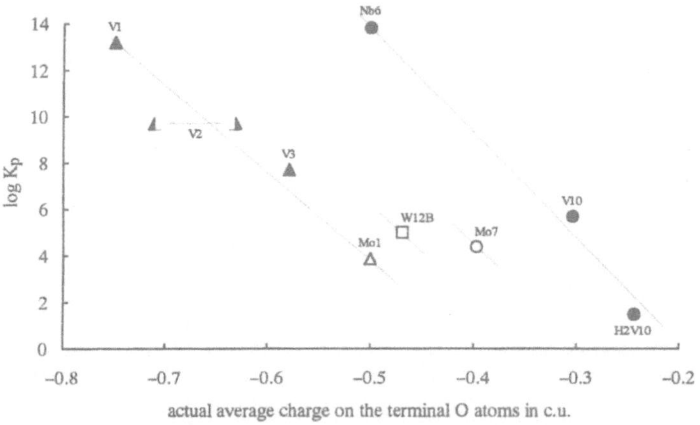

Fig. 33. Basicity of mono- and polyoxometalate ions, expressed by their protonation constants K_p for aqueous media, in dependence on the average negative charge on the terminal oxygen atoms (in the solid state). ▲ △ tetrahedral chain species, ● (highly compact) ○ (compact) octahedral block species, □ octahedral cage-type species; ▲ ● M^V species, △ ○ □ M^{VI} species; V1 VO_4^{3-}, W12B $W_{12}O_{40}(OH)_2^{10-}$, H2V10 $H_2V_{10}O_{28}^{4-}$ etc. Ionic media for the protonation constants: V and Mo species 3 M $Na(ClO_4)$ [14u, 14v, 286], Nb6 3 M K(Cl) [287], W12B 1 M NaCl [288]. (The charge of the protonated H2V10 species is corrected for hydrogen bonds of the OH groups according to the crystal structure data.)

are understood in principle. There will surely be a certain change of the charge on the terminal O atoms due to hydration of the anion or to hydrogen bonding with participation of solvent water molecules. This is analogous to the view that H_2O, coordinated to a cation, acts as a "bond valence transformer" (compare Sect. 4.4.3.2; see also Refs. 39a and 56b). Stabilizing or destabilizing effects of the protonation reactions on the M–O frameworks (as mentioned above) will also influence the protonation constants. The splitting of the data is most probably a consequence of the density of the terminal oxygen atoms as the main charge carriers at the surface of the polyanions. (The protonation constants have been taken from Refs. 14u, 14v, and 286–288.) A relationship between protonation constants and charge on any of the bridging O types could not be detected (and is not to be expected for theoretical reasons).

In Fig. 34 we have plotted the available reliable protonation constants of polyoxometalate species [14u, 14v, 286–290] versus the structural parameter

$$C = -m/(t + \alpha b) \tag{13}$$

where m is the formal ionic charge of the (poly)oxometalate ion, t the number of the terminal O atoms, b the number of the relevant bridging O atoms (assuming that the octahedral species are protonated at bridging and the tetrahedral species at terminal O atoms, see Sect. 4.6.7), and α an (average) factor characterizing the ability of the relevant bridging O atoms to accept negative charge in comparison to the terminal O atoms. The relevant bridging

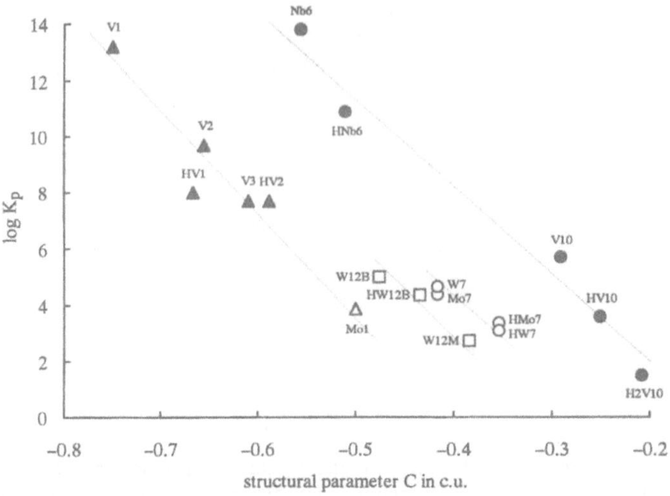

Fig. 34. Relationship between protonation constant K_p and a structural parameter C, derived from the structures according to $C = -m/(t + \alpha b)$. For the octahedral species protonation of bridging O atoms, the number of Opa + O2a + O3a atoms (= b), and $\alpha = 0.7$ for the highly compact M^V species, $\alpha = 0.3$ for the compact and cage-type M^{VI} species has been considered. For the tetrahedral species protonation of terminal O atoms, the number of O2l atoms (= b), and $\alpha = 0.1$ has been taken into consideration (see text). ▲ △ tetrahedral chain-like species, ● (highly compact) ○ (compact) octahedral block-type species, □ octahedral cage-type species; ▲ ● M^V species, △ ○ □ M^{VI} species; V1 VO_4^{3-}, HV10 $HV_{10}O_{28}^{5-}$, W12B $W_{12}O_{40}(OH)_2^{10-}$, etc. Ionic media with the protonation constants: 3 M at 25 °C (V [286] and Mo [14u, 14v] species: $NaClO_4$, Nb species [287]: KCl, W species [290]: LiCl) except HW12B (1 M NaCl at 20 °C) [288]

O atoms in the octahedral structures lie at the surface of the polyanions and are the Opa, O2a, and O3a atoms ($\alpha = 0.7$ for the highly compact M^V species and 0.3 for the compact and cage-type M^{VI} species) and in the tetrahedral structures the O2l atoms ($\alpha = 0.1$); the a values have been estimated such that there is a good overall correspondence between Figs. 33 and 34. The relationships can be used to estimate the protonation constants of species whose structures are merely known in respect of the connectivities of the atoms (e.g., for $H_2W_7O_{24}^{4-}$, $W_{10}O_{32}^{4-}$).

4.6.7
Protonation Sites: Protonation of Bridging Oxygen Atoms in Octahedral Polyoxometalate Ions

In the literature the protonation behavior of the polyoxometalate ions built up of MO_6 octahedra is rather contradictorily discussed. The following main argumentation lines, which differ only in the first step, are found [14a, 14g]:

(A) The negative ionic charge is mainly located on the terminal oxygen atoms [1, 2, 4–6] (A1). These are therefore the most basic oxygen atoms [2, 4–6]

(A2). Protonation sites are the most basic (the terminal [291–294]) oxygen atoms [2, 4, 5] (A3). (In the solid state other protonation sites are also possible, compare $Mo_8O_{26}(OH)_2^{6-}$, $H_2V_{10}O_{28}^{4-}$, $H_3V_{10}O_{28}^{3-}$ (A4).)

(B) The negative ionic charge is mainly located on the bridging oxygen atoms according to their coordination number (B1). These are therefore the most basic oxygen atoms [7, 8, 10–12, 13a] (B2). Protonation sites are the most basic oxygen atoms [7, 8, 10, 13a, 261b, 267b, 295] (B3). (This statement has been somewhat modified in Ref. 220.) (In the solid state other protonation sites are also possible [219] (B4).)

The results of the present investigation referring to the ionic charge, summarized in Tables 41 and 42, show that the first step in argumentation line (A), (A1), is correct, and that of argumentation line (B), (B1), obviously incorrect; hence statement (B2) is also incorrect. Consequently the occurrence of protonated bridging oxygen atoms in $H_2V_{10}O_{28}^{4-}$ and $H_3V_{10}O_{28}^{3-}$ in (aqueous) solution [220] (see Sects. 3.13 and 3.14) is a proof of *the incorrectness of the statements (A3) and (B3), the protonation sites to be the most basic oxygen atoms*. This surprising result requires an explanation.

Protonation of an O atom (formation of an O-H bond of 1 v.u.[13]) causes a reduction of the (negative) charge on that O atom, in the case of the bridging O2a2 and O3a atoms of the decavanadate ion by $\Delta c = 0.40$ to 0.52 c.u. (see Table 45), which requires no structural changes (bond lengths and bond valence alterations) in the M–O framework. The rest of the protonic charge is

Table 45. Bond valence alterations Δs in the neighboring M–O bonds and charge alterations Δc on the protonated oxygen atoms as compensation for the formation of an O-H bond of 1 v.u.

	Δs^a in v.u.	$\Sigma(\Delta s)^b$ in v.u.	Δc^a in c.u.	$\Sigma(\Delta s) - \Delta c$ in v.u./c.u.
$H_2V_{10}O_{28}^{4-}$:				
O2a2hftr	−0.249			
	−0.293	−0.542	0.458	−1.000
O2a2hbul	−0.249			
	−0.293	−0.542	0.458	−1.000
$H_3V_{10}O_{28}^{3-}$:				
O2a2hftr	−0.227			
	−0.255	−0.482	0.518	−1.000
O2a2hftl	−0.253			
	−0.270	−0.524	0.476	−1.000
O3ahft	−0.164			
	−0.147			
	−0.291	−0.603	0.397	−1.000

[a] From Tables 16 and 18.
[b] $\Sigma(\Delta s)$ corresponds to the sum of the changes of the charge on the unprotonated oxygen atoms.

[13] The fact that – due to hydrogen bonding – the O-H bond valences are actually close to 0.8 v.u. is for the following considerations of no relevance since all effects of this type of hydrogen bonding are intermolecularly compensated.

balanced, as a consequence of the valence sum rule for the protonated O atom, through alterations (*weakenings*) of the two or three, respectively, M–O(H) bonds of the protonated bridging O atom by altogether $\Sigma(\Delta s) = 0.60$ to 0.48 v.u. (Table 45). These primary structural changes, which amount for the individual M–O(H) bonds to $\Delta s = 0.15$ to 0.29 v.u. (Table 45), lead, as a consequence of the valence sum rule for the M atoms forming the M–O(H) bonds (and subsequently for other M atoms), via two or three, respectively, paths to a propagation of the bond valence alterations in alternating, preferably "*trans*" bond strengthenings and weakenings (bond shortenings and lengthenings) and cause also corresponding alterations of the charge (see Sects. 3.13 and 3.14). In this way the primary structural changes caused by the protonation are distributed over the whole molecule by which local changes are kept small. The advantage in advance for the protonation of terminal oxygen atoms due to their higher negative charge (and hence smaller necessary changes of bond lengths and bond valences) is on an average ca. 0.11 c.u./v.u. for $V_{10}O_{28}^{6-}$. However, if the protonated oxygen atom is a terminal one, the necessary primary changes can only distribute over the M–O framework via one path (and in the first step not via a "*trans*" route, see the next paragraph), i.e., the resulting distribution of the bond lengths and charge alterations is less well-balanced.

Apart from the consideration of the number of paths for the distribution of the necessary alterations (and hence of the size of the alterations in the individual paths) over the whole structure, there is a second line of argument allowing a substantiation of the protonation site. The propagation of the bond lengths, bond valence, and charge alterations as the result of a protonation occurs in alternate (a consequence of the valence sum rule), preferably "*trans*" (a consequence of the simple mechanism of the alterations for this case – the central M atom has only to shift a little within the MO_6 octahedron along the corresponding approximately linear O–M–O axis without substantial alterations of the O positions) bond weakenings and strengthenings of M–O bonds (Sects. 3.13 and 3.14). In *trans* position to the terminal oxygen atoms, however, are usually found high-coordinate and by the single bonds weakly bonded oxygen atoms which require, due to the small slope of the bond valence-bond length function in the region of weak (long) M–O bonds, great alterations of the bond lengths and hence make the inner bond length-bond valence compensation (Sect. 4.3.4.2) much more difficult if terminal oxygen atoms are protonated.

A third line of argument shows more directly the necessary strong structural changes coupled with the protonation of terminal oxygen atoms: The strong (short) terminal M–O bonds would be converted into middle-strong (medium-sized) bonds, i.e., type II MO_6 octahedra would be changed into type I octahedra, and type I octahedra even in the direction of regular octahedra. This would mean a complete reconstruction of the M–O frameworks. Thus we conclude that protonation of octahedral polyoxometalate ions occurs always at bridging oxygen atoms at the surface of the structures (i.e., at O2a and O3a atoms).

A second and third protonation is, concerning the protonation sites, surely influenced by the preceding protonations to keep the necessary structural changes as small as possible. In the literature [220] it has been proposed for the di- and triprotonation of $V_{10}O_{28}^{6-}$ that the mutual protonation sites lie such that the "*trans*" bond valence alternations occur in V_4O_4 rings which do not coincide in order that two protons do not compete for electron density from any of the same oxygen atoms (see Sects. 3.13 and 3.14).

It has been shown that protonation of bridging oxygen atoms weakens the M–O–M bridges slightly and thus the stability of the polyanions (their polymeric character). This effect is, however, much weaker than expected according to the stoichiometric proceedings since the simultaneous rearrangement processes permit the "activation" of negative charge on other bridging O atoms (see Sects. 3.13 and 3.14 and Table 46). Only for the case that the protonated oxygen atoms form exclusively "*trans*" M–O bonds and hence would bear a very high negative charge without protonation is there nearly no weakening of the polymeric character since most of the charge compensation can take place via Δc. This applies to $W_{12}O_{40}(OH)_2^{10-}$, $W_{12}O_{38}(OH)_2^{6-}$, and $Mo_{36}O_{112}(H_2O)_{16}^{8-}$.

Contrary to the above-treated protonation of bridging oxygen atoms there are some structures with protonated terminal (unshared) oxygen atoms: in $H_3W_6O_{22}^{5-}$ [58], $Mo_8O_{26}(OH)_2^{6-}$, $Mo_{36}O_{112}(H_2O)_{16}^{8-}$, "($\alpha$-)$MoO_3 \cdot H_2O$", and "$MoO_3 \cdot 2H_2O$". Common features of these species are the non-existence of their unprotonated forms and the occurrence of MO_6 octahedra with three unshared oxygen atoms (with three "free vertices"), one of which is (mono- or di-)protonated in each case. A more precise view reveals the reason for this behavior. Without protonation of one of the terminal oxygen atoms the integration of the corresponding MO_6 octahedron within the M–O framework would be very weak and reduce from ca. 2 (2.38 to 1.71) v.u. to ca. 1.6 v.u. or even ca. 1.0 v.u. (see Sect. 4.6.9, in particular Table 47 and Scheme (18)). Hence again the mode of protonation leads to the more stable polyoxo species.

Summing up we can state that the basicity of the polyoxometalate ions is without doubt determined by the high charge density of the terminal oxygen atoms. Protonation, however, occurs at bridging oxygen atoms at the surface of the octahedral polyoxometalate structures since in this case the necessary structural (bond lengths) changes are considerably smaller. The H atoms of protonated terminal oxygen atoms (unshared OH groups), certain OH groups in the interior of polyoxometalate structures, and diprotonated oxygen atoms (coordinated H_2O molecules) are non-acidic. These H atoms are not susceptible to deprotonation/protonation reactions: the structures do not exist in their absence.

The above discussion refers to polyoxometalate ions built up of MO_6 octahedra (MO_k polyhedra with $k > 4$) in which case there is an interdependence of the M–O bond lengths (bond valences) for geometrical reasons due to edge-sharing of the MO_k polyhedra. The polyoxometalate ions composed merely of corner-sharing MO_4 tetrahedra, i.e., without an interdependence of the M–O bond lengths (bond valences) for geometrical reasons, are obviously protonated at terminal oxygen atoms [275b, 296] as the most highly charged ones. In any case protonation of the bridging oxygen atoms in these structures would strongly destabilize their polymeric character.

4.6.8
The Contribution of the Different Factors Stabilizing the Polyoxometalate Ions by Bond Valence of Inner M–O Bonds

4.6.8.1
Structure-Stabilizing Factors

Several factors have been named in the literature as effecting stabilization of polyoxometalate ions [14r], for instance the condensation of r H_2O molecules in their formation reaction from q $M^nO_4^{w-}$ and p H^+ ions (an entropy effect) [2–4] and the build-up of MO_6 octahedra (increase of the number of M–O bonds by[14] meshing of the MO_4 units in the polyoxometalate structures through their simultaneous action as polydentate ligands and as acceptor of ligands; the "meshing effect" resembles the chelate effect) [2–4, 19] (see also Sect. 3.3 in Ref. 16). There are still other factors that can modify the stability of the polyoxometalate structures [14ff] $M_q^nO_{u-h}(OH)_h(H_2O)_o^{m-}$ (see also Sect. 3.6.7 in Ref. 16).

The bond valence approach allows a quantitative treatment of the circumstances. Table 46 summarizes some data on polyoxometalate ion

Table 46. Stabilization of the polymeric character of the polyoxometalate structures by "normal" bonding (S_n), by the meshing effect (S_m), by special structural adaptations (S_a), and by the overall bond valence of the inner M–O bonds (S_o), related to 1 M atom

Polyoxometalate ion	Z^+	S_n/q^a in v.u.	S_m/q^b in v.u.	S_a/q^c in v.u.	S_o/q^d in v.u.
M^{VI} species:					
MO_4^{2-}	0	0	0	0	0
$\langle HW_5O_{19}^{7-e}$	0.60	0.40	2.06	0.29	2.74$\rangle$
$Mo_2O_7^{2-}$	1	1.00	0	0.11	1.11
$[Mo_4O_{14}^{4-}]_\infty$	1	1.00	1.67	0.16	2.83
$Mo_7O_{24}^{6-}$	1.14	1.14	1.94	0.17	3.25
$\langle H_3W_6O_{22}^{5-e}$	1.17	0.67	1.98	0.18	2.83$\rangle$
$W_{12}O_{40}(OH)_2^{10-}$	1.17	1.00	2.42	0.29	3.70
$Mo_{10}O_{34}^{8-}$ (I)	1.2	1.20	1.37	0.22	2.79
$Mo_{10}O_{34}^{8-}$ (II)	1.2	1.20	1.37	0.25	2.82
$Mo_8O_{26}(OH)_2^{6-}$	1.25	1.00	1.72	0.23	2.95
$[Mo_8O_{27}^{6-}]_\infty$	1.25	1.25	1.73	0.18	3.16
$[Mo_6O_{20}^{4-}]_\infty$ (o)	1.33	1.33	1.17	0.14	2.64
$[Mo_6O_{20}^{4-}]_\infty$ (o,tp)	1.33	1.33	1.17	0.20	2.70
$(\beta\text{-})Mo_8O_{26}^{4-f}$	1.5	1.50	1.35[f]	0.19	3.04
$\alpha\text{-}Mo_8O_{26}^{4-}$	1.5	1.50	1.35	0.17	3.02
$W_{12}O_{38}(OH)_2^{6-}$	1.5	1.33	2.85	0.15	4.34
$W_{10}O_{32}^{4-}$	1.6	1.60	2.57	0.16	4.33
$Mo_6O_{19}^{2-}$	1.67	1.67	2.48	0.10	4.25
$W_6O_{19}^{2-}$	1.67	1.67	2.48	0.12	4.27
$Mo_{36}O_{112}(H_2O)_{16}^{8-}$	1.78	1.78	1.48	0.22	3.48

[14] From the point of view presented in this section we have to revise this statement by ourselves [14ii, 19, 20] according to 'increase *of the bond valence of inner M–O bonds by means of an increase* of the number of *inner* M–O bonds by ...'.

Table 46 (Continued)

$[Mo_5O_{15}(OH)(H_2O) \cdot H_2O^-]_\infty$ [g]	1.8				3.75
$[Mo_2O_6]_\infty$	2	2.00	2.00	0.20	4.20
$[Mo_2O_6(H_2O)_2]_\infty$	2	2.00	0.00[h]	0.22	2.22
$\{[Mo_{16}O_{48}(H_2O)_{16}] \cdot 16H_2O_\infty\}$	2	2.00	2.00	−0.06	3.94
M^V species:					
VO_4^{3-}	0	0	0	0	0
$V_2O_7^{4-}$ [i]	1	1.00	0	−0.10	0.90
$V_2O_7^{4-}$ [i]	1	1.00	0	0.13	1.13
$V_3O_{10}^{5-}$	1.33	1.33	0	−0.13	1.21
$Nb_6O_{19}^{8-}$	1.67	1.67	1.90	−0.07	3.50
$Ta_6O_{19}^{8-}$	1.67	1.67	1.90	0.04	3.61
$[VO_3^-]_\infty$	2	2.00	0	−0.04	1.96
$[V_2O_6^{2-}]_\infty$	2	2.00	0.00[h]	0.06	2.06
$V_5O_{14}^{3-}$	2.4	2.40	0	0.02	2.42
$V_{10}O_{28}^{6-}$	2.4	2.40	1.30	−0.06	3.64
$Nb_{10}O_{28}^{6-}$	2.4	2.40	1.30	0.02	3.72
$H_2V_{10}O_{28}^{4-}$	2.6	2.40	1.20	0.02	3.62
$H_3V_{10}O_{28}^{3-}$	2.7	2.40	1.15	0.06	3.61
$[V_4O_{10}]_\infty$	3	3.00	0.00[h]	0.19[j]	3.19[j]

[a] S_n: bond valence of the inner M–O bonds produced by "normal" bonding according to Eq. (15).
[b] S_m: bond valence of the inner M–O bonds produced by the meshing effect according to Eq. (16).
[c] S_a: bond valence of the inner M–O bonds produced by the special adaptations based on the preferential acceptance of the ionic charge by the terminal oxygen atoms and on the charge separation processes and diminished by the effects of the harmonization of the bond lengths, bond valences, and charges according to Eq. (17).
[d] S_o: total bond valence of the inner M–O bonds (stabilization of the polyoxometalate structure by bond valence of inner M–O bonds) according to Eq. (14).
[e] These species have only recently been obtained under extreme preparation conditions [57, 58] and therefore are not treated in Sect. 3.
[f] S_m and S_m/q for other $M_8O_{26}^{4-}$ block-type structures (for the structures see Refs. 5, 14ee, or 20; V is the (β-)$Mo_8O_{26}^{4-}$ structure, II is the only structure directly accessible via $Mo_7O_{24}^{6-}$):

structure	I	II	III, IV	V, VI
S_m in v.u.	18.00	16.20	14.40	10.80
S_m/q in v.u.	2.25	2.03	1.80	1.35

[g] Due to the statistical non-occupancy of Mo positions in the three-dimensional network structure [47] all MO_6 octahedra have no doubt identical bond lengths but there are four types of MO_6 octahedra characterized by different bridging (+) M–O bonds in them:

d in Å:	1.701	1.698	1.955	1.955	2.190	2.373	
s in v.u.:	1.669	1.682	0.904	0.904	0.512	0.329	(d_{0i}=1.9131Å)
type 1 (2×):	(Ot)	+	+	+	+	+	
type 2 (1×):	(Ot)	+	+	+	+	(Otw)	
type 3 (1×):	(Ot)	+	+	(Oth)	+	+	
type 4 (1×)	(Ot)	(Ot)	+	+	+	+	

[h] For the reasons of the absence of a meshing effect see text.
[i] The structures with the weakest and strongest V–O–V bridges have been selected.
[j] The very long M···O bonds have been neglected (compare Section 3.28).

stabilization by inner M–O bond valence. For a comparative view the quantities have been related to one M atom.

The overall stabilization of a structure by bond valence of inner M–O bonds corresponds to the sum of the bond valence of the bridging M–O bonds (ΣS_b) and amounts to

$$S_o = \Sigma s_b = qn - \Sigma s_u \quad (14)$$

(qn: total bond valence in the structure; Σs_u: sum of the bond valence of the outer M–O bonds, i.e., of the unshared O atoms). This quantity is based on experimental data and hence free of any assumptions.

The effects of three factors have been elaborated to contribute to S_o:

- a contribution of "normal" bonding,
- a contribution of the meshing effect in the case of most structures with extended coordination polyhedra (k > 4), and
- the favorable contribution by the special adaptations based
  - on the preferential acceptance of the ionic charge by the terminal oxygen atoms and
  - on the charge separation processes in the case of (approximately) linear M–O–M bridges and of bridging coordinated H_2O molecules and OH groups

  and diminished by the effects of
  - the harmonization of the bond lengths, bond valences, and charges, required by the interdependence of the M–O bond lengths,
  - the packing of the constituents of the salts, and
  - a high density of the negative charge on the terminal O atoms

  (factors treated in Sects. 4.3.4.2–4.3.4.4 and 4.4.3).

In the following paragraphs terms for the contribution of these factors are derived.

The contribution of "normal" bonding in the M–O–M bridges of the polyoxometalate structures, S_n, corresponds to the proceeding protonation reactions. Each proton reacting with an oxometalate species leads primarily to an M–O–H group and subsequently by condensation of an H_2O molecule from two M–O–H groups to an M–O–M bridge (at least formally), if mere protonation reactions (h) of the polyanions to give an OH group remain unconsidered. Since two protons result in an M–O–M bridge, each proton produces an M–O bond of 1 v.u. Consequently the inner bond valence due to normal bonding in a polyoxometalate ion is

$$S_n = 2r = p - h = qZ^+ - h \quad (15)$$

(r: number of eliminated O atoms by condensation of H_2O). This contribution corresponds to the bold bonding dashes in Fig. 1 in Ref. 16 and in Fig. 35, below, i.e., it characterizes the fictive $M_\xi O_{3\xi+1}$, $M_\xi O_{3\xi}$, $M_qO_{\leqq 3q}$, etc. entities. Coordinated H_2O molecules remain unconsidered in this way. This factor requires no knowledge about the structure, only the stoichiometric coefficients of the formation reaction of the polyoxo species (see Sect. 4.3.3, particularly footnote 8) have to be known. (See also Sect. 3.3 in Ref. 16.)

The determination (estimation) of the contribution by the meshing effect in the case of the structures with extended coordination polyhedra, S_m, requires additionally knowledge about the connectivity of the atoms in the polyoxometalate framework (knowledge of the number and types of unshared oxygen atoms). S_m can be calculated by subtracting from the total bond valence in the structure (qn) the bond valence of the unshared oxygen atoms (as a matter of course, without use of bond *length* data, i.e., of Σs_u!) and that of the inner M–O bonds by normal bonding, for the terminal O atoms and unshared OH groups[15] assuming a charge corresponding to an equidistribution of the ionic charge about the terminal oxygen atoms and unshared OH groups of the fictive $M_\xi O_{3\xi+1}$ chains, $M_\xi O_{3\xi}$ rings, and $M_qO_{\leqq 3q}$ entities, i.e., about all oxygen atoms of the structure except those forming the bridges in the fictive $M_\xi O_{3\xi+1}$ chains, $M_\xi O_{3\xi}$ rings, and $M_qO_{\leqq 3q}$ entities:

$$S_m = qn - (t + h_u/2)\left(2 - \frac{m}{u - (p - h)/2}\right) - S_n \tag{16}$$

(t: number of terminal O atoms in the structure, h_u: number of OH groups in the structure as unshared O atoms). The contribution of the meshing effect corresponds to the lean inner bonding dashes in Fig. 1 in Ref. 16. Note, however, when counting them that some of them are double bonds which has not been considered in the Figure. The effect of the meshing can plastically be seen from that Figure: Terminal oxygen atoms of the fictive $M_\xi O_{3\xi+1}$, $M_\xi O_{3\xi}$, or $M_qO_{\leqq 3q}$ entities become bridging ones by which the *stability* of the polymeric species is strongly increased and the number of terminal oxygen atoms is strongly reduced, i.e., their *basicity* is also strongly increased (see Sect. 3.6.7 in Ref. 16).

The determination (estimation) of the contribution by the special adaptations based on the preferential acceptance of the ionic charge by the terminal oxygen atoms and on the charge separation processes (factors named in Sect. 4.4.2), diminished by the effects of the harmonization of the bond lengths, bond valences, and charges, required by the interdependence of the M–O bond lengths, the packing of the constituents of the salts, and a high density of the negative charge on the terminal O atoms (factors named in Sects. 4.3.4.2–4.3.4.4 and 4.4.3), S_a, demands, in addition to the knowledge of the stoichiometric quantities and of the connectivity of the atoms in the structure, knowledge of the M–O bond lengths of the unshared oxygen atoms. If we assume for the reference state for the charge on the unshared oxygen atoms the same situation as for the determination of S_m, its contribution is

$$S_a = (t + h_u/2)\left(2 - \frac{m}{u - (p - h)/2}\right) - \Sigma s_u \tag{17}$$

The structures composed of MO_4 tetrahedra sharing corners show, as a matter of course, no stabilization by the meshing effect. For the

[15] The coordinated H_2O molecules remain unconsidered (see the preceding paragraph).

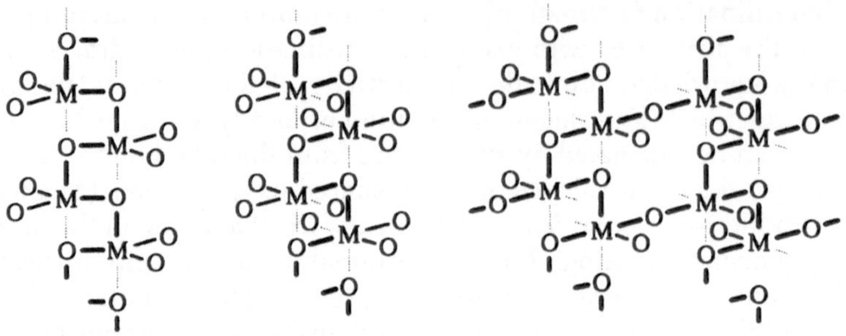

$[V_2O_6^{2-}]_\infty$ $[Mo_2O_6(H_2O)_2]_\infty$ $[V_4O_{10}]_\infty$

Fig. 35. Meshing and extension of the MO_4 tetrahedral building groups only by bridging oxygen atoms in $[V_2O_6^{2-}]_\infty$, $[Mo_2O_6(H_2O)_2]_\infty$, and $[V_4O_{10}]_\infty$ (representation as used in Fig. 1 of Ref. 16). The figured structures each represent one of an extremely large number of resonance structures. ▬, —— bonds of bond order one or two so that their sum about M amounts to the oxidation number of M (the *bold dashes* serve only for the accentuation of the $M_\xi O_{3\xi}$ chains and $M_\xi^V O_{2.5\xi}$ layers); ······ no bond in the figured resonance structure. In resonance formulas other combinations of the M atoms can form the chains ($[V_2O_6^{2-}]_\infty$, $[Mo_2O_6(H_2O)_2]_\infty$), or other bond patterns for the same combination of the M atoms can occur ($[V_2O_6^{2-}]_\infty$, $[Mo_2O_6(H_2O)_2]_\infty$, $[V_4O_{10}]_\infty$).

$[Mo_2O_6(H_2O)_2]_\infty$, $[V_2O_6^{2-}]_\infty$, and $[V_4O_{10}]_\infty$ structures, however, there is also no stabilization by a meshing effect although there are extended MO_k polyhedra. The reason can be seen from a more detailed view on these structures (Fig. 35): There are no terminal O atoms of fictive $M_\xi O_{3\xi+1}$ chains, $M_\xi O_{3\xi}$ rings, $M_qO_{\leqq 3q}$ entities, etc. which are converted into bridging ones by the extension of the MO_4 tetrahedra; only *bridging* O atoms of infinite $M_\xi O_{3\xi}$ chains ($[Mo_2O_6(H_2O)_2]_\infty$, $[V_2O_6^{2-}]_\infty$) and $M_q^VO_{2.5q}$ layers ($[V_4O_{10}]_\infty$) perform the extension by which no new inner M–O bonding is produced.

4.6.8.2
Contribution of the Different Factors to the Stabilization of the Different Structure Types

For all polyoxo species the sum of the stabilization by S_n, S_m, and S_a agrees, as expected (see a balance of Eqs. (14)–(17)), with the overall stabilization S_o.

The results for the *average* stabilization of the MO_k polyhedra by bond valence of inner M–O bonds (S_o/q), i.e., for the average bond valence proceeding from the M to the O atoms of neighboring MO_k polyhedra, are, in general, as expected (Fig. 36): it is greater for higher coordination numbers of M; it increases with the Z^+ value; it is for the M^V compounds generally somewhat smaller than for the M^{VI} compounds (the latter does, however, not apply to the structures composed of corner-sharing MO_4 tetrahedra, as expected).

The short chains consisting of corner-sharing MO_4 tetrahedra ($Mo_2O_7^{2-}$, $V_2O_7^{4-}$, $V_3O_{10}^{5-}$) show a small stabilization of approximately 1 (0.9 to 1.2) v.u.

per M atom, increasing to 2.0 v.u. per M atom for the infinitely long chains of corner-sharing MO_4 tetrahedra ($[VO_3^-]_\infty$) and to 2.4 v.u. per M atom for the cage-like structure with this connection mode of the MO_4 tetrahedra ($V_5O_{14}^{3-}$).

The structures composed of MO_k polyhedra with $k>4$ are stabilized according to the number of MO_k polyhedra having three, two, one or no free vertices, the least stabilization occurring with the polyhedra having three and the greatest stabilization occurring with those having no free vertices. Hence the structures composed of MO_6 octahedra (MO_k polyhedra) with one free vertex are the most stabilized ones ($Mo_6O_{19}^{2-}$, $W_6O_{19}^{2-}$, $W_{12}O_{38}(OH)_2^{6-}$, $W_{10}O_{32}^{4-}$, "MoO_3": 4.3 to 4.2 v.u. per M atom; $Nb_6O_{19}^{8-}$, $Ta_6O_{19}^{8-}$, $V_{10}O_{28}^{6-}$, $Nb_{10}O_{28}^{6-}$, $H_2V_{10}O_{28}^{4-}$, $H_3V_{10}O_{28}^{3-}$, $[V_4O_{10}]_\infty$: 3.7 to 3.2 v.u. per M atom). The structures containing most of the polyhedra with two free vertices are somewhat less stabilized ($W_6O_{19}(OH)_3^{5-}$, $Mo_7O_{24}^{6-}$, $Mo_{10}O_{34}^{8-}$(I), $Mo_{10}O_{34}^{8-}$(II), $Mo_8O_{26}(OH)_2^{6-}$, $[Mo_8O_{27}^{6-}]_\infty$, $[Mo_6O_{20}^{4-}]_\infty$(o), (β-)$Mo_8O_{26}^{4-}$, $W_{12}O_{40}(OH)_2^{10-}$, α-$Mo_8O_{26}^{4-}$, $Mo_{36}O_{112}(H_2O)_{16}^{8-}$, $[Mo_4O_{14}^{4-}]_\infty$, $[Mo_6O_{20}^{4-}]_\infty$ (o,tp), "$MoO_3 \cdot 2H_2O$": 3.9 to 2.6 v.u. per M atom;[16] $[V_2O_6^{2-}]_\infty$: 2.1 v.u. per M atom). The stabilization is independent of the fact whether the structures are of the block-, cage-, or of any other type; in particular the cage-type structures, appearing more loosely constructed than the block-type structures, are as strongly stabilized as the latter.

The average stabilizations of the MO_k polyhedra by normal bonding (S_n/q) are fixedly defined by the stoichiometry of the compounds [$= Z^+ - h/q = (p-h)/q = 2r/q$] and lie for the investigated M^{VI} polyoxo compounds in the range 0.40 ($HW_5O_{19}^{7-}$) to 2.00 v.u. ("MoO_3", "$MoO_3 \cdot H_2O$", "$MoO_3 \cdot 2H_2O$") per M atom and for the corresponding M^V species in the range 1.00 ($V_2O_7^{4-}$) to 3.00 v.u. ("V_2O_5") per M atom. $S_n/q < 1$ v.u. per M atom indicates species with extended MO_k polyhedra in the starting phase of the condensation mechanism in which H atoms originating from the addition mechanism are still present.

The average stabilizations of the MO_k polyhedra by the meshing effect (S_m/q) are also firmly defined, by the stoichiometry of the compounds and by the connectivity of the atoms (definition of the terminal O atoms and unshared OH groups), and lie for the M^{VI} polyoxo compounds with extended MO_k polyhedra in the range 1.17 to 2.85 v.u. per M atom and for the corresponding M^V compounds in the range 1.15 to 1.90 v.u. per M atom. The absence of a meshing effect for the $[Mo_2O_6(H_2O)_2]_\infty$, $[V_2O_6^{2-}]_\infty$, and $[V_4O_{10}]_\infty$ structures has been explained in Sect. 4.6.8.1. For the tetrahedral species $S_m/q = 0$, as expected.

The average stabilizations of the MO_k polyhedra by the special adaptations, based on the preferential acceptance of the ionic charge by the terminal oxygen atoms and on the charge separation processes and diminished by the effects of the harmonization of the bond lengths, bond valences, and charges

[16] The wide range of values in this group is well explained. Apart from the presence of MO_6 octahedra (MO_k polyhedra) with zero, one, two and/or three free vertices in varying amounts there is also the possibility of unshared O atoms, OH groups, and/or coordinated H_2O molecules.

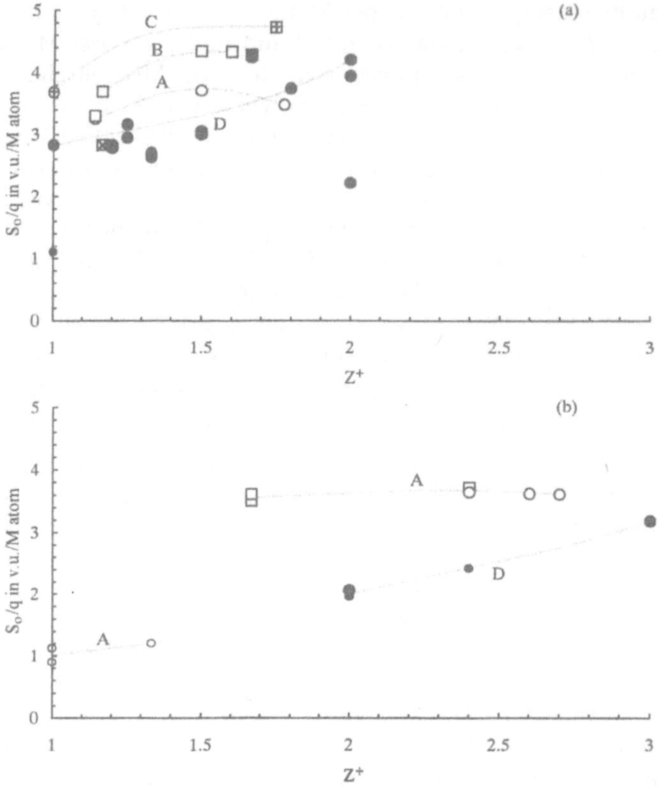

Fig. 36a,b. Stabilization of the polyoxometalate ions by bond valence of inner M–O bonds. (a) M^{VI} polyoxo species of Mo (● ○) and W (■ □). (b) M^V polyoxo species of V (● ○) and Nb, Ta (■ □). ○, □ species existing as equilibrium component in aqueous solution; ● ■ species existing only in the solid state; × species most probably existing in solution but far away from $P \approx Z = |Z^+|$ ($H_3W_6O_{22}^{5-}$ at $Z \approx 0.15$ and $HW_5O_{19}^{7-}$ at $Z \approx 0.29$). ● ○ polyoxo species exclusively built by MO_4 tetrahedra; the other symbols characterize octahedral polyoxo species (polyhedral species with $k > 4$). + Two heteropolyoxometalate ions as additional examples. A (Relatively) rapidly formed polyoxometalate ions existing in aqueous solution as equilibrium component; B slowly formed polyoxometalate ions existing in aqueous solution as equilibrium component; C heteropolyoxometalate ions existing in aqueous solution; D solid state polyoxo species obtained, as a rule, from aqueous solution. The polyoxo species can be identified by Table 46; additionally included species: $W_7O_{24}^{6-}$ ($Z^+ = 1.14$) and $Mo_8O_{26}^{4-}$ II ($Z^+ = 1.5$). The observed stabilization sequence (C > B > A > D) corresponds to theoretical expectations.

etc. (S_a/q), are for the M^{VI} polyoxo compounds with one exception (-0.06 v.u. per etc M atom for "$MoO_3 \cdot 2H_2O$") in the range 0.10 to 0.29 v.u. per M atom and for the corresponding M^V species with one exception (0.19 v.u. per M atom for "V_2O_5") in the range -0.13 to 0.13 v.u. per M atom. The smaller (and negative) values in the M^V cases reflect in particular the greater density of the negative charge for the M^V species (see Sects. 4.4.3.5 and 4.4.3.6, especially Table 43); hence the deviating value for "V_2O_5", which carries no formal ionic charge, is well explained. The deviating (negative) value for "$MoO_3 \cdot 2H_2O$" is due to the absence of any (true) charge separation effects (see Table 43). Other relationships could not be recognized. Note that the stabilization of the polyoxometalate structures by inner M–O bond valence based on this factor is generally only small compared to those based on normal bonding and on the meshing effect. Nevertheless the discussion in Sects. 4.2–4.5 deals nearly exclusively with these additional adaptations; the reason is that the stabilization by normal bonding and by the meshing effect is invariably determined by the stoichiometric coefficients and by the connectivity of the atoms. The small part of the additional adaptations and their simultaneously great structural influence make understandable the sensitivity with which M–O frameworks are changed with the cation (see Sect. 4.2).

The concept of the stabilization of the polyoxometalate species by bond valence on inner M–O bonds is confirmed by a relevant inspection of their structures. The data of Table 46 have been investigated according to the feature whether the polymetalate species occur as an equilibrium component in acidified aqueous metalate solutions or whether they exist only in a solid state structure. The structures of the former must be more strongly stabilized by bond valence of inner M–O bonds. The latter form only because of the unfavorable packing conditions for the polymetalate ions present as equilibrium components, cations, and molecules of water of crystallization and are ultimately precipitated because of their infinite chain, layer, or three-dimensional network structure or, in the case of the discrete polymetalate ions, because of a high lattice energy of the salt, low hydration energy of the ions, and/or a three-dimensional network formed by hydrogen bridges [253a, 254] (see also Sect. 4.2) when obtained from aqueous solution. Fig. 36 indicates that the polyoxometalate ions existing in solution ($Mo_7O_{24}^{6-}$, $W_7O_{24}^{6-}$, $Mo_{36}O_{112}(H_2O)_{16}^{8-}$, $Mo_8O_{26}^{4-}$ structure II [see Sect. 4.6.8.3]; $W_{12}O_{40}(OH)_2^{10-}$, $W_{12}O_{38}(OH)_2^{6-}$, $W_{10}O_{32}^{4-}$; $V_2O_7^{4-}$, $V_3O_{10}^{5-}$; $Nb_6O_{19}^{8-}$, $Ta_6O_{19}^{8-}$, $V_{10}O_{28}^{6-}$, $Nb_{10}O_{28}^{6-}$, $H_2V_{10}O_{28}^{4-}$, $H_3V_{10}O_{28}^{3-}$) show indeed the greatest stabilization by bond valence of inner M–O bonds (species of regions A and B). For the high Z^+ values of the species existing in solution, Fig. 36 shows a tendency for a decrease of the stabilization by bond valence of inner M–O bonds. For the MoO_4^{2-}/H^+ system it has indeed been shown that decomposition of the polymeric to give dimeric (and monomeric) species takes place [14hh, 253f]; for the VO_4^{3-}/H^+ system decomposition to give a monomer is proven [274a] (formation of species capable of consumption of more protons, see Sect. 3.6.7 in Ref. 16). The reason for the extraordinary stabilization of the insoluble $M_6^{VI}O_{19}^{2-}$ hexametalates and hence its fall from region D is discussed as follows. The species do not form in aqueous solution in detectable quantities because on the one hand for a molybdate species with $Z^+ = 1.67$ the optimal pH is ≈ 2 and on the other hand the triprotonated monomer, necessary for the formation of $M_6O_{19}^{2-}$ [20], requires a pH < 0 (pH values derived from diagrams in Ref. 14jj). The formation of insoluble salts, however, is possible because the small equilibrium concentration of $Mo_6O_{19}^{2-}$ in the solution is sufficient for their precipitation with suitable cations and hence rejection from the equilibrium

[254]. On the other side, $Mo_{36}O_{112}(H_2O)_{16}^{8-}$ is obviously a border case. Long standing of its crystals under the mother liquor leads to the formation of the so-called decamolybdates, containing the polymolybdate ion $[Mo_5O_{15}(OH)(H_2O) \cdot H_2O^-]_\infty$ [15c]. This behavior is expressed by the close neighborhood of both species in the stability diagram (Fig. 36(a)). The preparation conditions of "(α-)$MoO_3 \cdot H_2O$" are characterized by a very narrow range of acidification (P = 5.9 to 7.7 [245], 6.5 to 7.1 [297]) of the molybdate solution which might explain the rather small overall stabilization of this structure.

There is another feature of the polymetalate ions that can be validated by Fig. 36. Some of the structures existing as equilibrium components in the polymetalate solutions form rather rapidly (those defined by regions A), others form only slowly (region B) at the expense of rapidly formed species. The slowly formed species must be the more stable species which is indeed confirmed by Fig. 36 (region B lies above region A).

For further demonstration of the operation of this concept and its proof, in Fig. 36 we have given additionally two heteropolyoxometalate ions with central heteroatoms, $TeMo_6O_{24}^{6-}$ and $PW_{12}O_{40}^{3-}$, which are structurally related to $Mo_7O_{24}^{6-}$ and $W_{12}O_{38}(OH)_2^{6-}$, respectively. The bond valence of the (inner) X–O bonds has to be added to that of the M–O bonds; the Z^+ values have been calculated on the basis of $Te(OH)_6$ and H_3PO_4, respectively, as the heteroatom species that would exist at the pH values of the solutions under the formation conditions. The heteropolymetalate ions show an enhanced stability (region C lies above regions A and B); see also the discussion on the preferential formation of heteropolymetalate compared to isopolymetalate ions in Sect. 4.6.11.2.

The results presented in the foregoing three paragraphs represent the core of this investigation. Thermodynamic and kinetic features [the stability sequence *heteropolyoxometalate ions in aqueous solution > slowly formed isopolyoxometalate ions in aqueous solution > rapidly formed isopolyoxometalate ions in aqueous solution > (slowly formed) isopolyoxometalates of certain cations*] of the polyoxometalate ions (structures) confirm the overall stabilization of the structures as characterized by the sum of the bond valences of the bridging M–O (and X–O) bonds. Investigations of this kind are necessary to clarify the reasons for the formation of the mixed-valence heteropolyoxometalate ions containing "unusual" heteroelement "ions" or molecules briefly discussed in Sect. 4.6.11.2 and of the macropolyoxometalates recently [298–300] prepared.

4.6.8.3
Application: About the Structure of $Mo_8O_{26}^{4-}$ in Aqueous Solution

The structure of the $Mo_8O_{26}^{4-}$ equilibrium component in acidified aqueous molybdate solutions is rather controversially discussed in the literature [14gg]. Our main arguments against β-$Mo_8O_{26}^{4-}$ (Fig. 9), a structure observed in a number of solid-state polymolybdates, as an equilibrium component in solution are

- the insufficient agreement of the solid-state and solution Raman spectra of the species [4, 5, 254];
- the impossibility to prepare salts with common inorganic cations (only β-octamolybdates of ammonium and of cations of organic bases are known [254]; compare Sect. 4.2) and the long time for the crystallization process of the ammonium salt [254, 262c];
- the existence of more favorable, i.e., more basic (compare Sect. 3.6.7 in Ref. 16), isomeric $Mo_8O_{26}^{4-}$ structural possibilities (the β-structure is unfavorable in this respect due to its great number of terminal oxygen atoms and hence weak basicity according to the criteria discussed in Sect. 4.6.6; isomeric $Mo_8O_{26}^{4-}$ block-type structures [5, 14ee, 20] are by far more basic and should occur in solution [5, 262c]).

The present study supplies additional arguments against the existence of the β-form in aqueous solution: The stabilization by the meshing effect (see footnote f of Table 46) for β-$Mo_8O_{26}^{4-}$ (structure V) is comparatively small, $S_m/q = 1.35$ v.u. per M atom, whereas for $Mo_8O_{26}^{4-}$ structure II, the only $Mo_8O_{26}^{4-}$ structure directly accessible via $Mo_7O_{24}^{6-}$ [262d] (which is the equilibrium partner of $Mo_8O_{26}^{4-}$ in aqueous solution), it is 2.03 v.u. per M atom. Thus $Mo_8O_{26}^{4-}$ structure II has the advantage of a ca. 22% greater stabilization of the polymeric character by inner M–O bonding ($S_o/q = 3.72$ v.u. per M atom; since S_a/q for $Mo_8O_{26}^{4-}$ structure II is unknown, the value for β-$Mo_8O_{26}^{4-}$ was informally assumed) and of a greater rate of formation compared with β-$Mo_8O_{26}^{4-}$ ($S_o/q = 3.04$ v.u. per M atom only). Hence β-$Mo_8O_{26}^{4-}$ is clearly a species of region D whereas $Mo_8O_{26}^{4-}$ structure II belongs to region A.

4.6.9
MO_6 Octahedra with Three Unshared Oxygen Atoms: Strength of the Integration of the MO_k (k > 4) Polyhedra Within the M–O Frameworks

In the literature [13i, 281b, 284, 301] it is sometimes claimed that MO_6 octahedra with three unshared oxygen atoms cannot exist in polyoxometalate ions due to the 'Lipscomb principle'. Lipscomb [301] recognized in 1965 that no polyoxometalate ion *(known at that time)* contained MO_6 octahedra with more than two unshared oxygen atoms and argued that otherwise a very large number of polyoxometalate structures should exist, in contrast to the only small number of polyoxometalate types (structures) observed. The 'Lipscomb principle' [13i] was explained by a less good connection of the MO_6 octahedra in question to the M–O frameworks and hence the possibility for their easy split-off from the rest of the M–O frameworks [13i, 281b] (minor kinetic inertness of those structural parts [5, 14y, 20, 205]).

This argumentation is only partly correct. The *inner* (conditions defined by the metal M, mainly its size and oxidation number: attainable coordination number against oxygen; basicity as defined by the protonation constants; maximally attainable degree of protonation of the mono- and intermediary occurring polyoxometalate species [14e, 14l, 14o, 14dd]) and *outer* (conditions in the solution: metalate concentration; availability of protons, defined by the degree of acidification P or degree of protonation Z; pH value; reaction time; temperature; and others [14bb]) formation conditions for the polyoxometalate species determine the play for possible polyoxometalate structures, and of these in each case the most stable one(s) form. Hence, if according to the inner and outer formation conditions it is impossible to build octahedral polyoxometalate species with two or one unshared oxygen atoms only, less stable structural elements must be capable of existence. This refers primarily to the first acidification phase [first aggregation products; see, for instance, the formation of the tetrahedral polyvanadate [14cc] (compare footnote 8 in Ref. 16) and octahedral W_5O_{19} penta- [57] and W_6O_{22} hexatungstate [58] structures (see Sect. 4.6.12)]. Similarly, in the late acidification phase, at high degrees of acidification, it is necessary to have species in the solution that can consume the excessively offered protons [14x] (see, for instance, the formation of the highly protonated dimeric and monomeric, cationic Mo^{VI} species which come up to $Z^+ = 4$ [14cc]). Therefore in our group we argued (primarily for solute polyoxometalate species)

- that the parts of a polyoxometalate structure with MO_6 octahedra having three unshared oxygen atoms are not only kinetically less inert but, above all, also thermodynamically less stable since
  - the formation of such partial structures is connected with the condensation of only one H_2O molecule compared to the condensation of two or even three H_2O molecules with the formation of MO_6 octahedra having two or one, respectively, unshared oxygen atoms (stabilization through an entropy effect) [2, 4, 5, 14w, 20, 262a, 302];
  - the number of inner M–O bonds [14ii, 19, 20] and hence the inner bond valence is strongly reduced (Sect. 4.6.8);
  - according to a mechanism discussed by our group [5, 14k, 20] the kinetic inertness of a structure has also repercussions on its thermodynamic stability.
- that MO_6 octahedra having three unshared oxygen atoms occur if merely the monoprotonated monometalate ion is available for the aggregation reactions (i.e., at very low degrees of acidification and relatively high pH values of the metalate solution – an M^{VI} hexametalate ion under these conditions as $HM_6^{VI}O_{20}(OH)_2^{5-}$ [2, 4, 14e, 14s, 20]) or, if at high degrees of acidification of the metalate solution it becomes necessary to have on the pH level attained proton consuming species in the solution to consume further on the excessively offered protons [5, 14m].

The definite preparation and structural characterization of the polyoxo species "$MoO_3 \cdot H_2O$" [245, 246], $Mo_8O_{26}(OH)_2^{6-}$ [59, 303], $Mo_{36}O_{112}(H_2O)_{16}^{8-}$ [304, 305], $H_3W_6O_{22}^{5-}$ [58], and $HW_5O_{19}^{7-}$ [57] since 1973 shows that MO_6 octahedra with three unshared oxygen atoms do exist. Their usually (in the case of the $W4O_6$ octahedron in $H_3W_6O_{22}^{5-}$ modified) type II character is guaranteed by the mono- or diprotonation of one of the three unshared oxygen atoms (see also Sect. 4.6.7). The $M3O_6$ and $M5O_6$ octahedra in $HW_5O_{19}^{7-}$ are of type III (three short and three long M–O bonds). The penta- and hexatungstate ions were indeed obtained (as potassium and sodium salts, respectively) from solutions with a very low degree of acidification ($\approx$ 0.29 and 0.15, respectively) and high (2 M) tungstate concentrations [57, 58] and correspondingly at relatively high pH values (in weakly alkaline solution [58]; pH $\approx$ 9 according to data given in Ref. 290), i.e., in a range of very low concentrations of diprotonated monometalate ions. The 36-molybdate ion [5, 14m, 14x] and "$MoO_3 \cdot H_2O$" [245, 246] are examples for structures that consume a large number of protons for their formation at high degrees of acidification. (The dihydrogenoctamolybdate ion was obtained as an insoluble isopropylammonium salt by long storage of a heptamolybdate solution in the dark [15b, 59, 303, 306].)

It is of interest to study the strength of the integration of the MO_6 octahedra within the M–O frameworks in dependence on the number of unshared oxygen atoms and for the MO_6 octahedra with three unshared O atoms also on their degree of protonation. A measure for this integration is the sum of the bond valences of the bridging M–O bonds of these octahedra (Tables 47 and 48). (A rough, quick estimation for a given type of $M^{VI}O_k$ polyhedron can be made by summing up the valences of the M^{VI}–O bonds to the unshared oxygen atoms and formation of the difference against the oxidation number of M, assuming average bond valences for M^{VI}–Ot of 1.65 v.u., for M^{VI}–OH of 1.0 v.u., and for M^{VI}–OH_2 of 0.35 v.u.) Note that for the kinetic inertness of a polyoxometalate species as a whole the weakest site in the structure is decisive [14y, 262b].

Of particular interest is the integration of the MO_6 octahedra with three unshared O atoms, usually one of which is mono- or diprotonated. The protonation, in particular the diprotonation, leads to a stronger integration (greater stability, greater kinetic inertness) of the MO_6 octahedra in question. The reason is that monoprotonated terminal O atoms form middle-strong outer M–O bonds (structures (18c) and (18d)) compared to the stronger bonds of unprotonated terminal O atoms (structures (18a) and (18b)) which leaves more bond valence for the inner M–O bonds and thus stabilizes the structures. This effect is even greater for diprotonated terminal O atoms (coordinated H_2O molecules) which form weak outer M–O bonds.

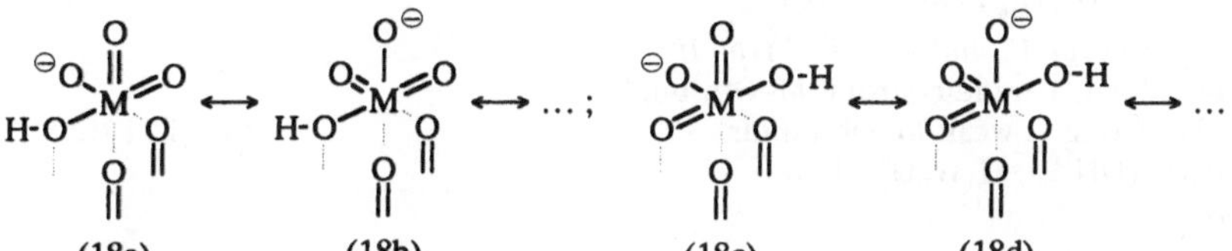
(18a) (18b) (18c) (18d) (18)

The rare MO_6 octahedra without unshared oxygen atoms are the most strongly integrated polyhedra, the sum of the M–Ob bond valences amounts to 6 (M^{VI}species) or 5 v.u. (M^V species). The standard type I $M^{VI}O_6$ octahedra are integrated by 4.63 to 4.23 v.u., those of the M^VO_6 octahedra by 3.61 to 3.19 v.u. The standard type II $M^{VI}O_6$ octahedra are integrated by 2.95 to 2.49 v.u. (only M^{VI} species containing this type are described; the M^V case exists only as a trigonal MO_5 bipyramid integrated by 2.06 v.u.). MO_6 octahedra with three unshared oxygen atoms (only M^{VI} species containing this

Table 47. Strength of the integration of the MO_k polyhedra ($k > 4$) in the M^{VI}–O frameworks as a measure for the kinetic inertness of the corresponding structural parts of polyoxometalate(VI) ions

Type of MO_k polyhedron	Sum of the M–Ob bond valences in v.u.[a]
Octahedra with 0 unshared oxygen atom:	
type II:	
2 strong pseudoterminal, 2 middle-strong, 2 weak M–Ob bonds	5.95 (5.87–6.03)
in: $Mo_7O_{24}^{6-}$, $Mo_{36}O_{112}(H_2O)_{16}^{8-}$	
Polyhedra with 1 unshared oxygen atom:	
type I (standard type):	
1 strong M–Ot bond;	
4 middle-strong, 1 weak M–Ob bond[b]	4.37 (4.23–4.63)
in: $M_6O_{19}^{2-}$, $W_{12}O_{40}(OH)_2^{10-}$, $W_{12}O_{38}(OH)_2^{6-}$, ...	
type II:	
1 strong M–Ot bond;	
1 strong pseudoterminal, 2 middle-strong, 2 weak M–Ob bonds	4.30 (4.20–4.38⟨4.53⟩)
in: (β-)$Mo_8O_{26}^{4-}$, $Mo_8O_{26}(OH)_2^{6-}$, $Mo_{10}O_{34}^{8-}$ (I), $HW_5O_{19}^{7-}$ [57], ...	
Polyhedra with 2 unshared oxygen atoms:	
type II:	
1 strong M–Ot, 1 weak M–Otw bond;	
1 strong pseudoterminal, 2 middle-strong, 1 weak M–Ob bond	3.92 (3.90–3.94)
in: $Mo_{36}O_{112}(H_2O)_{16}^{8-}$, "$MoO_3 \cdot 2H_2O$"	
type II (standard type):	
2 strong M–Ot bonds;	
2 middle-strong, 2 weak M–Ob bonds[b]	2.75 (2.49–2.95⟨3.14⟩)
in: $Mo_7O_{24}^{6-}$, (β-)$Mo_8O_{26}^{4-}$, $Mo_8O_{26}(OH)_2^{6-}$, $H_3W_6O_{22}^{5-}$ [58], ...	
Octahedra with 3 unshared oxygen atoms:	
type II:	
2 strong M–Ot, 1 weak M–Otw bond;	
2 middle-strong, 1 weak M–Ob bond	2.30 (2.22–2.38)
in: $Mo_{36}O_{112}(H_2O)_{16}^{8-}$, "$MoO_3 \cdot H_2O$"	
type II, quasi type II, and modified type II[c]:	
2 strong M–Ot, 1 middle-strong M–Oth bond;	
1 middle-strong, 2 weak M–Ob bonds	1.84 (⟨1.71–1.95⟩)
in: $Mo_8O_{26}(OH)_2^{6-}$, $H_3W_6O_{22}^{5-}$ [58]	
type III[d]:	
3 strong M–Ot bonds;	
3 weak M–Ob bonds	⟨1.63 (1.61–1.65)⟩
in: $HW_5O_{19}^{7-}$ [57]	
special type[e]:	
3 strong M–Ot bonds;	
3 (middle-strong and) weak M–Ob bonds	1.00
in: $M_4O_{12}(OH)_4^{4-}$, $HM_6O_{20}(OH)_2^{5-}$, ...	

[a] Unweighted average and range of the individual values according to Tables 8 to 11, 20 to 31, and 33 to 35; ⟨⟩: data of the borderline cases $W_6O_{19}(OH)_3^{5-}$ [58] and $HW_5O_{19}^{7-}$ [57].
[b] There are small modifications for polyhedra others than octahedra.

Table 47 (Continued)

[c] Quasi type II: average for two positions of Oth; modified type II: the two strong, two middle-strong, and two weak M–O bonds are in each case in *cis* position to each other (see Fig. 1 and Sect. 4.3.2).
[d] Type III: three strong, fac M–Ot and three weak (*fac*) M–Ob bonds in *trans* position to the strong bonds; note that our definition of type III refers like the definition of type I and type II to *MO_6 octahedra* with a corresponding number (I: 1, II: 2, III: 3) of strong (short) M–O bonds whereas in the literature [63] 'type III' refers to *polyoxometalate structures* consisting of type I *and* type II MO_6 octahedra.
[e] This type approximates the preceding case and has been postulated to occur in early intermediates formed according to certain formation mechanisms for polyoxometalate(VI) ions ($M_4O_{12}(OH)_4^{4-}$, $HM_6O_{20}(OH)_2^{5-}$, and others) [2, 4] under the assumption that the (non-acidic) H atoms (those guaranteeing the maintaining of the octahedral coordination of M, formulated in parentheses) occupy preferably bridging O atoms. Considering a maximum negative charge of 1 c.u. on the MO_6 octahedra and hence an average M–Ot bond valence of 1.67 v.u., the result for the sum of the inner M–O bond valences of the MO_6 octahedra in question is 1 v.u. In the light of the investigations in this section we now prefer to assume that the non-acidic H atoms form a 'terminal' M–OH (M–Oth) bond so that the intermediates belong to the next but one higher group. This type has, however, been observed in the compound $[(n\text{-}C_4H_9)_4N]_2[Mo_3O_7(CH_3C(CH_2O)_3)_2]$ [285] with a sum of the 3 weak M–Ob bond valences of 1.11 v.u.

type should exist and are described) are integrated by 2.38 to 1.00 v.u., depending on the degree of protonation of one of the unshared oxygen atoms (see also Sect. 4.6.7).

4.6.10
Relationship Between M–O Bond Valence and O–M–O Angle

In the literature the question has been raised whether the bond angles have an influence on the bond valences (as determined from the bond lengths). Unfortunately it is not always clear what the authors mean: whether there is an influence of the bond angle on the bond length-bond valence relationship (function) or merely an influence on the bond lengths or bond valence (as, for example, discussed in this article for $p\pi$–$d\pi$ bonding in approximately linear M–O–M bridges). Trömel [37] observed no significant influence of the bond angles on the bond lengths or bond valences. Allmann [29] considers it possible that there is a small influence. In the calculations of the present article no dependence of the O–M–O angles on the bond valences (on the bond length-bond valence *function*) was taken into account. We cannot see how a reference state in respect of the angles could be defined without arbitrariness. Also, the question arises why just the O–M–O and not the M–O–M angles should be of relevance. The correspondence of the standard deviations for the sum of the bond valences about M, for the individual M–O bond valences, and for the bond lengths with the standard deviation for the bond lengths given in the literature (see Sect. 4.1, above, and Sect. 3.1 in Ref. 35) indicates that there is no influence of the bond angles on the bond length-bond valence functions.

Table 48. Strength of the integration of the MO_k polyhedra ($k > 4$) in the M^V–O frameworks as a measure for the kinetic inertness of the corresponding structural parts of polyoxometalate(V) ions

Type of MO_k polyhedron	Sum of the M–Ob bond valences in v.u.[a]
Octahedra with 0 unshared oxygen atom:	
type II:	
2 strong pseudoterminal, 2 middle-strong, 2 weak M–Ob bonds	4.97 (4.89–5.01)
in: $M_{10}O_{28}^{6-}$, $H_2V_{10}O_{28}^{4-}$, $H_3V_{10}O_{28}^{3-}$	
Polyhedra with 1 unshared oxygen atom:	
type I (standard type):	
1 strong M–Ot bond;	
4 middle-strong, 1 weak M–Ob bond[b]	3.28 (3.19–3.61)
in: $M_6O_{19}^{8-}$, $M_{10}O_{28}^{6-}$, "V_2O_5", ...	
type II:	
1 strong M–Ot bond;	
1 strong pseudoterminal, 2 middle-strong, 2 weak M–Ob bonds	3.33 (3.31–3.35)
in: $H_3V_{10}O_{28}^{3-}$	
Polyhedra with 2 unshared oxygen atoms:	
type II:	
2 strong M–Ot bonds;	
2 middle-strong, 1 weak M–Ob bond	2.06
in: $[V_2O_6^{2-}]_\infty$	

[a] Unweighted average and range of the individual values according to Tables 12 to 16, 18, 32, and 36.
[b] Modifications for polyhedra others than octahedra.

On the other hand there is clearly a correlation between the average bond valence of a *cis*-pair of bonds and enclosed bond angle. There is the *rule* that great bond valences lead to enlarged bond angles *at the metal centers* [38, 307, 308]. The reason is seen in bond repulsion [308], in a repulsion between the electrons in the bonds [307], or in ligand-ligand repulsion [38]. Another possibility are simply the space requirements of the oxygen ligands as defined by their cone angle: For purely geometrical reasons the ligands in short (strong) bonds require large cone angles, those in long (weak) bonds need smaller cone angles at the metal centers if the same size of the ligand atoms is assumed in each case. The decisiveness of the angles at the metal centers and not so much of those at the oxygen atoms concerns the, on average, greater coordination numbers and hence greater crowding at the metal atoms.

The dependence between O–M–O angle and bond valence is most pronounced (no contradiction to the above rule) for the bond pairs with the strongest, namely the *cis*-terminal M–O bonds [307] since their oxygen atoms do not participate in additional (inner) M–O bonds and hence can rather independently adjust to any requirements. The other bond pairs show more or less great deviations. The following reasons for deviations come into force:

- The different MO_k polyhedra show already different (average) angles at the central M atoms.
- Even for an isolated MO_k polyhedron the O–M–O angles cannot be adjusted independently of each other.
- The additional (up to five) M atoms to which the oxygen atoms of the M–O bonds in question can be coordinated make the adjustment of optimal angles in the M–O framework still more difficult since the other M and O atoms also try to adopt optimal O–M–O and M–O–M angles.
- Special geometric conditions can hinder the formation of optimal angles. For instance the O–M–O angles between two medium-sized M–O bonds in the M_6O_{19} hexametalate ions can never exceed 90° although such angles are usually greater than 90°.

4.6.11
Investigations on Heteropolyoxometalates (and Mixed-Metal Isopolyoxometalates)

The description of heteropolyoxometalate ions will be given in a separate article. Here we will only briefly compare iso- and heteropolyoxometalate ions.

4.6.11.1
"Conventional" Heteropolyoxometalate Ions

In the M^{VI}–O and M^{V}–O frameworks of all types of heteropolyoxometalate ions and in the M^{VI}–V^{V}–O frameworks of mixed-metal isopolyoxometalate ions the distribution of the bond valence (bond lengths) and of the charge corresponds absolutely to that in the M–O frameworks of the isopolyoxo species [270, 272, 309, 310]. This includes in particular that the formal ionic charge of the heteropoly anions resides mainly on the terminal oxygen atoms, at least, however, on oxygen atoms at the surface of the structures (possibility for the necessary contacts with the cations).

In the XO_k heteroatom polyhedra of the central primary [253g] heteroatoms (e.g., $P^{V}O_4$, $Ni^{IV}O_6$, $U^{IV}O_{12}$) the oxygen atoms bear small to medium-sized negative or positive charge, according to the requirements of the geometry of the X–M–O frameworks (connectivity of the atoms), of the bond length-bond valence relationships, and of the valence sum rule. If, however, there is a negative charge it is either at the surface of the structure (possibility for contacts with cations) or intramolecularly compensated by positive charge on other oxygen atoms, or both. On no account can the heteropolyoxometalate ions with tetrahedrally coordinated P^{V}, As^{V}, Si^{IV}, and Ge^{IV} heteroatoms be described [13l, 311–314] as a phosphate, arsenate, etc. ion surrounded by an $M_{12}O_{36}$ cage (see also Sect. 4.8.1 and Ref. 253b).

4.6.11.2
(Mixed-Valence) Heteropolyoxovanadate(V/IV) Ions Containing "Unusual" Heteroelement "Ions" or Molecules

The mixed-valence[17] $V^{V/IV}$ heteropolyoxovanadate ions containing "unusual" heteroelement "ions" or molecules can be described like the other (iso- and hetero)polyoxometalate ions (i.e., the ionic charge is mainly located on the terminal oxygen atoms, etc.). The main difference as against the "conventional" heteropolyoxometalate ions is on the one hand the rather small bond valence which is "exchanged" between the $V^{V/IV}$–O frameworks (cages etc.) and the central heteroelements (Cl, Br, etc.: 1 v.u.), heteroelement molecules (H_2O, CH_3CN, etc.: formally 0 v.u.; see, however, the example given below), or heteroelement-oxo groups (CO_3: 2 v.u., ClO_4: 1 v.u., etc.) and on the other hand the great number of M atoms to which the heteroelements, heteroelement molecules, or heteroelement-oxo groups make bonds ($V_{15}O_{36}$ frameworks: 15 V atoms, $V_{18}O_{42}$ frameworks: 18 V atoms, $V_{22}O_{54}$: 22 V atoms, etc.). Hence there are in principle only very small (average) bond valences possible between the M atoms of the cage and the central heteroatom or oxygen atoms of the heteroatom sphere making the bond: $ClV_{15}O_{36}^{6-}$: 1/15 = 0.067 v.u.; $(CO_3)V_{15}O_{36}^{7-}$: 2/15 = 0.13 v.u.; $(ClO_4)HV_{22}O_{54}^{6-}$: 1/22 = 0.045 v.u.; etc. In the conventional heteropolyoxometalate ions the central M–O(X) bond valences are considerably greater [$(PO_4)Mo_{12}O_{36}^{3-}$ (Keggin type): 3/12 = 0.25 v.u. [253b]; $(Ni^{IV}O_6)Mo_6O_{18}^{8-}$ (Anderson type): 8/12 = 0.67 v.u.; $(U^{IV}O_{12})Mo_{12}O_{30}^{8-}$ [formally: $(U^{IV}O_{12}^{20-})(M_{12}O_{30}^{12+})$] (Dexter-Silverton type): 20/36 = 0.56 v.u.], however, *qualitatively* (in terms of the building principle) there is no difference between heteropolyoxovanadate(V/IV) ions with "unusual" heteroelement "ions" or molecules and conventional heteropolyoxometalate ions. The distinction according to the charge of/on the M–O cage [314] is artificial, already formally does not apply to some of the conventional heteropolyoxometalate ions (see above), and contradicts the real circumstances. In the following we discuss briefly an example each for a

[17] The d_o and B or N parameters of Eq. (1) are said to vary with their oxidation numbers for some elements M in M–O bonds, whereas for other elements M there is no such dependence. However, for Mo–O and V–O bonds there is no consensus on this point in the literature. Some authors [32, 50–52, 315] claim or assume that there is no difference, others [27, 33, 66, 49] state that there are small differences. Our observations for V–O bonds [35] revealed that there are small differences in the bond length-bond valence parameters for V^V and V^{IV} which are too large to be neglected [35]. We solved this problem of application to mixed-valence compounds in the following way. Since the bond length-bond valence functions do not respond sensitively to moderate changes of N [26, 29, 270] or B [33, 40, 270] if d_o is optimally fitted, it is possible to use a uniform B [35] or N for all oxidation states of M, in particular since the range of the oxidation states in the compounds in question is only small (in the literature sometimes [27] for V^V–O and V^{IV}–O bonds *both* bond length-bond valence parameters are given with different values). For the evaluation of a mixed-valence polyoxovanadate structure we chose B = 0.851 Å (see Table 1) and determined in a first approximation cycle a uniform d_{oi} and the (approximate) oxidation numbers of the different V atoms. In a second and, if necessary, in further approximation cycles we applied for each V atom a different d_{oi}, according to the particular oxidation numbers of the V atoms and the (as linear assumed) $d_{oi}(n)$ function with the slope $\Delta d_{oi}/\Delta n$ = 0.02 Å.

structure with a central "oxo anion", a monoatomic "anion", and a neutral molecule.

In the heteropolyanion *$(CO_3)V_{15}O_{36}^{7-}$* [316], −5.10 to −5.64 c.u. of the ionic charge are located on terminal oxygen atoms, additional −1.52 to −0.98 c.u.[18] on bridging oxygen atoms at the surface of the more or less spherical $V^{V/IV}$–O cage (summed up: −6.62 c.u.), and only −0.38 c.u. on the CO_3-oxygen atoms [310]. Hence the description of the polyanion as a carbonate ion surrounded by a polyoxometalate ion (as "anion in an anion") [202, 284, 314, 316–319] is an improper procedure. Two items are of particular interest in this respect:

(1) The range of the – related to the connectivities (coordination modes) reported in the literature – very long V···O distances is rather wide and starts already at 2.597 Å.
(2) The oxidation numbers of the V atoms differ considerably.

The very long V···O distances sum up to 0.89 v.u. (1.3% of the bond valence sum of the V atoms) in this case. They complete the VO_k polyhedra described in the literature as tetragonal VO_5 pyramids to give type I VO_6 octahedra with the very long V···O bonds inside the V–O cage in *trans* position to the terminal V–O bonds. The "*trans*" bond valences of the six VO_k polyhedra near to the "poles" of the approximately spherical cage amount on an average to 0.04 v.u. and those of the six VO_k polyhedra between the "poles" and "equator" on an average to 0.11 v.u.; the three VO_k polyhedra in "equatorial" position have 0.24 v.u. in the "*trans*" V–O bonds and are described already in the literature as VO_6 octahedra. These values must be judged on the one hand in comparison with the small (but definite) M–O bond valences of other polyoxo species, e.g., estimated from the oxidation numbers and the coordination modes of the relevant elements (M–O6l in $M_6O_{19}^{w-}$: 0.33 v.u.; M–O4f in $X^VM_{12}^{VI}O_{40}^{3-}$: 0.25 v.u. [253b]) or estimated according to Eq. (10), assuming an equidistribution of the available bond valence (M–O6l in $M_{10}O_{28}^{6-}$: 0.30 v.u.; M–O6l in $M_6O_{19}^{8-}$: 0.26 v.u.; see Table 40) or actually observed under consideration of the coordination modes generally accepted in the literature for the structures in question [M1–O6l in $M_{10}O_{28}^{6-}$: 0.24 v.u. (M = V), 0.19 v.u. (M = Nb); M1–Opa in $Mo_7O_{24}^{6-}$: 0.23 v.u.; V1br–O6lr in $H_3V_{10}O_{28}^{3-}$: 0.20 v.u.], and on the other hand in comparison with the corresponding (average) values for the case in question (the formal ionic charge of the carbonate ion divided by the number of vanadium atoms coordinated to the CO_3-oxygen atoms = 2/15 = 0.13 v.u.).

The bond valence sums for the different V atoms in $(CO_3)V_{15}O_{36}^{7-}$ specify their (fractional) oxidation numbers which must not be rounded. They range from 4.00 v.u. (the three type I VO_6 octahedra in "equatorial" position) via 4.67 v.u. (the six type I tetragonal VO_5 pyramids or VO_6 octahedra between

[18] The values depend upon assumptions about the origin of the positive charge on one sort of the O3l atoms, i.e., whether the positive charge follows primarily from a structural stabilization by bond strengthening of inner M–O bonds and hence is compensated by negative charge on terminal oxygen atoms (charge separation) or primarily from a better adjustment of the bond lengths to overcome restrictions caused by the rigidity of the X–M–O framework and hence may also be compensated by negative charge on bridging oxygen atoms.

"equator" and the "poles") to 4.50 v.u. (the six type I tetragonal VO_5 pyramids or VO_6 octahedra near to the "poles"); average: 4.467 v.u. This refers to the average values of the symmetry-equivalent V–O bonds; the individual values differ considerably (packing effects in the crystal).

In $XV_{15}O_{36}^{6-}(X = Cl, Br)$ [316] there is a similar situation. Of the formal ionic charge −4.82 to −5.20 c.u. (X = Cl) and −4.91 to −5.12 c.u. (X = Br) are located on the terminal oxygen atoms, additional −0.81 to −0.44 c.u. (X = Cl) and −0.97 to −0.76 c.u. (X = Br) on bridging oxygen atoms at the surface of the polyanions (summed up: −5.63 and −5.88 c.u., respectively), and only −0.37 and −0.12 c.u., respectively, on the halogen atoms (parameters of the bond length-bond valence functions for V^V–X and V^{IV}–X bonds: B = 0.852 Å, d_o = 2.16 Å for X = Cl and 2.30 Å for X = Br [49]). [Note that the (rest ?) charge on the central halogen atoms can be caused by an error of the X–V bond lengths of 0.02 and 0.01 v.u., respectively, only since this is a typical and extreme case for the occurrence of the "arithmetic" error mentioned in Sect. 4.1.] Hence again the description of the polyanions as anions (in this case halide ions) surrounded by negatively charged $V^{V/IV}$–O cages ("cluster shells" [284]) ("an anion in the interior of an anion" [320]) is improper (for a reason leading to this misinterpretation see also Sect. 4.8.1 and footnote 20). It contradicts additionally the principle of local valence balance (see Sect. 2.1 in Ref. 35). There is no problem to formulate resonance structures with negatively charged, uncharged, and even positively charged halogen atoms to accomodate for the different size of the halogen atoms.

In $(H_2O)V_{18}^{IV}O_{42}^{12-}$ [321, 322] the H_2O molecule would use on the anionic (oxygen) side 0.4 v.u. ([56a]; see Sect. 4.4.2.3) for bonds to ca. 9 V atoms of the V–O cage (average: 0.4/9 = 0.044 v.u. per M–O bond) and on the cationic (hydrogen) side correspondingly 0.4 v.u. for bonds to ca. 21 O atoms of the V–O cage (hydrogen bonds; average: 0.4/21 = 0.019 v.u. per O–H bond) if the usual O–H bond valence of 0.8 v.u. in H_2O [36, 39e, 40b, 56a] is retained.

Items (1) and (2) above are both to be interpreted in the same direction. The variability of the oxidation number of M is a fourth possibility to adjust bond valence to the geometrical requirements of the (X–)M–O frameworks (for the other possibilities see Sect. 4.6.3). The shallow slope of the bond length-bond valence function in the range of small bond valences together with the possibility of fractional and variable oxidation numbers for M allows a fine adjustment of the M–O cage to the more or less randomly large and randomly shaped molecular parts of only small bonding potential to be included in the cages, i.e., the mixed-valence M–O framework can rather differentially approach the enclosed molecular part to produce the necessary bond valence between them. Hence it is obviously on the one hand the pronounced property of vanadium to accept fractional and varying oxidation numbers and on the other hand the ease with which the monomeric V^V and V^{IV} species[19] – due to

[19] The relationships between the structures of the V–O frameworks of many heteropolyoxovanadates as sections of the structure of "V_2O_5" and the latter itself, pointed out in Ref. 321, are of no relevance for the formation of the heteropolyoxovanadates since the latter do not (directly) form from solid "V_2O_5".

their high basicity (based on the lower oxidation number as compared to the M^{VI} species) and hence high degree of protonation (V^V: $VO(OH)_3$ or other suitable species, e.g., VO_2^+,aq; the V^{IV} monomers should/could be deprotonated forms of $VO(H_2O)_5^{2+}$ [274b], e.g., $VO(OH)_2(H_2O)_3$) - can undergo aggregation reactions in the formation reaction under elimination of oxygen as H_2O, including the simple stripping off of coordinated H_2O molecules [14aa, 253d], that is responsible for the success to prepare the large number of mixed-valence polyoxovanadates with the great diversity of the included molecular parts. Otherwise there is no fundamental difference between the conventional and the new, "unusual" heteropolyoxometalate ions from the viewpoint of the bond valence model or of the building principle. Rather, there are only the gradual differences with regard to the bond valences. Hence we cannot agree to the operation of a new type of self-organization process [323]. Remarkable, however, is the smallness of the valence interchange which is sufficient to initiate the formation of the heteropolyoxometalates in question and is indeed for the single M–O(X) or M–X bond in the range of the bond valences usually occurring in host-guest interactions. These very small bond valences belong to the structure-determining factors and must not be neglected in the structural descriptions of the polyanions in question like bond valences <0.08 v.u. with the isopolyoxometalate ions. See also Sects. 4.6.2 and 4.6.3.

The driving forces and directing factors as well as the geometrical course with the formation of (iso- and) heteropolyoxo anions have systematically been discussed on a simple, conventional basis [14z, 205, 253e, 262g]. The M–O frameworks are successively built up around the heteroelement oxospecies (heteroelement polyhedron) as the "nucleus of aggregation". The driving forces for the formation of the heteropolymetalate ions are in the main (as for the isopolymetalate ions) the build-up of MO_6 octahedra (meshing effect etc.; see Sect. 4.6.8) and the liberation of H_2O molecules (favorable entropy change; compare Sect. 4.6.8). The formation of the heteropolymetalate ions is preferred to the formation of isopolymetalate ions and, hence, starts at higher pH values due to the earlier formation of MO_6 octahedra and earlier and stronger (more) liberation of H_2O molecules in the course of the aggregation processes. Thus, the formation of heteropolymetalate ions is energetically more favorable in all stages of the aggregation processes [2a, 205, 253b, 253e, 262e, 262f] (see also Sect. 3.6.7 in Ref. 16). Compared to the rather fast formation of the Keggin-type heteropolytungstate structures, the extremely slow formation of the structurally related metatungstate ion is assumed to take place in a quite different way since there is no "support" (the heteroatom tetrahedron) around which the aggregation reactions can readily proceed in a series of reaction steps according to the addition mechanism and condensation mechanism [253e, 262f] (for these mechanisms see also Sect. 3.1 in Ref. 16). Hence the generation of the Keggin structure in form of the metatungstate ion in the absence of tetrahedral heterooxo anions cannot be used in proof of different formation mechanisms [314] for the conventional (e.g., the "archetypal" Keggin structure) heteropolyanions and those containing "unusual" heteroelement "ions" or molecules.

4.6.12
The Special (Untypical Borderline) Case of the Isopolyoxometalate Ions $HW_5O_{19}^{7-}$ and $H_3W_6O_{22}^{5-}$

Salts of these isopolyoxometalate ions have only recently been prepared from aqueous tungstate solutions under extreme conditions in very small yields: rather high tungstate concentrations ($\approx$ 2M); rather small degree of acidification (P) or degree of protonation (Z) of the tungstate solution ($K_7[HW_5O_{19}] \cdot 10H_2O$: $Z \approx P = 0.29$; $Na_5[H_3W_6O_{22}] \cdot 18H_2O$: $Z \approx P = 0.15$), i.e., rather distant from the Z^+ values of the polyoxo species (W_5: $Z^+ = 3/5 = 0.60$; W_6: $Z^+ = 7/6 = 1.17$); high pH value ($\approx$ 9); very long reaction time [57, 58]. Apart from the reaction time, such conditions have been named [324] to be the optimum for a proof of the first aggregation products in the course of the formation of polyanions with extended MO_k polyhedra, i.e., for species occurring prior to the "parametalate" stage under conditions of scarcity of protons [the parametalate stage is assigned to the first polyoxometalate(VI) species forming in successively acidified aqueous M^{VI} solutions, provable by static methods of investigation, and can be characterized as the first aggregation product of M^{VI} requiring for its formation according to the "condensation mechanism" [2, 4, 20] a *di*protonated monometalate ion ($M_7O_{24}^{6-}$, $M_{12}O_{40}(OH)_2^{10-}$) and thus forming a "closed" structure having no octahedra with three unshared O atoms, as compared to the "open" structures characterized by the presence of MO_6 octahedra having three unshared O atoms]. Accordingly, the formation of the W_5 and W_6 species in question needs far-reaching absence of diprotonated monometalate ions. This is accomplished by the high tungstate concentration and the small degree of acidification (or degree of protonation) of the tungstate solution and leads to high pH values far from the pH range of existence of H_2MO_4 (pH $\approx$ 3; compare distribution, predominance, and/or existence area diagrams [290] like those in Ref. 14jj for the molybdate case). [Note also that the formation and precipitation of the para-B dodecatungstate as a main product [57] with its – compared to the Z value of the solution – high Z^+ value (1.17) leads to an additional reduction of the Z value of the mother liquor from which the W_5 compound crystallizes.]

In Ref. 2 two (**3, 4**) of three (**3** to **5**) M_5O_{19} and four (**9, 11, 13, 16**) of 15 (**7** to **21**) geometrically possible M_6O_{22} frameworks have been named as candidates for penta- and hexametalate ions that require only *mono*protonated monometalate ions for their formation according to the "condensation mechanism" (the bold-type numbers refer to the structures shown in Ref. 2). $HW_5O_{19}^{7-}$ [57] and $H_3W_6O_{22}^{5-}$ [58] each have one of the contracted structures (**4** and **13**), however, the pentatungstate ion is less strongly protonated than assumed in Ref. 2.

The less strong protonation of the pentatungstate ion means actually that (two) *un*protonated monometalate ions participate in the aggregation reaction to give the polymetalate ion which hence is characterized by $Z^+ < 1$. This is only the second example for such a case (the other example is the $W_4O_{16}^{8-}$ building group in the compound $7Li_2WO_4{\cdot}4H_2O$ [325, 326] which is formally a "monotungstate", i.e., a tungstate with $Z^+ = 0$).

The very slow formation of the (solid) W_5 compound (ca. 1 year [57]) is obviously due to the extreme formation and stability conditions.

The structural evaluation according to the bond valence model (not performed in Sect. 3) revealed - due to the borderline nature of the structures - some deviations from the normal case. In the W_6 structure the $W4O_6$ octahedron is a modified type II octahedron in which the two short, two medium-sized, and two long M–O bonds are in *cis* positions each (Fig. 1). In the W_5 structure the $W3O_6$ and $W5O_6$ octahedra are of a new type for polyoxometalate ions (type III) with three short, *fac* and three long (*fac*) M–O bonds in *trans* position to the short ones (Fig. 1). All the other MO_6 octahedra are of type II. Both structure determinations lie just beyond the reliability ($2\sigma_{\Sigma i}$) border.

The structures of $HW_5O_{19}^{7-}$ and $H_3W_6O_{22}^{5-}$ extend the bond valence (and bond lengths) ranges found for the other (poly)oxometalates of M^{VI} (Table 38) with regard to the M–Ot and M–Op bonds to 1.37 v.u. (1.78 Å) and with regard to the M–O* bonds to 0.69 v.u. (2.07 Å) and consequently the range of the charge on the Ot atoms to −0.63 c.u. (see Table 41). In the case of the M–Ot and M–Op bonds and Ot atoms this is a consequence of the extremely high charge of the W_5 species and in the case of the M–O* bonds obviously a consequence of the modified type II character of the $W4O_6$ octahedron of the W_6 species.

The results for the stabilization of the polymeric character of the structures (Table 46 and Fig. 36) show as a special feature that the stabilization by normal bonding is <1 v.u. per M atom. This is due to the fact that the structures belong to the starting phase of the condensation mechanism where only few O atoms have been eliminated as H_2O molecules ($r/q < 0.5$) and the principles of the foregoing addition mechanism [2, 4, 14q] are still governing the structural relationships (see also Fig. 1 in Ref. 16).

The results for the integration of the different kinds of MO_6 octahedra show as a special feature an extension of the types of MO_k polyhedra with three unshared oxygen atoms; for details see Table 47 (Sect. 4.6.9).

4.7 Conformity of the Results with Our Bond Models for Polyoxometalate Ions

The occurrence of significant bond strength (bond valence) between all neighboring M and O atoms in the M–O frameworks of the *polyoxometalate ions built up of MO_6 octahedra (MO_k polyhedra with $k>4$)* is in accordance with resonance schemes (5) and (6) of the bond model recently proposed [16]. The bond valences for the terminal and pseudoterminal M^{VI}–O bonds range from 1.80 to 1.40 v.u., for the medium-sized bridging M^{VI}–O bonds to usually low-coordinate oxygen atoms from 1.33 to 0.67 v.u., and for the long bridging M^{VI}–O bonds to usually high-coordinate oxygen atoms (oxygen atoms in *trans* position to terminal and pseudoterminal ones) from 0.62 to 0.23 v.u. The terminal and pseudoterminal M^V–O bonds range from 1.78 to 1.24 v.u., the medium-sized bridging M^V–O bonds from 1.05 to 0.43 v.u., and the long bridging M^{V}–O bonds from 0.50 to 0.19 v.u. (see Sect. 4.3.1). They are of the expected magnitude, see Sect. 3.4 in Ref. 16.

The occurrence of negative charge on nearly all types of oxygen atoms is mainly explained by resonance scheme (6) and the occurrence of positive charge on certain types of the bridging oxygen atoms by resonance scheme (7) of that bond model [16].

The acceptance of the negative ionic charge by mainly the terminal and pseudoterminal oxygen atoms is explained in the bond model by a number of reasons which could directly be taken over to explain the ascertained charges (see Sects. 3.4, 3.6.3–3.6.5, 3.6.7, and 3.6.8 in Ref. 16 and Sect. 4.4.2.1, above).

The acceptance of positive charge on the O atoms of linear or approximately linear M–O–M bridges is explained in the bond model by some reasons which could again directly be taken over to explain the ascertained charges (see Sects. 3.6.7 and 4.4 in Ref. 16 and Sect. 4.4.2.2, above). Their occurrence is obviously favored by the electronic configurations of the oxygen and metal atoms. The (multicenter [68, 69]) $p\pi$–$d\pi$ multiple bond model [67] does very well explain the observed dependence of the bond valences and of the positive charge upon the M–O–M angle. [Another effect that can be assigned to the electronic configuration of the metal (and oxygen) atoms is the *cis* configuration of the terminal oxygen atoms [13j, 64, 65, 261a, 327] in the type II MO_6 octahedra. However, this feature has unvariedly come over from the monometalate to the polymetalate ions and hence cannot have a stabilizing effect on the *polymeric* structures [14n]; see also Sect. 4.6.2 in Ref. 16.]

The occurrence of negative residual charges on angularly bridging oxygen atoms and even on linear M–O–M bridges (e.g., the five-coordinate O atoms in (β-)$Mo_8O_{26}^{4-}$) and the occasional occurrence of positive charge on oxygen atoms of angular M–O–M bridges (e.g., in (β-)$Mo_8O_{26}^{4-}$, α-$Mo_8O_{26}^{4-}$, and $W_{10}O_{32}^{4-}$) are also in accordance with the bond model and are caused by geometric requirements of the more or less rigid M–O frameworks in these cases (see Sect. 4.4.3.1). Negative residual charges on (angularly bridging) oxygen atoms at the surface of the M–O frameworks are also necessary for the charge balance in the packing of cations, polyoxometalate ions, and molecules of water of crystallization in the solid state (see Sect. 4.4.3.2).

Hence the formation of the structures is made possible by the ability of oxygen to accept varying amounts of (negative and positive) charge which allows the adjustment of the necessary bond lengths and charge.

Furthermore, the observed charge distributions in polyoxometalate ions with a high density of the formal ionic charge (unprotonated octahedral polyoxometalate ions of M^{V}) show indeed the acceptance of higher partial amounts of the formal ionic charge by bridging oxygen atoms and the suppression of charge separation processes (absence of positively charged oxygen atoms, see Sects. 4.4.3.5 and 4.4.3.6), as expected according to the bond model.

It has been discussed [16] for $M = Mo^{VI}$ and W^{VI} that certain bridging oxygen atoms cannot bear a negative charge (no resonance structures with a negative charge on these bridging oxygen atoms can be formulated except with O^{2-} – see resonance scheme (5), Sect. 3.11 – or by violation of coordination mode (V) in Ref. 16). In general the experimental results support this view.

Only for two structures are there small differences: For the $W_{12}O_{38}(OH)_2^{6-}$ ion a very small negative charge on the central, three-coordinate oxygen atoms (−0.05 c.u.) has been observed. This might be due to the poor quality of the structural data which is very strongly substantiated in this case (σ_d = 0.09 Å! [230]). For the $W_{10}O_{32}^{4-}$ ion a small negative charge on the central, five-coordinate oxygen atoms has been found (−0.07 c.u.). This might also be due to the generally less good quality of the structural data of polytungstates which is in this case also substantiated (σ_d = 0.03 Å [231]; see Sect. 3.19).

The bond valences and the charge distributions established above (Sects. 3.5–3.8) for the *tetrahedral polyoxometalate ions* also confirm our previous statements [23, 24] for such species. For the $V_2O_7^{4-}$, $V_3O_{10}^{5-}$, and $[VO_3^-]_\infty$ ions with a negative charge on the bridging oxygen atoms the additional resonance type (4c) must be assumed.

Similarly, the bond valences and the charge distributions established above (Sect. 3.4) for the unprotonated *tetrahedral monometalate ions* agree with previous statements [23].

4.8 Discussion in Connection with Statements in the Literature

The results discussed in Sects. 4.3–4.5, above, and the theoretical background treated in Sects. 3.6, 4.1, 4.3, and 4.4 of Ref. 16 indicate the existence of a very complex interplay of a large number of factors that determine structure and bonding of polyoxometalate ions. In contrast, statements in the literature on structure and bonding of these compounds name usually only one or two (ostensible) factors to be responsible for an observed (main) feature with a clear preference of "electronic" explanations. Beyond that the relations to reference (e.g., parental) states are usually neglected. In the following the main statements in the literature about structure and bonding in polyoxometalate ions are comparatively discussed.

4.8.1 *Previous Investigations on Bond Valences and Charge Distributions about Polyoxometalate Ions Based on Bond Length-Bond Valence Considerations. Description of Iso- and Heteropolyoxometalate Ions as Composed of Small Anionic Species and Polymeric Parts?*

Investigations on the distribution of the negative ionic charge over the various types of oxygen atoms, based on bond length-bond valence considerations, have already been described in the literature. The results are in direct contrast to such presented above.

The $M_6O_{19}^{2-}$ ion has been described as an O^{2-} anion surrounded by a neutral (= uncharged) M_6O_{18} cage [9, 13b, 197, 198, 294]; our result is a positively charged central oxygen atom and the negative ionic charge on the oxygen atoms of the surrounding M–O framework, mainly the terminal ones (see Sect. 3.11). Similarly, the α-$Mo_8O_{26}^{4-}$ ion has been viewed as an Mo_6O_{18} ring to which MoO_4^{2-} ions are attached on either side of the ring, etc. [9, 328, 232]. The differences of the interpretation of the same structural data (!) are caused by a number of reasons.

First, the bond length-bond valence functions used in the literature do not give such good descriptions of the structural data as those [35] applied in the present study. The basis for the determination of the parameters d_o and B in our investigations [35] have been species that contained only the metal (Mo^{VI}, W^{VI}, V^{V}, Nb^{V}, Ta^{V}) and oxygen atoms, whereas many of the other authors used also compounds containing additional elements and/or the metal in lower oxidation states. Furthermore, most of the other authors used doubtful evaluation methods instead of a fitting procedure with a least-squares treatment of the data together with a reliability test by which inaccurate structural data of the reference compounds could be detected and eliminated; etc. [35]. To demonstrate the differences and errors we can take the $Mo_6O_{19}^{2-}$ ion as an example (Table 49). The incorrectness of the published bond valences ("bond strengths" [13b]) can clearly be recognized if one considers the (not calculated or at least not published) ionic charge and the sum of the bond valences about the molybdenum atoms as control indices. They deviate strongly from the nominal values of −2 c.u. for the ionic charge and 6 v.u. for the (average) sum of the bond valences about molybdenum.

Secondly, the authors do not realize that in inorganic compounds fractional bond valences are the normal state [25, 36, 37] and must not be rounded to give integers as has more or less tacitly been done by some authors [9, 13b].

Table 49. Clarification of an incorrect interpretation of the $Mo_6O_{19}^{2-}$ structure in the literature

	d in Å	s in v.u. and c in c.u. in					
		Ref. 13b "wrong"[a]		Ref. 13b "correct"[b]		This work[c]	
Mo–Ot	1.68	*1.98*	−0.02	1.97	−0.03	1.75	−0.25
Mo–O2a	1.93	*0.97*	−0.06	0.87	−0.26	0.97	−0.06
Mo–O6 1	2.32	*0.28*	−0.32	0.29	−0.26	0.38	+0.25
Formal ionic charge in c.u.:		−1.16		−3.56		−2.00	
Σs_i about Mo in v.u.:		6.14		5.74		6.00	

[a] Published [13b] data in italics. An arithmetical error or a misprint [s(Mo–O2a) = 0.97 v.u.] and a poor bond length-bond valence function, out of date [$s = (1.882/d)^{6.0}$ [27]], led to wrong bond valences for the M–O bonds and for the charges on the oxygen atoms as is indicated by wrong values for the formal ionic charge and for the sum of the bond valences about the M atoms.

[b] Recalculation by the present authors employing the bond lengths and the bond length-bond valence function used in Ref. 13b. The influence of the poor bond length-bond valence function is still greater than appears according to the foregoing columns as is indicated by the still more deviating values for the formal ionic charge and for the sum of the bond valences about the Mo atoms.

[c] Use of a better bond length-bond valence function and in particular of an individually fitted d_0 parameter, d_{0i}.

Thirdly, the (potential) accuracy of bond length-bond valence functions is strongly underrated [7, 13b, 219], and hence together with the preceding item the authors regard themselves as authorized to round the decimals to integers [9, 13b]. Thus the (incorrect) MO–Ot bond valence of 1.98 v.u. and the (fortuitously correct) MO–Ob bond valence of 0.97 v.u. for $Mo_6O_{19}^{2-}$ (Table 49) have been interpreted as 2 v.u. and 1 v.u., respectively, which leads to a central O^{2-} ion [9]. Indeed, use of published values for the parameters d_o and B (or d_o and N) gives such different results [35] that the authors [9, 13b] could believe that they do not underrate the accuracy of the bond length-bond valence functions.

Fourthly, the authors [9, 13b] do not realize that for a given E–O bond length within an EO_k polyhedron there is no difference whether there is a negative charge on the oxygen atom or bonds to other E atoms (e.g., an E–O–M bridge) [253b]. For example, in SiO_4^{4-}, $Si(OH)_4$, and $SiM_{12}O_{40}^{4-}$ the Si–O bond lengths and hence bond valences are identical (or expected to be identical). This item leads also directly to the wrong interpretation that in polyoxometalate structures negatively charged species (O^{2-}, MO_4^{2-}, etc. ions) contact (uncharged) polymeric $(MO_3)_\zeta$ partial structures.

Fifthly, the authors [9, 13b] do not take into account that the cation-oxygen(oxometalate ion) and cation-oxygen(water) distances have to fulfill Eqs. (1) and (2) as well, which is not possible for distant, i.e., inner and unapproachable (e.g., O6l) negatively charged oxygen atoms (see Sect. 4.4.3.2).

In particular the last two items have also been neglected by a large number of authors [9, 13l, 311–313, 329] who concluded that the Keggin-type structures of heteropoly anions are composed of an inner PO_4^{3-}, SiO_4^{4-}, etc. ion surrounded by an (uncharged) $M_{12}O_{36}$ cage; see also Ref. 253b. Other authors [253b, 267c, 330] reject such a structural view.

The same applies to the heteropolyoxometalate (in particular -vanadate) ions containing "unusual" heteroelement "ions" or molecules. These cannot be described as "anions in the interior of anions" [202, 284, 314, 316, 317–320] since the negative ionic charge is virtually exclusively localized on oxygen atoms at the surface of the polyanions, mainly the terminal ones. The only difference as against the "conventional" heteropolyoxometalate ions is a gradual one with regard to the central M–O(X) or M–X bond valences (bond lengths) which are necessarily [due to the small charge of the initial species forming the "hetero parts" of the structures and the large number of M (or O) atoms in the surrounding cage] much smaller (longer[20]) than for the "conventional" heteropolyanions (see Sect. 4.6.11.2). In particular, no difference is to be expected with regard to the course of the formation of the "conventional" and the new, "unusual" heteropolyoxometalate ions. Template-controlled formation mechanisms [314] have also been discussed for the heteropolyanion types $XM_6O_{24}^{m-}$, $XM_6O_{24}H_6^{m-}$, $XM_9O_{32}^{m-}$, $X_2M_{10}O_{38}H_4^{m-}$ [205, 253e], $XM_{12}O_{40}^{m-}$, the precursor type $XM_{11}O_{39}^{m-}$ [253e], and others [253e]. The formation of the "archetypal" Keggin structure in form of the metatungstate ion in spite of the

[20] This great distance between the atoms of the "heteroelement grouping" (CO_3^{2-}, ClO_4^-, Cl^-, Br^-, CH_3CN, etc.) and surrounding M–O cage is just the crucial point which suggests the wrong idea of anions which appear to be suspended in the interior of anions.

absence of a template cannot be used to claim different formation mechanisms [314] for the "conventional" and the new, "unusual" heteropolyoxometalate ions since the Keggin frameworks of the heteropolyanions and of the metatungstate ion are obviously formed according to quite different mechanisms as is indicated by the extremely different reaction rates [253e].

The basicities and basicity sequences for the different types of oxygen atoms stated for some polyoxometalate ions in the literature [7, 8, 13b, 13j] are based on the considerations described above and therefore are also incorrect (see also Sects. 4.6.6 and 4.6.7). In particular it is surprising that terminal oxygen atoms are described as essentially non-basic, discouraging extensive protonation and further polymerization [13j, 331, 332], although protonation and aggregation (elimination of H_2O) reactions of mono- and polyoxometalate ions are the basis for the formation of the polyoxometalate ions (compare Sects. 1 and 4.6.6). (According to X-ray electron-spectroscopic studies the differences of the charge on the bridging and terminal oxygen atoms are slight [11].)

4.8.2
Derivation of "Bond Orders" for Certain M–O Bonds by Inspection of the Symmetry Properties of the Bonding Orbitals?

Attempts have been made to derive "bond orders" for certain, well describable oxometalate species (e.g., for MoO_4^{2-} [66] and $H_2V_{10}O_{28}^{4-}$ [216]) by an inspection of the symmetry properties of the bonding orbitals following a procedure of Cotton and Wing [333] and to relate them to the bond lengths. In this way bond orders greater than two for MO–Ot bonds were obtained [66, 216]. This contradicts our results for bond valences ("bond strengths", "bond orders"; see Sect. 2.1) and those of other authors [37]. Bond orders (bond multiplicities) greater than two for M–Ot bonds have also occasionally been discussed by other authors [247, 268].

4.8.3
Justification of the Term "Trans Influence" for the Explanation of the Occurrence of the Weak M–O Bonds?

Part of the inner (bridging) M–O bonds of the M–O frameworks built by MO_6 octahedra are weak (long) bonds in *trans* position to strong (short) monooxo-terminal (case of type I MO_6 octahedra) or *cis* dioxo-terminal or -pseudoterminal (case of type II MO_6 octahedra) multiple M–O bonds. The weakness of these bonds is ascribed to the "*trans* influence" of the M–O multiple bonds [13j, 64, 65, 268, 281b, 294, 332] (in the literature erroneously often also designated as "trans effect"). This concept, applied to monomeric metal or metalate complexes, may be of some use. For the case of the polyoxometalate species it is a rather useless concept: It explains nothing, suggests an incorrect assessment of the bonding situation (weak integration of the inner, high-coordinate oxygen atoms within the M–O frameworks; an influence proceeding from the terminal oxygen atoms), and blocks the view on other, relevant features.

A certain positive charge, v⊕, on an inner oxygen atom indicates a total bond valence of (2 + v) v.u. proceeding from that oxygen atom; a certain negative charge, v⊖, on a terminal oxygen atom indicates correspondingly a total bond valence of (2 − v) v.u. (Eq. (3)). Thus in reality the inner, high-coordinate oxygen atoms are altogether more strongly bound to M atoms (observed maximum for M^{VI}: 2.25 v.u. in the case of $Mo_6O_{19}^{2-}$) than the "strongly" ("multiple"-)bound terminal oxygen atoms (observed minimum for M^{VI} species: 1.47 v.u. in the case of $W_{12}O_{40}(OH)_2^{10-}$ and $W_{10}O_{32}^{4-}$). Even stronger bonds to the inner oxygen atoms would lead to yet more positive charge on them. Due to their highly negative charge, terminal oxygen atoms are generally the least strongly bound oxygen atoms if all M–O bonds of the oxygen atoms are taken into account.[21]

The final reasons for the typical appearance of the MO_6 octahedra, characterized as type II and type I, are on the one hand simply the requirements of the valence of the elements and of the geometry of the M–O frameworks (the network of bonds between the atoms) [16], that means the resulting distribution of the available bond valence over the molecules fulfills simultaneously the bond length-bond valence relationship(s), the valence sum rule, the geometric relationships (the connectivity of the atoms), and some additional stabilizing features, see Sect. 4.3.2, and on the other hand the space requirements of the M atoms which are fulfilled by the differentiation of the bond lengths, see Sect. 4.6.1. The mixture of covalent and coordinate bonds (due to the mutual action of the MO_4 building units as acceptors and as donors of ligands) described in our bond model (Ref. 16, in particular Sects. 3.1 and 4.1) is ideally suited to fulfill these requirements. Other factors have only modifying effects. In particular, as is shown by the great variety of the metal and oxygen coordination modes realized in polyoxometalate ions (see Sect. 3.2, Figs. 1 and 2), a decisive influence of the electronic configurations of the metal and oxygen atoms (of the symmetry properties of the bonding orbitals) cannot be detected[22] except for the atoms of linear or approximately linear M–O–M bridges. These tend to induce a charge separation – generation of positive charge on the bridging oxygen atoms in question and of negative charge on terminal oxygen atoms – and exert in this way a weakening influence on the terminal M–O bonds of the polymetalate ions, compare Sects. 4.3.4.1 and 4.4.2.2. It has also been shown that the bond valences of the terminal M–O bonds are smaller than expected for a uniform distribution of the bond valences over the M–O bonds (Sect. 4.3.3). The reason for all this is that the acceptance of negative charge by bridging oxygen atoms has an unfavorable effect on the stability (on the polymeric character) of the structures (Sects. 4.3.4.1 and 4.4.2.1) and on the space available for the central M atoms (Sect. 4.6.1). These are entirely different things than the above "trans influence".

[21] Another concern is that M–O bonds of multiply bridging oxygen atoms – due to the weaker individual bonds – may be kinetically less inert.

[22] This means, the possibilities of the bonding orbitals correspond perfectly to the geometrical requirements of the M–O frameworks and exert no own influence.

4.8.4
Type II and Type I MO_6 Octahedra as a Consequence of Favorable Orbital Symmetries?

It has been claimed [64, 65, 216] that the occurrence of type II and type I MO_6 octahedra in polyoxometalate ions is an effect of the favorable electronic configuration of the metal (and oxygen) atoms for these cases (see also Ref. 39b). The following arguments speak against this view as a *sole* decisive factor:

- The appearance of the polyoxometalate ions as structures with a definite surface and formal ionic charge, defined by the connectivity of the atoms and their valence (oxidation number), and the space requirements of M request the occurrence of strong (multiple-bonded) outer (terminal) and of weaker inner (bridging) M–O bonds (compare Sects. 4.3.2, 4.6.4, and 4.8.3).
- the great variability of the coordination modes of the M atoms in the MO_k polyhedra, which comprises
  - apart from MO_6 octahedra of type II and type I MO_4 tetrahedra with every kind of distortion, tetragonal MO_5 pyramids of type II and type I, trigonal MO_5 bipyramids of type II, and pentagonal MO_7 bipyramids of type I;
  - apart from type II and type I MO_6 octahedra also type III octahedra;
  - in one case (W4 in $H_3W_6O_{22}^{5-}$ [58]) even a bond lengths (bond valence) pattern different from type II and type I;
  - a rather continuous sequence of intermediates between type II and type I MO_6 octahedra;
- the real existence of regular-octahedrally coordinated M atoms (e.g., in MoF_6 [269]).

In the oxides and oxide hydrates there occur also MO_6 octahedra or other MO_k polyhedra with type II or type I characteristics. In this case the distortions of the MO_6 octahedra (MO_k polyhedra) result from the construction of the M–O frameworks as layers or chains of closely packed oxygen atoms by which an "inner surface" is generated; otherwise the necessary (minimum) average M–O bond lengths for the octahedral coordination of M cannot be realized. Charge separation processes enable the necessary adjustment of the bond lengths and bond valences and thus the M–O layer or chain frameworks with terminal (negatively charged) and bridging or water- (positively charged) oxygen atoms to be formed (see Sect. 4.6.4). Hence again one cannot definitely conclude from the occurrence of type II or type I MO_k polyhedra that the electronic configuration of the metal atoms is a decisive factor. (For other explanations of distortions of MO_k polyhedra see Refs. 334 and 335.)

Nevertheless there may indeed be an influence of the electronic configuration of the metal atoms *as one of the many other factors* discussed in Sects. 4.3 and 4.4 to influence M–O bond lengths and bond valences and charges on the oxygen atoms.

Remarkably, the frequent formulations of the protonated monomolybdate ions as $MoO(OH)_5^-$ and $Mo(OH)_6$ (see Ref. 23) have only occasionally [13k] been criticized under this point of view.

4.8.5
Maximization of the Terminal MO and MO_2 pπ-dπ Interactions for the Stabilization of Polyoxo Anions?

A number of authors assume that terminal MO and MO_2 pπ-dπ interactions stabilize polyoxometalate structures [13j, 53, 64, 65, 268, 281a, 284, 332, 336, 337]. Their maximization requires in the case of the MO_2 groups avoidance of competition for the same p and d (t_{2g}) orbitals and hence demands *cis* arrangement of the oxygen atoms [13j, 64, 65, 261a, 268, 281a, 294, 332, 337, 338].

The increased bond valences of the inner M–O bonds (Sects. 4.3.3 and 4.3.4.1) and the reduced terminal M–O bond valences (Sect. 4.3.3), the occurrence of positive charge on bridging oxygen atoms (Sects. 4.4.1 and 4.4.2.2) and the increased negative charge on the terminal oxygen atoms (Sects. 4.4.1 and 4.4.2.1) show that there is by no means a maximization of the terminal MO and MO_2 pπ-dπ interactions but rather the reverse. The reasons for these effects are the stabilization of the polymeric character of the polyoxometalate structures and the increase of their basicity (see also Sect. 4.6.6, Fig. 33). Both factors lead under the conditions of the formation of the polyoxometalate ions to an enhanced consumption of protons and thus favor their formation (law of mass action) [16]. Strengthening of the terminal M–O bonds, i.e., reduction of the charge on the terminal O atoms can neither stabilize the polymeric structures nor enhance their basicity; such effects would be directed to a destabilization of the system.

The occurrence of terminal MO and MO_2 *multiple* bonds (in pπ-dπ interactions) in the polyoxometalate ions is an emanation of the appearance of their structures with a definite surface and formal ionic charge, defined by the connectivity of the atoms and their oxidation number, as has already been mentioned (Sects. 4.3.2, 4.8.3, and 4.8.4). The same, however, applies to the parent monomeric metalate ions [14n]. Thus the above view about a stabilization of the polyoxometalate structures by MO and MO_2 pπ-dπ interactions has nothing to do with *poly*oxometalate structures but only with the character of metal-oxygen bonding in metal and oxygen containing compounds (monomeric oxometalate ions, polyoxometalate ions, chelate complexes of monomeric and aggregated metalate ions, etc.); see also Sect. 4.6.2 in Ref. 16.

4.8.6
Deviations of Bond Valence Sums for O and M Atoms from their Oxidation Numbers?

In Sect. 3.1 of Ref. 35 it has been summarized that adequately derived and applied bond length-bond valence relationships describe the bond valence of the M–O bonds and the charge distribution over the O atoms of oxometalate species with an accuracy that corresponds virtually to the accuracy of the bond lengths data of the structures. In the literature, however, there are stated examples (that do not refer to polyoxometalates) with large deviations for bond valence sums about individual O and M atoms. For a discussion of these deviations see Sects. 3.4 and 3.5 in Ref. 35.

4.8.7
Chemical and Steric Constraints in Inorganic Solids

After having formulated the main results of the present investigation [55, 270, 271], a paper was published [36] in which the conflicting requirements of chemical bonding and three-dimensional geometry (i.e., of connectivity of the atoms, bond lengths, bond length-bond valence relationship, valence sum rule) in inorganic solids are treated. Whereas the main aspects of that paper correspond with our view (Sects. 4.3.4.2 and 4.4.3.1), there are some details that are not in agreement with our conclusions. The discrepancies are due to some generalizations not applicable to polyoxometalates and polyoxometalate ions.

The question for an "ideal structure" ("ideal bonding geometry", "ideal bond lengths") [36] can rather differently be answered for polyoxometalate ions: assuming an equidistribution of the M–O bond valences (average bond valences according to Eq. (10)) and of the charge over the oxygen atoms (no systematic influence on the distribution of the bond valences over the M–O bonds and/or on the distribution of the ionic charge over the oxygen atoms of the polymetalate ion and/or no charge separation are present) (see Sect. 4.3.3); assuming that the (negative) ionic charge is exclusively accepted by the terminal oxygen atoms (maximum stabilization of the polymeric character of the polyoxo anions) and that there is either no or a full charge separation (see Sect. 4.4.2.4); assuming that all resonance formulas for the polyoxometalate ions have the same weight and the bond valences correspond to the "resonance bond numbers" (see Sect. 3.4 in Ref. 16). Assumptions of coordination spheres being as regular as possible and the equal-valence rule ("each atom shares its valence as equally as possible among the bonds that it forms") [36, 39c], however, clearly contradict the occurrence of type II and type I MO_6 octahedra (MO_k polyhedra) and other coordination modes observed for the polyoxometalate structures and *in particular the relevant theoretical background.*

The dependence of the importance of steric effects in a given structure on the strength of the bonds is also seen differently. The strongest bonds, the terminal M–O bonds having bond valences greater than 1.47 v.u. (M^{VI}–O) or greater than 1.39 v.u. (M^{V}–O), adapt primarily to the requirements of the inner bonds of the M–O frameworks and to those of the packing of (poly)oxometalate ions, cations, and molecules of water of crystallization in the crystal (see Sect. 4.8.3). According to Brown [36] these bonds are rigid and cannot be adapted to the steric constraints.

4.8.8
Compact Structures Obeying the 'Lipscomb Principle' as the Most Favorable Building Principle of Polyoxometalate Ions?

It has been claimed [339] that the most favorable structures obey, besides some other factors, in particular the "principle of compactness" (as defined in Ref. 200) and the 'Lipscomb principle' [13i, 301].

Using the term "compactness", the authors [339] actually mean block-type structures without further distinction. The restriction to block-type structures largely neglects the cage-type (e.g., $W_{12}O_{40}(OH)_2^{10-}$, $Mo_{36}O_{112}(H_2O)_{16}^{8-}$) and certain chain-like structures (e.g., $[Mo_4O_{14}^{4-}]_\infty$) which do not appear as compact ones; the structures composed of corner-sharing MO_4 tetrahedra are neglected in any case. The above-cited cage-type structures belong to the most stable species, existing even in aqueous solution, compare Fig. 36.

Referring to the 'Lipscomb principle', this principle can be largely overcome by (mono- or, still better, di-)protonation of one of the three unshared oxygen atoms of the MO_6 octahedra in question. Meanwhile, a number of polyoxo species built according to this principle have been prepared and/or identified, among others the $Mo_{36}O_{112}(H_2O)_{16}^{8-}$ ion (see Sect. 4.6.9) existing even in aqueous solution and hence showing a particular stability.

Considering the discussion in Sect. 4.6.9 we can state that the question of this section must be negated on principle for the range of low degrees of acidification of the metalate solution at relatively high pH values and for the range of high degrees of acidification. Only for the intermediate range can the question, with some reservations, be answered affirmatively.

5
Summary

Refined and individually fitted bond length-bond valence functions have been applied to the structural data (bond lengths) of isopolyoxometalate ions, oxides, and oxide hydrates of Mo^{VI}, W^{VI}, V^V, Nb^V, and Ta^V to calculate the bond valences of the different types of M–O bonds and the charge on the different types of oxygen atoms. The sum of the bond valences about the M atoms, which amounts to the oxidation number of M, shows a standard deviation of 0.07 v.u. (1.2 to 1.3%) only. The estimated standard deviation for the M–O bond valences amounts to 0.012 v.u., corresponding to a standard deviation for the bond lengths of 0.010 Å, and for the charge on the O atoms to 0.020 c.u.

Data for the same polyoxo species (but with different cations, numbers of water of crystallization, and/or investigated by different authors) show a small standard deviation for the charge on the O atoms, on an average of 0.015 c.u. for the bridging and 0.032 c.u. for the terminal O atoms (these standard deviations refer to the average of the symmetry-equivalent O atoms). For the terminal M–O bonds of a polyoxo compound as a whole in individual cases great deviations (up to 0.26 v.u./c.u.) and in the case of the monomeric species even greater deviations (up to 0.36 v.u./c.u.) occur, obviously to comply with the geometrical and charge-wise packing conditions of cations, oxometalate ions, and molecules of water of crystallization.

From the relatively small differences for the inner bond lengths of a polyoxometalate type it was concluded that the structures in the free (e.g., dissolved) state resemble strongly those in the crystal and have the ideal symmetry. In many cases different cations induce the formation of quite different isopolyoxometalate structures in *insoluble* compounds under otherwise the same preparation conditions which is obviously due to the better packing conditions for cations, *changed* polyoxo anions, and molecules of water of crystallization.

The pattern of the bond valences in the different types of MO_k polyhedra is as follows:

- In the $M^{VI}O_6$ octahedra the strong, terminal or pseudoterminal M–O bonds range from 1.80 to 1.40 v.u.; the middle-strong M–O bonds to usually low-coordinately bridging oxygen atoms from 1.33 to 0.67 v.u.; the weak M–O bonds to usually high-coordinately bridging oxygen atoms in *trans* position to the terminal or pseudoterminal M–O bonds from 0.62 to 0.23 v.u. For the M^VO_6 octahedra the corresponding ranges are 1.78 to 1.24 v.u.; 1.05 to 0.43 v.u.; 0.50 to 0.19 v.u. Type II and type I MO_6 octahedra, from which intermediate stages exist, show no significant differences.
- The other, rarely occurring MO_k polyhedra with extended ($k > 4$) coordination sphere (tetragonal MO_5 pyramids, trigonal MO_5 bipyramids, pentagonal MO_7 bipyramids) exist also as type II or type I analogues and show similar bond valences.
- In the $M^{VI}O_4$ (MoO_4) tetrahedra of all types of isopolyoxo anions the terminal M–O bonds range from 1.70 to 1.44 v.u. and the bridging M–O bonds from 1.56 to 1.11 v.u. For the M^VO_4 (VO_4) tetrahedra the corresponding ranges are 1.68 to 1.29 v.u. and 1.14 to 0.81 v.u.
- In isolated $M^{VI}O_4$ tetrahedra the (terminal) M–O bond valences range from 1.86 to 1.19 (average: 1.50) v.u., in isolated M^VO_4 (V^VO_4) tetrahedra from 1.53 to 1.00 (average: 1.25) v.u.

The reasons for the general bond valence (bond lengths) differentiation in strong (short), middle-strong (medium-sized), and weak (long) M–O bonds in the MO_k polyhedra with extended coordination sphere are on the one hand simply the realization of the valence sum rule and of the bond length-bond valence relationships for the elements of the M–O frameworks with given connectivity of their atoms and on the other hand the spatial requirements of the M atoms.

Comparison of the M–O bond valences (calculated from the bond lengths according to the bond length-bond valence functions) with average bond valences, assuming an equidistribution of the bond valence over the M–O bonds of the structures, shows the following relations:

- Terminal M–O bonds are always weaker (by down to 0.41 v.u.) than expected for an equidistribution of the bond valence over the M–O bonds.

Accordingly the sum of the bond valence of the inner, bridging M–O bonds is always greater than that expected for the occurrence of equidistributed M–O bonds.

- Pseudoterminal M–O bonds are somewhat weaker than terminal ones, but are always considerably stronger (by 0.29 to 0.66 v.u.) than expected for the M–O bonds of an average two-coordinate O atom. The complementary M–O bonds are always considerably weaker (by 0.27 to 0.64 v.u.) than expected for the M–O bonds of an average two-coordinate O atom.
- Bridging M–O bonds of MO_4 tetrahedra are always stronger (by up to 0.77 v.u.) than expected for an average M–O bond of this type.
- The other bridging M–O bonds (those of the MO_6 octahedra and other MO_k polyhedra with $k > 4$) are usually stronger (by as much as 0.54 v.u.), in rare cases somewhat weaker (by down to 0.30 v.u.) than expected for an average M–O bond of a correspondingly coordinated oxygen atom, if this oxygen atom is not in a *trans* position to a terminal or pseudoterminal one; they are usually weaker (by down to 0.47 v.u.), in rare cases somewhat stronger (by up to 0.20 v.u.) than expected for an average M–O bond of a correspondingly coordinated oxygen atom, if this oxygen atom is in a *trans* position to a terminal or pseudoterminal one.

Some operating factors for the occurrence of the above bond valence (bond lengths) alterations are discussed as follows:

- The generally enhanced sum of the M–O bond valences of the bridging O atoms favors the holding together of the MO_k building units and thus stabilizes the polymeric character of the structures with its favorable consequences.
- The geometrically conditioned interdependence of the inner (bridging) M–O bond lengths and the simultaneous necessity to fulfill the bond length-bond valence relationship and the valence sum rule lead to restricted ranges for possible M–O bond lengths and hence bond valences. The restrictions are the more pronounced the more compact and/or the more symmetrical, that means the more rigid the M–O frameworks are.
- Coincidence of several middle-strong M–O bonds about an oxygen atom causes reduced bond valences and hence again restricted ranges for M–O bond valences and bond lengths.
- Coincidence of a few and weak M–O bonds about an oxygen atom causes likewise restricted ranges for M–O bond valences and bond lengths.

The pattern of the charge distribution over the O atoms of the polyoxometalate ions (and oxides or oxide hydrates) is as follows:

- The terminal O atoms bear by far most of the negative ionic charge (even the terminal oxygen atoms of the oxides and oxide hydrates, which have no formal ionic charge, bear a negative charge). It depends approximately on the Z^+ value of the polyoxo species and ranges for those of M^{VI} from (average values) −0.47 to −0.20 c.u. (individual values for a given species differ from the respective average by up to 0.20 c.u.) and for the M^V species from (average values) −0.66 to −0.19 c.u. (individual values differ from the respective average by up to 0.26 c.u.).

- The pseudoterminal O atoms, as a sort of terminal but actually two-coordinate O atoms, bear a somewhat smaller negative ionic charge. Angular bridges with an M–O–M angle smaller than ca. 125° show, as a rule, a "normal" behavior. For the M^{VI} species the charge ranges from −0.24 to −0.09 c.u. Linear or approximately linear bridges with an M–O–M angle from 180° to ca. 125° bear a somewhat smaller negative charge, in the case of the M^{VI} species from −0.09 to +0.01 c.u. The rare pseudoterminal oxygen atoms of the M^V species belong virtually to the group of the angularly bridging two-coordinate oxygen atoms since their second bonds are middle-strong M–O bonds (and not weak "*trans*" bonds as in all M^{VI} cases).
- The angularly bridging two- and three-coordinate oxygen atoms (all M–O–M angles are smaller than ca. 125°) bear yet smaller negative and even small positive charges (M^{VI} species: −0.24 to +0.11 c.u.; M^V species: −0.40 to +0.05 c.u.).
- Oxygen atoms that are part of a linear or approximately linear M–O–M bridge (at least one M–O–M angle in the bridge is greater than ca. 125°) bear, as a rule, a positive charge (up to +0.25 c.u. for the M^{VI} and up to +0.27 c.u. for the M^V species), if the basic negative charge density, as expressed by the quotient −m/t (m: number of charges, t: number of terminal oxygen atoms of the polyoxo species), is not too high (−m/t > −0.5 to −0.7 c.u.). If it is great, the corresponding oxygen atoms bear a negative charge (case of the highly compact M^V species $M_6O_{19}^{8-}$ and $M_{10}O_{28}^{6-}$).
- Coordinated bridging H_2O molecules bear a positive charge, in the only example ($Mo_{36}O_{112}(H_2O)_{16}^{8-}$) ranging from +0.16 to +0.23 c.u., if the "standard" O–H bond valence of 0.8 v.u. is assumed. Bridging OH groups bear also positive charge, in the protonated decavanadate ions under the same conditions ranging from +0.01 to +0.07 c.u. (In the central, secluded cavities of the para-B and metatungstate ions there is no such charge on the OH groups.)

The occurrence of positive charge on some types of O atoms is one of the most remarkable results of the calculations and requires the simultaneous appearance of an equal amount of negative charge ("charge separation") apart from that representing the formal ionic charge.

The above charge distribution leads to the formulation of the following *rules*:

(1) Negative charge is preferentially accepted by terminal oxygen atoms.
(2) Pseudoterminal (two-coordinate) oxygen atoms behave approximately as terminal ones and bear somewhat less negative charge.
(3) Angularly bridging oxygen atoms bear only small negative and occasionally positive charge.

(4) Oxygen atoms participating in linear or approximately linear M–O–M bridges bear usually positive charge.
(5) Oxygen atoms forming an OH bridge or a coordinated H_2O molecule in a bridge are positively charged.
(6) Oxygen atoms on which two of the structural features superimpose bear an intermediate charge.
(7) Bridging oxygen atoms of polyoxo species with a high negative charge density bear increased negative and reduced positive charges.

The reasons for the operation of rules (1) to (6) are discussed as follows:

- The preferential acceptance of the negative ionic charge by the terminal (and pseudoterminal) oxygen atoms is caused and conducted by
  - the high statistical weight of the Lewis-type resonance structures with charged terminal (and pseudoterminal) oxygen atoms;
  - the stabilization of the polymeric structure if the negative charge on the bridging oxygen atoms is as small as possible (full utilization of the bonding power of the bridging oxygen atoms for the holding together of the MO_k building groups of the species);
  - the enhanced basicity of the polyoxometalate ions as a consequence of the more highly negatively charged terminal oxygen atoms and hence in acidified solutions increased consumption of protons according to Le Chatelier's principle;
  - the space requirements of unshared electron pairs (cf. the VSEPR model) in which the charge electrons are merged in;
  - the large distances between the negative ionic charges and hence small Coulombic repulsion between them;
  - the necessary contacts to the cations in the crystal.
- The occurrence of positive charge on linearly or approximately linearly bridging oxygen atoms is caused by charge separation processes and conducted by
  - the additional stabilization of the polymeric structures through bond strengthening of inner M–O bonds by multicenter $p\pi$–$d\pi$ multiple bonds in the respective M–O–M bridges (for solutions in combination with Le Chatelier's principle) if the simultaneously generated negative charge is accepted by terminal oxygen atoms;
  - the ionic interaction between positively and negatively charged oxygen atoms which leads to a further additional stabilization of the structures if the geometric conditions for an interaction are given;
  - the enhanced basicity of the polyoxometalate ions as a consequence of the more highly negatively charged terminal oxygen atoms and hence in acidified solutions increased consumption of protons according to Le Chatelier's principle;
  - (in the case of the oxides:) the possibility to obtain for the M atoms octahedral coordination (gain of space) by differentiation of the bond lengths through variation of the bond types.

- The occurrence of positive charge on *bridging* OH groups and *bridging* coordinated H_2O molecules is also caused by charge separation processes and conducted by
  - the additional stabilization of the polymeric structures through bond strengthening of inner M–O bonds (for solutions in combination with Le Chatelier's principle) if the simultaneously generated negative charge is accepted by terminal oxygen atoms;
  - the ionic interaction between positively and negatively charged oxygen atoms, as discussed only just;
  - the enhanced basicity of the polyoxometalate ions as discussed only just;
  - (in the case of the oxides:) the possibility to obtain for the M atoms octahedral coordination (gain of space) by differentiation of the bond lengths through variation of the bond types, as discussed only just.
- Coincidence of the factors coming into force leads to a superimposition of the effects (e.g., a pseudoterminal oxygen atom in an approximately linear M–O–M bridge is more or less uncharged).

Hence the stability of the polyoxometalate species (or of the polyoxometalate systems) becomes optimal if the (negative) ionic charge and the negative charge generated by charge separation are exclusively spread over the terminal (and pseudoterminal) oxygen atoms and if much positive charge is generated by charge separation and accepted by suitable bridging oxygen atoms, OH groups, and/or coordinated H_2O molecules.

The reasons for the deviations from rules (1) to (6) and for the operation of rule (7) are discussed as follows:

- The interdependence of the *inner, bridging* M–O bond lengths of the polyoxometalate structures containing MO_k polyhedra with $k > 4$ and the simultaneous necessity to fulfill the bond length–bond valence relationship and the valence sum rule lead to restricted bond lengths and hence restricted ranges for possible M–O bond valences. The restrictions are the more pronounced the more compact and/or the more symmetrical the M–O frameworks are (the more rigid they are). This *can* lead to changes of the charge in either direction.
- The geometrical and charge-wise requirements of the packing of cations, oxometalate ions, and molecules of water of crystallization in the solid salts can be considerably improved by a corresponding adaptation of the bond lengths and charges of the *outer, terminal* oxygen atoms. Changes on the inner, bridging oxygen atoms are very limited; if the necessary changes for different cations are too great, different polyoxometalate types form.
- Coincidence of several medium-sized M–O bonds about a bridging oxygen atom leads to positive charge on the corresponding oxygen atom and causes restricted ranges for the M–O bond valences (and bond lengths).
- Coincidence of only a few and weak M–O bonds about a bridging oxygen atom leads to negative charge on the corresponding oxygen atom and causes also restricted ranges for the M–O bond valences (and bond

lengths). (Virtually, even high-coordinate O atoms can belong to this case if the bond valences are sufficiently small.)

- A high formal ionic charge and a small number of terminal oxygen atoms in a structure (as applies to the highly compact, octahedral isopolyoxometalate ions of M^V) lead to very highly charged terminal oxygen atoms. The charge-related stress, apparently due to the space requirements of the charge cloud, can be reduced by the acceptance of a partial amount of the charge by bridging oxygen atoms and hence leads to more negatively charged bridging oxygen atoms. Additionally, polyoxo structures with an $M^V_qO_{<3q}$ framework require Lewis structures with some charge on bridging O atoms.
- A generally high negative charge density on the terminal oxygen atoms acts, moreover, suppressive on the charge separation process since the negative charge generated along with the positive charge would increase the negative charge density on the terminal oxygen atoms still more.

One of the most important properties of oxygen for the realization of the great variety of the oxometalate structures through adjustment of bond lengths and bond valences to the structural requirements (connectivity of the atoms) is its ability to accept in a wide range variable amounts of negative and even positive charge. Moreover, there is the possibility to adjust bond lengths, bond valences, or charges by formation of more or less unsymmetrical M–O–M bridges on the basis of the hyperbola-like form of the bond length-bond valence function.

Octahedral coordination of M (or E) by oxygen requires a minimum average M–O (E–O) bond length of 1.92 Å. Factors having an influence upon that are

- a sufficient size of M in itself, e.g., as characterized by the single-bond distance d_o;
- a suitable oxidation number of M: smaller oxidation numbers cause smaller bond valences and hence longer M–O bonds;
- a protonation of the monometalate ion and in the later aggregation process the substitution of H by M, which leads to a diversification of the M–O bond types and hence bond lengths and in particular, as a consequence of the non-linearity of the bond length-bond valence function, to an increase of the average bond lengths.

The extension of the coordination sphere of the monovanadate ions requires (at least) their *di*protonation by which geometrical conditions comparable to those of the *mono*protonated monomolybdate and monotungstate ions are achieved. Te^{VI} and I^{VII} form more regular EO_6 octahedra in the monomeric species $Te(OH)_6$ and $IO(OH)_5$ and hence also in the few polymeric species derived from them since the tetrahedral species TeO_4^{2-} and IO_4^-, due to their great μ value (μ: index in the conjugate oxoacid $EO_\mu(OH)_\nu$) and consequently low basicity, cannot be protonated and thus cannot undergo an extension of their coordination sphere in the way observed for MoO_4^{2-}, WO_4^{2-}, and VO_4^{3-}.

The occurrence of type II and type I MO_6 octahedra (MO_k polyhedra) and of positively and negatively charged oxygen atoms in the oxides and oxide hydrates is explained by the formation of "inner surfaces" through chain or layer structures under differentiation of the M–O bond lengths and bond valences by the operation of charge separation processes and hence the possibility to obtain for the M atoms octahedral coordination (gain of space) by differentiation of the bond lengths through variation of the bond types. It explains simultaneously the amphoteric character of the oxides and oxide hydrates.

Very long M···O distances beyond those that define the commonly assumed coordination number and coordination polyhedron for M (average M···O distances $\gtrapprox$ 2.5 Å for Mo^{VI} and W^{VI}, $\gtrapprox$ 2.4 Å for V^V; average bond valences $\lessapprox$ 0.2 v.u.) can rest on the following reasons:

- Bond valences up to 0.03 v.u., in rare, special cases 0.05 v.u. are the inevitable consequence of the accumulation, i.e., of the connection modes of the MO_k polyhedra in the polyoxometalate structures.
- Greater bond valences (above 0.03 or 0.05 up to ≈0.20 v.u.) occur only in some of the structures containing MO_k polyhedra smaller than MO_6 octahedra. Their occurrence may be regarded
  - as occasion to obtain yet an increased coordination number for M;
  - as a consequence of the packing conditions for cations, (poly)oxometalate ions, and molecules of water of crystallization in the solid-state structure;
  - as – in the case of the heteropolyoxometalate ions with "unusual" heteroatoms and fractional oxidation numbers of M (preferably V) – special adjustment of the M–O bond lengths to the geometrical conditions of the X–M–O frameworks and to the requirements of the bond length-bond valence function and valence sum rule.

The strength of the integration of the MO_6 octahedra (MO_k polyhedra with $k > 4$) within the M–O frameworks depends upon the number of unshared oxygen atoms (free vertices) of the MO_6 octahedra and for the MO_6 octahedra with three unshared oxygen atoms additionally upon their protonation state. MO_6 octahedra without unshared oxygen atoms are maximally integrated, by 6 v.u. (M^{VI} species) or 5 v.u. (M^V species). The standard type I $M^{VI}O_6$ octahedra are integrated by 4.63 to 4.23 v.u., the M^VO_6 octahedra by 3.61 to 3.19 v.u. The standard type II $M^{VI}O_6$ octahedra are integrated by 2.95 to 2.49 v.u.; the M^V case does only exist as trigonal MO_5 bipyramid, integrated by 2.06 v.u. MO_6 octahedra with three unshared oxygen atoms (only M^{VI} species containing this type are described; M^V species are not to be expected because of the generally higher basicity of the M^V species and hence easy availability of the necessary diprotonated monomers) are integrated by 2.38 to 1.00 v.u., depending on the degree of protonation of one of the unshared oxygen atoms (decreasing integration in the sequence diprotonation > monoprotonation > no protonation). The weakest integrated type of MO_6 octahedra determines the kinetic inertness of the respective structure.

The overall stabilization of the polyoxometalate ions by bond valence of bridging M–O bonds (stabilization of the polymeric character of the structures) can be expressed by the average sum of the inner M–O bond valence per M atom. The structures composed of MO_6 octahedra with one free vertex are the most strongly stabilized structures (M^{VI} case: 4.3 to 4.2 v.u. per M atom; M^{V} case: 3.7 to 3.2 v.u. per M atom). The structures containing most of the MO_6 octahedra (MO_k polyhedra) with two free vertices are somewhat less stabilized (M^{VI} case: 3.9 to 2.6 v.u. per M atom; M^{V} case: 2.1 v.u. per M atom). (The cage-type structures, appearing more loosely constructed than the block-type structures, are as strongly stabilized as the latter.) The least-stabilized structures are the corner-shared tetrahedral chains (2.0 to 0.9 v.u. per M atom, according to the chain length).

According to another classification, polyoxometalate ions existing in aqueous solution equilibria are in general the most stabilized species by bond valence of inner M–O bonds – the slowly, at the expense of the rapidly formed species appearing structures showing the greater stabilization. $Mo_{36}O_{112}(H_2O)_{16}^{8-}$ is a border case. Solid state structures of species that do not exist in aqueous solution as equilibrium component and hence being less stabilized by bond valence of inner M–O bonds are, as a rule, formed in evasive reactions; a well understandable exception is $M_6^{VI}O_{19}^{2-}$. Soluble heteropolyoxometalate ions are yet better stabilized by bond valence of inner M–O bonds than the soluble isopolyoxometalate ions otherwise existing in the solutions.

Three factors contribute to the overall stabilization by inner M–O bonds: "normal" bonding (which would result if no expansion of the coordination sphere of M would take place); the "meshing effect" in most structures with extended MO_k ($k > 4$) polyhedra (which is a result of the expansion of the coordination sphere of M); the bond strengthening of the inner M–O bonds and the harmonization of the bond lengths, bond valences, and charges (which result from a variety of special adaptation processes). Normal bonding corresponds to the stoichiometry of the polyoxo species and results in on an average 0.40 to 2.00 v.u. per M atom for the M^{VI} ($Z^+ = 0.60$ to 2) and in on an average 1.00 to 3.00 v.u. per M atom for the M^{V} ($Z^+ = 1$ to 3) species investigated. Values <1 v.u. per M atom for polyoxometalate ions indicate structures in the starting phase of the condensation mechanism where H atoms originating from the addition mechanism are still present (which maintain the octahedral coordination of M and hence should be non-acidic). The contribution by the meshing effect in the case of the polyoxo species with extended MO_k polyhedra corresponds to a fictitious conversion of terminal into bridging oxygen atoms during the extension of the MO_k polyhedra and amounts to 1.17 to 2.85 v.u. per M atom for the M^{VI} and to 1.15 to 1.90 v.u. per M atom for the M^{V} species of the above Z^+ ranges; for $[V_2O_6^{2-}]_\infty$, "(α-) $MoO_3 \cdot H_2O$", and "V_2O_5" there is no fictitious conversion of terminal into bridging oxygen atoms and hence no meshing effect since only bridging oxygen atoms are used for the extension of the MO_k polyhedra. The contribution by the special adaptations which result from the preferential acceptance of the (negative) ionic charge and of the negative charge generated through charge separation by the terminal and of the simultaneously generated positive charge by bridging oxygen atoms, diminished by destabilizing effects of the geometrical requirements of the M–O frameworks (interdependence of the M–O bond lengths), of the packing of the constituents of the salts, and of the density of the negative charge in the polyoxo species results for the M^{VI} case with one exception ("$MoO_3 \cdot 2H_2O$": −0.06 v.u. per M atom) in 0.10 to 0.29 v.u. per M atom and for the M^V case with one exception ("V_2O_5": 0.19 v.u. per M atom) in −0.13 to 0.13 v.u. per M atom. The smaller (and negative) values in the M^V case reflect the in general greater density of the negative charge of the M^V species. The deviating, large positive value for "V_2O_5", which carries no formal ionic charge, is hence well explained; the deviating value for "$MoO_3 \cdot 2H_2O$" is also well explained since there is no ("true") charge separation in this case. The importance of this special adaptation in spite of its smallness is due to the fact that normal bonding and the meshing effect are unvariably determined by the stoichiometric coefficients and by the connectivity of the atoms.

Application of the bond valence model to the problem of the structural identity of the $Mo_8O_{26}^{4-}$ ion as equilibrium component in aqueous solution by calculation of the sum of the inner bond valences for the different block-type isomers speaks so far (in addition to other arguments) against the (β-)$Mo_8O_{26}^{4-}$ structure: The only structure directly accessible via the equilibrium component $Mo_7O_{24}^{6-}$ provides ca. 22% more bond valence for the inner M–O bonds.

The basicity of the polyoxometalate ions (as characterized by their protonation constants) is clearly determined by the (average) negative charge on the terminal (as the most basic) oxygen atoms and by the density of the terminal oxygen atoms at the surface of the structures.

In the only available example of *octahedral* structures ($V_{10}O_{28}^{6-}$) that is characterized both in unprotonated and protonated form, protonation occurs (proven for the solution and for the solid state) on bridging oxygen atoms at the surface of the polyanion (O2a and O3a atoms) and not on the most basic, i.e., terminal oxygen atoms. The reason is obviously the better realization of the necessary structural changes which should be as small as possible: The requisite bond valence and hence bond lengths alterations can better distribute over the entire structure due to

- the better "coupling" of the protonated bridging oxygen atoms to the M–O frameworks (efficacy of two or three, respectively, paths for the propagation of the alterations compared to only one path in the case of protonation of terminal oxygen atoms) and
- the better response of bridging M–O bonds to "*trans*" bond valence and hence bond lengths alterations which need only a little shift of the central M atoms in the MO_6 octahedra along the corresponding approximately linear O–M–O axis ("*trans*" bonds of terminal O atoms are weak, long bonds which can only incompletely compensate the necessary bond valence alterations because of their weakness in itself and because of the small slope of the bond valence-bond length function in the region of weak (long) M–O bonds and hence great necessary alterations of the bond lengths).

Moreover, protonation of the short, strong terminal M–O bonds would result in medium-sized, middle-strong M–O bonds which would change the types of MO_6 octahedra and hence mean a complete reconstruction of the M–O frameworks.

In the *tetrahedral* structures, which exhibit no interdependence of bond lengths due to the rigidity of M–O frameworks and hence present in this respect no structural problems, protonation occurs on terminal oxygen atoms as the most basic oxygen atoms. The resulting protonation products are again the most stable species.

In the case of the species $W_{12}O_{40}(OH)_2^{10-}$, $W_{12}O_{38}(OH)_2^{6-}$, and $Mo_{36}O_{112}(H_2O)_{16}^{8-}$, which do not exist in the unprotonated form, the protonated bridging oxygen atoms form exclusively "*trans*" M–O bonds and hence would bear a very high negative charge without protonation.

In the case of the species $H_3W_6O_{22}^{5-}$, $Mo_8O_{26}(OH)_2^{6-}$, $Mo_{36}O_{112}(H_2O)_{16}^{8-}$, "$MoO_3 \cdot H_2O$", and "$MoO_3 \cdot 2H_2O$", containing protonated unshared oxygen atoms, the protonation occurred on MO_6 octahedra with three unshared oxygen atoms, and the unprotonated species do not exist. In these examples the reason for the protonation is obviously the stronger integration of the protonated MO_6 octahedra within the M–O frameworks which would otherwise reduce from the range 2.38 to 1.71 v.u. to ca. 1.0 v.u. and hence destabilize the structure.

The heteropolyoxometalate structures adhere to the same principles of building and bonding, those with "unusual" heteroelement "anions" or molecules included. That means in particular that the inner structural (hetero) parts obey the principles of the bond valence model too. Due to the only small bonding potential of the hetero parts (e.g., because of a small ionic charge of the initial hetero atom species) and the great number of partner atoms in the surrounding M–O framework (cage) there is necessarily only little bond valence for the individual bonds between atoms of the hetero part and such of the surrounding M–O framework. The shallow slope of the bond length-bond valence function in the range of small bond valences and the possibility for fractional and variable oxidation numbers for the addenda atoms M (in particular for V) allow a fine adjustment of the bonding parameters (bond lengths, bond valences, charge distribution).

The above results are in full accordance with bond models for polyoxometalate species proposed by our group. Both, the above results and the bond models contradict a number of statements about polyoxometalate ions in the literature; for example:

- The $M_6O_{19}^{2-}$ ions have positively charged central oxygen atoms, and most of the ionic charge is located on the terminal oxygen atoms. Hence these species cannot be described as O^{2-} ions surrounded by neutral (= uncharged) M_6O_{18} cages [9, 13b, 197, 198, 294]. The same is true for many other iso- [9, 232, 253b, 328] and in particular heteropolyoxometalate (e.g., the Keggin structure [9, 140, 311–313, 329]) ions, those with "unusual" heteroelement "anions" or molecules included; i.e., the description of heteropolyoxometalate ions as "anions in the interior of anions" [202, 284, 314, 316–320] is incorrect. The reasons for the incorrect interpretation of the structural data in the literature have been allocated.

- The statements that the ionic charge is mainly located on the bridging oxygen atoms [7–12, 13a] contradict the results of the present investigation and of theoretical considerations [16], in particular stability arguments based on the bonding power of bridging oxygen atoms and electrostatic arguments (electrostatic attraction between cations and polyanions in the salts).
- The characterization of terminal oxygen atoms as having little or no basic character [9, 13n, 261a, 284, 332, 340] contradicts the charge distributions established in this investigation and such derived from theoretical studies [16].
- The characterization of the most basic oxygen atoms as those undergoing protonation [2, 4, 5, 7, 8, 10, 13a, 261b, 267b, 295] is obviously not true in the case of the structures composed of MO_k polyhedra with $k > 4$.
- Bond orders (bond valences) greater than two for terminal M–O bonds, as derived from inspections of the symmetry properties of the bonding orbitals [66, 216, 247, 268], have never been observed in the present investigation.
- The term "trans influence" (of multiple-bonded, terminal oxygen atoms), frequently used to "explain" the pattern of the bond lengths (bond valences) in the different types of MO_k ($k > 4$) polyhedra [13j, 64, 65, 268, 281b, 294, 332], suggests an incorrect assessment of the bonding situation of the inner oxygen atoms (weak integration of the inner, high-coordinate oxygen atoms in *trans* position to the strong, terminal or pseudoterminal oxygen atoms within the M–O frameworks; an influence proceeding from the terminal oxygen atoms). In reality the oxygen atoms in *trans* positions to the terminal M–O bonds are the altogether most strongly bound oxygen atoms, and because of the interdependence of the inner M–O bond lengths in the M–O frameworks it is the inner M–O bonds that have primarily to be adjusted and hence influence the terminal M–O bond lengths. Strong terminal and weak inner M–O bonds (but weaker respectively – on an average – stronger than corresponds to an equidistribution of the available M–O bond valence) are an inherent emanation of the connectivity of the atoms in the structures of the polyoxo species, i.e., there is no other way to divide the available bond valence among the different M–O bonds, simultaneously fulfilling the interdependence of the bond lengths in the M–O frameworks, the bond length-bond valence relationship, and the valence sum rule, and the mixture of covalent and coordinate bonds (due to the mutual action of the MO_4 building units as acceptor and as donor of ligands as described in our bond model) is ideally suited to fulfill these requirements.
- The many statements that *poly*oxo anions are stabilized by a maximization of the terminal MO and MO_2 $p\pi$–$d\pi$ interactions [13j, 53, 64, 65, 268, 281a, 284, 332, 336, 337] are obviously incorrect. A maximization of the terminal MO and MO_2 $p\pi$–$d\pi$ interactions does actually not take place and would destabilize the polymeric structures.

Narrow thinking leads to naive answers. They are comfortable because they are not complicated by consideration of indirect consequences and interactions. But they are also misleading and thus dangerous since you can in this way never realize an appropriate picture of the intermeshed reality.

Narrow thinking not only limits our horizons but also leads us astray. You may consider a problem as precisely as you can but as long as you do not consider all its interactions as a unit you are approaching the problem in the wrong way. You will come to apparent answers that are expensive and may even be contrary to the actual situation.

Frederic Vester [342]

6 List of Symbols

A	fictive basic charge per O atom in the starting material for polyoxometalate ion formation
B	parameter of the exponential bond length-bond valence function for a definite element pair
B	additional charge per O atom, caused by the elimination of O (condensation of H_2O) in the course of formation of the polyoxometalate ions
b	number of the relevant bridging O atoms in a structure, besides the terminal O atoms also somewhat determining the basicity of the polyoxometalate ions
C	structural parameter determining the basicity of the polyoxometalate ions
C	further additional charge per terminal O atom, resulting from the preferential acceptance of the charge by the terminal O atoms in the polyoxometalate ions
c	charge (generally)
D	additional charge per terminal O atom, produced by charge separation in the polyoxometalate ions
d	E–O, M–O, or X–O distance (bond length)
d_o	parameter of the bond length-bond valence function for a definite element pair (equals the "single-bond" distance)
d_{oi}	individually (for a given structure determination) fitted d_o parameter of the bond length-bond valence function
E	element with a positive oxidation number
E	charge per terminal O atom if the formal ionic charge would be exclusively localized on the terminal O atoms of the polyoxometalate ions
h	number of OH groups in a polyoxometalate structure (species)
h_u	number of OH groups as unshared O atoms (of Oth atoms) in the structure
K_a	acid constant
K_p	protonation constant
k	coordination number

M	group VI or group V transition metal ("addenda" atom)
M–O	(in text passages, not in formulas) short, medium-sized, and long M–O bonds in the range 2 to 0.2 v.u., characterizing the commonly assumed tetrahedral, tetragonal pyramidal, trigonal-bipyramidal, octahedral, and pentagonal-bipyramidal coordination of the MO_k polyhedra
M···O	(in text passages, not in formulas) additional, very long bonds with <0.2 v.u., extending the commonly assumed coordination sphere of M
m	number of charges of the polyoxo species $M_q^nO_u^{m-}$ or, in a more general form, $M_q^nO_{u-h}(OH)_h(H_2O)_o^{m-}$
N	parameter of the bond length-bond valence power function for a definite element pair
n	oxidation number, (atom) valence of the element considered; in this article n means usually n_M
Ot	terminal O atom ($M–O^{\ominus}$, M=O); the number of terminal O atoms in a structure or in a structural part is one of the main parameters determining the basicity of the structure or of the structural part (in the literature 'terminal' and 'unshared' O atoms are often equalized)
Ou	unshared (outer) O atom ($M–O^{\ominus}$, M=O, M–OH, M···OH_2); the number of unshared O atoms (the number of free vertices) of an MO_k polyhedron is the main parameter determining the kinetic inertness of the corresponding structural part
Ob	bridging (inner) O atom without any distinction
Obl	bridging O atom in a long ("*trans*") M–O bond
Obm	bridging O atom in a medium-sized M–O bond
Op	pseudoterminal (two-coordinate) O atom without further distinction
Opa	pseudoterminal (two-coordinate) O atom in an angular M–O–M bridge
Opl	pseudoterminal (two-coordinate) O atom in an (approximately) linear M–O–M bridge
O2a	2-coordinate O atom in an angular M–O–M bridge
O3a	3-coordinate O atom in an angular M–O–M bridge
O2l	2-coordinate O atom in an (approximately) linear M–O–M bridge
O3l	3-coordinate O atom in an (approximately) linear M–O–M bridge; the simultaneous presence of an angular M–O–M bridge is irrelevant
O4l	4-coordinate O atom in an (approximately) linear M–O–M bridge; the simultaneous presence of angular M–O–M bridges is irrelevant
O5l	5-coordinate O atom, participating in two (approximately) linear M–O–M bridges; the simultaneous presence of angular M–O–M bridges is irrelevant
O6l	6-coordinate O atom, participating in three (approximately) linear M–O–M bridges; the simultaneous presence of angular M–O–M bridges is irrelevant
O3f	3-coordinate O atom in an 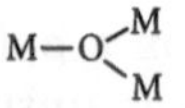 or 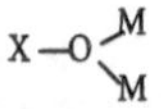 furcation
O4f	4-coordinate O atom in an 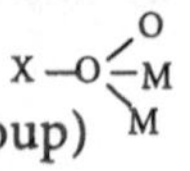 furcation
Oh	protonated O atom (OH group)
Ow	coordinated water molecule

O*	(bridging) O atom in trans position to a terminal or pseudoterminal oxygen, usually participating in a long M–O bond
O[*,b]c	complementary (second), weak [* in the case of M^{VI}] or middle-strong [b in the case of M^{V}] bond of a pseudoterminal (two-coordinate) O atom
☞	The additions h, w, *, and c are usually coupled with other additions. For the further additions b, f, t, u, l, and r in final positions see footnote a of Table 16. For the additions ′, ″, and ‴ see Sect. 3.3
o	number of coordinated H_2O molecules in a structure
P	degree of acidification of a metalate solution: molar ratio of H^+ ions *introduced* to MO_4^{w-} ions *introduced*
p	stoichiometric coefficient of H^+ in the overall equation for formation of the polyoxo species
pH	negative decadic logarithm of the H^+ *concentration*
pK_a	negative decadic logarithm of the acid constant
pK_b	negative decadic logarithm of the base constant
q	stoichiometric coefficient of MO_4^{w-} in the overall equation for formation of the polyoxo species; degree of aggregation
r	stoichiometric coefficient of eliminated (condensed) H_2O in the overall equation for formation of the polyoxo species
S_a	sum of the inner bond valence in a structure, caused by special adaptations due to the preferential acceptance of the (negative) ionic charge and of the negative charge generated through charge separation by terminal oxygen atoms while the simultaneously generated positive charge is accepted by bridging oxygen atoms, diminished by destabilizing effects due to the harmonization of the bond lengths, bond valences, and charges of the M–O frameworks
S_n	sum of the inner bond valence in a structure, caused by "normal bonding"
S_m	sum of the inner bond valence in a structure, caused by the "meshing effect"
S_o	overall bond valence of the inner M–O bonds in a structure
s	bond valence (in the older literature: bond strength, bond order) (generally)
s_i	bond valence of the ith M–O or X–O bond about an M, X, or O atom considered
s_u	bond valence of an unshared O atom
s_b	bond valence of a bridging O atom
$\bar{s}(k_O)$	average bond valence of the M–O bonds about all oxygen atoms with a given coordination number against M and a given protonation state
T	total negative charge on the terminal oxygen atoms of a structure
t	number of terminal oxygen atoms in a polyoxo structure
u	number of O atoms in a structure (those of coordinated H_2O molecules excluded)
v	an assumed positive or negative charge on an O atom (Sect. 4.8.3)
w	charge number of the unprotonated monomeric metalate ion

X	heteroelement (conventional heteroelements like P^V, Te^{VI}, Ni^{IV}, etc. in the case of Mo^{VI}, W^{VI}, V^V, etc. as addenda elements; "unusual" heteroelements like Cl^{-I}, Br^{-I}, etc., Cl^{VII}, S^{VI}, etc., OH_2, CH_3CN, etc. in the case of $V^{V/IV}$ as addenda element)
x	additional (average) charge per terminal oxygen atom, due to the preferential acceptance of the ionic charge by the terminal oxygen atoms
Y	positive charge in a structure produced by charge separation
Z	degree of protonation of a metalate solution: molar ratio of *reacted* H^+ ions to MO_4^{w-} ions *introduced*
Z^+	ratio of the stoichiometric coefficients of H^+ and $M^nO_4^{w-}$ ($Z^+ \equiv p/q$) in the overall equation for formation of the isopolyoxometalate ion $M_q^nO_u^{m-}$ or, generalized, $M_q^nO_{u-h}(OH)_h(H_2O)_o^{m-}$
α	factor characterizing the ability of the relevant bridging O atoms to accept negative charge in comparison to terminal O atoms (Eq. (13))
Δ	difference
ζ	fictive degree of aggregation of partial structures ($1 < \zeta \leqq q$)
μ	index in the oxoacid $EO_\mu(OH)_\nu$
ν	index in the oxoacid $EO_\mu(OH)_\nu$
ξ	fictive, non-committed stoichiometric quantity for M ($\xi = 1$ to q)
Σ	sum
Σs_i	$\equiv \sum_{i=1}^{k} s_i$
σ_d	standard deviation of the bond lengths (also used for those stated in the literature)
σ_Σ	standard deviation for the sum of the bond valences about M
$\sigma_{\Sigma i}$	standard deviation for the sum of the bond valences about M, related to individually fitted d_{oi} values
σ_{si}	estimated standard deviation for the valence of the M–O bonds, related to individually fitted d_{oi} values
σ_{ci}	estimated standard deviation for the charge on the O atoms, related to individually fitted d_{oi} values
σ_{di}	estimated standard deviation for the bond lengths, based on σ_{si} and $\sigma_{\Sigma i}$

Running index:

i	for the k O atoms about an M atom (or k M atoms about an O atom)

Acknowledgement. The authors are indebted to the Fonds der Chemischen Industrie for financial support. Discussions with Dr. H. J. Lunk (Humboldt Universität, Berlin) are gratefully acknowledged.

7 References

1. Linnett JW (1961) J Chem Soc 3796–3803
2. Tytko KH, Glemser O (1971) Z Naturforsch B 26: 659–678; (a) p 678

3. Tytko KH (1971) Angew Chem 83: 935–936; Angew Chem Int Ed Engl 10: 860–861
4. Tytko KH, Glemser O (1976) Advan Inorg Chem Radiochem 19: 239–315
5. Tytko KH, Baethe G, Hirschfeld ER, Mehmke K, Stellhorn D (1983) Z Anorg Allgem Chem 503: 43–66
6. Druskovich DM, Kepert DL (1975) Australian J Chem 28: 2365–2372
7. Klemperer WG, Shum WJ (1977) J Am Chem Soc 99: 3544–3545
8. Klemperer WG, Shum WJ (1978) J Am Chem Soc 100: 4891–4893
9. Day VW, Klemperer WG (1985) Science 228: 533–541
10. Kazanskii LP, Spitsyn VI (1975) Dokl Akad Nauk SSSR 223: 381–384; Dokl Phys Chem Proc Acad Sci USSR 220/225: 721–723
11. Kazanskii LP, Spitsyn VI (1976) Dokl Akad Nauk SSSR 227: 140–143; Dokl Phys Chem Proc Acad Sci USSR 226/231: 225–227
12. Kazanskii LP, Fedotov MA, Spitsyn VI (1977) Dokl Akad Nauk SSSR 233: 152–155; Dokl Phys Chem Proc Acad Sci 232/237: 250–253
13. Pope MT (1983) Heteropoly and Isopoly Oxometalates. Springer, Berlin. (a) pp 11, 21, 37, 128, (b) pp 20–21, (c) p 18, (d) p 17, (e) p 20, (f) pp 23–24, (g) pp 5, 48, 81, (h) pp 46–47, (i) pp 19, 130, (j) pp 128–132, (k) p 43, (l) p 72, (m) p 129, (n) p 128
14. Tytko KH (1987) Oxomolybdenum(VI) Species in Aqueous Solution, In: Gmelin Handbook of Inorganic Chemistry, 8th edn, Molybdenum Suppl Vol B 3a, pp 67–358. (a) pp 253–254, 258, 316–321, (b) pp 273–340, 344–352, (c) pp 289–291, 298–299, (d) p 280, (e) pp 295–297, (f) p 320, (g) p 72, (h) p 78, (i) pp 215–216, (j) pp 209–211, (k) pp 212–213, (l) pp 330–331, (m) pp 321–322, (n) pp 309–310, (o) pp 316–321, (p) pp 325, 346, (q) pp 291–295, (r) pp 305–309, (s) p 347, (t) pp 288–289, (u) pp 230–233, (v) pp 257–258, (w) pp 307–308, (x) pp 333–334, (y) pp 311–312, (z) pp 305–340, (aa) pp 330–332, 351, (bb) pp 335–337, (cc) pp 344–352, (dd) pp 274–277, (ee) pp 286–287, (ff) pp 305–311, (gg) pp 188–190, 261–262, 215–216, (hh) pp 196–200, 269–272, (ii) pp 306–307, (jj) pp 201–208, (kk) pp 334–335
15. Tytko KH (1985) Molybdate Hydrates with Alkali Metals and Ammonium and with Alkaline Earth Metals, In: Gmelin Handbook of Inorganic Chemistry, 8th edn, Molybdenum Suppl Vol B 4, pp 2–213. (a) p 37, (b) pp 151–152, (c) pp 67, 69
16. Tytko KH (1999) Structure and Bonding 93: 67–127
17. Freedman ML (1958) J Am Chem Soc 80: 2072–2077
18. Kepert DL (1962) Progr Inorg Chem 4: 199–274, pp 260–263
19. Tytko KH, Glemser O (1969) Chimia 23: 494–502
20. Tytko KH (1983) Chem Scr 22: 201–208
21. Pauling L (1944) General Chemistry, Edward Bros, Ann Arbor, Michigan
22. Ricci JE (1948) J Am Chem Soc 70: 109–113
23. Tytko KH (1986) Polyhedron 5: 497–503
24. Tytko KH, Mehmke J (1983) Z Anorg Allgem Chem 503: 67–86
25. Brown ID (1978) Chem Soc Rev 7: 359–376
26. Brown ID, Shannon RD (1973) Acta Crystallogr A 29: 266–282
27. Brown ID, Wu KK (1976) Acta Crystallogr B 32: 1957–1959
28. Donnay G, Donnay JDH (1973) Acta Crystallogr B 29: 1417–1425
29. Allmann R (1975) Monatsh Chem 106: 779–793
30. Donnay G, Allmann R (1970) Am Mineral 55: 1003–1015
31. Pyatenko YuA (1972) Kristallografiya 17: 773–779; Sov Phys-Crystallogr 17: 677–682
32. Trömel M (1983) Acta Crystallogr B 39: 664–669; (1984) B 40: 338–342; (1986) B 42: 138–141
33. Brown ID, Altermatt D (1985) Acta Crystallogr B 41: 244–247
34. Pauling L (1947) J Am Chem Soc 69: 542–553
35. Tytko KH, Mehmke J, Kurad D (1999) Structure and Bonding 93: 1–66
36. Brown ID (1992) Acta Crystallogr B 48: 553–572
37. Trömel M (1992) Z Kristallogr 200: 177–187
38. Brown ID (1974) J Solid State Chem 11: 214–233

39. Brown ID (1994) Bond-Length-Bond-Valence Relationships in Inorganic Solids, In: Bürgi HB, Dunitz JD (eds), Structure Correlation, vol 2, VCH, Weinheim, pp 405–429. (a) p 418, (b) p 424, (c) p 412–413, (d) pp 414–415, (e) p 421
40. Brown ID (1981) The Bond Valence Method: An Empirical Approach to Chemical Structure and Bonding, In: O'Keeffe M, Navrotsky A (eds), Structure and Bonding in Crystals, vol 2, Academic Press, New York, pp 1–30. (a) pp 2–3, (b) pp 11–12
41. Böschen I, Buss B, Krebs B (1974) Acta Crystallogr B 30: 48–56
42. Vivier H, Bernard J, Djomaa H (1977) Rev Chim Minerale 14: 584–604
43. D'Amour H, Allmann R (1972) Z Kristallogr 136: 23–47
44. Allmann R (1971) Acta Crystallogr B 27: 1393–1404
45. D'Amour H, Allmann R (1973) Z Kristallogr 138: 5–18
46. Allmann R, D'Amour H (1975) Z Kristallogr 141: 161–173
47. Krebs B, Paulat-Böschen I (1976) Acta Crystallogr B 32: 1697–1704
48. Perloff A (1970) Inorg Chem 9: 2228–2239
49. Brese NE, O'Keeffe M (1991) Acta Crystallogr B 47: 192–197
50. McCarley RE (1986) Polyhedron 5: 51–61
51. Bart JCJ, Ragaini V (1979) Inorg Chim Acta 36: 261–265
52. Zachariasen WH (1978) J Less-Common Metals 62: 1–7
53. Evans HT Jr, Gatehouse BM, Leverett P (1975) J Chem Soc Dalton Trans 504–514
54. Cotton FA, Wilkinson G (1982) Anorganische Chemie, 4th edn, Verlag Chemie, Weinheim; p 867
55. Krebs B, Stiller S, Tytko KH, Mehmke J (1991) Eur J Solid State Inorg Chem 28: 883–903
56. Hawthorne FC (1992) Z Kristallogr 201: 183–206; (a) p 193, (b) pp 183–184, 195–199, (c) pp 187–189
57. Fuchs J, Palm R, Hartl H (1996) Angew Chem 108: 2820–2822; Angew Chem Int Ed Engl 35: 2651–2653
58. Hartl H, Palm R, Fuchs J (1993) Angew Chem 105: 1545–1547; Angew Chem Int Ed Engl 32: 1492
59. Isobe M, Marumo R, Yamase T, Ikawa T (1978) Acta Crystallogr B 34: 2728–2731
60. Förster A, Kreusler HU, Fuchs J (1985) Z Naturforsch B 40: 1139–1148
61. Bergerhoff G, Hundt R, Sievers R, Brown ID (1983) J Chem Inf Comput Sci 23: 66–69
62. Allen SH, Bellard S, Brice MD, Cartwright BA, Doubleday A, Higgs H, Hummelink T, Hummelink-Peters BG, Kennard O, Motherwell WDS, Rodgers JR, Watson DG (1979) Acta Crystallogr B 35: 2331–2339
63. Pope MT (1972) Inorg Chem 11: 1973–1974
64. Porai-Koshits MA, Atovmyan LO (1975) Koord Khim 1: 1271–1281; Sov J Coord Chem 1: 1065–1074
65. Shustorovich EM, Porai-Koshits MA, Buslaev YuA (1975) Coord Chem Rev 17: 1–98, pp 67–81
66. Schröder FA (1975) Acta Crystallogr B 31: 2294–2309
67. Cruickshank DWJ (1961) J Chem Soc 5486–5504
68. Kepert DL (1973) Isopolyanions and Heteropolyanions, In: Comprehensive Inorganic Chemistry, Vol 4, Pergamon Press, Oxford, pp 607–672; pp 620–624
69. Kepert, DL (1972) The Early Transition Metals, Academic Press, London-New York, pp 52, 54
70. Clark GM, Morley R (1976) Chem Soc Rev 5: 269–295
71. Tytko KH (1979) Chem uns Zeit 13: 184–194
72. Auray M, Quarton M, Tarte P (1986) Acta Crystallogr C 42: 257–259
73. Bensch W, Hug P, Emmenegger R, Reller A, Oswald HR (1987) Mater Res Bull 22: 447–454
74. Serezhkin VN, Efremov VA, Trunov VK (1987) Zh Neorg Khim 32: 2695–2699; Russ J Inorg Chem 32: 1568–1570
75. Chen HY, Sleight AW (1986) J Solid State Chem 63: 70–75
76. Clearfield A, Moini A, Rudolf PR (1985) Inorg Chem 24: 4606–4609

77. Coquerel G, Gicquel-Mayer C, Mayer M, Perez G (1983) Acta Crystallogr C 39: 1602–1604
78. Efremov VA, Kudin OV, Velikodnyi YuA, Trunov VK, Makarevich LG (1981) Zh Neorg Khim 26: 2112–2116; Russ J Inorg Chem 26: 1138–1141
79. Gatehouse BM, Leverett P (1969) J Chem Soc A 849–854
80. Gicquel-Mayer C, Mayer M, Perez G (1981) Acta Crystallogr B 37: 1035–1039
81. Launay S, Rimsky A (1980) Acta Crystallogr B 36: 910–912
82. Mueller M, Hildmann BO, Hahn T (1987) Acta Crystallogr C 43: 184–186
83. Solodovnikov SF, Klevtsova RF, Kim VG, Klevtsov PV (1986) Zh Strukt Khim 27(6): 100–106; J Struct Chem 27: 928–933
84. Lazoryak BI, Efremov VA (1986) Kristallografiya 31: 237–243; Sov Phys Crystallogr 31: 138–142
85. Bars O, Le Marouille JY, Grandjean D (1977) Acta Crystallogr B 33: 1155–1157
86. Peytavin S, Philippot E, Maurin M (1974) J Solid State Chem 11: 71–77
87. Le Marouille JY, Bars O, Grandjean D (1980) Acta Crystallogr B 36: 2558–2560
88. Khazheeva ZI, Simonov VI, Mokhosoev MV, Smirnyagina NN, Kozhevnikova NM, Alekseev FP (1987) Kristallografiya 32: 79–82; Sov Phys Crystallogr 32: 43–44
89. Klevtsova RF, Glinskaya LA (1982) Zh Strukt Khim 23(5): 176–179; J Struct Chem 23: 816–818
90. Gonschorek W, Hahn T (1973) Z Kristallogr 138: 167–176
91. Matsumoto K, Kobayashi A, Sasaki Y (1975) Bull Chem Soc Japan 48: 1009–1013
92. Nagano O (1979) Acta Crystallogr B 35: 465–467
93. Ozeki T, Ichida H, Sasaki Y (1987) Acta Crystallogr C 43: 2220–2221
94. Clearfield A, Sims MJ, Gopal R (1976) Inorg Chem 15: 335–338
95. Kobtsev BM, Kharitonov YuA, Pobedimskaya EA, Belov NV (1968) Dokl Akad Nauk SSSR 179(1): 84–87; Sov Phys Dokl 13: 193–195
96. Jeitschko W, Sleight AW (1972) Acta Crystallogr B 28: 3174–3178
97. De Boer JJ (1974) Acta Crystallogr B 30: 1878–1880
98. Koster AS, Kools FXNM, Rieck GD (1969) Acta Crystallogr B 25: 1704–1708
99. Okada K, Ossaka J (1980) Acta Crystallogr B 36: 657–659
100. Okada K, Ossaka J, Iwai S (1979) Acta Crystallogr B 35: 2189–2191
101. Shannon RD, Calvo C (1973) J Solid State Chem 6: 538–549
102. Kato K, Takayama-Muromachi E (1985) Acta Crystallogr C 41: 1415–1417
103. Krishnamachari N, Calvo C (1971) Can J Chem 49: 1629–1637
104. Süsse P, Buerger MJ (1970) Z Kristallogr 131: 161–174
105. Lohmüller G, Schmidt G, Deppisch B, Gramlich V, Scheringer C (1973) Acta Crystallogr B 29: 141–142
106. Baglio JA, Sovers OJ (1971) J Solid State Chem 3: 458–465
107. Robertson B, Kostiner E (1972) J Solid State Chem 4: 29–37
108. Coing-Boyat J (1982) Acta Crystallogr B 38: 1546–1548
109. Touboul M, Toledano P (1980) Acta Crystallogr B 36: 240–245
110. Sleight AW, Chen HY, Ferretti A, Cox DE (1979) Mater Res Bull 14: 1571–1581
111. Liu Jian-Cheng, Chen Jia-Ping, Li De-Yu (1983) Wuli Xuebao [Acta Phys Sin] 32: 1053–1060
112. Dreyer G, Tillmanns E (1981) Neues Jahrb Mineral Monatsh (4) 151–154
113. Rice CE, Robinson WR (1976) Acta Crystallogr B 32: 2232–2233
114. Brusset H, Madaule-Aubry F, Mahe R, Boursier C (1971) C R Hebd Seances Acad Sci Ser C 273: 455–458
115. Fuess H, Kallel A (1972) J Solid State Chem 5: 11–14
116. Touboul M, Toledano P (1981) J Solid State Chem 38: 386–393
117. Abrahams SC, Marsh P, Ravez J (1983) Acta Crystallogr C 39: 680–683
118. Salmon R, Parent C, Le Flem G, Vlasse M (1976) Acta Crystallogr B 32: 2799–2802
119. Velikodnyi YuA, Trunov VK, Zhuravlev VD, Makarevich LG (1982) Kristallografiya 27: 226–228; Sov Phys Crystallogr 27: 138–140
120. Calestani G, Andreetti GD (1984) Z Kristallogr 168: 41–51

121. Olazcuaga R, Reau JM, Le Flem G, Hagenmuller P (1975) Z Anorg Allgem Chem 412: 271–280
122. Gopal R, Calvo C (1973) Z Kristallogr 137: 67–85
123. Gopal R, Calvo C (1971) Can J Chem 49: 3056–3059
124. Granzin J, Pohl D (1984) Z Kristallogr 169: 289–294
125. Cox DE, Moodenbaugh AR, Sleight AW, Chen HY (1980) National Bureau of Standards (US) Special Publication 567, 189–201
126. Will G, Schaefer W (1971) J Phys C 4: 811–819
127. Schaefer W (1972) Ber Kernforschungsanlage Jülich, Juel 830-RX 70
128. Vlasse M, Salmon R, Parent C (1976) Inorg Chem 15: 1440–1444
129. Parent C, Fava J, Salmon R, Vlasse M, Le Flem G, Hagenmuller P, Antic-Fidancev E, Lemaitre-Blaise M, Caro P (1979) Nouv J Chim 3: 523–527
130. Holt E, Drai S, Olazcuaga R, Vlasse M (1977) Acta Crystallogr B 33: 95–98
131. Quarton M, Kahn A (1979) Acta Crystallogr B 35: 2529–2532
132. Balko VP, Rylov GM, Bakakin VV (1974) Zh Strukt Khim 15(1): 155–157; J Struct Chem 15: 143–147
133. Wichmann R, Müller-Buschbaum H (1986) Rev Chim Minerale 23: 1–7
134. Day VW, Fredrich MF, Klemperer WG, Shum WJ (1977) J Am Chem Soc 99: 6146–6148
135. Kato K, Takayama E (1983) Acta Crystallogr C 39: 1480–1482
136. Kato K, Takayama-Muromachi E (1985) Acta Crystallogr C 41: 1411–1413
137. Kato K, Takayama-Muromachi E (1985) Acta Crystallogr C 41: 1413–1415
138. Gopal R, Calvo C (1974) Acta Crystallogr B 30: 2491–2493
139. Konnert JA, Evans HT Jr (1975) Acta Crystallogr B 31: 2688–2690
140. Baglio JA, Dann JN (1972) J Solid State Chem 4: 87–93
141. Hawthorne FC, Calvo C (1978) J Solid State Chem 26: 345–355
142. Dorm E, Marinder BO (1967) Acta Chem Scand 21: 590–591
143. Sauerbrei EE, Faggiani R, Calvo C (1974) Acta Crystallogr B 30: 2907–2909
144. Calvo C, Faggiani R (1975) Acta Crystallogr B 31: 603–605
145. Mercurio-Lavaud D, Frit B (1973) Acta Crystallogr B 29: 2737–2741
146. Gopal R, Calvo C (1973) Can J Chem 51: 1004–1009
147. Au PKL, Calvo C (1967) Can J Chem 45: 2297–2302
148. Shannon RD, Calvo C (1973) Can J Chem 51: 70–76
149. Björnberg A (1979) Acta Chem Scand A 33: 539–546
150. Kato K, Takyama-Muromachi E (1985) Acta Crystallogr C 41: 163–165
151. Masse R, Averbuch-Pouchot MT, Durif A, Guitel JC (1983) Acta Crystallogr C 39: 1608–1610
152. Kato K, Takayama-Muromachi E (1985) Acta Crystallogr C 41: 1409–1411
153. Kato K, Takayama-Muromachi E (1985) Acta Crystallogr C 41: 647–649
154. Fuchs J, Mahjour S, Pickardt J (1976) Angew Chem 88: 385; Angew Chem Int Ed Engl 15: 374
155. Heath E, Howarth OW (1981) J Chem Soc Dalton Trans 1105–1110
156. Pettersson L, Hedman B, Andersson I, Ingri N (1983) Chem Scr 22: 254–264
157. Shannon RD, Calvo C (1973) Can J Chem 51: 265–273
158. Hawthorne FC, Calvo C (1977) J Solid State Chem 22: 157–170
159. Marumo F, Isobe M, Iwai S, Kondo Y (1974) Acta Crystallogr B 30: 1628–1630
160. Idler KL, Calvo C, Ng HN (1978) J Solid State Chem 25: 285–294
161. Ng HN, Calvo C, Idler KL (1979) J Solid State Chem 27: 357–366
162. Ganne M, Piffard Y, Tournoux M (1974) Can J Chem 52: 3539–3543
163. Day VW, Klemperer WG, Yaghi OM (1989) J Am Chem Soc 111: 4518–4519
164. Sjöbom K, Hedman B (1973) Acta Chem Scand 27: 3673–3691
165. Don A, Weakley TJR (1981) Acta Crystallogr B 37: 451–453
166. Ohashi Y, Yanagi K, Sasada Y, Yamase T (1982) Bull Chem Soc Japan 55: 1254–1260
167. Román P, Gutiérrez-Zorilla JM, Martínez-Ripoll M, García-Blanco S (1985) Z Kristallogr 173: 283–292

168. Román P, Gutiérrez-Zorilla JM, Martínez-Ripoll M, García-Blanco S (1986) Trans Met Chem 11: 143–150
169. Fuchs J, Flindt EP (1979) Z Naturforsch B 34: 412–422
170. Burtseva KG, Chernaya TS, Sirota MI (1978) Dokl Akad Nauk SSSR 243: 104–107; Sov Phys Dokl 23: 784–786
171. Román P, Jaud J, Galy J (1981) Z Kristallogr 154: 59–68
172. Román P, Martínez-Ripoll M, Jaud J (1982) Z Kristallogr 158: 141–147
173. Román P, González-Aguado ME, Esteban-Calderón C, Martínez-Ripoll M, García-Blanco S (1983) Z Kristallogr 165: 271–276
174. Román P, González-Aguado ME, Esteban-Calderón C, Martínez-Ripoll M (1984) Inorg Chem Acta 90: 115–120
175. Román P, Gutiérrez-Zorilla JM, Martínez-Ripoll M, García-Blanco S (1985) Z Kristallogr 173: 169–178
176. Román P, Gutiérrez-Zorilla JM, Esteban-Calderón C, Martínez-Ripoll M, García-Blanco S (1985) Polyhedron 4: 1043–1046
177. Román P, Gutiérrez-Zorilla JM, Martínez-Ripoll M, García-Blanco S (1986) Polyhedron 5: 1799–1803
178. Román P, Gutiérrez-Zorilla JM, Martínez-Ripoll M, García-Blanco S (1987) J Crystallogr Spectrosc Res 17: 109–119
179. Román P, Gutiérrez-Zorilla JM, Martínez-Ripoll M, Garçía-Blanco S (1987) Trans Met Chem 12: 159–167
180. Weakley TJR (1982) Polyhedron 1: 17–19
181. Wilson AJ, McKee V, Penfold BR, Wilkins CJ (1984) Acta Crystallogr C 40: 2027- 2030
182. Bharadwaj PK, Ohashi Y, Sasada Y, Sasaki Y, Yamase T (1984) Acta Crystallogr C 40: 48–50
183. Fuchs J, Knöpnadel I (1982) Z Kristallogr 158: 165–179
184. Arzoumanian H, Baldy A, Lai R, Odreman A, Metzger J, Pierrot M (1985) J Organomet Chem 295: 343–352
185. Román P, Vegas A, Martínez-Ripoll M, García-Blanco S (1982) Z Kristallogr 159: 291–295
186. Tytko KH (1989) Oxomolybdenum(VI) Species in Nonaqueous (Organic) Solvents, In: Gmelin Handbook of Inorganic Chemistry, 8th edn, Molybdenum Suppl Vol B 3b, pp 209–266. (a) p 239, (b) pp 214–221, 227–230, (c) pp 115–117
187. Allcock HR, Bissel EC, Shawl ET (1973) Inorg Chem 12: 2963–2968
188. Garner CD, Howlader NC, Mabbs FE, McPhail AT, Miller RW, Onan KD (1978) J Chem Soc Dalton Trans 1582–1589
189. Nagano O, Sasaki Y (1979) Acta Crystallogr B 35: 2387–2389
190. Dahlstrom P, Zubieta J, Neaves B, Dilworth JR (1982) Cryst Struct Commun 11: 463–469
191. Clegg W, Sheldrick GM, Garner CD, Walton IB (1982) Acta Crystallogr B 38: 2906–2909
192. Fuchs J, Freiwald W, Hartl H (1978) Acta Crystallogr B 34: 1764–1770
193. Henning G, Hüllen A (1969) Z Kristallogr 130: 162–172
194. Goiffon A, Philippot E, Maurin M (1980) Rev Chim Minerale 17: 466–476
195. LaRue WA, Liu AT, San Filippo J (1980) Inorg Chem 19: 315–320
196. Kirillova NI, Kolomnikov IS, Zolotarev YuA, Lysyak TV, Struchkov YuT (1977) Koord Khim 3: 1895–1899; Sov J Coord Chem 3: 1488–1492
197. Fuchs J (1973) Z Naturforsch B 28: 389–404
198. Mattes R, Bierbüsse H, Fuchs J (1971) Z Anorg Allgem Chem 385: 230–242
199. Kepert DL (1969) Inorg Chem 8: 1556–1558
200. Goiffon A, Spinner B (1975) Rev Chim Minerale 12: 316–327
201. Goiffon A, Spinner B (1975) Bull Soc Chim France 2435–2441
202. Pope MT (1994) Polyoxoanions, In: King RB (ed), Encyclopedia of Inorganic Chemistry, vol 6, Wiley, Chichester, 3361–3371; (a) p 3361
203. D'Amour H (1976) Acta Crystallogr B 32: 729–740
204. Lange G, Hahn H, Dehnicke K (1969) Z Naturforsch B 24: 1498–1507
205. Tytko KH (1974) 16th Int Conf Coord Chem Proc, Dublin, Lecture and Ref R8

206. Durif A, Averbuch-Pouchot MT, Guitel JC (1980) Acta Crystallogr B 36: 680–682
207. Evans HT Jr (1966) Inorg Chem 5: 967–977
208. Swallow AG, Ahmed FR, Barnes WH (1966) Acta Crystallogr 21: 397–405
209. Saf'yanov YuN, Kuz'min A, Belov NV (1978) Dokl Akad Nauk SSSR 242: 603–605; Sov Phys Dokl 23: 639–641
210. Saf'yanov YuN, Belov NV (1976) Dokl Akad Nauk 227: 1112–1115; Sov Phys Dokl 21: 176–178
211. Rivero BE, Rigotti G, Punte G, Navaza A (1984) Acta Crystallogr C 40: 715–718
212. Rivero BE, Punte G, Rigotti G, Navaza A (1985) Acta Crystallogr C 41: 817–820
213. Graeber EJ, Morosin B (1977) Acta Crystallogr B 33: 2137–2143
214. Saf'yanov YuN, Kuz'min EA, Belov NV (1978) Kristallografiya 23: 697–702; Sov Phys Crystallogr 23: 390–392
215. Kempf JY, Rohmer MM, Poblet JM, Bo C, Bénard M (1992) J Am Chem Soc 114: 1136–1146
216. Shao M, Wang L, Zhang Z, Tang Y (1984) Sci Sin Ser B (Engl Ed) 27: 137–148
217. Capparelli MV, Goodgame DML, Hayman PB, Skapski AC (1986) J Chem Soc Chem Commun 776–777
218. Debaerdemaeker T, Arrieta JM, Amigo JM (1982) Acta Crystallogr B 38: 2465–2468
219. Evans HT Jr, Pope MT (1984) Inorg Chem 23: 501–504
220. Day VW, Klemperer WG, Maltbie DJ (1987) J Am Chem Soc 109: 2991–3002
221. Garin JL, Costamagna JA (1988) Acta Crystallogr C 44: 779–782
222. Bharadwaj PK, Ohashi Y, Sasada Y, Sasaki Y, Yamase T (1986) Acta Crystallogr C 42: 545–547
223. Kreusler HU, Förster A, Fuchs J (1980) Z Naturforsch B 35: 242–244
224. Toraya H, Marumo F, Yamase T (1984) Acta Crystallogr B 40: 145–150
225. Cruywagen JJ, van der Merve IFY, Nassimbeni LR, Niven ML, Symonds EA (1986) J Crystallogr Spectrosc Res 16: 525–535
226. Evans HT Jr, Rollins OW (1976) Acta Crystallogr B 32: 1565–1567
227. Tsay YH, Silverton JV (1973) Z Kristallogr 137: 256–279
228. Averbuch-Pouchot MT, Tordjman I, Durif A, Guitel JC (1979) Acta Crystallogr B 35: 1675–1677
229. Evans HT Jr, Prince E (1983) J Am Chem Soc 105: 4838–4839
230. Asami M, Ichida H, Sasaki Y (1984) Acta Crystallogr C 40: 35–37
231. Fuchs J, Hartl H, Schiller W, Gerlach U (1976) Acta Crystallogr B 32: 740–749
232. Day VW, Fredrich MF, Klemperer WG, Shum W (1977) J Am Chem Soc 99: 952–953
233. Tze-Chen Hsieh, Shaikh SN, Zubieta J (1987) Inorg Chem 26: 4079–4089
234. Fuchs J, Hartl H (1976) Angew Chem 88: 385–386; Angew Chem Int Ed Engl 15: 375
235. Krebs B, Paulat-Böschen I (1982) Acta Crystallogr B 38: 1710–1718
236. Armour AW, Drew MGB, Mitchell PCH (1975) J Chem Soc Dalton Trans 1493–1496
237. Knöpnadel I, Hartl H, Hunnius W, Fuchs J (1974) Angew Chem 86: 894–895; Angew Chem Int Ed Engl 13: 823
238. Magarill SA, Klevtsova RF (1971) Kristallografiya 16: 742–745; Sov Phys Crystallogr 16: 645–648
239. Gatehouse BM, Leverett P (1968) J Chem Soc A 1398–1405
240. Seleborg M (1966) Acta Chem Scand 20: 2195–2201
241. Björnberg A, Hedman B (1977) Acta Chem Scand A 31: 579–584
242. Kato K, Takayama E (1984) Acta Crystallogr B 40: 102–105
243. Sedlacek P, Dornberger-Schiff K (1965) Acta Crystallogr 18: 407–410
244. Kihlborg L (1963) Arkiv Kemi 21: 357–364
245. Böschen I, Krebs B (1974) Acta Crystallogr B 30: 1795–1800
246. Oswald HR, Günter JR, Dubler E (1975) J Solid State Chem 13: 330–338
247. Krebs B (1972) Acta Crystallogr B 28: 2222–2231
248. Åsbrink S, Brandt BG (1971) Chem Scr 1: 169–181
249. Enjalbert R, Galy J (1986) Acta Crystallogr C 42: 1467–1469
250. Byström A, Wilhelmi KA, Brotzen O (1950) Acta Chem Scand 4: 1119–1130

251. Bachmann HG, Ahmed FR, Barnes WH (1961) Z Kristallogr 115: 110–131
252. Filowitz M, Klemperer WG, Shum W (1978) J Am Chem Soc 100: 2580–2581
253. Tytko KH (1989) Reactions of Oxomolybdenum(VI) Species in Aqueous Solution, In: Gmelin Handbook of Inorganic Chemistry, 8th edn, Molybdenum Suppl Vol B 3b, pp 1–207. (a) pp 18–22, (b) pp 118–119, (c) p 33, (d) p 117, (e) pp 112–117, (f) pp 99–101, (g) 36–40
254. Tytko KH, Schönfeld B (1975) Z Naturforsch B 30: 471–484
255. Richter H, Fuchs J (1984) Z Naturforsch B 39: 623–627
256. Noltemeier M, Sheldrick GM, Tytko KH, Ichinose N (1985) unpublished; see also Ref. 253a
257. Tolédano P, Touboul M (1978) Acta Crystallogr B 34: 3547–3551
258. Benchrifa R, Leblanc M, De Pape R (1989) Eur J Solid State Inorg Chem 26: 593–601
259. Klemperer WG, Shum W (1976) J Am Chem Soc 98: 8291–8293
260. Fuchs J, Brüdgam I (1977) Z Naturforsch B 32: 853–857
261. Pope MT (1992) Progr Inorg Chem 39: 181–257; (a) pp 182–186, (b) pp 186, (c) pp 189–193, (d) p 213
262. Tytko KH (1977) Habilitationsschrift, Göttingen. (a) pp 112–115, 145–150, (b) pp 113–115, (c) pp 132–135, 154, (d) pp 117–123, 132–135, (e) 162–163, (f) 145–148, (g) pp 78–163, (h) p 152
263. Cordis V, Tytko KH, Glemser O (1975) Z Naturforsch B 30: 834–841
264. Strandberg R (1977) Acta Crystallogr B 33: 3090–3096
265. Strandberg R (1975) Acta Chem Scand A 29: 350–358
266. Ichida H, Sasaki Y (1983) Acta Crystallogr C 39: 529–533
267. Spitsyn VI, Kazanskii LP, Torchenkova EA (1981) Soviet Sci Rev B 3: pp 111–196. (a) pp 113–114, 131, (b) pp 148–149, (c) p 120
268. Porai-Koshits MA, Atovmyan LO (1981) Zh Neorg Khim 26: 3171–3180; Russ J Inorg Chem 26: 1697–1703
269. Gmelin Handbook of Inorganic Chemistry, 8th edn (1990) Molybdenum Suppl Vol B 5, pp 132–133
270. Mehmke J (1990) Dissertation, Göttingen
271. Tytko KH, Mehmke J (1990) 28th Int Conf Coord Chem, Gera
272. Menge R (1992) Diplomarbeit, Göttingen
273. Evans HT Jr (1974) Acta Crystallogr B 30: 2095–2100
274. Baes CF Jr, Mesmer RE (1976) The Hydrolysis of Cations, Wiley, New York. (a) pp 201–210, (b) p 200, (c) pp 249–252, (d) pp 392–396
275. Greenwood NN, Earnshaw A (1985) Chemistry of the Elements, Pergamon Press, Oxford. (a) p 1144, (b) pp 588–589, 596, (c) pp 914–915, (d) pp 1022–1025
276. Trömel M (1994) Z Kristallogr 209: 18–21
277. Gmelin Handbook of Inorganic Chemistry, 8th edn (1987) Molybdenum Suppl Vol B 3a, p 19
278. Tytko KH, Baethe G, Mehmke K (1987) Z Anorg Allgem Chem 555: 98–108
279. Tytko KH, Baethe G (1987) Z Anorg Allgem Chem 555: 85–97
280. Wiberg N (1985) Holleman-Wiberg: Lehrbuch der Anorganischen Chemie, 91th–100th edn, de Gruyter, Berlin; p 1097–1098
281. Pope MT (1987) In: Wilkinson G et al (eds) Comprehensive Coordination Chemistry, vol 3, Pergamon Press, Oxford, pp 1023–1058. (a) p 1025, (b) p 1024
282. Cruywagen JJ, Heyns JBB (1981) S Afr J Chem 34: 118–120
283. Richardson E (1960) J Inorg Nucl Chem 12: 349–353
284. Pope MT, Müller A (1991) Angew Chem 103: 56–70; Angew Chem Int Ed Engl 30: 34–48
285. Ma L, Liu S, Zubieta J (1989) Inorg Chem 28: 175–177
286. Pettersson L, Andersson I, Hedman B (1985) Chem Scr 25: 309–317
287. Neumann G (1964) Acta Chem Scand 18: 278–280
288. Schwarzenbach G, Geier G, Littler J (1962) Helv Chim Acta 45: 2601–2612
289. Tytko KH, Cordis V, Mehmke K, Hirschfeld ER (1985) US-Japan Semin Catal Activ Polyoxoanions, Shimoda, Abstr pp 35–39

290. Mehmke K (1988) Dissertation, Göttingen
291. Howarth OW, Jarrold MJ (1978) J Chem Soc Dalton Trans 503–506
292. Griffith WP, Lesniak PJB (1969) J Chem Soc A 1066–1071
293. Corigliano F, Di Pasquale S (1975) Inorg Chim Acta 12: 99–101
294. Freeman A, Schultz FA, Reilley CN (1982) Inorg Chem 21: 567–576
295. Barcza L, Pope MT (1975) J Phys Chem 79: 92–93
296. Griffith WP, Wickins TD (1966) J Chem Soc A 1087–1090
297. Böschen I (1974) Dissertation, Kiel; pp 70–71, 77
298. Müller A, Meyer J, Krickemeyer E, Diemann E (1996) Angew Chem 108: 1296–1299; Angew Chem Int Ed Engl 35: 1206–1208
299. Müller A, Krickemeyer E, Meyer J, Bögge H, Peters F, Plass W, Diemann E, Dillinger S, Nonnenbruch F, Randerath M, Menke C (1995) Angew Chem 107: 2293–2295; Angew Chem Int Ed Engl 34: 2122
300. Müller A, Reuter H (1995) Angew Chem 107: 2505–2239; Angew Chem Int Ed Engl 34: 2311
301. Lipscomb WN (1965) Inorg Chem 4: 132–134
302. Tytko KH (1975) Chemiedozententagung Düsseldorf, Referateband p A 40
303. Yamase T, Ikawa T (1977) Bull Chem Soc Japan 50: 746–749
304. Tytko KH, Schönfeld B, Buss B, Glemser O (1973) Angew Chem 85: 305–307; Angew Chem Int Ed Engl 12: 330–332
305. Paulat-Böschen I (1979) J Chem Soc Chem Commun 780–782
306. Yamase T, Hayashi H, Ikawa T (1974) Chem Lett. 1055–1056
307. Porth D (1991) Diplomarbeit, Göttingen
308. Stiefel EI (1977) Progr Inorg Chem 22: 1–223, p 27
309. Keck D (1992) Staatsexamensarbeit, Göttingen
310. Fischer S (1994) Staatsexamensarbeit, Göttingen
311. Clark CJ, Hall D (1976) Acta Crystallogr B 32: 1545–1547
312. Boeyens JCA, McDougal GJ, Smit J van R (1976) J Solid State Chem 18: 191–199
313. Day VW, Klemperer WG, Yaghi OM (1991) Nature 352: 115–116
314. Pope MT (1992) Nature 355: 27
315. Fink L, Trömel M (1992) Z Kristallogr 200: 169–175
316. Müller A, Penk M, Rohlfing R, Krickemeyer E, Döring J (1990) Angew Chem 102: 927–929; Angew Chem Int Ed Engl 29: 926
317. Müller A, Krickemeyer E, Penk M, Rohlfing R, Armatage A, Bögge H (1991) Angew Chem 103: 1720–1722; Angew Chem Int Ed Engl 30: 1674
318. Reuter H (1992) Angew Chem 104: 1210–1213; Angew Chem Int Ed Engl 31: 1185
319. Müller A, Reuter H, Dillinger S (1995) Angew Chem 107: 2505–2539; Angew Chem Int Ed Engl 34: 2311–2327
320. Penk M (1991) Dissertation, Bielefeld, p 62
321. Klemperer WG, Marquart TA, Yaghi OM (1992) Angew Chem 104: 51–53; Angew Chem Int Ed Engl 31: 49
322. Johnson GK, Schlemper EO (1978) J Am Chem Soc 100: 3645–3646
323. Müller A (1991) Nature 352: 115
324. Tytko KH, Glemser O (1970) Z Naturforsch B 25: 429–430
325. Hüllen A (1964) Naturwissenschaften 51: 508; Angew Chem 76: 588
326. Jahr KF, Fuchs J (1966) Angew Chem 78: 725–735; Angew Chem Int Ed Engl. 5: 689
327. Griffith WP (1970) Coord Chem Rev 5: 459–517
328. Klemperer WG (1990) Inorg Synth 27: 71–135; pp 71–74
329. Fuchs J, Thiele A, Palm R (1982) Z Naturforsch B 37: 1418–1421
330. Brown GM, Noe-Spirlet MR, Busing WR, Levy HA (1977) Acta Crystallogr B 33: 1038–1046
331. Kwak W, Pope MT, Scully TF (1975) J Am Chem Soc 97: 5735–5738
332. Pope MT (1983) 29th IUPAC Congr Abstr Papers, Cologne, p 22
333. Cotton FA, Wing RM (1965) Inorg Chem 4: 867–873
334. Megaw HD (1968) Acta Crystallogr B 24: 149–153

335. Dunitz JD, Orgel LE (1960) Advan Inorg Chem Radiochem 2: 1–60
336. Evans HT Jr (1971) Perspect Struct Chem 4: 1–59
337. Wilson AJ, McKee V, Penfold BR, Wilkins CJ (1984) Acta Crystallogr C 40: 2027–2030
338. Stiefel E (1987) Molybdenum(VI), In: Wilkinson G et al (eds), Comprehensive Coordination Chemistry, vol 3, Pergamon Press, Oxford, pp 1375–1420; p 1380
339. Kahn MI, Zubieta J (1995) Progr Inorg Chem 43: 1–149; pp 25
340. Baker LCW, Lebioda L, Grochowski J, Mukherjee HG (1980) J Am Chem Soc 102: 3274–3276
341. Braitenberg V (1984) Vehicles: Experiments in Synthetic Psychology, MIT Press, Cambridge, Mass
342. Vester F (ca. 1982) at an exhibition in Cassel

AUTHOR INDEX VOLUMES 1–93